Cloud Control Systems

Analysis, Design and Estimation

Emerging Methodologies and Applications in Modelling, Identification and Control

Cloud Control Systems

Analysis, Design and Estimation

Magdi S. Mahmoud
King Fahd University of Petroleum and Minerals
Systems Engineering Department
Dhahran, Saudi Arabia

Yuanqing Xia
Beijing Institute of Technology
School of Automation
Beijing, China

Series Editors
Stephen Ison
Lucy Budd

ELSEVIER
ACADEMIC PRESS
An imprint of Elsevier

Academic Press is an imprint of Elsevier
125 London Wall, London EC2Y 5AS, United Kingdom
525 B Street, Suite 1650, San Diego, CA 92101, United States
50 Hampshire Street, 5th Floor, Cambridge, MA 02139, United States
The Boulevard, Langford Lane, Kidlington, Oxford OX5 1GB, United Kingdom

Library of Congress Cataloging-in-Publication Data
A catalog record for this book is available from the Library of Congress

British Library Cataloguing-in-Publication Data
A catalogue record for this book is available from the British Library

ISBN: 978-0-12-818701-2

For information on all Academic Press publications
visit our website at https://www.elsevier.com/books-and-journals

Publisher: Mara Conner
Acquisition Editor: Sonnini R. Yura
Editorial Project Manager: John Leonard
Production Project Manager: Nirmala Arumugam
Designer: Matthew Limbert

Typeset by VTeX

This book is dedicated to our families. With tolerance, patience, and wonderful frame of mind they have encouraged and supported us for many years.

Magdi S. Mahmoud, Yuanqing Xia

Contents

8. Secure group consensus

9. Cybersecurity for the electric power system

About the authors

Magdi S. Mahmoud is Distinguished Professor at KFUPM, Dhahran, Saudi Arabia. He obtained a BSc (Honors) in Communication Engineering, an MSc in Electronic Engineering, and a PhD in systems engineering from Cairo University in 1968, 1972, and 1974, respectively. He has been a Professor of Engineering since 1984. He has been on the faculty at different universities worldwide including Egypt (CU, AUC), Kuwait (KU), UAE (UAEU), UK (UMIST), USA (Pitt, Case Western), Singapore (NTU), and Australia (Adelaide). He has lectured in Venezuela (Caracas), Germany (Hanover), UK (Kent), USA (Texas, UoSA), Canada (Montreal), and China (BIT, Yanshan, USTB). He is the principal author of 23 books and 18 book chapters, and is the author/co-author of more than 600 peer-reviewed papers.

He is the recipient of the Science State Incentive Prizes for outstanding research in engineering (Egypt) in 1978 and 1986, the Abdul-Hameed Showman Prize for Young Arab Scientists in engineering sciences (Jordan) in 1986, the Prestigious Award for Best Researcher at Kuwait University (Kuwait) in 1992, the State Medal of Science and Arts-first class (Egypt) in 1979, and the State Distinguished Award-first class (Egypt) in 1995. He was listed in the 1979 edition of Who's Who in Technology Today (USA). He was the vice-chairman of the IFAC-SECOM working group on large-scale systems methodology and applications (1981–86), and an associate editor of the LSS Journal (1985–88) and editor-at-large of the EEE series, Marcel-Dekker, USA. He has been an associate editor of the International Journal of Parallel and Distributed Systems of Networks (IASTED) since 1997. He is a member of the New York Academy of Sciences. He is currently actively engaged in teaching and research in the development of modern methodologies to distributed control and filtering of networked-control systems, cyber-physical systems, and secure control of renewable-energy systems.

He is a fellow of the IEE; a senior member of the IEEE; a member of Sigma Xi, the CEI (UK), the Egyptian Engineers society, and the Kuwait Engineers society; and a registered consultant engineer of information engineering and systems (Egypt).

Yuanqing Xia was born in Anhui Province, China in 1971, and graduated from the Department of Mathematics, Chuzhou University, China in 1991. He received a MSc in Fundamental Mathematics from Anhui University, China, in 1998, and a PhD in Control Theory and Control Engineering from Beijing University of Aeronautics and Astronautics, China, in 2001. From 1991 to 1995 he was with Tongcheng Middle-School, China, where he worked as a teacher. From January 2002 to November 2003 he was a postdoctoral research associate at the Institute of Systems Science, Academy of Mathematics and System Sciences, Chinese Academy of Sciences, China, where he worked on navigation, guidance, and control. From November 2003 to February 2004 he was with the National University of Singapore as a Research Fellow, where he worked on variable structure control. From February 2004 to February 2006 he was with the University of Glamorgan, UK, as a Research Fellow, where he studied networked control systems. From February 2007 to June 2008 he was a guest professor with Innsbruck Medical University, Austria, where he worked on biomedical signal processing. Since July 2004 he has been with the School of Automation, Beijing Institute of Technology, Beijing, first as an Associate Professor and then since 2008 as Professor. In 2012 he was appointed Xu Teli Distinguished Professor at the Beijing Institute of Technology, and then in 2016 he was made Chair Professor. In 2012 he obtained the National Science Foundation for Distinguished Young Scholars of China; in 2016 he was honored as the Yangtze River Scholar Distinguished Professor and was supported by the National High Level Talents Special Support Plan ("Million People Plan") by the Organization Department of the CPC Central Committee. He is now the Dean of the School of Automation, Beijing Institute of Technology. He has published ten monographs in Springer, John Wiley, and CRC, and more than 200 papers in international scientific journals. He is a deputy editor of the Journal of Beijing Institute of Technology and an associate editor of Acta Automatica Sinica; Control Theory and Applications; the International Journal of Innovative Computing, Information, and Control; and the International Journal of Automation and Computing. He obtained the Second Award of the Beijing Municipal Science and Technology (No. 1) in 2010 and 2015, the Second National Award for Science and Technology (No. 2) in 2011, and the Second Natural Science Award of the Ministry of Education (No. 1) in 2012 and 2017. His research interests include networked control systems, robust control and signal processing, active disturbance rejection control, cloud control systems and flight control.

Preface

This volume lays down the basic definitions and essential ingredients, and provides some new developments in the growing area of cloud control systems (CCSs). On the one hand, a CCS essentially contains a cyber-physical system (CPS) and a cyber-physical control system (CPCS). In this regard, the field of CCSs embraces the idea of "*control as a service*," that is, control algorithms can be scheduled as a kind of resource. In CCSs there are three closed loops: the control loop, followed by a scheduling loop and a decision-making loop.

On the other hand, the term cyber-physical system (CPS) is a generic term for a variety of modern control systems in real-life use, such as supervisory control and data acquisition (SCADA) systems, industrial control systems (ICSs), building control systems (BCSs), and the global electrical smart grid (SG). These control systems are made up of computers, electrical and mechanical devices, and manual processes overseen by humans. CPSs perform automated or partially automated control of physical equipment in manufacturing and chemical plants, electric utilities, distribution and transportation systems, and many other industries. In addition, CPSs integrate computational resources, communication capabilities, sensing, and actuation in an effort to monitor and control physical processes. There are several types of CPS found in critical infrastructure such as transportation networks, unmanned aerial vehicles (UAVs), nuclear power generation (NPG), electric power distribution networks (EPDNs), water and gas distribution networks (GDNs), and advanced communication systems (ACSs).

A key difference between CPSs and traditional information technology (IT) systems is that CPSs interact strongly with the physical environment, and the availability of the physical devices is the most important security aspect. However, CPSs are also cyber systems and are therefore vulnerable to cyber attacks. This connection with the physical world, however, presents unique challenges and opportunities. The security of the physical machines in a networked environment depends on the security of the electronic control systems, but cybersecurity is not typically the main design concern. The main concern for CPSs is the availability of the physical machines governing the operations. As CPS owners continue to install remote network control devices and incorporate an increasing number of insecure Internet of Things (IoT) devices in their industrial processes, the underlying security of their operations becomes increasingly vulnerable.

This volume is dedicated to *control systems* undergoing *cyber-physical* attacks, which categorizes an essential portion of future industrial automation systems. It therefore focuses on current cybersecurity issues of CPSs, develops secure (sufficiently safe) control and estimation algorithms, and alleviates potential concerns for future CPS designers and operators. Guaranteeing secure future CPSs is an integral factor for keeping our critical infrastructure safe.

From this perspective the pedagogical objectives of the book are the following:

1. To introduce a coherent and unified framework for studying CPCSs with particular emphasis on the analysis, design, and estimation (detection/identification) in relation to security issues and different types of attacks;
2. To acquaint students with the control theory background required to read and contribute to the research literature on CPCSs;
3. To present the main ideas and demonstrations of the major results of safe operation of CPSs;
4. To provide a modest coverage of cloud-based approaches to control systems, and of secure control methodologies to CPSs against various types of malicious attacks.

- **Chapter 1:** (An overview)
  This chapter provides a guided tour into the key ingredients of cloud control systems (CCSs) and their prevailing features under normal operating environments and when subjected to cyber-physical attacks.
- **Chapter 2:** (Cloud control systems venture)
  This chapter focuses on the issues underlying the analysis, design, and estimation methods of cloud control systems (CCSs) with particular emphasis on workflow and security objectives under different attacks.
- **Chapter 3:** (Distributed denial-of-service attacks)
  This chapter critically examines the impact of distributed denial-of-service (DoS) attacks in cyber-physical control systems (CPCSs).
- **Chapter 4:** (Distributed cloud control systems)
  This chapter discusses further the paradigm of cyber-physical control systems (CPCSs) and considers several approaches.
- **Chapter 5:** (Secure stabilization of distributed systems)
  This chapter examines the construction of stabilization methods that guarantee secure (sufficiently safe) operation of cyber-physical control systems (CPCSs).
- **Chapter 6:** (False data injection tacks)
  This chapter introduces some typical practical case studies.
- **Chapter 7:** (Stabilization schemes for secure control)
  This chapter examines networked control systems in the presence of attacks that prevent transmission over the network. We characterize the frequency and duration of the denial-of-service (DoS) attacks under which input-to-state

stability (ISS) of the closed-loop system can be preserved. Then a secure observer-based controller for discrete-time cyber-physical systems subject to both cyber attacks and physical attacks is presented

- **Chapter 8:** (Secure group consensus)
  In this chapter, couple-group consensus of multi-agent systems under denial-of-service (DoS) attacks is studied. Specifically, we study couple-group consensus problems involving DoS attacks within subgroups.
- **Chapter 9:** (Cybersecurity for the electric power system)
  This chapter highlights the significance of cyber infrastructure security in conjunction with power application security to prevent, mitigate, and tolerate cyber attacks. A layered approach is introduced to evaluating risk based on the security of both the physical power applications and the supporting cyber infrastructure.
- **Chapter 10:** (Resilient design under cyber attacks)
  In this chapter, we consider a data injection attack on a cyber-physical system (CPS). We propose a hybrid framework for detecting the presence of an attack and operating the plant in spite of the attack. Additionally, an improved observer-based stabilizing controller will be proposed for CPSs including random measurements and actuation delays and affected by denial-of-service (DoS) and deception attacks.
- **Chapter 11:** (Safety assurance under stealthy cyber attacks)
  In this chapter, we examine the performance of stealthy deception attacks from the system's perspective. We investigate three kinds of stealthy deception attacks according to the attacker's ability to compromise the system.
- **Chapter 12:** (A unified game approach under DoS attacks)
  We consider in this chapter the problem of resilient control of a networked control system (NCS) under a denial-of-service (DoS) attack via a unified game approach. The DoS attacks lead to extra constraints in the NCS, where the packets may be jammed by a malicious adversary.
- **Chapter 13:** (Secure estimation subject to cyber stochastic attacks)
  In the first section of this chapter, a secure estimator for discrete-time delayed nonlinear systems considering both denial-of-service (DoS) and deception attacks will be presented. The DoS and deception attacks will be considered Bernoulli distributed white sequences with variable probabilities. Next we consider the problem of resilient dynamical state estimation in the presence of integrity attacks. We conduct resilience and performance analysis for a convex optimization-based estimator.
- **Chapter 14:** (Cloud-based approach in data centers)
  Data centers play an important role in modern information technology infrastructures. A data center is a home to computational power, storage, and applications necessary to support an enterprise or business. In this chapter, we treat data centers from a cyber-physical system (CPS) perspective. Current methods for controlling information technology (IT) and

cooling technology (CT) in data centers are classified according to the degree to which they take into account both cyber and physical considerations.

Magdi S. Mahmoud
KFUPM, Dhahran, Saudi Arabia

Yuanqing Xia
BIT, Beijing, China

March 2019

Acknowledgments

Special thanks are due to the Elsevier team, particularly Acquisitions Editor Sonnini R. Yura, Editorial Project Manager John Leonard, and Copyrights Coordinator Ashwathi Aravindakshan for their guidance, assistance, and dedication throughout the publishing process. We are grateful to all the anonymous referees for carefully reviewing and selecting the appropriate topics for the final version during this process. Portions of this volume were developed and upgraded while offering the graduate courses **SCE-612-171, SCE-612-172, SCE-701-172, SCE-701-181, SCE-515-182** at KFUPM, Saudi Arabia. The support afforded by the Deanship of Scientific Research (DSR) at KFUPM through project no. **BW 181004**, the National Key Research and Development Program of China under Grant **2018YFB1003700**, and the National Natural Science Foundation under Grant 61836001 are gratefully acknowledged.

Magdi S. Mahmoud
Yuanqing Xia

March 2019

Chapter 1

An overview

Contents

1.1 Preliminaries

In the conventional design of control systems, all of the system components, including sensors, controller, actuator, and plant, are installed within a single facility. This arrangement can lead to a high cost of construction, imposes communication constraints, and yields a lack of flexibility. The recent development of Information and Communication Technologies (ICTs) have greatly facilitated the integration of advanced technologies into already designed control systems [1]. Today's Networked Control Systems (NCSs) have incorporated several functionalities, including reduced size, speed, and the ability to work for a long time, to name a few. In turn, these functionalities demand that NCSs possess huge flexible computational resources of smaller size, which are difficult to achieve with the conventional design of these systems.

With the development of Cloud Computing Technologies (CCTs) the issues of resource constrained NCSs are nearly solved. It turns out that the combination of CCTs and NCSs makes it possible to save energy, which reduces the processing energy used in the modern design of NCSs to enhance the system's lifetime. In addition, it reduces the size of NCSs by shifting the core processing unit to a remote "cloud" server. The favorable achievement of this integration is the massive parallel computation on the cloud server.

In another research direction, critical infrastructure sites and facilities are becoming increasingly dependent on interconnected physical and cyber-based real-time distributed control systems (RTDCSs). A mounting cybersecurity threat results from the nature of these ubiquitous and sometimes unrestrained communications interconnections. Much work is under way in numerous organizations to characterize cyber threats, determine the means to minimize risk,

Cloud Control Systems. https://doi.org/10.1016/B978-0-12-818701-2.00009-3

and develop mitigation strategies to address potential consequences. While it seems natural that a simple application of cyber-protection methods derived from the corporate business IT domain should lead to an acceptable solution, the reality is that the characteristics of RTDCSs make many of these methods inadequate and unsatisfactory or even harmful. A solution lies in developing a defense-in-depth approach that ranges from protection at the communications interconnection level to the control system's functional characteristics that are designed to maintain control in the face of malicious intrusion. This paper summarizes the nature of RTDCSs from a cybersecurity perspective and discusses issues, vulnerabilities, candidate mitigation approaches, and metrics.

One real-world application of a CCS is the Google self-driving car, which determines its accurate position by sharing sensor information with satellite and the cloud [2]. Building on [1], Fig. 1.1 illustrates some of the applications of the cloud-based control system that use cloud computing resources for better performance.

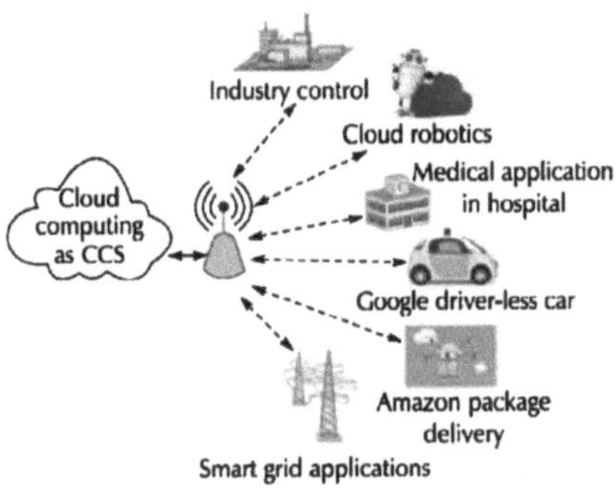

FIGURE 1.1 Representative of cloud-based applications.

Although there are many advantages of using the cloud for different resources, there are many security challenges related to the cyber and physical parts of the system. Local controllers and cloud servers transmit packets of sensor information and the feedback control signals to each other. This communication could be subject to different kinds of security attacks, including *eavesdropping* (where the attacker only watches the information, but does not modify it) and *modification attacks* (where the attacker modifies the message and sends the wrong solution to the local controller).

1.1.1 Real-time distributed control systems

Cyber-critical infrastructure is the junction of control systems and cyber systems. Control systems can be as simple as a self-contained feedback loop, or can be a very complex, networked system of interdependent, hierarchical control systems with multiple components physically distributed over a wide area (miles, counties, states, or larger). The key word in this description is "networked." In its truest sense the term means "an interconnected or interrelated group of nodes." The consequences of control failure and damage potential are proportional to the systems under direct control. Control systems must perform their critical functions without interruption. *Real-Time Distributed Control Systems* (RTDCSs) integrate computing and communication capabilities with monitoring and control of entities in the physical world. These systems are usually composed of a set of networked agents, including sensors, actuators, control processing units, and communication devices. While some forms of RTDCSs are already in use, the widespread growth of wireless embedded sensors and actuators is creating several new applications in areas such as medical devices, autonomous vehicles, and smart structures, and is increasing the role of existing applications such as supervisory control and data acquisition (SCADA) systems.

Currently, RTDCSs are ill prepared for the highly interconnected communications environment that is becoming standard practice. Originally the systems were used on stand-alone networks in physically protected locations without threat of subversion. With the use of data collection and control activation systems being set in remote, unattended locations connected to a public or shared network, this exposure allows intrusion if not properly protected from a perimeter aspect, and more importantly, from a resilient component aspect.

Many RTDCSs are safety critical: their failure can cause irreparable harm to the physical system being controlled and to the people who depend on it. SCADA systems, in particular, perform vital functions in national critical infrastructures, such as electric power distribution, oil and natural gas, water and waste water distribution systems, and transportation systems. The disruption of these control systems could have a significant impact on public health and safety and could lead to large economic losses. While most of the effort for protecting RTDCSs (and SCADA systems in particular) has been done in reliability (i.e., protection against random failures), there is an urgent and growing concern for protection against malicious cyber attacks.

Methods derived from a corporate business information technology IT domain would lead to an acceptable solution if the physical losses were limited to just data. The reality is that the characteristics of RTDCSs make many of these methods inadequate and unsatisfactory or even harmful. A solution lies in developing a defense-in-depth approach ranging from the protection of communications interconnection, to the functional characteristics of the control systems designed to ensure proper control under malicious intrusion, or to a fail-safe analog that includes intrusion tolerant capabilities that ensure critical functionality and survivability.

1.1.2 Synopsis of the security problem

This section provides a synopsis of the problem domain, a framework for defense-in-depth, mitigation methods, and metrics that codify RTDCS resilience to intrusion. We conclude that while the various fields currently used to solve the problem (using elements from information security, sensor network security, and control theory) can give the necessary mechanisms for the security of control systems, these mechanisms alone are not sufficient for the security of RTDCSs. Historically, control systems are in manned, protected environments and are under constant monitoring. Such perimeter isolation, or *fence-and-gate*, views of protection are impractical as control systems are frequently located at unmanned, unmonitored installations. Security of these sites is performed by a literal fence and lock. This security is easily subverted by a well-informed intruder who can gain physical access undetected and consequently leave these remote systems subject to control by hostile intruders. Extending perimeter security may be impractical, if not impossible. Furthermore, it is entirely possible that a trusted insider can become an adversary, which raises the risk of danger to the greater control system, to the equipment under its control, or both. RTDCSs have an additional complication of being responsible for operating critical infrastructures and facilities of great economic or strategic value. Examples include electric power distribution, telecommunications, public transportation, water supply and sewage, chemical plants, oil and gas pipelines, and military vessels.

Cyber control is considered fast, accurate, and able to optimize resources (e.g., energy efficiency) and delivery of services while minimizing overall cost. These advantages drive networked implementation. A recent example is the synchrophasor, which captures time-accurate current and voltage (phase) at critical points on the electric grid. Unprecedented knowledge of power flow and stability is obtained from this information. Installation of RTDCS elements in the power system base improves the information from the *smart grid* and if designed properly (i.e., if it is attack tolerant) improves the cybersecurity of the conglomerate of networked devices that make up the smart grid.

1.2 Basics of cloud control systems

This book introduces the basic definitions and some new developments in the growing area of cloud-based control systems or in short CCSs. In this regard there are different views on the topic including *Remote-Control Systems (RCSs); Wireless-Control Systems (WCSs); Internet-Control Systems (ICSs); NCSs*. Extending NCSs from a wide spectrum, one fundamental view is that a CCS essentially contains a CPS and a cyber-physical control system (CPCS). Here a CCS embraces the idea of "*control as a service (CaaS)*", that is, *control algorithms can be scheduled as a kind of resource*. In a CCS there are three closed loops:

1) the control loop;
2) the scheduling loop;
3) the decision-making loop.

However, it is not confined to a closed-loop control format. From this perspective a CCS focuses on managing the virtual control resources, especially the control algorithm, in addition to a local control that is necessary to satisfy a short time-delay tolerance system. A cloud control plus a local control scheme will not only achieve a powerful process of industrial automation (IA), but also an accurate and real-time control.

A generic infrastructure of a cloud data center is depicted in Fig. 1.2 where a self-managed, dynamic, and dependable infrastructure constantly delivers the expected quality of service with reasonable operation costs and an acceptable carbon footprint for large-scale services with sometimes dramatic variations in capacity demands.

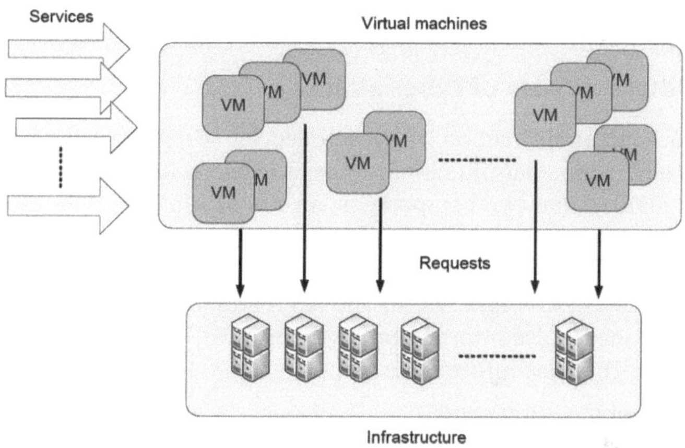

FIGURE 1.2 Basic concept of the cloud.

The current challenges for clouds include

- traffic performance monitoring of large distributed systems;
- workload models;
- scalability effects;

Instead, security challenges normally arise when

(a) the computation is corrupted by false sensor information or
(b) the control centers send malicious control actions to the physical process.

We want to emphasize that *CaaS* gives the control research community a suitable platform by providing services like Software-as-a-Service (*SaaS*), Platform-as-a-Service (*PaaS*), and Infrastructure-as-a-Service (*IaaS*) on the "pay as you go" scheme.

1.2.1 Cloud control security

The security of a CCS includes

- service security;

- storage security;
- management security;
- network security as in NCSs.

Looking ahead, we can say that a CCS represents a significant class of modern control systems that possesses the following features:

- It embodies the fundamental features of cloud computing and introduces *CaaS* as an additional layer of cloud computing hierarchy;
- It shares the major constituents of CPSs in terms of realization of secure control and estimation methods against cyber attacks, and guaranteeing safe computational processing by employing particularly structured private platforms;
- It extends NCSs further to ensure deploying advanced technologies of IA.

1.2.2 Different types of cyber attacks

Consider a general abstraction of a CCS and let y represent the sensor measurements and u the control commands sent to the actuators. A controller can usually be divided into two components: an estimation algorithm used to track the state of the physical system given y, and the control algorithm that selects a control command u given the current estimate. Attacks to a CCS can be summarized as follows (see Fig. 1.3): **A1** and **A3** represent *deception attacks*, where the adversary sends false information $\tilde{y} \neq y$ or $\tilde{u} \neq u$ from one or more sensors or controllers. The false information can include

(1) an incorrect measurement;
(2) the incorrect time when the measurement was taken;
(3) the incorrect sender identification (ID).

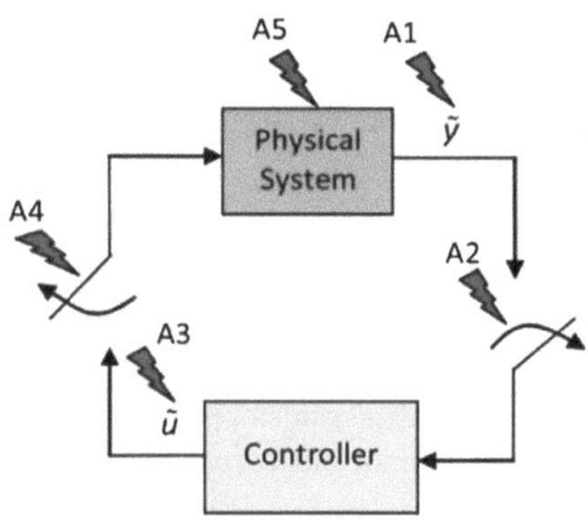

FIGURE 1.3 Different types of cyber attacks.

The adversary can launch these attacks by obtaining the secret key or by compromising some sensors (**A1**) or controllers (**A3**). **A2** and **A4** represent *denial-of-service (DoS) attacks*, where the adversary prevents the controller from receiving sensor measurements. To launch a DoS attack the adversary can jam the communication channels, compromise devices to prevent them from sending data, attack the routing protocols, etc. Finally, **A5** represents a direct attack against the actuators or an external physical attack on the plant. From an algorithmic perspective, we cannot provide solutions to these attacks (other than detecting them). Therefore, significant efforts must be placed on deterring and preventing the compromise of actuators and other direct attacks against the physical system, for example by securing the physical system, employing monitoring cameras, etc. Although these attacks are more devastating, we believe that a risk-averse adversary will launch cyber attacks **A1–A4** because it is more difficult to identify and prosecute the perpetrators, it is not physically dangerous for the attacker, and the attacker may not be constrained by geography or distance to the network. The reader is advised to consult Chapter 2 for further analysis and detailed discussions.

1.2.3 Passive versus active attacks

Violations of the desired security properties typically arise through known attack mechanisms. A taxonomy developed by the National Institute of Standards and Technology (NIST) is segregated into passive attacks, which require nothing more than an ability to eavesdrop on wireless communications, and active attacks, which require active interference. Passive attacks are difficult to detect as they involve no alteration or introduction of data; there are two types, both of which are attacks on confidentiality. Active attacks allow an attacker to be more intrusive; there are four types.

Passive attacks:

- *Eavesdropping*
  An attacker acquires data by passive interception of information transactions. If encryption is used, cracking the encryption and decrypting the traffic counts as a passive eavesdropping attack.
- *Traffic analysis*
  Deduction of certain properties regarding information transactions based on the participants, duration, timing, bandwidth, and other properties that are difficult to disguise in a packet-encrypted wireless environment allow an attacker to examine a network by observing its transmissions.

Active attacks:

- *Masquerades*
  An attacker fraudulently impersonates an authorized entity to gain access to information resources. A "man-in-the-middle" attack involves a double masquerade—the attacker convinces the sender that she is the authorized recipient, and convinces the recipient that she is the intended sender. Man-in-the middle attacks on WiFi networks using

a counterfeit access point are common. Successful masquerades can compromise all aspects of security.

- *Replay*
  An attacker is able to rebroadcast a previous message and elicit a reaction. This reaction either allows the attacker to force the information system into a vulnerable state (e.g., a system reset) or to collect information to enable further attacks (such as WEB encrypted packets). Replays are most directly a compromise of integrity, but also compromise authentication, access control, and non-repudiation. Selected replay attacks can also impinge on availability and confidentiality.
- *Message modification*
  Modification of transmitted packets by delaying, inserting, reordering, or deleting en route changes a message. In a wireless network, man-in-the-middle attacks are the most direct route to message modification. Message modification is a violation of integrity, but can potentially affect all aspects of security.
- *Denial of service*
  DoS occurs when an attacker compromises the availability of an information system. In a wireless environment the most direct routes to DoS are disabling one of the communication partners or jamming the wireless channel itself.

Jamming: Traditionally the term jamming refers to the disruption of communications systems by the use of intentional electromagnetic interference. Jamming targets corrupt the desired signals from expected users or blocks communications between users by keeping the communications medium busy. Jamming can originate from a single attacker or multiple attackers in coordination, and can target a specific user or the entire shared medium. The result is a DoS attacks, which can vary from simple to sophisticated. An attacker can send a signal with considerably higher signal strength than the usual signal levels in the system, and then flood the channel so that no user can communicate through it. A more sophisticated way is for the attacker to gain access to the system and violate the network protocol for sending packets, thereby causing many more packet collisions. In the context of electric power grids, jamming can result in a security breach in the form of DoS for communications systems by blocking the on and off activation of remote generating sites or the opening and closing of transmission line switches in response to load demands. In particular, wireless communications systems are more vulnerable to jamming because of their potential for access from covert locations.

1.2.4 Fundamental requirements

The ultimate purpose of using cyber infrastructure (including sensing, computing, and communication hardware and software) is to intelligently monitor (from physical to cyber) and control (from cyber to physical) our physical world. A system with a tight coupling of cyber and physical objects is called a CPS [3], [4], which has become one of the most important and popular computer applications today. Table 1.1 lists the major differences between cyber resources and physical objects [5], [6]. A CPS often relies on sensors and actuators to implement tight interactions between cyber and physical objects. The sensors (cyber objects) can be used to monitor the physical environments, and the actuators or controllers can be used to change the physical parameters. Regarding the interactions between sensors and controllers and extending the work of [7], we depict the wind power system in Fig. 1.4, in which there exist three types of communications between sensors and controllers:

TABLE 1.1 Cyber and physical properties of CPSs.

	Cyber	Physical
Method of ensuring proper order	Sequences	Real time
Event synchronization	Synchronous	Asynchronous
Time properties	Discrete	Continuous
Structure	Computing abstractions	Physical laws

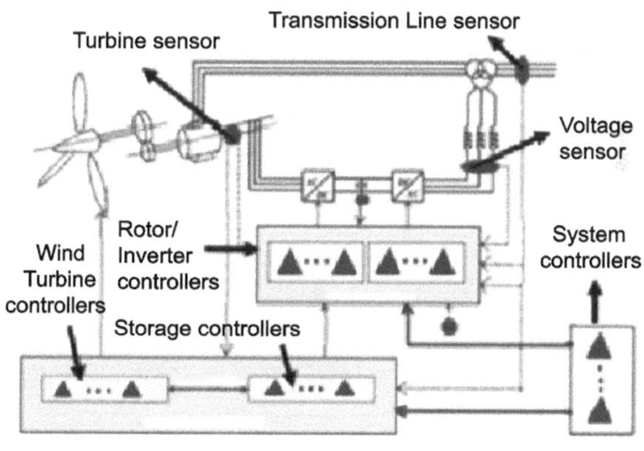

FIGURE 1.4 Locations of sensors/controllers in a smart grid.

1. *Sensor-to-sensor (S–S) coordination*: The sensors in a power cluster (with hundreds of wind turbines) need to communicate with each other to find an electromagnetic distribution map for power flow analysis;
2. *Sensor-to-controller (S–C) coordination*: A controller makes decisions based on the collected sensor data. A controller may need data from both local and remote sensors;
3. *Controller-to-controller (C–C) coordination*: A controller may need to coordinate with other controllers to make a coherent decision.

Typically a storage controller needs to work with other controllers (that control loads and renewable sources) to decide whether the storage unit should be charged or discharged and how much electricity load it should handle.

As treated in [8] and [9], the sensor and controller relationship can be represented as a NCS with inputs (sensor data) and outputs (control commands), see Fig. 1.5, where a wireless sensor and controller (WSCN) with delay and packet loss can be used to describe a CPS. It has state transitions based on the control results.

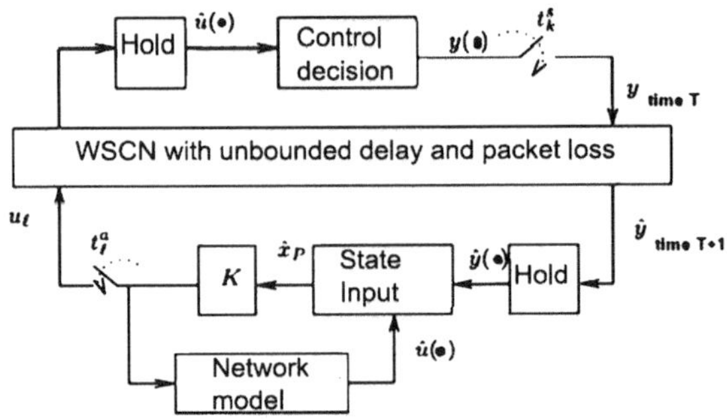

FIGURE 1.5 Schematic CPS state transition.

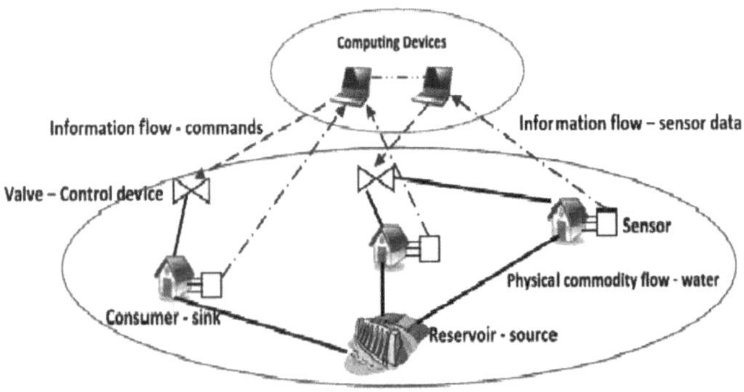

FIGURE 1.6 A water distribution system as a CPS.

In Fig. 1.6 an intelligent water distribution network is presented. Among the physical components there are pipes, values, and reservoirs. Using this system, researchers are able to track water use. They are also able to predict where most of the water will be consumed. It has a multilayer architecture. One layer is the actual water flow, such as a reservoir of a sink. This layer has cyber objects (sensors) that communicate to the higher level cyber objects, such as computer devices, how much water will be used and when. This allows the computers to send the water where it will be needed at the correct times. It also allows monitoring of the maintenance side of the water flow. It achieves this by monitoring what amount is being used at a house and how much water is being sent to that section. If more water is being sent to a section than is being used, the computers will know there is leak or malfunction.

1.2.5 Design consideration

A CCS is a "*system of systems*" where complex and heterogeneous systems interact in a continuous manner, and proper regulation of the system necessitates careful co-design of the overall architecture. Since a CPS lies at the core of a CCS, we focus in the sequel on the consideration of CPS.

A resilient CPS design includes three features (3S):

(1) *Stability:* no matter how the environment generates noise and uncertain factors the control system should always eventually reach a stable decision result;

(2) *Security:* the system should be able to detect and countermeasure the cyber-physical interaction attacks;

(3) *Systematicity:* the components of cyber and physical should be seamlessly integrated into a systematic design.

To achieve such a resilient CPS the following challenges should be addressed:

(A) *Dependability;*

(B) *Consistency;*

(C) *Reliability;*

(D) *Cyber–physical mismatch;*

(E) *Cyber–physical coupling security.*

Briefly stated, dependability is an important quality for any CPS, where in some applications adaptability brings higher dependability. Here raw physical process (RPP) data is collected, and the system is controlled by an intelligent computational world.

To achieve consistency, each component in the CPS can be accounted for in a base architecture (BA), and every path of communication and physical connection between elements is allowed in the BA by connectors. This means that the system should know all the possible paths. If an incorrect connection or assumption is made, it will not be in the BA. To show this multiview consistency, additional tools are needed. Fig. 1.7 shows the design flow in the water system used to check consistency.

A disconnection often lies between program execution and physical requirements. Programs essentially have 100% reliability in the sense that a program will go through the exact same set of commands in exactly the same order every time it is run.

In a CPS the interaction and coordination between the physical elements and the cyber elements of a system are key aspects. In the physical world one of the most dominant characteristics is its dynamics, that is, the state of the

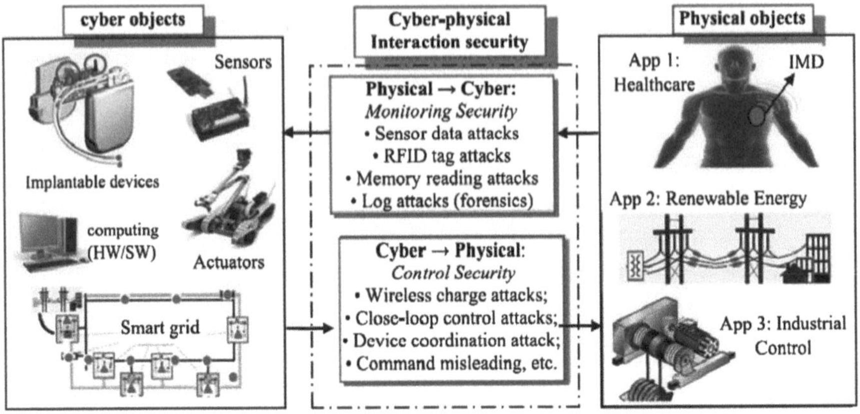

FIGURE 1.7 A water distribution system as a CPS.

system constantly changes over time. On the other hand, in the cyber world these dynamics are more appropriately defined as a series of sequences that do not have temporal semantics. There are two basic approaches to analyzing this problem: cyberizing the physical (CtP), where cyber interfaces and properties are imposed on a physical system, and physicalizing the cyber (PtC), where software and cyber components are represented dynamically in real time.

A CPS should be resilient to both natural faults and malicious attacks. In particular, we will describe how we can use a suitable control model and corresponding security scheme to build a resilient CPS. In CPSs the physical systems are susceptible to the cybersecurity vulnerabilities from a monitoring and control security perspective (see Fig. 1.7).

Over the last 10 years the concept of CPSs has emphasized the integrated modeling and analysis of computational platforms and the physical processes that are controlled by such platforms. One typical class of CPS is made up of embedded control systems. In such a system, physical processes are controlled by a piece of software running on an embedded platform. These systems are commonly found in automotive, avionics, IA, and medical devices.

Typical design layouts are the following:

(I) *Separate and iterative design*, as shown in Fig. 1.8;
(II) *Platform design for control applications*, as shown in Fig. 1.9;
(III) *Control platform for co-design synthesis*, as shown in Fig. 1.10.

1.3 A view on modeling cloud control systems

It is important to have a suitable CCS model with quantitative cyber-physical interaction descriptions in order to understand different types of control and security designs. Here we will explain some modeling issues in CCSs. Recall that physical processes are made up of a combination of different processes

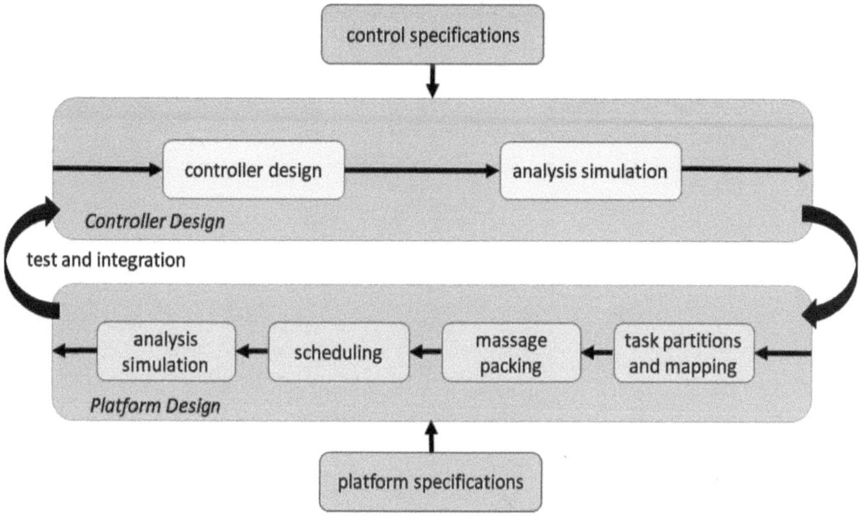

FIGURE 1.8 Separate and iterative design layout.

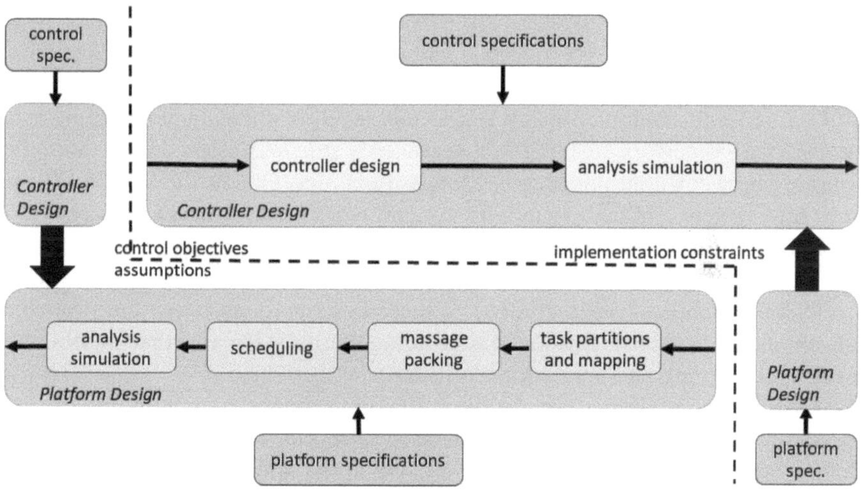

FIGURE 1.9 Layout of platform design for control applications.

that run in parallel. The job of measuring and controlling these processes by orchestrating actions that have an influence on the processes is a very important task performed in an embedded system. Models are a major stepping stone in the development of CCSs.

Generally, models can show how the design process has evolved, and help form the specifications that govern a system. In addition, models allow a CCS design to be tested in a safe environment, which will allow engineers to determine if any design defects exist. To model a CCS, engineers will have to include

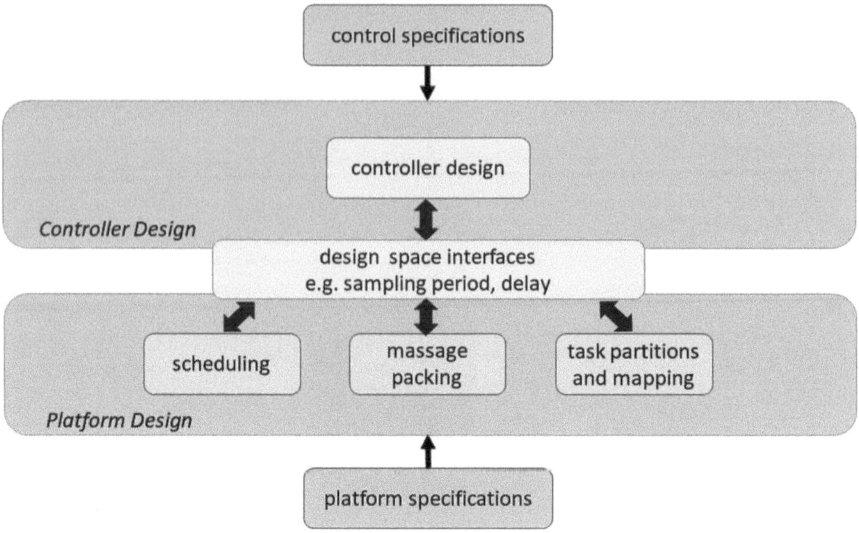

FIGURE 1.10 Control platform for co-design synthesis layout.

the models of the physical processes, and models of the software, computation platforms, and networks.

One general standpoint is that at a certain level, mathematical modeling and an analysis framework for a CPS is necessarily *a hybrid* issue due to the tight coupling between continuous and discrete dynamics. One of the simplest form of hybrid system, called a switching system, is one that switches between different operation modes to adapt to various changes.

Since computing and networking systems in CCSs interact with the physical world, predictability (or timeliness) of these systems is an important property that should be provided. Real-time scheduling theory is the area that studies this issue in computing and networking systems.

Along with advances in real-time scheduling theory, computing platforms for real-time and embedded systems have been developed and used successfully in many application areas [27]. However, due to the scale, structure, and behavioral complexities of current and future CCSs, it is an important challenge to develop extensible, scalable, and adaptable software platforms that can operate in distributed, heterogeneous, time-critical, and safety-critical environments.

1.3.1 Development and activities

Scientific CCS composition is a new system architecture pattern that is composed of hierarchical systems including components and subsystems, service quality theory, agreements, modeling language, and tools that can analyze, integrate, and simulate different components. Computation theory should be able to handle feedback control of real-time systems based on event-driven strategies that suit the asynchronous dynamic event processing on different timescales. Research into CCSs is just at their beginning worldwide. Since a CCS is the integration of multidisciplinary heterogeneous systems, without a unified global model CCS research is carried out by experts in various areas from the perspective of applications in their own fields. At present, CCS research mainly focuses on studies of system architecture, information processing, and software design.

1.3.2 Architecture of cloud control systems

Modeling can be considered the technology used to describe the target system before completion. CCS architecture is the base of research and development, and CCS models must be modified and integrated on the basis of existing structures of physical systems, network systems, and computer systems. Abstraction and modeling of communication, computation, and physical dynamics on different timescales are also needed to accommodate the development of CCSs. We propose a kind of cloud control system structure model that can be divided into three layers: a user layer, an information system layer, and a physical system layer. The physical system is composed of a large number of embedded systems, sensor networks, smart chips, etc., that take charge of the collection and transmission of information and the execution of control signals; it is the foundation of the CCS. The information system layer is mainly responsible for the transmission and processing of the data collected from the physical system, which is the core of the CCS. The user layer mainly completes the work, such as data query, strategy and safety protection, under a human-computer interaction environment that should be guaranteed by regular CCS operations. CCSs run in the form of a closed-loop control. The architecture of a CCS is shown in Fig. 1.11.

The function of each part in Fig. 1.11 is as follows:

1) Sensor Networks: They use a variety of sensors and real-time embedded systems for real-time data acquisition; conduct analog-to-digital conversion of collected data and other processes including data encryption and data integration through collection nodes; protect the security of data transmission (privacy, integrity, and non-repudiation); reduce the network energy consumption by energy management; apply real-time data protection technology to real-time processing.

2) Next Generation Network Systems: They use antihacking and defense technology against a variety of network attacks; use high-performance encryption algorithms and CA authentication technology to ensure the safety of data transmission; realize the rapid exchange of data transmission by optimizing existing routing algorithms; change the existing network system structure with the "best effort" to provide real-time network transmission services for the system.

3) Data Center: The sensor network transmits data to the data center for storage through next-generation network systems. The data center checks the authentication and integrity of received data and stores the data if they pass the

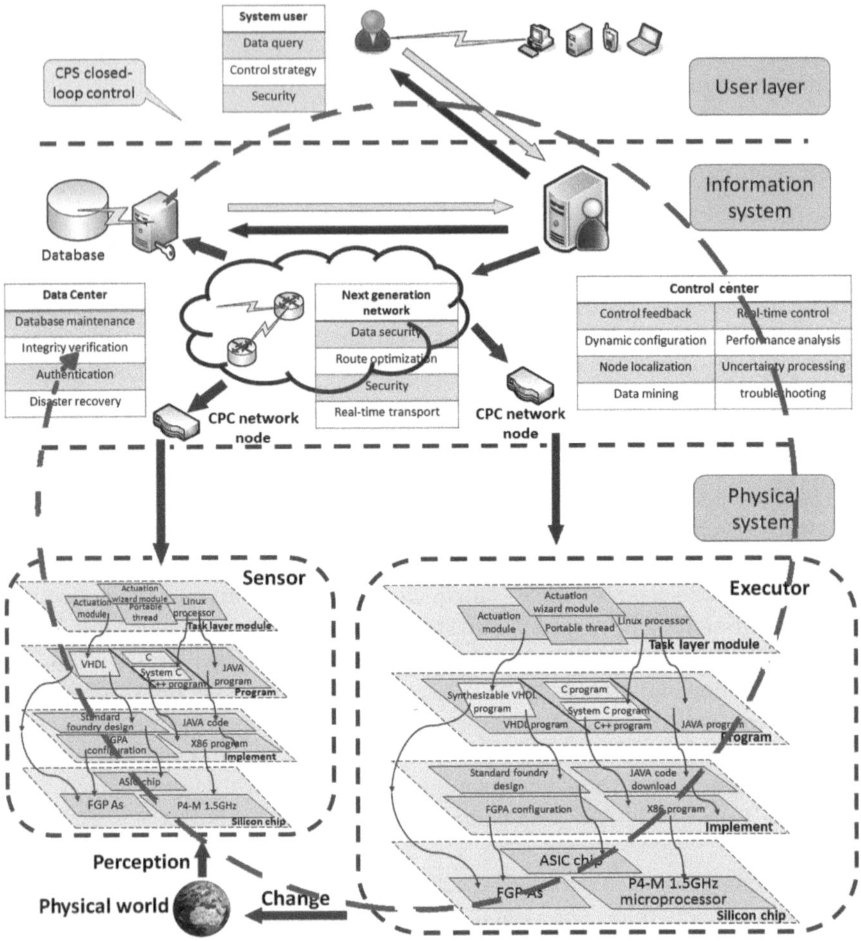

FIGURE 1.11 A CCS architecture.

inspection, otherwise it sends a message to the control center. Then the control center sends the control signals to the actuator, which notifies the sensor network nodes to collect data again. The data center is also responsible for routine maintenance of the database and quick response to instructions sent by the control center (e.g., queries). Regular emergency treatments are also needed to prevent the database from collapse.

4) Control Center: The control center is the most important part of a CCS. It receives the inquiry instructions sent by users and then sends query command to the data center after identity authentication. It categorizes the query results according to control strategies, reports back to the user if they meet the requirements, otherwise finds out the location of the node-by-node positioning technology, and sends control instructions to the actuators for corresponding

processing. The control center configuration policy can be dynamically adjusted according to the users' needs. It conducts forecast analysis and performance analysis of the CPS behavior through data mining technology and uncertainty processing technology, detects the network and node failure through fault diagnosis technology, conducts the corresponding processing, and ensures the real-time control processing of CPSs through real-time control technology.

5) Actuator Networks: They receive the control instructions from the control center and send the control instructions to the corresponding nodes.

6) System User: The system user includes a variety of WEB servers, individual hosts, and external devices. It is responsible for the communication with the CPS, sending inquiry instruction to the control center, and receiving feedback data. Users can send definitions and revised control strategies to the control center for execution.

In this model the CCS would run under closed-loop control, and the real-time capability, security, and system performance are fully considered so that it can preliminarily meet future CCS requirements. Some scholars have also conducted research on the system architecture of CCSs with different studied subjects and from different application perspectives.

An advanced power grid is a complex real-time system that contains network and physical components. Each part may function well independently, but not when they are combined together because the interference may cause errors, for instance the violation of the Nyquist rate in the frequency domain. Y. Sun et al. proposed using RT-PROMELA to build a model that can represent frequency interference and use the real-time interference of Real-Time Sensor Protocol for Information via Negotiation (RT-SPIN) detection to test the accuracy of CPS components. It solved the problem of multiple clock variables in collaboration processing caused by the lack of real-time and asynchronous interaction of components. [12], [14], [15] established a cloud control energy system dynamic model with distributed sensing and control, and discussed the process of information exchange between components in this model, and used the model to develop interactive protocols between the embedded system control terminal and the network system.

As a selective architecture of a CPS, and building on the effort of [11], the reader is referred to Fig. 1.11 and the results reported in [12], [13], [14], [15].

1.4 Notes

This chapter has briefly shed light on several topics including

- Introduction to cloud control systems (CCS);
- Requirements of the portion related to cyber physical systems (CPS);
- Issues of architecture, modeling, and simulation of CCS;
- Control and estimation methodologies.

On the one hand, the Internet of Things (IoT) particularly becomes a utility with increased sophistication in sensing, actuation, communications, control, and creating knowledge from vast amounts of data. On the other hand, cloud computing is revolutionizing access to distributed information and computing resources that can facilitate future data and computation-intensive cloud control functions and improve the reliability and safety of the dynamics. This brought about the security attacks that are problematic in NCSs. The field of CPSs that, in the sense of generalized NCSs, is rapidly emerging due to a wide range of potential applications. However, there is a strong need for novel analysis and synthesis tools in control theory to guarantee safe and secure operation despite the presence of possible malicious attacks.

All of these topics and more will be discussed further in the following chapters. The interested readers are advised to consult the papers and books in the list of references for additional demonstrations, and particularly [16], [17], [18], [19], and [26].

Chapter 2

Cloud control systems venture

Contents

2.1 Introduction

A Cloud Control System (CCS) is a new type of system that integrates computation with physical processes and embodies sufficiently safe control methodologies to deal with cyber attacks. The components of cyber-physical control systems (e.g., controllers, sensors, actuators) transmit information to cyber space through sensing a real-world environment; they also reflect the policy of the cyber space back to the real world [28]. A cyber-physical system (CPS) is a physical and engineered system whose operations are monitored, coordinated, controlled, and integrated by a computing and communication core. This intimate coupling between the cyber and the physical is manifested from the nano-world to large-scale wide-area systems of systems. Recall that the Internet transformed how humans interact and communicate with one another, revolutionized how and where information is accessed, and even changed how people buy and sell products. Similarly, CCSs will transform how we interact with and

Cloud Control Systems. https://doi.org/10.1016/B978-0-12-818701-2.00010-X

control the physical world around us [29]. Cyber-physical systems may consist of many interconnected parts that must instantaneously exchange, parse, and act upon heterogeneous data in a coordinated way. This creates two major challenges when designing CPSs: the amount of data available from various data sources that should be processed at any given time and the choice of process controls in response to the information obtained. An optimal balance needs to be attained between data availability and its quality in order to effectively control the underlying physical processes [30].

2.1.1 Characteristics

A CCS has many and diverse characteristics: distributed management and control, a high degree of automation, real time performance requirements, reorganizing/reconfiguring dynamics, multiscale and multisystem control features, networking at multiple scales, wide geographic distribution with components in locations that lack physical security, integration at multiple temporal and spatial scales, and input and possible feedback taken from the physical environment. As emphasized in Chapter 1, a CPS represents the heart of a CCS. In this regard the objective of a CPS is to monitor the behavior of physical processes and to actuate actions that change its behavior in order to make the physical environment work correctly and more effectively.

2.1.2 Cloud control system venture

There are four main phases in a CCS venture [31]: *Monitoring*, *Networking*, *Computation, and Actuation*:

- *Monitoring* refers to giving feedback on any past actions that are taken by the CCS and ensuring correct operations in future. Monitoring physical processes and environments is the basic function of a CCS.
- *Networking* means that when there are multiple sensors in a CCS, they can all generate data in real time and many of them can generate a lot of data to be aggregated or diffused for further processing by analyzers, and at the same time different applications need to interact with communications.
- *Computing* refers to reasoning about and analyzing the data collected during monitoring, to cross-check whether the physical processes are satisfying the prescribed criteria. If not, then the corrective actions that were proposed earlier can be executed in order to ensure that the criteria are met.
- *Actuation* is used to execute the actions determined in the computing phase. It can correct the cyber behavior of a CCS and can change physical processes and many other forms of actions as needed.

A few examples based on these characteristics are included in [32], where modern power grid CCS, wind farms, and solar farms constitute the physical resources; data that are collected from the sensors of these resources constitute the cyber part of the system. Often a communication channel is involved to transmit the data that are used to monitor and control the physical resources.

On the cyber side, computations are carried out with the objective of maximizing utilization of renewable sources, and a suitable decision is taken based on which of the physical resources are further controlled. Another example is the body sensor network, which is a network of medical devices that can sense, actuate, and communicate with each other through a wireless network. An aircraft can also be seen as a CCS whose smart sensors and networking system enable it to monitor its operation while coordinating with ground stations.

Remark 2.1. From another angle, a CCS is a composition of independently interacting components, including computational elements, communications, and control systems. Applications of CPSs are instituted at different levels of integration, ranging from large-scale systems (mobile CCS, data centers, networking systems, social networking and gaming, surveillance, electric power grids and energy systems, power and thermal management, nation-wide power grids) to medium-scale systems (smart homes and buildings) and to small-scale systems (e.g., ubiquitous health care systems, including implantable medical devices). Cyber-physical systems primarily change how we interact with the physical world, with each system requiring different levels of security based on the sensitivity of the control system and the information it carries. Considering the remarkable progress in CPS technologies in recent years, advancement in security and trust measures is much needed to counter the security violations and privacy leakage of integration elements [33].

2.1.3 Security

Security is a must in CCS. It is necessary to ensure that the systems are trustworthy and secure, and that they protect the privacy of information. For example, patients depending on implanted medical devices want protection of their identity and critical health information that could be exposed when their devices are connected to monitoring networks. Industry requires protection of intellectual property and of sensitive business and demographic information. Ensuring the confidentiality of information and controlling the access and use of data are challenging, especially as the systems that collect, manage, and analyze information are rapidly evolving and in some cases need to operate in a distributed or relatively open environment.

2.2 Cloud control system security objectives

To ensure the security of a CCS the following security objectives must be achieved. Fig. 2.1 depicts the various security objectives of a CCS.

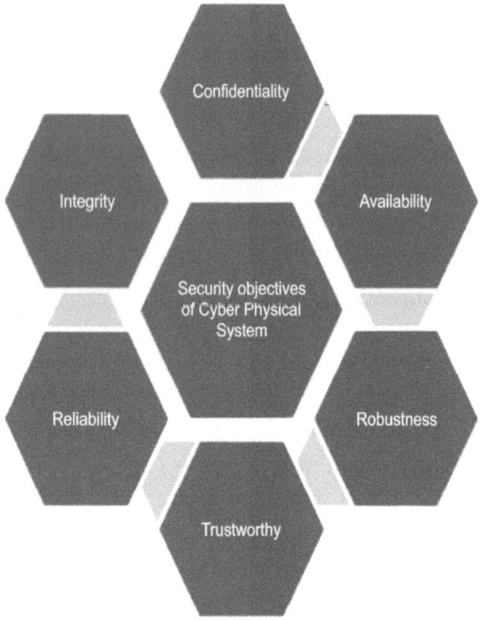

FIGURE 2.1 Security objectives of CCS.

2.2.1 Confidentiality

Confidentiality means that a CCS should have the ability to block information from being disclosed to unauthorized individuals or systems. For example, in a healthcare CCS the personal health records of any patient can be transmitted from a local repository or device to the clinician or analyzer center. The healthcare CCS should maintain confidentiality by securing the transmitted data, by restricting the places storing the patients' personal health records, and by limiting access to these storage places. Disclosure of health data in any way raises doubts regarding the system's confidentiality. If any unauthorized person accesses these records, a notification of confidentiality leak should be sent out. This ensures that all sensitive information generated within the system is disclosed only to those who are supposed to see it.

2.2.2 Integrity

Integrity means modification of any resource or data can only be possible after authorization. To ensure integrity of data a CPS requires the capability of detecting any changes introduced by unauthorized or malicious activity in the message being passed. It ensures that all information generated and exchanged during the system's operation is accurate and complete without any alterations.

2.2.3 Availability

Availability in a CPS refers to providing service every time it is requested by preventing computing, control, and communication corruption due to any failures in hardware, system upgrades, power outages, or attacks. It ensures that any entity that uses the data, services, and resources of the system are able to do so when required.

2.2.4 Reliability

Reliability in a CCS refers to ensuring that the data, transactions, and communications are genuine. In a CCS the reliability performs an originality check on all the related processes such as monitoring, networking, computing, and actuation.

2.2.5 Robustness

Robustness of a CCS refers to the level of system quality; it describes the degree to which a system is capable of working properly and effectively even in the presence of incorrect inputs, malfunctions, or disturbance [34].

2.2.6 Trustworthiness

Trustworthiness in a CCS refers to the extent that the system can be relied upon to correctly perform the system tasks under predefined operational and environment conditions over a predefined time. A threat is a violation of security [35]. A system needs to be guarded against threats in order to ensure its correct operation at all times. The execution of the threats is called an attack, while the entities that execute these threats are called attackers [36].

2.3 Types of attacks in cloud control system

Cyber-physical systems are defined as the integration of computation, communication, and control in order to achieve the desired performance of the physical processes [37]. Recently, CCS has become an interest of research due to the wide range of applications in a variety of fields: sustainable and blackout-free electricity generation and distribution, clean and energy-aware buildings and cities, smart medical and healthcare systems, transportation networks, chemical process control, smart grids, water/gas distribution networks, emergency management, etc. [38].

On the other hand, CCS has great potential security threats and can be affected by several cyber attacks without any signs of failure. These attacks can cause disruption to the physical system. So it is crucial to ensure the control system security, robustness, and resilience in practical CPS applications.

Security in safety-critical systems in which physical damage can be caused is a crucial issue. In CCS, a lot of research can be done regarding security issues.

One example of these issues is the cyber attack on a supervisory control and data acquisition (SCADA) system described in [39]. Other cyber attacks have targeted water systems [40], power utilities [41], trams [42], and natural gas pipeline systems [43]. The most famous is the Stuxnet worm that affected a SCADA system in Iran's nuclear program control systems causing substantial damage [44], [45], [46].

Several survey papers are available in the literature summarizing the updated results on security control of CPS applications. In [47] the authors discussed the security of networked control systems (NCSs) based on introducing the attack space which is defined by the system knowledge of the adversaries. Then the model and analysis of replay, bias injection, and zero dynamics attacks were presented. A conceptual model was proposed in [48] for NCSs with malicious adversaries, cyber security metrics for stealthy adversaries, and the detection of stealthy adversaries. Approaches of modeling data centers in CPS applications related to cloud computing were presented and their applications were surveyed [49]. A general introduction to the issue of CPS security in NCSs can be found in [50]. Literatures on both security filtering and security control of CPS applications were reviewed while focusing on the three main three types of cyber attacks mentioned above. [51] summarized the detection, design, and secure estimation and control of CPS cyber attacks.

The approaches of a Software-Defined Networking (SDN) controller which is able to establish paths between sensors and actuators were presented [52]. The differences between CPSs and the Internet of Things (IoT) were discussed, and several kinds of CPS attacks were reviewed [53]. The CPS is analyzed as a special case of IoT, and its characteristics and applications in control theory, computer sciences, and communication engineering were discussed in [54]. A general review including the application domain, security, privacy, and defense of CPS applications were provided in [55]. CPS security under a unified framework was presented in [56]; the CCS attacks were classified as cyber attacks, cyber-physical attacks, and physical attacks, and then the security control of CPSs was discussed. In [57] the security challenges at different layers of CPS architecture, and the risk assessment and methods of securing CPSs were analyzed. The concept and characteristics of CCS were introduced, and the development of CCS was presented from several points of view, such as software design, information processing technology, and system models [11]. In [58] the authors reviewed several issues of CPS security, such as modeling of CPS, describing several attacks, discussing the performance analysis and detection methods, and security control and estimation in CCS.

The vulnerability of the cyber parts a CCS allows injecting attacks into the system in a stealthy and unpredictable way. As an example, inserting some malware such as worms and viruses in medium-access control layers or the networked components' computerization could cause disruption of coordination packets. Moreover, an attacker can gain illegal access to the monitoring centers while obtaining the encryption key for the purposes of damaging normal operations [58]. This means that the attacker can either arbitrarily damage the dynamics of the system or induce any perturbation when there is a lack of security protection. The effect of attacks on networked systems was discussed, and a new security index to analyze the effect of such attacks by applying H_2 norms of attacks targeting outputs was presented in [59]. Additionally, optimization problems were solved by selecting inputs or outputs that pointed to attacks with maximum impact and minimum detectability.

In general, cyber attacks in the literature can be classified into three main types: denial-of-service (DoS) attacks, deception attacks, and replay attacks. The focus of this section will be on each of the attack types. The modeling and detection of each attack will be addressed, and the control of CCS under attack will be discussed in detail.

2.3.1 Detection of cyber attacks

The first step in building a secure CPS is to build a reliable attack detector. There are four main strategies proposed in the literature for attack detection as listed in Fig. 2.2.

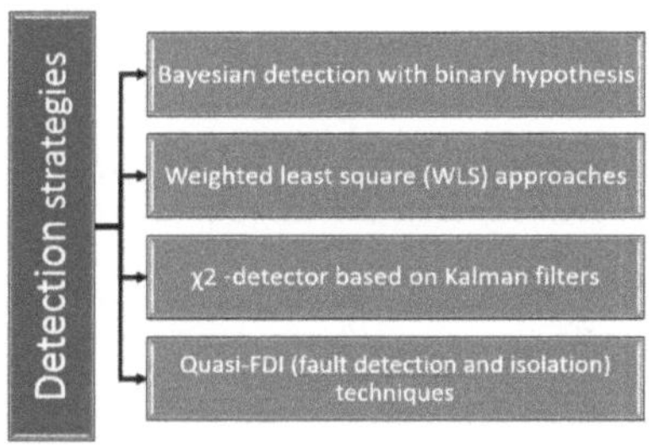

FIGURE 2.2 Detection strategies.

2.3.2 Bayesian detection with binary hypothesis

This is one of the detection methods that is used, especially in deception attacks in the sensor networks, and it consists of a hypothesis test with prior probabilities of two hypotheses [60,61]. The performance limit of cooperative spectrum sensing is analyzed subject to Byzantine attacks, while the fusion center is affected by false data which increases the probability of incorrect sensor output [62]. A likelihood ratio detector based on binary hypothesis is proposed to cope with the predetermined bounded error in sensor network data for the security of the smart grid [63]. In [64], a detector based on an improved likelihood ratio subject to colored Gaussian noise under false data injection attacks is proposed for both observable and unobservable scenarios in SCADA systems.

One example of this method was applied in [65]. The main assumption applied is that all packets transmitted from a node are independent. This means that finding a packet to be malicious will not affect the probability of the next packet being malicious. As a result the attacks could have several forms, affecting one or more packets. For the computation of trust values, it is assumed that the node is sending N packets, of which m packets are trusted to be normal. The distribution of observing $n(N) = m$ is described by the following binomial distribution:

$$P(n(N) = m|p) = C(N, m)p^m(1-p)^{N-m}, \tag{2.1}$$

where $P(n_i : \mathit{normal}) = p$ is the probability of the ith packet being normal, $n(N)$ is the number of normal packets, and n_i means that the ith packet is normal, $C(N, m)$ is a permutation matrix given by $N!/(m!(N-m)!)$ and $N!$ denoted the factorial of N.

The Bayesian model estimates the probability of $P(V_{N+1} = 1|n(N) = m)$, and determines whether the $N + 1$ packet is normal or not. The following probability distribution is formulated using the Bayesian theorem:

$$P(V_{N+1} = 1|n(N) = m) = \frac{P(V_{N+1} = 1, n(N) = m)}{P(n(N) = m)}. \tag{2.2}$$

Now the marginal probability distribution can be applied, leading to the following equations:

$$P(n(N) = m) = \int_0^1 P(n(N) = k|p)f(p) \cdot dp \tag{2.3}$$

$$P(V_{N+1} = 1, n(N) = m) = \int_0^1 P(n(N) = m|p)f(p)p \cdot dp. \tag{2.4}$$

Since there is no prior data for p, it is assumed to be determined by a uniform prior distribution $f(p) = 1$, where $p \in [0, 1]$. So, Eqs. (2.1)–(2.4) could be rewritten as follows:

$$\begin{aligned} P(V_{N+1} = 1|n(N) = m) &= \frac{\int_0^1 P(n(N) = m|p)f(p)p \cdot dp}{\int_0^1 P(n(N) = k|p)f(p) \cdot dp} \\ &= \frac{k+1}{N+2}. \end{aligned} \tag{2.5}$$

In conclusion, by solving Eq. (2.5) the number of normal packets m and the total number of packets N can both be determined in a wireless sensor network after collecting the traffic information, and a malicious node can be known accordingly by applying an appropriate threshold. So, a malicious node can be determined using this model.

2.3.3 Weighted least-squares approaches

A weighted least-squares (WLS) approach is an efficient and reliable attack detection method for the measurement of data. It is widely applied, especially in power systems [66,67] and smart grids [68,69]. The judgment on the presence of a bad measurement is carried out by comparing the constructed measurement residual with a predefined threshold.

Consider the following linearized system:

$$\mathbf{z} = \mathbf{H}\mathbf{x} + \mathbf{e}, \tag{2.6}$$

where $\mathbf{H} = \left[h_{ij}\right]_{m\times n}$ is the Jacobian matrix of the measurement with full column rank when $m > n$; $\mathbf{z}$ and $\mathbf{x}$ is the measurement and states vectors, respectively; and $\mathbf{e}$ is the noise affecting the system. The estimation problem is to calculate an estimate $\hat{\mathbf{x}}$ of the variables $\mathbf{x}$ that is the best fit of the meter measurements $\mathbf{z}$ with reference to (2.6). The difference between the observed measurements $\mathbf{z}$ and the estimated measurements $\hat{\mathbf{z}}$, which is the residual, is defined as $\mathbf{r} = \mathbf{z} - \hat{\mathbf{z}} = \mathbf{z} - \mathbf{H}\hat{\mathbf{x}}$. The WLS criterion problem is to calculate an estimate $\hat{\mathbf{x}}$ that minimizes the performance index $J\left(\hat{\mathbf{x}}\right)$, which is described by the formula

$$\min_{\hat{\mathbf{x}}} J\left(\hat{\mathbf{x}}\right) \triangleq \left(\mathbf{z} - \mathbf{H}\hat{\mathbf{x}}\right)^T \mathbf{W}\left(\mathbf{z} - \mathbf{H}\hat{\mathbf{x}}\right), \tag{2.7}$$

where the weight matrix $\mathbf{W} \triangleq \boldsymbol{\Sigma}^{-1}$. In order to calculate the first-order optimal condition, $J\left(\hat{\mathbf{x}}\right)$ is differentiated [70],

$$\hat{\mathbf{x}} = \left(\mathbf{H}^T\mathbf{W}\mathbf{H}\right)^{-1}\mathbf{H}^T\mathbf{W}\mathbf{z} \triangleq \mathbf{E}\mathbf{z}, \tag{2.8}$$

where $\mathbf{E}$ is the "pseudoinverse" of $\mathbf{H}$ and $\mathbf{E}\mathbf{H} = \mathbf{I}$.

2.3.4 χ^2 Detector based on Kalman filters

In this approach the characteristic of the Kalman filter residual is used instead of the calculation of the WLS making it suitable for use in the presence of bad or false data.

Let us consider the following linear-time invariant (LTI) model:

$$\begin{aligned} x_{k+1} &= Ax_k + Bu_k + w_k, \\ y_k &= Cx_k + v_k, \end{aligned} \tag{2.9}$$

where $x_k \in \mathbb{R}^n$, $u_k \in \mathbb{R}^p$, and $y_k \in \mathbb{R}^m$ are the state variables, control input, and system measurements, respectively; $w_k \in \mathbb{R}^n$ and $v_k \sim \mathcal{N}(0, R)$ are the process noise and the measurement noise, respectively. The optimal state estimate $\hat{x}_{k|k}$ can be calculated using the following Kalman filter:

$$\begin{aligned} \hat{x}_{0|-1} &= \bar{x}_0, \quad P_{0|-1} = \Sigma \\ \hat{x}_{k+1|k} &= A\hat{x}_k + Bu_k, \quad P_{k+1|k} = AP_kA^T + Q \\ K_k &= P_{k|k-1}C^T\left(CP_{k|k-1}C^T + R\right)^{-1} \\ \hat{x}_k &= \hat{x}_{k|k-1} + K_k\left(y_k - C\hat{x}_{k|k-1}\right) \end{aligned} \tag{2.10}$$

$$P_k = P_{k|k-1} - K_k C P_{k|k-1}. \tag{2.11}$$

It has been shown that the Kalman filter resides in $y_i - C\hat{x}_{i|i-1}$ for system (2.9) with the Kalman filter, and the linear-quadratic-Gaussian (LQG) controller is Gaussian independent and identically distributed (i.i.d.) with zero mean and covariance [71].

Let the weigted error signal be defined as

$$g_k \triangleq \sum_{i=k-\mathcal{T}+1}^{k} \left(y_i - C\hat{x}_{i|i-1}\right)^T \mathcal{P}^{-1} \left(y_i - C\hat{x}_{i|i-1}\right), \tag{2.12}$$

where $\mathcal{T}$ is the window size. The value of g_k has an χ^2 distribution with $m\mathcal{T}$ degrees of freedom during the normal operation, which means a lower probability of a larger g_k. The χ^2 detector at time k is

$$g_k \underset{H_1}{\overset{H_0}{\lesseqgtr}} \delta, \tag{2.13}$$

where δ is the threshold selected for a predetermined false alarm probability and H_1 denotes a triggered alarm [71]. This detector was implemented for SCADA systems in [71] and a similar result was presented in [72].

The χ^2 detector with the cosine similarity matching was applied for detecting false data injection attacks affecting smart grid [73]. A residue-based detection algorithm is presented that detects deception attacks in a remote state estimation application where a remote estimator receives data from smart sensors [74]. Another application of this approach was proposed in [75] to detect bias injection attacks for stochastic linear dynamical systems. The feature selection of the χ^2 approach and multiclass support vector machine (SVM) was applied to build an intrusion detection model [76].

2.3.5 Fault detection and isolation techniques

Fault detection and isolation (FDI) is a well-known technique, and is applied widely in NCSs. It consists of monitoring a system and identifying the occurrence of a fault and its type and location. This technique has attracted attention and has been applied to determine the existence of external attacks in CPS applications. In [77] the CPS subjected to several attacks was modeled as a descriptor system and the attacks were considered as unknown inputs affecting both the state and the measurements. Then the undetectable attacks were characterized using the graph theory; using the leveraging on tools from geometric control theory, centralized and distributed monitors were designed to detect and distinguish attacks. A fault detection method based on a geometric approach was implemented to detect faults and cyber attacks in power networks [78].

A general system was implemented to detect both sensor-actuator faults and deception attacks affecting a water distribution network [79]. Similarly, a model-free fault detection and diagnosis system was presented to detect and isolate faults in large-scale CPSs by using a model-free approach [80]. A model-based approach was also used to design a cyber attack detector for water distribution systems [81]. An Intelligent Generalized Predictive Controller (IGPC) was designed that is capable of detecting both faults and cyber attacks, and it can also differentiate between them [82]. One disadvantage of the FDI technique is that the cyber attacks may target a known weakness in the system; this is different from failures, which are normally independent or random. So this technique requires a careful examination in order to design the desired robust system.

2.4 Denial-of-service attacks

Denial-of-service attacks are strategies that are often used to occupy the communication resources in order to prohibit the transmission of measurement and/or control signals and that cause the maximum possible deterioration of the system performance.

The most dangerous type of DoS attack is the distributed DoS (DDoS), also called a coordinated attack, in which a large number of compromised machines are used to perform the attack [83]. Moreover, DDoS attacks frequently occur due to the simplicity of creating them, their low cost, and the high impact on systems, including the ability to completely disconnect an organization [84,1]. It has been shown that this kind of attack can cause instability in power grids [85] and can produce long delay jitter in NCS packets [86].

The DoS attacks in a radio frequency identification (RFID) system can be categorized based on the factors causing them as follows [87]:

1. System Jamming: Electromagnetic jamming done to prohibit tags from communicating with readers;
2. Desynchronization Attack: Destroying synchronization between the tag and the RFID reader, which causes a permanent disabling of the authentication capability of an RFID tag;
3. Tag Data Modification: Changing the data to a random number which cannot be identified by the reader;
4. Kill Command Attack: Sending a kill command with the hacked password causing a permanent disabling of the tag;
5. Random DoS Attack: Affecting the system by injecting short periods of noise signals.

2.4.1 Approaches of modeling a denial-of-service attack

There are two main methods for modeling DoS attacks in CPSs: the queueing model and the stochastic model.

2.4.1.1 Queuing model

Networking devices such as firewalls, end computers, and routers begin to perform badly under DoS attacks while dealing with high packet rate due to the constraints on memory resources, central processing units, interrupted processing, and input/output (I/O) processing. Thus, delay jitter and packet loss are highly common under attack, which may affect performance of the control system: rise and settling time, mean-squared error, and percentage overshoot.

The packet transmission of NCSs under DoS attacks is approximated by applying two simple models based on a multiple-input queue [88]:

- Type I: The attackers launch DoS attacks to an endpoint from computers in the local area close to the endpoint. This causes a loss of a large number of packets.
- Type II: The attackers launch DoS attacks remotely to service-provider-edge routers leading to a slowdown of the network links between a remote plant and a controller (see Fig. 2.3).

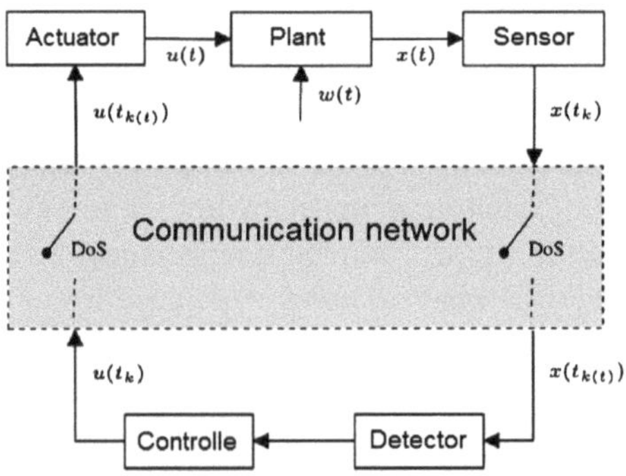

FIGURE 2.3 Block diagram of the closed-loop system with DoS attack [89].

The DoS is considered the phenomenon that may prevent the control signal from being treated at the desired time [89]. This means measurement and control channels can be affected separately, so it can be assumed that during a DoS attack, data can be neither sent nor received. Let us consider $\{h_n\}_{n\in\mathbb{N}_0}$, where $h_0 \geq 0$ is the sequence of DoS off/on transitions, i.e., the time instants at which a DoS changes a transition from zero to one (possible to interrupt communication). Then

$$H_n := \{h_n\} \cup [h_n, h_n + \tau_n[\tag{2.14}$$

with H_n is the time-interval of nth DoS attack, and its length is $\tau_n \in \mathbb{R}_{\geq 0}$, during that time the communication is not available. If $\tau_n = 0$, then the nth DoS attack is represented as a single pulse at time h_n.

The actuator generates an input based on the most recent data received from the controller during the DoS attack. Given $\tau,\ t \in \mathbb{R}_{\geq 0}$ with $t \geq \tau$, consider that

$$\Xi(\tau,t) := \bigcup_{n\in\mathbb{N}_0} H_n \bigcap [\tau,t] \tag{2.15}$$

$$\Theta(\tau,t) := [\tau,t] \setminus \Xi(\tau,t). \tag{2.16}$$

That means, for each interval $[\tau,t]$, that $\Xi(\tau,t)$ and $\Theta(\tau,t)$ refer to the sets of time instants where communication is prohibited and allowed, respectively. The control signal applied to the system at each $t \in \mathbb{R}_{\geq 0}$ can be represented as

$$u(t) = Kx\left(t_{k(t)}\right), \tag{2.17}$$

where

$$k(t) := \begin{cases} -1, & \text{if } \Theta(0,t) = \emptyset \\ \sup\{k \in \mathbb{N}_0 \mid t_k \in \Theta(0,t)\}, & \text{otherwise.} \end{cases} \tag{2.18}$$

That means, for each $t \in \mathbb{R}_{\geq 0}$, that $k(t)$ represents the most recent successful control update.

2.4.1.2 Stochastic model

A stochastic model could be either a Bernoulli model [90,91] or a Markov model [92]. The Bernoulli model can be seen from the following LTI system:

$$\begin{cases} x(k+1) = Ax(k) + \alpha(k)Bu(k) + w(k) \\ y(k) = \beta(k)Cx(k) + v(k) \end{cases}, \tag{2.19}$$

where $w(k)$ and $v(k)$ are the process and measurement noise, respectively, and are normally considered Gaussian i.i.d. random vectors with mean 0 and covariance Q, and $\alpha(k)$ and $\beta(k)$ are Bernoulli i.i.d. related to occurrences of the DoS attack on the process and measurement noise, respectively [91].

For the Markov model, consider the following system:

$$\begin{cases} x(k+1) = Ax(k) + \alpha(\xi(k+1))Bu(k) + w(k) \\ y(k) = Cx(k) + v(k) \end{cases}, \tag{2.20}$$

where $\alpha(\xi(k+1)) \in \{0,1\}$ is the Markov modulated DoS attack sequence that prevents transmission of the control signal packets to the actuator where $\xi(k)$ is related to the internal state of the attacker [92].

2.4.2 Secure estimation approaches

Compared with other types of attacks such as the replay attack, the DoS attack has received less attention from researchers. This is largely due to its nature, which focuses on controlling the CPS. The detection approaches mentioned in the following cover this area; listed here are some research on this topic.

Detection of DoS attacks and defending against them in a state estimation problem of a linear discrete-time system were both considered [93]. In this system the data of the sensors were sent to the estimator through a packet-dropping communication network. A modified Kalman filter, which was previously defined in [94], for state estimation in an unreliable communication network was applied in this problem. First, the detection problem was formulated as a hypothesis testing problem while considering the a priori knowledge of the statistics of the network. Second, two defending strategies were proposed, one using a secured packet-coding approach to compensate for the missing data, and the other based on raising of the transmission power to overcome the jamming effect of the attack.

The game theory approach was applied such that the interaction between the sensor and the attacker was modeled as a zero-sum stochastic game for a remote state estimation subject to DoS attacks [95]. The existence of a stationary Nash equilibrium was initially discussed for this game, and then optimal strategies were designed to adjust the transmission power of the sensor. In [96] a strategic form game was applied to calculate the asymptotic performance of the remote estimator, then was used to determine the duty cycle over an infinite time horizon to guarantee a predefined bounded error.

The problem of fault-tolerant control (FTC) for a nonlinear chaotic system with adaptive slide-mode control was presented while considering both network faults and DoS attacks [97]. It was considered that network faults consist of deterioration, perturbations of the nonlinear couplings, and signal attenuation. The compensation of the faulted and perturbed couplings were obtained by applying a slide-mode control strategy by means of adaptive estimations of the unknown parameters. After that the Lyapunov stability theory and mathematical analysis method were implemented to guarantee the asymptotic synchronization of the nonlinear chaotic systems.

2.4.3 Secure control approaches of denial-of-signal attack

Several approaches for controlling CPS subject to DoS attacks are discussed in the literature. Following are discussions of the main approaches, which are summarized in Fig. 2.4.

2.4.3.1 Stochastic time delay system approach

In this approach the DoS is modeled as a stochastic process with a delay in the signal. In [98], both DoS and deception attacks are considered to be randomly occurring, and they are modeled as two sets of Bernoulli distributed white sequences. Let us consider a discrete-time stochastic system with multiplicative

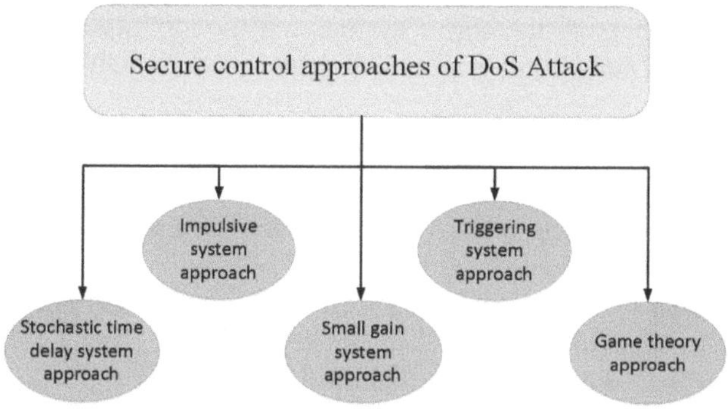

FIGURE 2.4 Secure control approaches of DoS attacks.

noise affecting the system and the measurement as follows:

$$\begin{cases} x_{k+1} = (A_0 + \sum_{i=1}^{r} \omega_{i,k} A_i) x_k + B u_k \\ \tilde{y}_k = (C_0 + \sum_{i=1}^{s} \bar{\omega}_{i,k} C_i) x_k \end{cases}, \tag{2.21}$$

where $x_k \in \mathbb{R}^{n_x}$ is the state vector, $\tilde{y}_k \in \mathbb{R}^{n_y}$ is the sensor measurement, and $u_k \in \mathbb{R}^{n_u}$ is the controller input; $A_i (i = 0, 1, \cdots, r)$, B, and $C_i (i = 0, 1, \cdots, s)$ are known constant matrices with appropriate dimensions; $\omega_{i,k} \in \mathbb{R} (i = 1, 2, \cdots, r)$ and $\bar{\omega}_{i,k} \in \mathbb{R} (i = 1, 2, \cdots, s)$ are multiplicative noise with zero means and unity variances, and are mutually uncorrelated in k; and i, r, and s are known positive integers. The rank of B is assumed to be n_u.

The following attack model is applied to study these problems:

$$y_{k_s} = \alpha_{k_s} (\tilde{y}_{k_s} + \gamma_{k_s} v_{k_s}) + (1 - \alpha_{k_s}) y_{k_{s-1}}, \tag{2.22}$$

where y_{k_s} is the data received by the controller and $v_{k_s} \in \mathbb{R}^{n_y}$ stands for the signals injected by attackers which is described by

$$v_{k_s} = -\tilde{y}_{k_s} + \xi_{k_s}, \tag{2.23}$$

where ξ_{k_s} is an arbitrary bounded energy signal satisfying

$$\|\xi_{k_s}\| \leq \delta_2. \tag{2.24}$$

The stochastic variables α_{k_s} and γ_{k_s} are Bernoulli distributed white sequences with values of 0 or 1 and with the following probabilities:

$$\begin{aligned} &\text{Prob}\{\alpha_{k_s} = 0\} = 1 - \bar{\alpha}, \qquad \text{Prob}\{\alpha_{k_s} = 1\} = \bar{\alpha}, \\ &\text{Prob}\{\gamma_{k_s} = 0\} = 1 - \bar{\gamma}, \qquad \text{Prob}\{\gamma_{k_s} = 1\} = \bar{\gamma}, \end{aligned} \tag{2.25}$$

where $\bar{\alpha} \in [0, 1)$ and $\bar{\gamma} \in [0, 1)$ are two known constants. The stochastic approach is applied and some sufficient conditions are obtained to ensure the security requirements of the above system and by solving certain linear matrix inequalities (LMIs) with nonlinear constraints for calculating the desired controller gain.

2.4.3.2 *Impulsive system approach, hybrid model*

In this approach the system under DoS attack is represented by a hybrid model, or in other words an impulsive system.

The design of resource-aware and resilient control strategies for NCSs affected by malicious DoS attacks were considered in [99] and [100]. In particular, an output-based event-triggered control scheme was applied to obtain the control and communication strategy in a class of nonlinear feedback systems affected by exogenous disturbances. The existence of a robust strictly positive lower bound on the inter-event times was guaranteed by implementing this framework even in the presence of disturbances and DoS attacks.

Consider the following plant $\mathcal{P}$:

$$\mathcal{P}: \begin{cases} \dot{x}_P = f_P(x_P,\ u,\ w) \\ y = g_P(x_P) \end{cases}, \tag{2.26}$$

where $w \in \mathbb{R}^{n_w}$ is a disturbance input, $x_p \in \mathbb{R}^{n_p}$ is the state vector, $u \in \mathbb{R}^{n_u}$ is the control input, and $y \in \mathbb{R}^{n_y}$ is the measured output of plant $\mathcal{P}$, and controller $\mathcal{C}$:

$$\mathcal{C}: \begin{cases} \dot{x}_c = f_c(x_c, \hat{y}) \\ u = g_c(x_c, \hat{y}) \end{cases}, \tag{2.27}$$

where $x_c \in \mathbb{R}^{n_c}$ denotes the controller state, $\hat{y} \in \mathbb{R}^{n_y}$ is the most recently received measurement, and $u \in \mathbb{R}^{n_u}$ is the controller output. The performance output is given by $z = q(x)$, where $z \in \mathbb{R}^{n_z}$ and $x = (x_p, x_c)$. The DoS attacks interval is denoted by $\{H_n\}_{n\in\mathbb{N}} \in \mathcal{I}_{DoS}$, which is the period in time at which communication is not available between the sensor and the controller because of the attack. So the collection of times of DoS attacks are given by

$$\mathcal{T} := \bigcup_{n\in\mathbb{N}} H_n. \tag{2.28}$$

By applying the hybrid modeling framework, the jump/update of $\hat{y}$ and the update of the transmission error $e := \hat{y} - y$, can be written as

$$\hat{y}^+ = \begin{cases} y, & \text{when } t_j \notin \mathcal{T}, \\ \hat{y}, & \text{when } t_j \in \mathcal{T}, \end{cases} \tag{2.29}$$

$$e^+ = \begin{cases} 0, & \text{when } t_j \notin \mathcal{T}, \\ e, & \text{when } t_j \in \mathcal{T}, \end{cases} \tag{2.30}$$

for each $j \in \mathbb{N}$; the maximal allowable transmission interval bound τ_{miet} is characterized by

$$\tau_{miet} = \begin{cases} \frac{1}{Lr} \arctan\left(\frac{r(1-\lambda)}{2\frac{\lambda}{1+\lambda}(\frac{\gamma}{L}-1)+1+\lambda}\right), & \gamma > L \\ \frac{1}{L}\frac{1-\lambda}{1+\lambda}, & \gamma = L \;, \\ \frac{1}{Lr} \arctan\left(\frac{r(1-\lambda)}{2\frac{\lambda}{\lambda+1}(\frac{\gamma}{L}-1)+1+\lambda}\right), & \gamma < L \end{cases} \tag{2.31}$$

where $r = \sqrt{|(\gamma/L)^2 - 1|}$; $L \geq 0$ is a constant; $\lambda \in (0, 1)$ represents the information locally available at the event-triggering mechanism (ETM); and γ is obtained from the following condition:

$$\langle \nabla V(x), f(x, e, w)\rangle \leq -\rho(|x|) - \rho(|y|) - H^2(x, w) \\ -\sigma_1(W(e)) + \gamma^2 W^2(e) + \theta^2|w|^2. \tag{2.32}$$

The details of this condition are described in [100]. Finally, by the normal consideration that DoS attacks are restricted in terms of frequency and duration, the desired stability and performance criteria in terms of induced L_∞-gains are also guaranteed [100].

2.4.3.3 Small-gain approach

In [101] the stabilization problem of distributed systems subjected to DoS attack, DoS frequency characterization, and duration of preserved stability are investigated. A hybrid communication strategy is also considered to save communication resources. It was shown that communication load can be reduced effectively and Zeno behavior can be prevented by using the hybrid transmission strategy.

As an example of this approach a large-scale system that consists of N interacting subsystems is considered with the following model:

$$\dot{x}_i(t) = A_i x_i(t) + B_i u_i(t) + \sum_{j \in N_i} H_{ij} x_j(t), \tag{2.33}$$

where A_i, B_i, and H_{ij} are matrices with appropriate dimensions; $t \in \mathbb{R}_{>0}$; and $x_i(t)$ and $u_i(t)$ are the state and control input of subsystem i, respectively. The control input applied to subsystem i is

$$u_i(t) = K_i x_i(t_k^i) + \sum_{j \in N_i} L_{ij} x_j(t_k^j), \tag{2.34}$$

where L_{ij} is the coupling gain in the controller. Let $\{h_n\}n \in N_0$, $h_0 \geq 0$, denote the sequence of DoS *off/on* transitions, i.e., the time instants at which DoS exhibits a transition from zero to one. So

$$H_n := \{h_n\} \cup [h_n, h_n + \tau_n[\tag{2.35}$$

represents the n-th DoS time-interval, of a length $\tau_n \in \mathbb{R}_{\geq 0}$, over which the network is under a DoS attack. Also, let

$$\Xi(\tau, t) := \bigcup_{n \in \mathbb{N}_0} H_n \bigcap [\tau, t] \tag{2.36}$$

be the subset of $[\tau, t]$ where the network is under a DoS attack.

Assumption 2.1 (DoS frequency). There exist $\Xi(\tau, t) := \bigcup_{n \in \mathbb{N}_0} H_n \bigcap [\tau, t]$ constants $\eta \in \mathbb{R}_{\geq 0}$ and $\tau_D \in \mathbb{R}_{>0}$ such that

$$n(\tau, t) \leq \eta + \frac{t - \tau}{\tau_D} \tag{2.37}$$

for all $\tau, t \in \mathbb{R}_{\geq 0}$ with $t \geq \tau$.

Assumption 2.2 (DoS duration). There exist constants $\kappa \in \mathbb{R}_{\geq 0}$ and $T \in \mathbb{R}_{>1}$ such that

$$|\Xi(\tau, t)| \leq \kappa + \frac{t - \tau}{T} \tag{2.38}$$

for all $\tau, t \in \mathbb{R}_{\geq 0}$ with $t \geq \tau$.

Assumption 2.3 (intersampling of round-robin). In the absence of DoS attacks, there exists an intersampling interval Δ such that

$$||e_i(t)|| \leq \sigma_i ||x_i(t)|| \tag{2.39}$$

holds, where σ_i is a suitable design parameter.

Theorem 2.1. *For a distributed system (2.33) with a control input (2.34), the plant-controller communicates over a shared network with a round-robin communication protocol with sampling interval Δ, as in Assumption 2.3. The large-scale system is asymptotically stable for any DoS attack sequence satisfying Assumption 2.1 and 2.2 with arbitrary η and κ, and with τ_D and T if*

$$\frac{1}{T} + \frac{\Delta_*}{\tau_D} < \frac{\omega_1}{\omega_1 + \omega_2}, \tag{2.40}$$

in which $\Delta_* = N\Delta$, $\omega_1 := min\{\frac{l_i - \sigma_i^2 j_i}{\lambda_{max}(P_i)\mu_i}\}$, and $\omega_2 := \frac{4max\{j_i\}}{min\{\mu_i \lambda_{min}(P_i)\}}$. The parameters l_i, j_i, μ_i, and σ_i are as in [101], Lemma 1.

The detailed proof of Theorem 2.1 is provided in [101].

2.4.3.4 Triggering strategy

A plant-jammer-operator setup is considered where the communication channel between the operator and the plant is affected by a periodic jammer [102]. An event-triggering time-sequence was adopted to reduce the communication in the system. This triggering time-sequence has the ability of defending against the jammer attack and also rendering the system asymptotically stable under some circumstances as presented in [102]. Let $x \in \mathbb{R}^n$ and $u \in \mathbb{R}^m$ be the state vector and the input vector, respectively. The following system is considered:

$$\begin{aligned} \dot{x}(t) &= Ax(t) + Bu(t), \\ u(t) &= Kx(t_k), \quad \forall t \in [t_k, t_{k+1}[, \end{aligned} \tag{2.41}$$

where A, B, and K are matrices of proper dimensions, and $\{t_k\}_{k\geq 1}$ is the triggering time-sequence. Let $e(t) = x(t_k) - x(t)$, $\forall t \in [t_k, t_{k+1}[$. The system is asymptotically stable if the Lyapunov function $V(x) = x^T Px$ associated with $||Q|| > 1$ is considered and the control $u(t)$ is updated at times t_k according to the following triggering law:

$$|e(t_k)|^2 = \sigma \frac{\|Q\| - 1}{\|PBK\|^2} |x(t_k)|^2, \quad k \geq 1. \tag{2.42}$$

The triggering time-sequence is given by

$$\begin{aligned} t^*_{k,n} = \; & \{t_l \text{ satisfying (2.42) } | t_l \in [(n-1)T, (n-1)T + T^{\text{cr}}_{\text{off}}]\} \\ & \cup \{nT\}. \end{aligned} \tag{2.43}$$

Theorem 2.2. *The system (2.41) with the triggering law (2.43) is asymptotically stable if the following conditions are satisfied:*

$$\frac{(1-\sigma)T^{\text{cr}}_{\text{off}}(\|Q\| - 1)}{2} > \|P\| \ln(\alpha), \tag{2.44}$$

where

$$\begin{aligned} \alpha \triangleq \; & \exp((T - T^{\text{cr}}_{\text{off}})\mu(A+BK)) + \frac{\|BK\|}{\mu(A+BK)} \times \\ & \left(\frac{\|BK\|}{\|A\|} + 1\right)(1 - \exp((T - T^{\text{cr}}_{\text{off}})\|A\|)) \times \\ & (1 - \exp((T - T^{\text{cr}}_{\text{off}})\mu(A+BK))), \end{aligned} \tag{2.45}$$

and

$$\mu(A+BK) < 0. \tag{2.46}$$

The detailed proof of Theorem 2.2 is provided in [102].

The triggering method was applied in the literature, for example, the control strategy was designed for linear [103] and nonlinear [104] systems subjected to DoS attacks based on the analysis of the input-to-state stability (ISS)-Lyapunov function. In these systems the maximal percentage of time of losing feedback data without leading the system to instability was characterized and an event-based controller for which the existence of a minimal intersampling time is guaranteed was proposed.

2.4.3.5 Game theory approach

Game theory deals with strategic interactions among multiple decision makers, named players [105]. The preference ordering of each player among several options is augmented in an objective function for that player. And each player tries to optimize an objective function. The objective function of a player depends on the alternatives of at least one other player and in general of all the players in any nontrivial game. So the optimization process for each player depends on the choices of the other players [105]. For more information about game theory and its application in networks, we refer the interested reader to [106], [107], [108], and [109].

This approach was applied to achieve secure control in a wide range of research. In [110] the dropout caused by DoS attacks is modeled as a Markov process based on the game between attack and defense strategies. After that, four theorems were derived using the Lyapunov theory to ensure the system stability. A Nash Q-learning algorithm is presented to handle the computation complexity of the optimal strategies problem for both players [121]. The sensor data are sent to a remote estimator over a multichannel network, which may be affected by a malicious attacker. The sensor requires a single channel to be selected from among these paths to transmit data packets with less probability of being attacked. On the other hand, the attacker must select which channel to attack. A two-player zero-sum stochastic game framework is proposed and solved to model an interactive decision-making problem [121].

Other examples of applying this approach can be found in the literature: a hybrid game-theory framework where the occurrence of unanticipated events is represented by stochastic switching, and deterministic uncertainties described by the known range of disturbances was used to build a robust secure system [105]; a unified game approach was applied to design a resilient control of an NCS [112]; a multistage hierarchical game with a corresponding hierarchy of decisions was implemented to achieve a resilient control system [113].

2.4.4 Jamming attack

A special type of DoS attack called a jamming attack refers to the situation when an attacker occupies one channel to prevent other nodes from using it and that causes communication stoppage.

In [114] the stochastic game theory was applied to obtain an optimal defense mechanism for NCSs subject to jamming attacks. The dynamic interactions between the attacker and the sensor transmitter in the NCS were formulated as a two-player zero-sum stochastic game. The cost function in this stochastic game includes the resource costs used to conduct cyber-layer defense and attack actions, and the possible degraded dynamic performance of the NCS. The effects of the interactions between the attacker and the defender on the dynamic performance of the NCS were considered by this cost function. In the end, a stochastic dynamic programming (SDP) problem was solved to obtain the optimal defense mechanism.

The security issues in remote state estimation of CPSs were discussed in [115]. The communication between a sensor node and a remote estimator takes place through a wireless channel, which may be jammed by an attacker. The interactive decision making process of both communicating and attacking was studied while considering energy constraints for both the sensor and the attacker. A game-theoretic problem was formulated, and the optimal strategies for both sides constituting a Nash equilibrium of a zero-sum game were proved. A constraint-relaxed problem was designed and the Markov chain theory was used to obtain the corresponding solutions.

The optimal jamming attack that maximizes the linear quadratic Gaussian (LQG) control cost function while considering energy constraint was considered in [116]. The optimal jamming attack schedule and the corresponding cost function were derived after analyzing the properties of the cost function under an arbitrary attack schedule. The impact of jamming attacks on broadcasting by introducing a new analytical model was investigated [117]. The feasibility of the existing threshold-based methods to detect jamming in real-time applications was also discussed and a real-time Medium Access Control-based (MAC-based) detection method was proposed to meet the requirements of safety applications in vehicular networks.

In [118], an optimal energy efficient jamming attack schedule against remote state estimation through wireless channels under energy constraints of jamming attacker was proposed. The proposed schedule was derived with the objective of maximizing the remote estimation error covariances, and an optimal jamming attack schedule for a multisystem case was derived using the optimal algorithm.

The event-based controller synthesis problem for NCSs under the resilient event-triggering communication scheme (RETCS) and periodic jamming attacks was discussed [119]. The jamming attacks imposed by power-constrained pulsewidth-modulated jammers are considered partially identified, that is, the period of the jammer and a uniform lower bound on the jammer's sleeping periods are known. In the end, the piecewise Lyapunov functional was implemented to ensure the exponential stability of the system.

2.5 Deception attack

A deception attack, also called a false data injection (FDI) attack or a malicious attack, is defined as the modification of the data integrity for the transmitted packets among some cyber parts in a CPS [120,121]. Stuxnet, for example, is a famous malicious computer worm that has the ability to reprogram the code running in programmable logic controllers (PLCs) in the SCADA systems causing a deviation from the required behavior. Another example is found in the transmission systems of power grids; attacks can be launched by adversaries by hacking remote terminal units (RTUs) like sensors placed in substations [98]. An example of this kind of attack that has affected SCADA water systems hierarchically consists of several attacks including several objectives in different cyber layers and was discussed in [122].

2.5.1 Modeling the deception attack

Generally speaking, a deception attack can occur in one of two forms: random attacks in which arbitrary measurements are modified, and targeted attacks in which specific states are affected [123]. From a control engineering point of view, a deception attack is modeled as a stochastic process [120] and [124]. To clarify the idea, consider the following system:

$$\begin{cases} x(k+1) = Ax(k) + Bu(k) \\ \bar{y}(k) = Cx(k) \\ y(k) = \bar{y}(k) + \alpha(k)v(k) \end{cases}, \tag{2.47}$$

where $x(k) \in \mathbb{R}^{n_x}$, $u(k) \in \mathbb{R}^{n_u}$, $\bar{y}(k) \in \mathbb{R}^{n_y}$, and $y(k) \in \mathbb{R}^{n_y}$ are the states, control input, the measured output, and the signal received by the controller, respectively. The stochastic variable $\alpha(k)$ is a Bernoulli distributed white sequence denoting the possibility of the occurrence of the deception attacks by taking values on 1 and 0 such that

$$\begin{cases} \text{Prob}\{\alpha(k) = 1\} = \bar{\alpha} \\ \text{Prob}\{\alpha(k) = 0\} = 1 - \bar{\alpha} \end{cases}, \tag{2.48}$$

and the deception attack here is described as

$$v(k) = -\bar{y}(k) + \eta(k), \tag{2.49}$$

where $\eta(k)$ is bounded by

$$\sum_{k=0}^{\infty} \eta^T(k)\eta(k) \le \bar{\eta}. \tag{2.50}$$

The variance-constrained distributed filtering problem for a class of time-varying systems subject to multiplicative noise, unknown but bounded disturbances, and deception attacks over sensor networks was discussed in [125]. The available measurements at each sensing node are collected from both of the individual sensors and their neighbors. The attacker inserts some deception signals into the true signals of the control input u_k and the measurement outputs $y_{i,k}$ during the process of data transmission, as shown in Fig. 2.5.

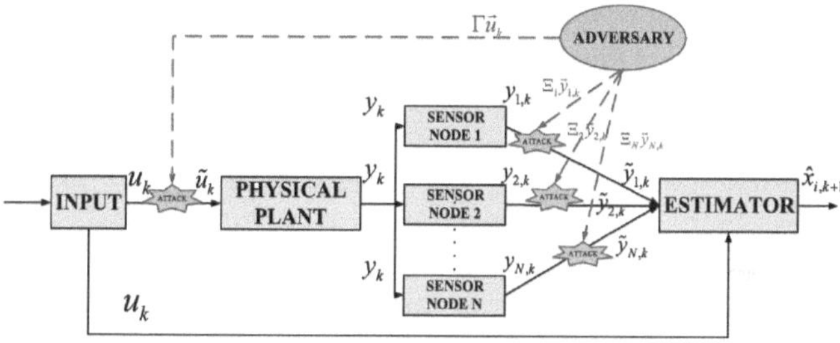

FIGURE 2.5 Model of the deception attack [125].

A novel model for the deception attack was presented where the malicious signals are injected by the adversary into both control and measurement data during the process of information transmission via the communication network. The attacker affects the system by the following signals [125]:

$$\begin{cases} \vec{u}_k = -u_k + \delta_k \\ \vec{y}_{i,k} = -y_{i,k} + \theta_{i,k}, \quad i = 1, 2, \dots, N \end{cases}, \tag{2.51}$$

where δ_k and $\theta_{i,k}$ $(i = 1, 2, \dots, N)$ are the unknown but bounded signals belonging to the following ellipsoids:

$$\begin{aligned} \delta_k \in \mathbb{E}(0, S_k, m) &\triangleq \{\delta_k \in \mathbb{R}^m : \delta_k^{\mathrm{T}} S_k^{-1} \delta_k \leq 1\}, \\ \theta_{i,k} \in \mathbb{E}(0, R_{i,k}, p) &\triangleq \{\theta_{i,k} \in \mathbb{R}^p : \theta_{i,k}^{\mathrm{T}} R_{i,k}^{-1} \theta_{i,k} \leq 1\}, \end{aligned} \tag{2.52}$$

with S_k and $R_{i,k}$ $(i = 1, 2, \dots, N)$ being positive definite matrices of compatible dimensions.

Now the actual control input $\tilde{u}_k$ and the actual measurement outputs $\tilde{y}_{i,k}$ are described by

$$\begin{cases} \tilde{u}_k = u_k + \Gamma \vec{u}_k \\ \tilde{y}_{i,k} = y_{i,k} + \Xi_i \vec{y}_{i,k}, \quad i = 1, 2, \dots, N \end{cases}, \tag{2.53}$$

where the matrices Γ and Ξ_i represent the physical constraints imposed on the attack signals and are considered diagonal matrices, and each element inside it has known lower and upper bounds. By resorting to the recursive linear matrix inequality approach, a sufficient condition is derived for the existence of the desired filter satisfying the predetermined characteristic of the estimation error variance.

The interested reader is referred to [126] for a comprehensive review on the models used in the literature of false data attacks in power systems.

2.5.2 Secure estimation approaches of the deception attack

The estimation of deception attacks is a serious issue since this kind of attack can be designed to avoid any detection mechanism in the system in addition to the original goal of affecting the stability of the system [74], [77], [122], [127], [128]. In [75] the issue of bias injection attacks targeting the Kalman filter in a system including the χ^2 detector was studied. It was proved that the problem of worst-case scenario can be reduced to a quadratically constrained program obtaining the criterion that is used to choose which sensors to secure and the condition on the number of sensors required to maintain the effect of the attack within a predetermined threshold.

The problem of centralized security for linear time-invariant stochastic systems with multirate-sensor fusion subject to deception attacks was discussed [129]. The data sent by adversaries on each sensor are considered as additional signals that satisfy boundary conditions similar to (2.50). Then the lifting technique was used to formulate single-rate discrete-time systems. After that the sufficient conditions were obtained using the stochastic analysis techniques to achieve the predetermined security level of the original system.

It is known that the deception attack in addition to the effect of a uniform quantization can be considered within a problem of the distributed recursive filtering of a discrete-time delayed stochastic system. To show how it works, we consider the following system:

$$\begin{aligned} x(k+1) &= \left(A_0(k) + \sum_{s=1}^{r} \omega_s(k) A_s(k) \right) x(k) \\ &\quad + \left(A_0^d(k) + \sum_{s=1}^{r} \omega_s(k) A_s^d(k) \right) x(k-\tau) \\ &\quad + B(k)\omega(k), \end{aligned} \tag{2.54}$$

with n sensors described by

$$\begin{aligned} \tilde{y}_i(k) &= (C_0(k) + \bar{\omega}_i(k) C_i(k))\, x(k) + D(k) v_i(k), \\ &\qquad i = 1, 2, \cdots, n, \end{aligned} \tag{2.55}$$

where $x(k) \in \mathbb{R}^{n_x}$ is the state that cannot be observed directly; $\tilde{y}_i(k) \in \mathbb{R}^{n_y}$ is the output of sensor i without quantization; $\omega(k) \in \mathbb{R}^s$ and $v_i(k) \in \mathbb{R}^p$ $(i = 1, 2, \cdots, n)$ are the white noise with zero-mean and unity covariance, and are mutually uncorrelated in k and i; $\omega_s(k) \in \mathbb{R}(s = 1, 2, \cdots, r)$ and $\bar{\omega}_i(k) \in \mathbb{R}$ are multiplicative noise with zero-mean and unity variances, and are mutually uncorrelated in k; r and τ are two known positive integers; and $A_s(k)$, $A_s^d(k)$ and $C_i(k)$ are known constant matrices with compatible dimensions. Here the effect of the deception attack is considered to be similar to Eqs. (2.47)–(2.50). Note that the upper bound for the filtering error covariance is characterized and was used to obtain the gain matrices of the Kalman-type recursive filter by solving Riccati-like difference equations.

The event-triggered scheme was applied to design a distributed state estimator for a system of wireless sensor networks subject to false data injection attack [130]. The estimate of each sensor is checked to see if it is attacked at each time step before transmitting the data to its neighboring sensors, and if it is attacked it will be stopped. An optimal estimator gain is proposed using the event-triggered scheme by minimizing the mean-squared estimation error covariance, and the stability of the designed distributed estimator is guaranteed by deriving a sufficient condition.

This Bayesian method was applied for both the detection and estimation of states for CPSs subject to switching signal attacks and faked measurements [131]. The problem was formulated and solved as a hybrid Bernoulli filter that updates in real time the joint posterior density of the detection attack Bernoulli set and of the state vector.

In [132], an algorithm based on a Kalman filter that guarantees a secure state estimation for stochastic dynamic systems was presented. It is considered that the adversary affected arbitrary subset of sensors in this problem and an upper bound on the number of sensors under attack was characterized to maintain an acceptable state estimation error. In [133] the conditions of insecure estimation was derived for an NCS subject to false data injection attacks including a χ^2 detector. Additionally, the attack is generated using a specific algorithm and a scheme for protecting a few communication channels instead of protecting all of them is presented.

The filtering problem of a nonlinear stochastic discrete-time delay system affected by randomly sensor saturation and randomly deception attacks were discussed in [134]. Let us consider the following system:

$$\begin{aligned} x(k+1) &= Ax(k) + A_d x(k-d(k)) + Bf(x(k)) \\ &\quad + B_d f_d(x(k-d(k))) + D\omega(k), \end{aligned} \tag{2.56}$$

where $x(k) \in \mathbb{R}^{n_x}$ is the state vector; $\omega(k) \in \mathbb{R}$ is a zero-mean Gaussian white noise sequence with $\mathbb{E}[\omega^2] \leq \delta^2$; and A, A_d, B, B_d, and D are known real constant matrices with appropriate dimensions. The nonlinear functions f and f_d satisfy the following bounded conditions:

$$\begin{aligned} &\left[f(x) - K_1 x\right]^T \left[f(x) - K_2 x\right] \leq 0, \\ &\left[f_d(x) - T_1 x\right]^T \left[f_d(x) - T_2 x\right] \leq 0, \end{aligned} \tag{2.57}$$

where K_1, K_2, T_1, and T_2 are known real matrices of appropriate dimensions and $K = K_1 - K_2$ and $T = T_1 - T_2$ are symmetric positive definite matrices.

The following filter model is used in this system:

$$\hat{x}(k+1) = F\hat{x}(k) + Ny(k). \tag{2.58}$$

A sufficient condition is first derived to guarantee the desired security level in the filtering system by applying the stochastic analysis techniques. Then a linear matrix inequality with nonlinear constraints is solved to obtain the filter gain.

2.5.3 Secure control approaches of the deception attack

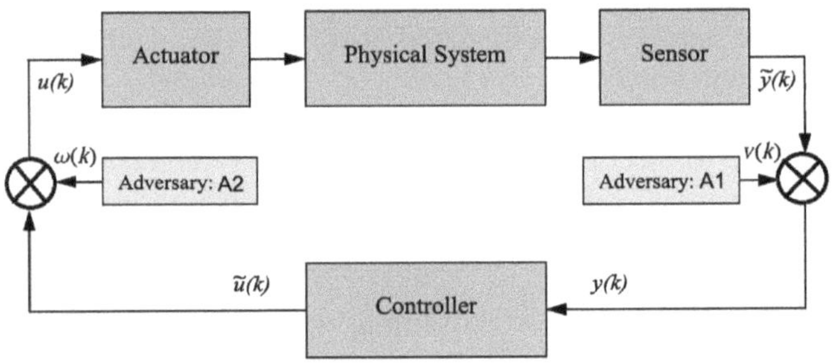

FIGURE 2.6 Schematic of a deception attack.

The security control problem with quadratic cost criterion for a class of discrete-time stochastic nonlinear systems affected by deception attacks was discussed in [120]. The measurement and actuating signals were both subjected to the deception attacks shown in Fig. 2.6 and modeled as

$$\begin{aligned} u(k) &= \tilde{u}(k) + \gamma(k)\omega(k), \\ y(k) &= \tilde{y}(k) + \delta(k)v(k), \end{aligned} \tag{2.59}$$

where $u(k)$ is the actuator input, $\tilde{u}(k)$ is the controller output subject to attacks, $y(k)$ is the received signal by the controller, $\tilde{y}(k)$ is the sensor measurement subject to attacks, $\omega(k)$ and $v(k)$ are the signals transmitted by the attacker, and $\gamma(k)$ and $\delta(k)$ are two mutually independent Bernoulli-distributed white sequences stochastic variables with values 0 or 1 with probabilities as follows:

$$\begin{cases} \text{Prob}\{\gamma(k) = 1\} = \bar{\gamma}, & \text{Prob}\{\gamma(k) = 0\} = 1 - \bar{\gamma}, \\ \text{Prob}\{\delta(k) = 1\} = \bar{\delta}, & \text{Prob}\{\delta(k) = 0\} = 1 - \bar{\delta}. \end{cases} \tag{2.60}$$

It is assumed that the attackers (A1) and (A2) in Fig. 2.6 insert false data in the system such that $\omega(k) = -\tilde{u}(k) + \zeta_1(k)$ and $v(k) = -\tilde{y}(k) + \zeta_2(k)$, where

$\zeta_1(k)$ and $\zeta_2(k)$ are the bounded energy signals satisfying $\|\zeta_1(k)\| + \|\zeta_2(k)\| \leq \sigma$, and σ is a known positive scalar.

The objective in this problem is to build a dynamic output feedback controller such that the prescribed security in probability is achieved while obtaining an upper bound of the preselected quadratic cost function. So a stochastic analysis approach is applied to derive the sufficient conditions with the form of matrix inequalities in the framework of the input-to-state stability in probability. In addition, the controller gain and the upper bound are obtained by applying the matrix inverse lemma.

A secure networked predictive control system (SNPCS) architecture is presented [135], which integrates the Data Encryption Standard (DES) algorithm, Message Digest (MD5) algorithm, timestamp strategy, and recursive networked predictive control (RNPC) method. The RNPC method based on round-trip time delays is applied to ensure the control system performance when it is affected by deception attacks such that it will compensate for the attacks effects and the imperfections in the network such as packet dropout, packet disorder, and time-varying delay.

In [136] the problem of consensus control for a class of discrete-time varying stochastic multiagent systems including stochastic nonlinearities and those affected by deception attacks was considered. A new definition of quasi-consensus was presented to represent the consensus process with a constraint on all agents to stay within a certain ellipsoidal region at each time instant. Based on the provided topology, the measurement output available for the controller from both the individual agent and the neighbors. In order to achieve the quasi-consensus, sufficient conditions are obtained by using a set of recursive matrix inequalities for the existence of the desired control scheme.

A resilient control strategy was proposed for NCSs affected by stealthy false-data injection attacks that are designed so that they cannot be detected using the control input and measurement data [137]. The consequence of a zero dynamic attack on the state variable of the plant is undetectable during attack and then it appears after the end of the attack. So, a resilient linear quadratic Gaussian controller was proposed such that the Kalman filter is updated online from information given by an active version of the generalized likelihood ratio detector with the ability to quickly recover the nominal behavior of the system after the attack is finished [137].

An adaptive controller of CPSs subjected to simultaneous sensor and actuator attacks were proposed in [138]. In addition, an improved adaptive resilient control scheme was presented to mitigate adversarial attacks in CPSs [139]. The adaptive bound estimation mechanism and a Nussbam function with faster growth rate, and a two-step backstepping method were implemented to mitigate the effects of unknown sensor and actuator attacks, and then state variables were constrained by applying an exponentially decaying barrier Lyapunov function.

The reliable and optimal control problems of data-driven CPSs subject to a class of actuator attacks was discussed in [140]. An unknown continuous-time linear physical system with the external disturbance was considered, and control input signals to be sent via network layers are assumed to be vulnerable to cyber attacks. A novel data-based adaptive integral sliding-mode control strategy was presented to eliminate the effect of the actuator attacks such that the stability and a nearly optimal performance of the CPSs can be obtained. In [141] the upper bound of the worst stealthy attacks was obtained using the detection mechanisms of the abnormal monitor, which is composed of the information of the detector's threshold, the attack's structure, and frequency characteristic for a class of frequency-constrained sensor and actuator attacks.

2.5.4 Replay attack

A special kind of deception attacks, called replay attacks, occur when the adversary succeeds in recording some of the data transmission, such as sensing data, and injects it in the CCS [142]. This kind of attack takes place in two phases. In the first phase the attacker records data from the system, as shown in Fig. 2.7; in the second phase the attacker injects this data into the system, which could then be directed to the physical system, as shown in Fig. 2.8 [47]. As an example, the attacker can create a communication link between two end points to insert replay messages observed in different regions; these are called wormhole attacks, and they are common in wireless sensor networks.

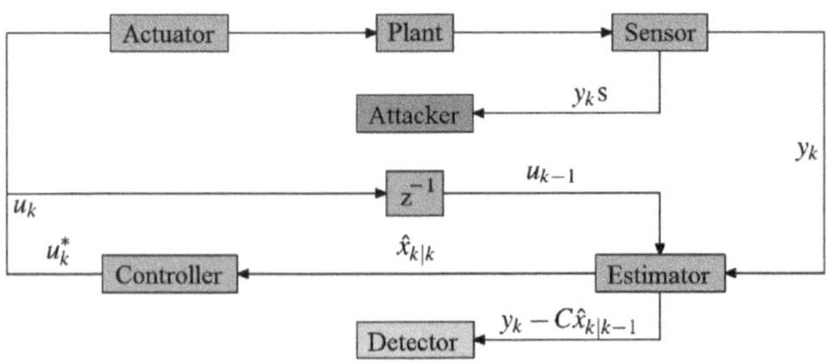

FIGURE 2.7 Schematic of the first stage of a replay attack.

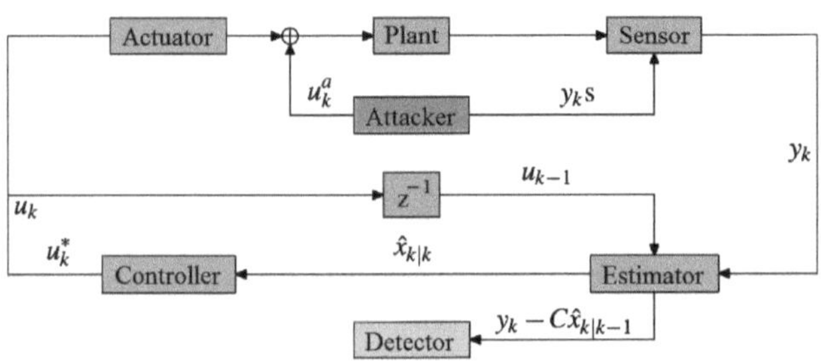

FIGURE 2.8 Schematic of the second stage of a replay attack.

Clearly, no system information is needed in this kind of attack, including information on the designed controllers or estimators. This behavior of the attack makes it very difficult to detect. One solution against this attack is to adopt timestamps or counters in the transmitted data. In modeling, this attack could be considered as variable delays with unknown data on the upper bounds and variable rates. However, from the scheduling point of view the admissible maximum upper bound can be calculated by applying the time-delayed system theory together with optimization approaches [58].

However, there is very little literatures that have addressed controlling of CCSs subject to replay attacks. As an example, a variation in the receding-horizon control subject to replay attacks was discussed in [143], which derives a simple and explicit relation among the infinite-horizon cost, the computing, and attacking horizons. After that the asymptotic exponential stability of systems is guaranteed by providing a set of sufficient conditions. Another example was provided for a discrete-time linear time invariant Gaussian system subject to replay attacks in [142]. To ensure the desired probability of detection by trading off between decreasing control accuracy and increasing control effort, an infinite horizon linear quadratic Gaussian controller has been proposed.

The conditions of the feasibility of the replay attack and suggesting countermeasures that optimize the probability of detection by conceding control performance was described in [71], where the effect of integrity attacks on the control systems is analyzed and countermeasures capable of exposing such attacks are proposed.

2.6 Notes

With the development of cloud computing and the processing techniques of big data, the era of CCSs is dawning. A preliminary structure and algorithm are proposed. Some new results on CCSs will be found in our future publications. We believe that there will be more interesting and important results that will be produced in this new research area.

Cyber-physical systems are almost everywhere; they can be accessed and controlled remotely. These features make them more vulnerable to cyber attacks. Since these systems provide critical services, having them under attack would have dangerous consequences. Unfortunately, cyber attacks may be detected, but only after the damage is done. Therefore, developing a cyber system that can survive an attack is a challenge. In this chapter the literature on security aspects of CPSs was surveyed. First, we presented some of the existing methods for detecting cyber attacks. Second, we focused on three main types of cyber attacks, namely DoS, deception, and replay attacks. In our discussion, we surveyed some exiting models of these attacks, approaches of filtering the CPSs subject to these attacks, and approaches of controlling the CPSs subject to these attacks.

One important factor in CPS security is that attacks might not only come from outside the system, but also from inside, for example from employees who do not need much additional knowledge about the target system. The knowledge that insiders possess often gives them unrestricted access to steal or modify data in the system or to deactivate that system. This makes it important to have a secure control system to maintain the stability of the system during such an attack.

Security control techniques for CPSs are still at an early stage in comparison with other control applications. The effects of a successful attack on an NCS are generally more serious in comparison with attacks on other systems due to the core of critical infrastructures. It is an inevitable challenge how to mitigate the impact of cyber attacks along with the other imperfections of the NCSs.

One of the ongoing challenging research issues is to design a secure filter based on attacked measurement outputs so as to achieve an acceptable index of security performance [129]. But the existing filtering schemes might not work properly to guarantee the security since it is difficult or even impossible for defenders to estimate when or how the system is affected by a cyber attack. The traditional Kalman filter for example has the capability of achieving the minimal variance of the filtering errors by considering the exact knowledge of noise statistics, while this assumption is usually not true for CPSs since the statistical characteristics of signals transmitted by the attacker cannot usually be obtained [129].

In an H_∞ filtering framework the disturbance noise is required to be energy-bounded. This means that the energy of the external signals tends to zero when time goes on. But this requirement may be too stringent for CPSs as the signals sent by attackers cannot be ensured to meet the L_2 gain condition. The filtering and control problem with security constraints is an emerging topic of research that is starting to attract some initial attention. As an example, in [143] a security-guaranteed estimation strategy has been formulated against integrity attacks by using the minimax optimization technology, where the estimator optimally minimizes the worst-case scenario of the expected cost considering all possible attacks launched by the adversary [129].

The estimation performance can be increased by using multiple sensor systems instead of one single sensor system. Data fusion is a process in which data are received and integrated from different sensors observing the same system, which often leads to better estimation accuracy. In the previous works considering data fusion, the sensor systems were implicitly considered to have equal sampling rates. This assumption is, unfortunately, quite restrictive since sensor devices tend to be asynchronous with different sampling rates because of the hardware constraints, and the resulting multirate fusion problem has drawn some preliminary research attention. Some examples are [144], [145], and [146].

Another issue is that CPSs may be subject to multiple attacks at the same time. An adaptive strategy compensating different types of attacks has not yet received adequate attention for industrial CPSs, and the impact on the system performance should be discussed in detail. In addition, in practical problems security requirements and resource constraints (communication bandwidth, limited energy, etc.) usually need to be considered simultaneously.

Chapter 3

Distributed denial-of-service attacks

Contents

3.1 Introduction

The Internet has become an important part of our society in numerous ways, such as in economics, government, business, and daily personal life. Furthermore, an increasing amount of critical infrastructure, for example, power grids or air traffic control, are managed and controlled via the Internet, in addition to traditional infrastructure for communication. However, today's cyberspace is full of attacks, such as Distributed Denial-of-Service (DDoS) attacks, information phishing, financial fraud, email spamming, and so on. DDoS attacks are global attacks and have become a serious problem of today's Internet. As depicted in Fig. 3.1, the results of [152] highlight the architecture and methods developed for network defense mechanisms, attack taxonomies, attack launching mechanisms, and their pros and cons.

DDoS attacks are adroit in nature and follows the same techniques as regular Denial-of-Service (DoS) attacks, but the attack are performed on a much larger scale using botnets [153], as shown in Figs. 3.2 and 3.3. A botnet is a wide chain of hundreds or thousands of remotely controlled compromised hosts (zombies,

Cloud Control Systems. https://doi.org/10.1016/B978-0-12-818701-2.00011-1

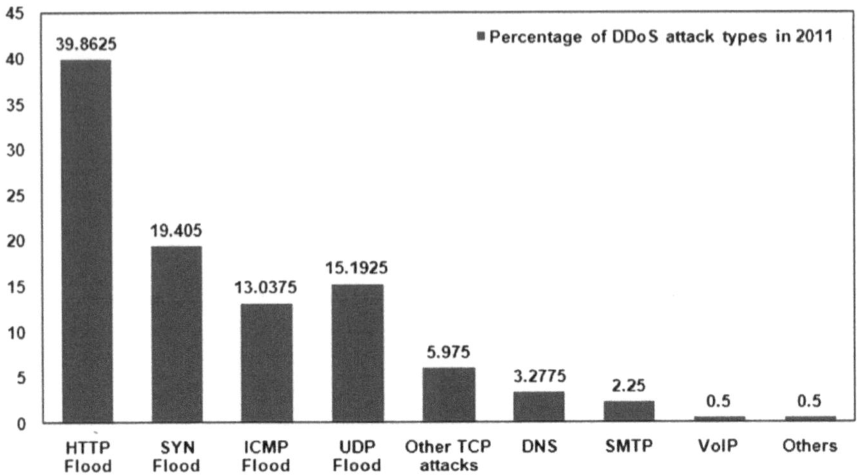

FIGURE 3.1 Distributed DoS attack by type [152].

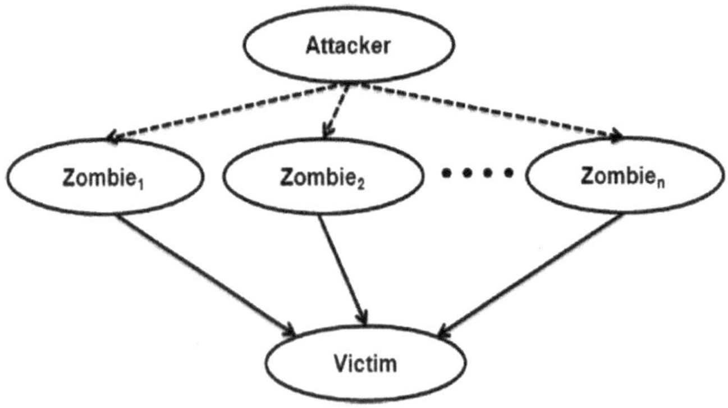

FIGURE 3.2 Direct distributed DoS attack.

bots, or slave agents) under the control of one or more intruders used to attack a particular victim.

3.2 Methods and tools

The first well-documented DDoS attack appears to have occurred in August 1999 when a DDoS tool called Trinoo was deployed in at least 227 systems, to flood a single University of Minnesota computer, which was knocked out for more than two days. The first large-scale DDoS attack took place in February 2001. On February 7, Yahoo! was the victim of a DDoS attack during which its Internet portal was inaccessible for three hours. On February 8, Amazon, Buy.com, CNN, and eBay were all hit by DDoS attacks that caused them to either stop functioning completely or slowed them down significantly

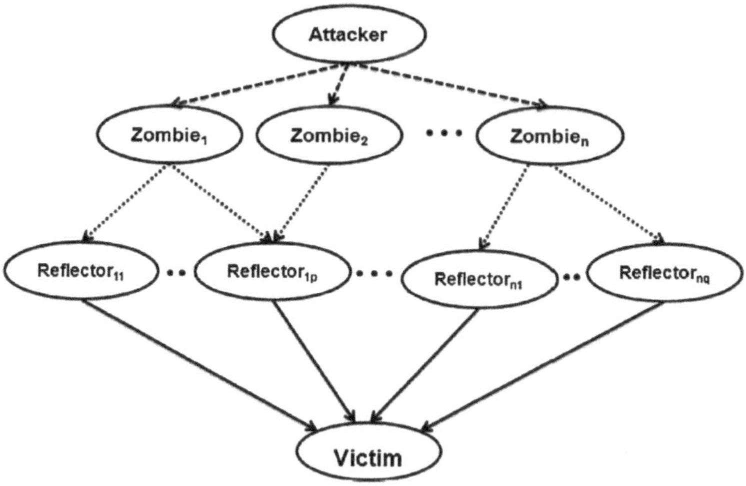

FIGURE 3.3 Indirect distributed DoS attack.

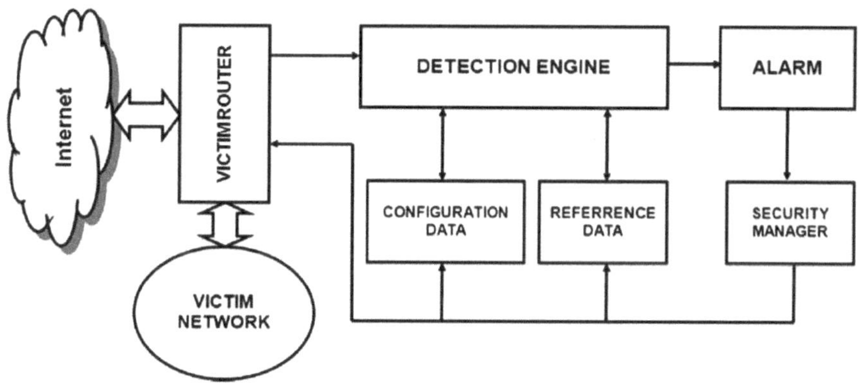

FIGURE 3.4 Generic architecture for victim-end DDoS defense mechanism.

http://www.garykessler.net/library/ddos.html. DDoS attack networks follow two types of architectures: the Agent-Handler architecture and the Internet Relay Chat (IRC)-based architecture, as discussed by [155]. The Agent-Handler architecture for DDoS attacks is comprised of clients, handlers, and agents (see Fig. 3.4). The attacker communicates with the rest of the DDoS attack system at the client systems. The handlers are often software packages located throughout the Internet that are used by the client to communicate with the agents. Instances of the agent software are placed in the compromised systems that finally carry out the attack. The owners and users of the agent systems are generally unaware of the situation. In the IRC-based DDoS attack architecture an IRC communication channel is used to connect the client(s) to the agents.

3.2.1 DDoS strategy

In general there are several steps in launching a DDoS attack, as summarized below:

1. *Selection of agents.* The attacker chooses the agents that will perform the attack. Based on the nature of vulnerabilities present, some machines are compromised to use as agents. Attackers victimize these machines, which have abundant resources, so that a powerful attack stream can be generated. In early years the attackers attempted to acquire control of these machines manually. However, with the development of advanced security tools, it has become easier to identify these machines automatically and instantly.
2. *Compromise.* The attacker exploits security holes and vulnerabilities of the agent machines and plants the attack code. In addition, the attacker also takes the necessary steps to protect the planted code from being identified and deactivated. As per the direct DDoS attack strategy shown in Fig. 3.2, the compromised nodes (i.e., the zombies between the attacker and the victim) are recruited by unwitting accomplice hosts from a large number of unprotected hosts connected through the Internet in high bandwidth. On the other hand, the DDoS attack strategy shown in Fig. 3.3 is more complex due to the inclusion of intermediate layer(s) between the zombies and victim(s). It further complicates the traceback mostly due to the following:
   (i) complexity in untangling the traceback information (partial) with reference to multiple sources, and/or
   (ii) having to connect a large number of routers or servers.

   Self-propagating tools such as the Ramen worm [156] and Code Red [157] automate this phase. Unless a sophisticated defense mechanism is used, it is usually difficult for the users and owners of the agent systems to realize that they have become a part of a DDoS attack system. Another important feature of such an agent system is that the agent programs are very cost effective both in terms of memory and bandwidth. Hence they affect the performance of the system minimally.
3. *Communication.* The attacker communicates with any number of handlers to identify which agents are up and running, when to schedule attacks, or when to upgrade agents. These communications between the attackers and handlers can be via various protocols, such as Internet Control Message Protocol (ICMP), Transmission Control Protocol (TCP), or User Datagram Protocol (UDP). Depending on the configuration of the attack network, agents can communicate with a single handler or multiple handlers.
4. *Attack.* The attacker initiates the attack. The victim, the duration of the attack, and special features of the attack such as the type, length, time-to-live (TTL), and port numbers can be adjusted. If there are substantial

variations in the properties of attack packets, it is beneficial to the attacker since it complicates detection.

Every computer connected to the Internet is an attractive target for attackers for making bots or zombies, even if the user does not know about it. Zombies are enrolled through the use of worms, backdoors, or Trojan horses by sending an e-mail content, a captivating link, or a trust-inspiring sender address to the vulnerable machines [154], [158]. Sometimes the data that originates from a single bot is very small, but the cumulative traffic from a sufficient number of bots emerging at the end user's system is so enormous that it exhausts the system resources. Therefore, Lowrate DDoS (LDDoS) attacks are devastating and harder to expose as the traffic appears to be normal and can be controlled by a particular link [151].

On the other hand, highrate DDoS (HDDoS) attacks are quickly recognized with the prevailing detection methods. Nowadays, DDoS attacks are conducted in the form of packet flooding and link flooding attacks. Such attacks have increased on the Internet because the attacker knows what information can be obtained where and how. Due to the presence of vulnerabilities in Internet protocols, web applications, and operating systems, it becomes easy for the attacker to launch such attacks. These attacks are performed with motives like hactivism (to generate media attention) or gaining profit through extortion.

3.2.2 Types of DDoS attacks

According to the records in [161], the largest DDoS attack in history was orchestrated in October 2016 using a new Mirai botnet against the servers of an American company called Dyn, that steers much of the Internet's Domain Name System (DNS) infrastructure. Mirai was the primary source of the pernicious attack traffic. Unlike other botnets, Mirai botnets had used the Internet of Things (IoT) devices such as digital cameras and DVR players to bring down websites (including Twitter, Netflix, the Guardian, CNN, Reddit, and many others) in Europe and the United States. According to the estimates of Dyn, the attack had a prodigious attack strength of 1.2 terabits (1200 gigabytes) per second and had intricate "100,000 malicious agents." The trend of the average strength of DDoS attacks is shown in Fig. 3.5.

Basically there are two types of DDoS attacks, namely flooding attacks and vulnerability attacks [154], as described in Fig. 3.6. In flooding attacks the attacker sets a zombie army to send junk or attack packets to the destination in order to raise the traffic to a level that a victim cannot handle, and the victim's system closes down or crashes [154]. On the basis of the attack mechanism, it categorizes the flooding attacks into direct and indirect (through reflectors) DDoS attacks. On the basis of the protocol level that is targeted, flooding attacks are grouped into Network/Transport level (Net-DDoS attacks) and Application level DDoS flooding attacks (App-DDoS attacks) [158]. Attacks like TCP, UDP, and ICMP flooding comes under

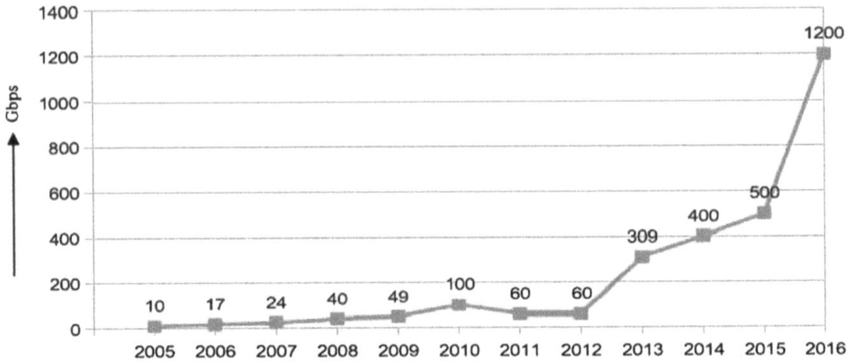

FIGURE 3.5 Strength of DDoS attacks in Gbps (Bhandari et al., 2015).

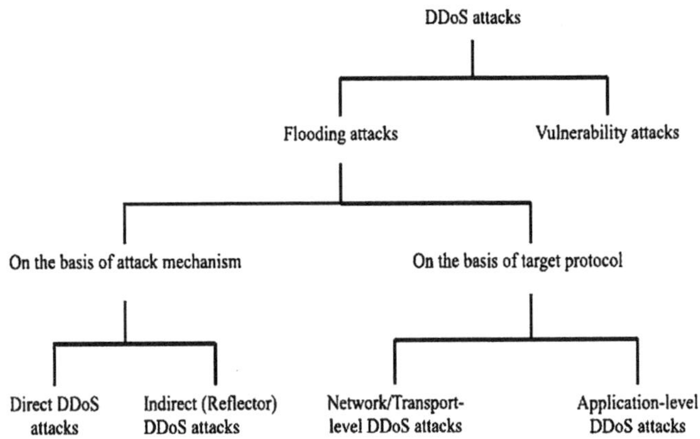

FIGURE 3.6 Types of DDoS attacks.

the category of Net-DDoS flooding attacks, while Hypertext Transfer Protocol (HTTP) flooding comes under App-DDoS flooding attacks. In [159], [160] the authors introduce the App-DDoS attacks and discuss the inability of network-level detection methods to catch the App-DDoS attacks. The number of these attacks is growing rapidly, they are becoming harder to detect, and they cause more severe problems in accessing a particular online service (or web server) compared to the Net-DDoS attacks. In vulnerability attacks the attacker browses for unprotected openings in the software implementation and exploits them to bring the system down or to recruit zombies for further attacks. These attacks use the exacted performance of different protocols (such as TCP and HTTP) to ravage the resources of the victim's server and prevent it from processing events or requests from the authorized users.

3.3 Detection techniques against DDoS attacks

The Internet has brought us cloud computing, which constitutes three major services, namely platform as a service, infrastructure as a service, and software as a service [162]. This increase in data and information storage within the cloud environment has raised cloud security concerns over the safety of data and information. It has also led to distributed attacks such as ICMP flood, the Ping of Death, the slowloris, the SYN flood attack, the UDP flood attack, malformed packet attacks, protocol vulnerability exploitation, and the HTTP flood molest [163], [164]. The choice on any attack type depends on the ease of the exploitation or its mastery by the attacker.

Previous researchers have expounded on how distributed attacks in the cloud can be detected, prevented, and mitigated. These techniques apply two major detection mechanisms of signatures or anomalies. They can use one, both, or be intelligent enough to learn new attacks based on set rules. The next section offers a review of various traditional intrusion detection techniques. Furthermore, it reviews the various classes of cloud computing-based detection methods and offers examples, the underlying purpose being to compare the various detection methods and point out the strengths and limitations they pose. Beyond the review, the chapter will show how specific techniques by specific scholars were successful or failed in the detection process against DDoS attacks in the cloud. In the analysis the performance of evaluation metrics used in a given technique will be shown. Additionally, the analysis will point out the various data sets and tools used by these techniques. It will thus be possible to decide which of the techniques is efficient or has potential for future enhancement.

3.3.1 Literature review

Existing techniques utilize different forms of algorithms to detect and determine attack levels within the cloud. HTTP-DoS and Extensible Markup Language (XML)-DoS attacks are known to lead to exhaustion of resources [165]. Cloud-based intrusion detection techniques are an improved version of traditional intrusion detection system. The first section of this chapter discusses various traditional intrusion detection techniques that are also applied in the cloud. The second section will show cloud-specific intrusion detection techniques.

3.3.2 Signature-based detection technique

This detection, also known as a misuse technique, compares known information to already captured signatures stored in the database. The technique is only suitable for the detection of known attacks. A common tool used in a signature detection technique is the SNORT tool [166]. SNORT, as an open source network intrusion prevention system, capable of performing real-time traffic analysis and packet logging on IP networks, is widely used in the technical media as it allows its users to set their own rules and use those rules in regulating attacks on either the training set or real data set of attack.

In the study conducted by Mazzariello, Canonico, and Bifulco, the authors deployed the network-based intrusion detection system (IDS) at separate cloud positions. By considering two scenarios in calculating the performance of the IDS, two results were depicted. First they inferred that the load on the controller increased, and the IDS detected the likelihood of the attack. Second, deploying an IDS close to the virtual machine resulted in the increase in the CPU load [167].

3.3.3 Anomaly-based detection technique

These techniques observe the behavior of an event and determine existing anomalies. The Shannon-Wiener index theory analyzes random data with an aim to unraveling existing uncertainty. [168] defines entropy as a measure of abnormal behavior or randomness. In a separate study, data from a single class proved to contain less entropy in contrast to statistics from multiple classes. Headers present in the sampled data were analyzed to determine the Internet Protocol (IP) and ports before computing their entropy. A certain threshold was then constituted to detect a DDoS attack: when the observed abnormality surpasses a set threshold, the IDS raises alarm alerts [169], [170]. An approach for detecting HTTP-based DDoS attacks was proposed by [171]. It entails a five-step filter tree approach of cloud defense. These steps include filtering of sensors and hop counts, diverging IP frequencies, double signatures, and puzzle solving [171]. The approach helped in determining anomalies with the various hop counts and treating the sources of these anomalies as attack sources.

3.3.4 Artificial neural network intrusion detection techniques

Techniques utilizing artificial neural networks (ANNs) to detect intrusions aim at generalizing incomplete data and classifying it as either intrusive or normal. An ANN IDS can either utilize a Multi-Layer Perceptron (MLP), Back propagation (BP), or a Multi-Layer Feed-Forward (MLFF) technique. An approach by Gradiega Ibarra, Ledesma, and Garcia compared the use of self-organization map (SOM) to MLP in determining intrusion rates and found that SOM provides higher accuracy rates of detection compared to ANN [172]. Cannady utilized a signature-based detection mechanism in a three-layer neural network as a way to detect any intrusions. He used a nine-network feature vector consisting of the source port, protocol identification, raw data, destination port, data length, source IP address, ICMP code, the type of ICMP, and the destination IP address to determine the intrusions [172].

3.3.5 Genetic algorithm intrusion detection systems

The use of genetic algorithms in the development of IDS helps incorporate various network features to determine the best possible parameters for improvement of accuracy and optimization of results. Gong, Zulkernine, and Abolmaesumi implemented seven network features used to analyze packets: Duration, Protocol, Source IP, Destination IP, Source Port, Destination Port, and Attack Name. By using fitness function frameworks that support confidence, the authors were able to detect and determine network intrusions with high accuracy levels.

In [172] a solution was proposed that combined both genetic algorithms and fuzzy logic to detect signature and anomaly attacks. Fuzzy logic helps in accounting for quantitative parameters, while genetic algorithms determine the best-fit parameters that are introduced by the fuzzy logic. This approach proved able to solve the best-fit problem in the cloud environment. It also showed that using the genetic algorithm in developing IDS is effective for cloud use because selecting optimal network features as the parameters for intrusion detection increases the IDS accuracy level [172].

3.4 Epilogue

DDoS attacks mainly take advantage of the architecture of the Internet and this is what makes them powerful. While designing the Internet, the prime concern was to provide functionality, not security. As a result, many security issues have been raised that are exploited by attackers. Some of the issues are given below:

- Internet security is highly interdependent. No matter how secure a victim's system may be, whether or not this system will be a DDoS victim depends on the rest of the global Internet [13,14].
- Internet resources are limited. Every Internet host has limited resources that sooner or later can be exhausted by a sufficiently large number of users.
- Many against a few. If the resources of the attackers are greater than the resources of the victims, the success of the attack is almost assured.
- Intelligence and resources are not collocated. Most of the intelligence needed for service guarantees is located at end hosts. At the same time, high-bandwidth pathways needed for high throughput are situated in the intermediate network. Such abundant resources present in parts of the network are covertly exploited by the attacker to launch a successful flooding attack.
- The handlers or masters, which are hosts with special programs running on them, are capable of controlling multiple agents.
- The attack daemon agents or zombie hosts are hosts that each run a special program and are responsible for generating a stream of packets towards the intended victim. These machines are commonly external to the victim's own network in order to disable an efficient response from the victim, and are external to the network of the attacker in order to preclude liability if the attack is traced back to the source.

3.5 Stabilization of distributed discrete systems

We learned that a cloud control system (CCS) is an integration of communication, computation, and control used to achieve the desired performance of physical systems. With its wide range of applications, including sustainable and blackout-free electricity generation and distribution, CCSs have attracted the interest of researchers [316]. Other applications for cyber-physical systems (CPSs) include clean and energy-aware buildings and cities, smart medical and healthcare systems, transportation networks, chemical process control, smart grids, water and gas distribution networks, and emergency management to new a few [38].

On the other hand, security issues increase the challenges in control of CPSs because CPSs have a high possibility of being affected by several cyber attacks without providing any notification of failure. These attacks can lead to a disruption to the physical system; for example, the disarrangement of coordination packets in medium access control layers could be a result of malware injected by an adversary. Moreover, in order to destroy the nominal operation, an attacker can illegally obtain access to the supervision centers while obtaining the encryption key. That means that the system dynamics can be disturbed arbitrarily by the attacker, and when there is a lack of security protection either in hardware or software strategies an attacker has the capability of inducing any perturbation [58].

3.5.1 Introduction

The communication among the items of control systems, i.e., sensors, actuators, and controllers, occurs through a common network medium. This network needs to be secured to prevent the vulnerability of attack by adversaries during data transmission. These attacks could lead the system to instability or drive the plant to undesired operations, as mentioned before. Thus, considering security issues is very important in designing the controllers for such a system [306].

DoS attacks are strategies that are often used to occupy the communication resources in order to block the transmission of measurement and/or control signals, and that cause the maximum possible deterioration of the system's performance. Several approaches for controlling systems affected by DoS attacks are applied in the literature, due to its high importance in CPSs [306].

The cyclic small-gain theorem was implemented to design an output-feedback controller for large-scale nonlinear systems subject to nonsmooth sensor noise [250]. A distributed output feedback control of linear-time invariant (LTI) systems in the presence of unreliable communication was designed by solving an optimization control problem [307]. The problem of lossy sensors and cyber attacks in discrete-time multiagent systems was discussed and a distributed observer-based consensus controller was proposed using an event-triggering method [121]. The backstepping adaptive approach was implemented for a large-scale stochastic nonlinear time-delay system in the presence of constrained outputs and saturation of actuators [308].

An observer-based controller was proposed for linear systems affected by process disturbances and false data injection attacks by implementing a controller gain scheme and a supervisory switching strategy [309]. A secure distributed controller for power systems subject to time-varying data injection attacks were proposed using a model predictive control approach [11]. A distributed controller is designed for networked control systems undergoing stochastic cyber attacks using an event-triggered approach [310]. Another implementation of an event-triggered approach was presented with the help of H_∞ optimization to achieve the stability of neural networks affected by cyber attacks and considering a constrained bandwidth of the network [311].

The small-gain approach was widely applied by researchers to solve the stabilization problem of distributed systems. An event-triggered sampling scheme was presented to ensure the stability of large-scale systems by distributed controllers in the presence of limited communication medium [285], [286]. A hierarchical game method was presented to solve the control problem of a wireless networked control system subject to DoS attack [312]. A robust pinning synchronization control problem was proposed to ensure that the initial state will be restored for a complex CPS subject to mixed attacks affecting independent transmitting channels [313].

The analysis problem of distributed systems subject to cyber attacks has attracted many researchers. The duration and frequency of the DoS attacks for CCSs with multiple transmission channels were characterized to ensure the stability of a switched system [314]. The bound of a DoS attack frequency and duration was discussed for distributed systems in the presence of pure round-robin communication [101]. The major drawback of these techniques is that the availability of the full state for all subsystems is considered, which is not true for most of the real practical CPSs. Below we will examine the stabilization problem of a discrete-time distributed CCS subject to a DoS attack, while considering that partial information of the states is available through the output of each system. The main contributions of our work are the following:

- Discussing the robustness problem in distributed CPSs by examining the stabilization of these systems in the face of DoS attacks and elaborating on the published work from various points of view;
- Considering a static output feedback control problem of a "nominal discrete-time" distributed CPS and designing an appropriate control law using the LMI technique to achieve closed-loop stability;
- Deriving a bound of attack frequency and duration to ensure the stability of the distributed CPS with partial information by means of a simple typical scenario where the communication sequence is purely round-robin;
- Demonstrating the feasibility of the proposed system through numerical simulation.

3.5.2 Distributed cloud control system (DCCS)

Let us consider the following discrete-time distributed system consisting of N interacting subsystems

$$\begin{aligned} x_i(k+1) &= A_i x_i(k) + B_i u_i(k) + \sum_{j \in N_i} H_{ij} x_j(k), \\ y_i(k) &= C_i x_i(k), \end{aligned} \tag{3.1}$$

where A_i, B_i, H_{ij}, and C_i are system matrices with appropriate dimensions; $x_i(k)$, $u_i(k)$, and $y_i(k)$ are state, control input, and output of each subsystem i, respectively; and N_i denotes the set of neighbors of subsystem i.

The distributed systems are controlled through a communication network that is used by each subsystem to send the outputs of the sensors to controllers. The controllers use these data to calculate the input signals and send them to the actuators of the systems. The outputs arrive in sample-and-hold fashion such as $y_i(k_i^{\prime})$, where $k_i^{\prime}$ represents the sequence of transmission instants of subsystem i.

Remark 3.1. We assume that there exists a feedback gain K_i such that matrix $\bar{A}_i = A_i + B_i K_i C_i$ is a Hurwitz matrix. So each control input u_i affecting subsystem i is written as

$$u_i(k) = K_i C_i x_i(k_i^{\prime}) + \sum_{j \in N_i} L_{ij} C_j x_j(k_i^{\prime}). \tag{3.2}$$

3.5.3 Characteristics of the denial-of-service attacks

In this chapter the effect of DoS attacks will be considered as a failure in the transmission of signals. In addition, this effect will be accumulated with the failure caused by channel unavailability. The communication attempts of all the subsystems are simultaneously affected by the DoS attacks because the network is shared.

As in [89] the model of the DoS attacks will be considered to have a limited frequency and duration. Let $\{H_n\}_{n \in \mathbb{N}_0}, h_0 \geq 0$, refer to the sequence of DoS off/on transitions such that the time instants at which DoS exhibits a transition from possible to impossible transmissions (or zero to one). So the n-th DoS time interval of length $\tau_n \in \mathbb{R}_{\geq 0}$ is given by

$$H_n := \{h_n\} \cup [h_n, h_n + \tau_{n-1}]. \tag{3.3}$$

If $\tau_n = 0$, then H_n takes the form of a single pulse at h_n. If $\tau_n \neq 0$, $[h_n, h_n + \tau_{n-1}]$ represents an interval from the instant h_n (include h_n) to $(h_n + \tau_{n-1})$. Similarly, $[\tau, k-1]$ represents an interval from τ to $k-1$. Given $\tau, k \in \mathbb{R} \geq 0$ with $k \geq \tau$, let $n(\tau, k)$ refer to the number of DoS off/on transitions over $[\tau, k-1]$, and let $\Xi(\tau, k)$ refer to the subset of $[\tau, k]$ during which the network affected by the DoS attack such that

$$\Xi(\tau, k) := \bigcup_{n\in\mathbb{N}_0} H_n \cap [\tau, k]. \tag{3.4}$$

In addition, $\Theta(\tau, k)$ refers to the interval where the attack does not exist as such, and is represented by

$$\Theta(\tau, k) := [\tau, k] \backslash \Xi(\tau, k). \tag{3.5}$$

Assumption 3.1 (frequency of the DoS attack). There exist constants $\eta \in \mathbb{R} \geq 0$ and $\tau_D \in \mathbb{R} > 0$ such that

$$n(\tau, k) \leq \eta + \frac{k-\tau}{\tau_D} \tag{3.6}$$

for all $\tau, k \in \mathbb{R} \geq 0$ with $k \geq \tau$.

Assumption 3.2 (duration of the DoS attack). There exist constants $\kappa \in \mathbb{R} \geq 0$ and $T \in \mathbb{R} > 1$ such that

$$|\Xi(\tau, k)| \leq \kappa + \frac{k-\tau}{T} \tag{3.7}$$

for all $\tau, k \in \mathbb{R} \geq 0$ with $k \geq \tau$.

Remark 3.2. Assumptions 3.1 and 3.2 constrain the average frequency and duration of the DoS attack signals. τ_D and η in Assumption 3.1 can be named as the average dwell-time between consecutive DoS off/on transitions and the chattering bound, respectively. Assumption 3.2 constrains the duration of the DoS attack such that it is limited by a certain fraction of time named $1/T$. The constant κ is used for regulation [101].

3.5.4 Nominal design results

The objective in this section is to find stability conditions for the distributed CPS affected by DoS attacks. We will first present an output feedback controller to ensure the stability of the system in the nominal case (absence of DoS). Then we will discuss the stabilization problem of distributed CPS under a digital communication channel in the nominal scenario.

The error between the value of the current state and the transmitted state is defined as $e_i(k)$, where i refers to the subsystem such that

$$e_i(k) = x_i(k_i') - x_i(k), \quad i = 1, 2, \ldots, N. \tag{3.8}$$

The dynamics of each subsystem i can be described by combining (3.1), (3.2), and (3.8) as

$$x_i(k+1) = \bar{A}_i x_i(k) + B_i K_i C_i e_i(k) + \sum_{j\in N_i} \bar{A}_j x_j(k) + B_i \sum_{j\in N_i} L_{ij} C_j e_j(k), \tag{3.9}$$

where $\bar{A}_j = B_i L_{ij} C_j + H_{ij}$. It should be noted that the interconnected neighbors $x_j(k)$ affect the dynamics of subsystem i in addition to $e_i(k)$ and $e_j(k)$.

Remark 3.3. It is clear from (3.9) that the stability can be accomplished in the case of small errors on e and weak couplings. Moreover, the "smallness" of e can be explained by the x-dependent bound $\|e_i(k)\| \leq \sigma_i \|x_i(k)\|$, with a suitable design parameter σ_i.

Our objective in this section is to design a static output feedback in the form of (3.2) to achieve the asymptotic stability for nominal distributed systems (3.1).

Theorem 3.1. *Let the controller gains K_i and L_{ij} of (3.2) be given. System (3.9) is asymptotically stable if there exist positive matrices P_i satisfying the following inequalities:*

$$\Lambda_i = \begin{bmatrix} \Lambda_{1i} & \Lambda_{2i} & \Lambda_{3i} & \Lambda_{4i} \\ * & \Lambda_{5i} & \Lambda_{6i} & \Lambda_{7i} \\ * & * & \Lambda_{8i} & \Lambda_{9i} \\ * & * & * & \Lambda_{10i} \end{bmatrix} < 0, \tag{3.10}$$

where

$$\begin{aligned}
\Lambda_{1i} &= \bar{A}_i^T P_i \bar{A}_i - P_i, & \Lambda_{2i} &= \bar{A}_i^T P_i B_i K_i C_i, \\
\Lambda_{3i} &= \bar{A}_i^T P_i \sum_{j\in N_i} \bar{A}_j, & \Lambda_{4i} &= \bar{A}_i^T P_i B_i \sum_{j\in N_i} L_{ij} C_j, \\
\Lambda_{5i} &= C_i^T K_i^T B_i^T P_i B_i K_i C_i, & \Lambda_{6i} &= C_i^T K_i^T B_i^T P_i \sum_{j\in N_i} \bar{A}_j, \\
\Lambda_{7i} &= C_i^T K_i^T B_i^T P_i B_i \sum_{j\in N_i} L_{ij} C_j, & \Lambda_{8i} &= \sum_{j\in N_i} \bar{A}_j^T P_i \bar{A}_j, \\
\Lambda_{9i} &= \sum_{j\in N_i} \bar{A}_j^T P_i B_i L_{ij} C_j, & \Lambda_{10i} &= \sum_{j\in N_i} C_j^T L_{ij}^T B_i^T P_i^T B_i L_{ij} C_j.
\end{aligned} \tag{3.11}$$

Proof. To establish the main theorem the following Lyapunov function is constructed:

$$V_i(k) = x_i^T(k) P_i x_i(k). \tag{3.12}$$

Evaluating the difference of $V_i(k)$, we have

$$\Delta V_i(k) \quad = \quad V_i(k+1) - V_i(k) < 0$$

$$
\begin{aligned}
\Delta V_i(k) &= x_i^T(k)\left(\bar{A}_i^T P_i \bar{A}_i - P_i\right)x_i(k) + 2x_i^T(k)\bar{A}_i^T P_i B_i K_i C_i e_i(k) \\
&+ 2x_i^T(k)\bar{A}_i^T P_i \sum_{j\in N_i} \bar{A}_j x_j(k) \\
&+ 2x_i^T(k)\bar{A}_i^T P_i \sum_{j\in N_i} B_i L_{ij} C_j e_j(k) \\
&+ e_i^T(k) C_i^T K_i^T B_i^T P_i B_i K_i C_i e_i(k) \\
&+ 2e_i^T(k) C_i^T K_i^T B_i^T P_i \sum_{j\in N_i} \bar{A}_j x_j(k) \\
&+ 2e_i^T(k) C_i^T K_i^T B_i^T P_i \sum_{j\in N_i} B_i L_{ij} C_j e_j(k) \\
&+ \sum_{j\in N_i}\left(x_j^T(k)\bar{A}_j^T\right) P_i \sum_{j\in N_i}\left(\bar{A}_j x_j(k)\right) \\
&+ 2\sum_{j\in N_i}\left(x_j^T(k)\bar{A}_j^T\right) P_i \sum_{j\in N_i}\left(B_i L_{ij} C_j e_j(k)\right) \\
&+ \sum_{j\in N_i}\left(e_j^T(k) C_j^T L_{ij}^T B_i^T\right) P_i \sum_{j\in N_i}\left(B_i L_{ij} C_j e_j(k)\right) \\
&< 0.
\end{aligned} \tag{3.13}
$$

This can be rewritten as

$$\Delta V_i(k) = \Omega_i^T(k)\Lambda_i\Omega_i(k) < 0, \tag{3.14}$$

where

$$\Omega_i^T(k) = \begin{bmatrix} x_i(k) & e_i(k) & \sum_{j\in N_i} x_j(k) & \sum_{j\in N_i} e_j(k) \end{bmatrix}^T, \tag{3.15}$$

and Λ_i as described in (3.10). □

Theorem 3.2. *If there exist positive definite matrices X_i and matrices Y_i and Y_{ij} such that the following inequality is satisfied*

$$\bar{\Lambda}_i = \begin{bmatrix} -X_i & 0 & 0 & 0 & Y_i^T B_i^T + X_i A_i^T \\ 0 & \epsilon_i I & 0 & 0 & Y_i^T B_i^T \\ 0 & 0 & \epsilon_i I & 0 & \sum_{j\in N_i} Y_{ij}^T B_i^T + X_i H_{ij}^T \\ 0 & 0 & 0 & \epsilon_i I & \sum_{j\in N_i} Y_{ij}^T B_i^T \\ * & * & * & * & -X_i \end{bmatrix} < 0, \tag{3.16}$$

then system (3.9) is asymptotically stable and the controller gains are given by

$$K_i = Y_i X_i^{-1}(C_i)^{\dagger}, \quad L_{ij} = Y_{ij} X_i^{-1}(C_j)^{\dagger}. \tag{3.17}$$

Proof. Λ_i in LMI (3.10) can be rewritten as

$$\Lambda_i = \bar{\Lambda}_{i11} + \bar{\Lambda}_{i12} P_i \bar{\Lambda}_{i12}^T < 0, \tag{3.18}$$

with $\bar{\Lambda}_{i11} = \text{diag}\{-P_i, 0, 0, 0\}$ and

$$\bar{\Lambda}_{i12}^T = \begin{bmatrix} \bar{A}_i & B_i K_i C_i & \sum_{j\in N_i} \bar{A}_j & \sum_{j\in N_i} B_i L_{ij} C_j \end{bmatrix}^T.$$

In turn, LMI (3.18) is formulated using Schur complements as

$$\begin{bmatrix} \bar{\Lambda}_{i11} & \bar{\Lambda}_{i12} \\ \bar{\Lambda}_{i12}^T & -P_i^{-1} \end{bmatrix} < 0. \tag{3.19}$$

Now, define $X_i = P_i^{-1}$, then multiply (3.19) from the right and left by

$$\text{diag}\{X_i, X_i, X_i, X_i, I\}.$$

Then, by recalling the values of $\bar{A}_i$ and $\bar{A}_j$, and by selecting $Y_i^T := X_i C_i^T K_i^T$ and $Y_{ij}^T := X_i C_j^T L_{ij}^T$, the LMI (3.16) can be obtained. □

3.5.5 A small-gain approach for distributed CPS

In this section a periodic sampling protocol will be implemented to update the law, e.g. round-robin protocol.

Assumption 3.3 (intersampling of round-robin). There exists an intersampling interval Δ in the absence of DoS attacks that satisfies

$$\|e_i(k)\| \leq \sigma_i \|x_i(k)\| \tag{3.20}$$

holds, where σ_i is a suitable design parameter.

Although we were able to find a value of Δ satisfying (3.20), σ_i should be selected appropriately to avoid losing the stability [101].

For $Q_i = Q_i^T > 0$, and P_i is the unique solution of the Lyapunov equation $\bar{A}_i^T P_i \bar{A}_i - P_i + Q_i = 0$. We consider the Lyapunov function $V_i(k) = x_i^T(k) P_i x_i(k)$ for each subsystem i satisfying

$$\lambda_{\min}(P_i)\|x_i(k)\|^2 \leq V_i(x_i(k)) \leq \lambda_{\max}(P_i)\|x_i(k)\|^2, \tag{3.21}$$

where $\lambda_{\min}(P_i)$ and $\lambda_{\max}(P_i)$ refer to the smallest and largest eigenvalue of P_i, respectively. The selection of σ_i to ensure the stability of the system is presented by Lemma 3.1.

Lemma 3.1. *For a distributed CPS described by (3.1) controlled by inputs described by (3.2), suppose that the spectral radius $r(A^{-1}B) < 1$. The distributed CPS is asymptotically stable if there exists σ_i such that*

$$\sigma_i < \sqrt{\frac{l_i}{j_i}}, \tag{3.22}$$

where l_i is the i-th entry of the row vector $L := \mu^T(A - B) = [l_1, l_2, \ldots, l_N]$ and j_i is the i-th entry of the row vector $J := \mu^T \Gamma = [j_1, j_2, \ldots, j_N]$. $\mu \in \mathbb{R}_+^N$ is an arbitrary column vector satisfying $\mu^T(-A + B) < 0$. The matrices A, B, and Γ are given by

$$A = \begin{bmatrix} \alpha_1 & & \\ & \ddots & \\ & & \alpha_N \end{bmatrix}, \tag{3.23}$$

$$B = \begin{bmatrix} 0 & \beta_{12} & \cdots & \beta_{1N} \\ \beta_{21} & 0 & \cdots & \beta_{2N} \\ \vdots & \vdots & \vdots & \vdots \\ \beta_{N1} & \beta_{N2} & \cdots & 0 \end{bmatrix}, \tag{3.24}$$

$$\Gamma = \begin{bmatrix} \gamma_{11} & \gamma_{12} & \cdots & \gamma_{1N} \\ \gamma_{21} & \gamma_{22} & \cdots & \gamma_{2N} \\ \vdots & \vdots & \vdots & \vdots \\ \gamma_{N1} & \gamma_{N2} & \cdots & \gamma_{NN} \end{bmatrix}, \tag{3.25}$$

with

$$\alpha_i = \lambda_{\min}(Q_i) - \delta\|\bar{A}_i^T P_i \bar{A}_i\|^2, \tag{3.26}$$

$$\beta_{ij} = \frac{\|\bar{A}_j^T P_i \bar{A}_j\|^2}{\delta} + \|C_i^T K_i^T B_i^T P_i \bar{A}_j\|^2, \tag{3.27}$$

$$\gamma_{ii} = \frac{\|\bar{A}_i^T P_i B_i K_i C_i\|^2}{\delta} + \|C_i^T K_i^T B_i^T P_i B_i K_i C_i\|, \tag{3.28}$$

$$\begin{aligned}\gamma_{ij} &= \frac{\|\bar{A}_i^T P_i B_i L_{ij} C_j\|^2}{\delta} + \frac{\|C_i^T K_i^T B_i^T P_i B_i L_{ij} C_j\|^2}{\delta} \\ &+ \frac{\|\bar{A}_j^T P_i B_i L_{ij} C_j\|^2}{\delta} + \|C_j^T L_{ij}^T B_i^T P_i B_i L_{ij} C_j\|,\end{aligned} \tag{3.29}$$

where δ is positive and real such that $\alpha_i > 0$ and $\lambda_{\min}(Q_i)$ is the minimum eigenvalue of Q_i for $i = 1, 2, \ldots, N$.

Proof. The difference equation (3.13) can be described by

$$\begin{aligned}
\Delta V_i(k) &\le -\lambda_{\min}(Q_i)\,\|x_i(k)\|^2 + \left\|2\bar{A}_i^T P_i B_i K_i C_i\right\| \|x_i(k)\|\,\|e_i(k)\| \\
&+ \sum_{j\in N_i} \left\|2\bar{A}_i^T P_i \bar{A}_j\right\| \|x_i(k)\|\,\left\|x_j(k)\right\| \\
&+ \sum_{j\in N_i} \left\|2\bar{A}_i^T P_i B_i L_{ij} C_j\right\| \|x_i(k)\|\,\|e_i(k)\| \\
&+ \left\|C_i^T K_i^T B_i^T P_i B_i K_i C_i\right\| \|e_i(k)\|^2 \\
&+ \sum_{j\in N_i} \left\|2C_i^T K_i^T B_i^T P_i \bar{A}_j\right\| \|e_i(k)\|\,\left\|x_j(k)\right\| \\
&+ \sum_{j\in N_i} \left\|2C_i^T K_i^T B_i^T P_i B_i L_{ij} C_j\right\| \|e_i(k)\|\,\left\|e_j(k)\right\| \\
&+ \sum_{j\in N_i} \left\|\bar{A}_j^T P_i \bar{A}_j\right\| \left\|x_j(k)\right\|^2 \\
&+ \sum_{j\in N_i} \left\|2\bar{A}_j^T P_i B_i L_{ij} C_j\right\| \left\|x_j(k)\right\|\,\left\|e_j(k)\right\| \\
&+ \sum_{j\in N_i} \left\|C_j^T L_{ij}^T B_i^T P_i B_i L_{ij} C_j\right\| \left\|e_j(k)\right\|^2 .
\end{aligned} \tag{3.30}$$

The Young's inequalities for any matrices E, F, and G with any positive real δ yield the following:

$$\|E\|\,\|F\|\,\|G\| \le \delta\,\|F\|^2 + \frac{1}{\delta}\,\|E\|^2\,\|G\|^2 . \tag{3.31}$$

Using (3.31), (3.30) can be rewritten as

$$\begin{aligned}
\Delta V_i(x_i(k)) &\le -\alpha_i\,\|x_i(k)\|^2 + \sum_{j\in N_i} \beta_{ij}\,\left\|x_j(k)\right\|^2 + \gamma_{ii}\,\|e_i(k)\|^2 \\
&+ \sum_{j\in N_i} \gamma_{ij}\,\left\|e_j(k)\right\|^2
\end{aligned} \tag{3.32}$$

with α_i, β_{ij}, γ_{ii}, and γ_{ij} as in (3.26)–(3.29). In addition, δ can be always found such that $\alpha_i > 0$ for $i = 1, 2, \ldots, N$.

By defining vectors

$$\begin{aligned}
V_{vec}(x_i(k)) &:= [V_1(x_1(k)), V_2(x_2(k)), \ldots, V_N(x_N(k))]^T \\
\|x(k)\|_{vec} &:= \left[\|x_1(k)\|^2, \|x_2(k)\|^2, \ldots, \|x_N(k)\|^2\right]^T \\
\|e(k)\|_{vec} &:= \left[\|e_1(k)\|^2, \|e_2(k)\|^2, \ldots, \|e_N(k)\|^2\right]^T .
\end{aligned}$$

The inequality (3.32) can be compactly written as

$$\Delta V_i(x_i(k)) \quad \leq \quad (-A+B)\left\|x(k)\right\|_{vec} + \Gamma \left\|e(k)\right\|_{vec} \tag{3.33}$$

with A, B, and Γ as in Lemma 3.1.

There exists a positive vector $\mu \in \mathbb{R}_+^n$ satisfying $\mu^T(-A+B) < 0$ if the spectral radius satisfies $r(A^{-1}B) < 1$ [258]. We choose the Lyapunov function to be $V(x(k)) := \mu^T V_{vec}(x_i(k))$. Then ΔV yields

$$\begin{aligned} \Delta V(x(k)) &= \mu^T \Delta V_{vec}(x_i(k)) \\ &\leq \mu^T(-A+B)\left\|x(k)\right\|_{vec} + \mu^T\Gamma \left\|e(k)\right\|_{vec}. \end{aligned} \tag{3.34}$$

By noticing that $\mu^T(-A+B) < 0$, we have

$$\Delta V(x(k)) \leq -L\left\|x(k)\right\|_{vec} + J\left\|e(k)\right\|_{vec}, \tag{3.35}$$

where $L := \mu^T(A-B)$ and $J := \mu^T\Gamma$ are row vectors.

Let l_i and j_i be the entries of vectors L and J, respectively. So we obtain

$$\begin{aligned} \Delta V(x(k)) &\leq \sum_{i\in N} l_i \left\|x_i(k)\right\|^2 + \sum_{i\in N} j_i \left\|e_i(k)\right\|^2 \\ &= -\sum_{i\in N}\left(l_i \left\|x_i(k)\right\|^2 - j_i \left\|e_i(k)\right\|^2\right), \end{aligned} \tag{3.36}$$

which implies asymptotic stability with $\sigma_i < \sqrt{\frac{l_i}{j_i}}$. □

Remark 3.4. The matrices P_i and Q_i, which are required to apply Lemma 3.1 and for calculating σ_i, can be calculated by solving LMI (3.16) of Theorem 3.2, such that

$$P_i = X_i^{-1}, \quad Q_i = P_i - \bar{A}_i^T P_i \bar{A}_i. \tag{3.37}$$

3.5.6 Stability analysis under denial-of-service attacks

We discussed in the previous section the selection of σ_i, which is related to the error bound to ensure the asymptotic stability of a distributed CPS using a round-robin protocol but under no attack. Unfortunately the stability of the system is not guaranteed when the DoS attack affects the system even with the same round-robin protocol [101]. Our objective in this section is to discuss the stability of the distributed CPSs subject to DoS attacks for a round-robin protocol.

Theorem 3.3. *For a distributed CPS (3.1) and a control input (3.2), a round-robin protocol is used in the network communication with sampling interval Δ as in Assumption 3.3. The distributed system subjected to a DoS attack with time* *and duration as in Assumptions 3.1 and 3.2 with arbitrary η and κ, and with τ_D and T is asymptotically stable if*

$$\frac{1}{T} + \frac{\Delta_*}{\tau_D} < 1; \quad \omega_1 < 1, \tag{3.38}$$

in which

$$\omega_1; = \min\{\frac{l_i - \sigma_i^2 j_i}{\lambda_{\max}(P_i)\mu_i}\}, \quad \omega_2; = \frac{4\max\{j_i\}}{\min\{\mu_i \lambda_{\min}(P_i)\}} \tag{3.39}$$

and $\Delta_* = N\Delta$, l_i, j_i, μ_i, and σ_i are as in Lemma 3.1.

Proof. The proof of Theorem 3.3 is shown in three steps:

Step 1. *Lyapunov function in periods without attacks.* Referring to Assumption 3.3, (3.20) holds true for σ_i satisfying Lemma 3.1 and (3.36) is negative. So the derivative of the Lyapunov function is described as

$$\begin{aligned} \Delta V(x(k)) &\leq -\sum_{i \in N}\left(l_i - j_i\sigma_i^2\right)\|x_i(k)\|^2 \leq -\sum_{i \in N}\frac{l_i - j_i\sigma_i^2}{\lambda_{\max}(P_i)\mu_i}\mu_i V_i \\ &\triangleq -\omega_1 V(x(k)), \end{aligned} \tag{3.40}$$

where $\omega_1 = \min\{\frac{l_i - \sigma_i^2 j_i}{\lambda_{\max}(P_i)\mu_i}\}$. Thus, for $k \in [h_n + \tau_n, h_{n+1}[$ (DoS-free time), the Lyapunov function yields

$$V(x(k)) \leq (1-\omega_1)^{k-h_n-\tau_n} V(x(h_n + \tau_n)). \tag{3.41}$$

Step 2. *Lyapunov function during attack periods.* Let us define z_m^i as the last successful sampling instant before the DoS attack occurs. By implementing $e_i(k)$ as defined before, we get

$$e_i(k) = x_i(z_m^i) - x_i(k) = x_i(h_n) - x_i(k) \tag{3.42}$$

and

$$\begin{aligned} \|e_i(k)\|^2 &\leq \|x_i(h_n)\|^2 + 2\|x_i(k)\|\,\|x_i(h_n)\| + \|x_i(k)\|^2 \\ &\leq 2\|x_i(h_n)\|^2 + 2\|x_i(k)\|^2 \end{aligned} \tag{3.43}$$

for $k \in H_n$. By summing up $\|e_i(k)\|^2$ for $i \in N$, we obtain

$$\sum_{i \in N}\|e_i(k)\|^2 \leq 2\sum_{i \in N}\|x_i(h_n)\|^2 + 2\sum_{i \in N}\|x_i(k)\|^2. \tag{3.44}$$

If $\sum_{i \in N}\|x_i(h_n)\|^2 \leq \sum_{i \in N}\|x_i(k)\|^2$, then we have

$$\sum_{i \in N}\|e_i(k)\|^2 \leq 4\sum_{i \in N}\|x_i(k)\|^2.$$

Otherwise, we have $\sum_{i\in N}\|e_i(k)\|^2 \le 4\sum_{i\in N}\|x_i(h_n)\|^2$.
Recalling (3.36), it is simple to see that

$$\Delta V(x(k)) \le \sum_{i\in N} j_i \|e_i(k)\|^2 . \tag{3.45}$$

Thus, for all $k \in H_n$ (DoS-active time) in the case that $\sum_{i\in N}\|x_i(h_n)\|^2 \le \sum_{i\in N}\|x_i(k)\|^2$, the difference of the Lyapunov function yields

$$\begin{aligned}\Delta V(x(k)) &\le \max\{j_i\}\sum_{i\in N}\|e_i(k)\|^2 \le 4\max\{j_i\}\sum_{i\in N}\|x_i(k)\|^2\\ &\le \frac{4\max\{j_i\}}{\min\{\mu_i\lambda_{\min}(P_i)\}}\sum_{i\in N}\mu_i V(x_i(k))\\ &\triangleq \omega_2 V(x(k))\end{aligned} \tag{3.46}$$

with $\omega_2 := \frac{4\max\{j_i\}}{\min\{\mu_i\lambda_{\min}(P_i)\}}$. On the other hand, for all $k \in H_n$ such that

$$\sum_{i\in N}\|x_i(h_n)\|^2 > \sum_{i\in N}\|x_i(k)\|^2 ,$$

we have

$$\Delta V(x(k)) \le \omega_2 V(x(h_n)). \tag{3.47}$$

Thus, (3.46) and (3.47) imply the Lyapunov function during H_n satisfies

$$V(x(k)) \le (1+\omega_2)^{k-h_n} V(x(h_n)). \tag{3.48}$$

Step 3. *Switching between stable and unstable modes.* Let us consider a DoS attack with period τ_n. The overall system has to wait an additional period of length $N\Delta$ at the end of this attack to have a full round of communications. So the period where at least one subsystem transmission is not successful can be upper bounded by $\tau_n + N\Delta$. For all $\tau, k \in \mathbb{R}_{\le 0}$ with $k \ge \tau$, the total length where communication is not possible over $[\tau, k[$, say $|\bar{\Xi}(\tau,k)|$, can be upper bounded by

$$|\bar{\Xi}(\tau,k)| \le |\Xi(\tau,k)| + (1+n(\tau,k))\Delta_* \le \kappa_* + \frac{k-\tau}{T_*}, \tag{3.49}$$

where $\Delta_* = N\Delta$, $\kappa_* := \kappa + (1+\eta)\Delta_*$, and $T_* = \frac{\tau_D T}{\tau_D + T\Delta_*}$.
In addition, we consider the additional waiting time caused by the protocol, and the Lyapunov function yields $V(x(k)) \le (1-\omega_1)^{k-h_n-\tau_n-N\Delta} \times V(x(h_n+\tau_n+N\Delta))$ for $t \in [h_n+\tau_n+N\Delta, h_{n+1}[$ and $V(x(k)) \le (1+\omega_2)^{k-h_n} V(x(h_n))$ for $t \in [h_n, h_n+\tau_n+N\Delta[$.

As a result, we can deal with the overall behavior of the closed-loop system as a switching system with two modes. Applying simple iterations to the Lyapunov functions in and out of DoS attacks status, we obtain

$$V(x(k)) \leq (1-\omega_1)^{\left[k-\kappa_*-(\frac{1}{T}+\frac{\Delta_*}{\tau_D})k\right]}(1+\omega_2)^{\left[\kappa_*+(\frac{1}{T}+\frac{\Delta_*}{\tau_D})k\right]}V(x(0)). \tag{3.50}$$

To ensure the stability of the last equation, (3.38) can be easily obtained and this completes the proof. □

Remark 3.5. The resilience of the distributed systems depends on the largeness of ω_1 and the smallness of ω_2. To achieve this, we can try to find K_i and L_{ij} such that $\|B_i C_i K_i\|$ and $\|B_i C_j L_{ij}\|$ are small. On the other hand, the sampling interval of round-robin also affects stability in the sense that it determines how fast the overall system can restore the communication. We can always apply a shorter round-robin intersampling time to reduce the left-hand side of (3.38) at the expense of higher communication load.

3.5.7 Illustrative example

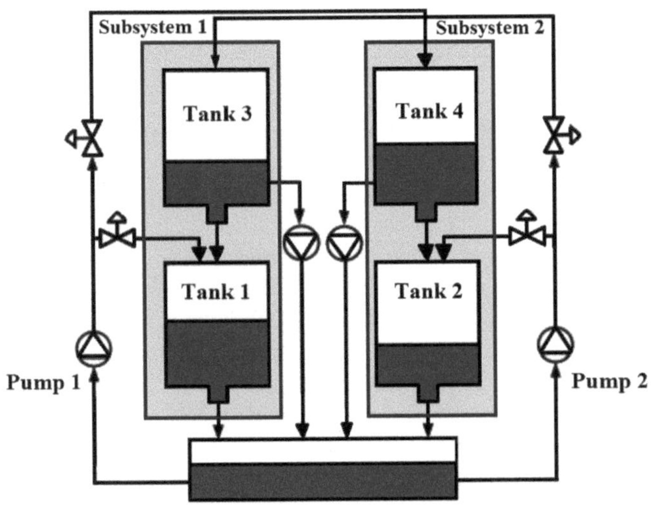

FIGURE 3.7 Schematic diagram of a quadruple-tank system.

The effectiveness of the proposed method presented in this chapter is shown by considering one of the common CPSs, the quadruple-tank process described in [315]. As shown in Fig. 3.7, the system consists of four tanks (two upper and two lower) and our objective is to control the level in the lower two tanks with two pumps. The process has two inputs (input voltage to the pumps) and two outputs (voltage from level measurement devices). Following are the system matrices for the linearized discrete-time state space model of the system:

$$\begin{aligned}
A_1 &= \begin{bmatrix} 0.9998 & 0 \\ 0 & 0.9998 \end{bmatrix}, \quad B_1 = \begin{bmatrix} 0.6359(10)^{-3} \\ 0.4559(10)^{-3} \end{bmatrix}, \\
H_{12} &= \begin{bmatrix} 0 & 0.0003 \\ 0 & 0 \end{bmatrix}, \quad C_1 = \begin{bmatrix} 1 & 0 \end{bmatrix}, \\
A_2 &= \begin{bmatrix} 0.9999 & 0 \\ 0 & 0.9997 \end{bmatrix}, \quad B_2 = \begin{bmatrix} 0.488(10)^{-3} \\ 0.6279(10)^{-3} \end{bmatrix}, \\
H_{21} &= \begin{bmatrix} 0 & 0.0002 \\ 0 & 0 \end{bmatrix}, \quad C_2 = \begin{bmatrix} 1 & 0 \end{bmatrix}.
\end{aligned}$$

Using the toolbox for modeling and optimization in MATLAB® (YLMIP), the controller gains were determined to be as follows:

$$\begin{aligned}
K_1 &= -0.8476, \quad K_2 = -1.8838, \\
L_{12} &= -0.4756, \quad L_{21} = -1.3312.
\end{aligned} \tag{3.51}$$

Furthermore, we found that

$$\begin{aligned}
P_1 &= 10^{-5} \times \begin{bmatrix} 0.6421 & -0.5143 \\ -0.5143 & 0.9261 \end{bmatrix}, \\
P_2 &= 10^{-5} \times \begin{bmatrix} 0.8421 & -0.4895 \\ -0.4895 & 0.6357 \end{bmatrix}, \\
Q_1 &= 10^{-8} \times \begin{bmatrix} 0.5514 & -0.1250 \\ -0.1250 & 0.3704 \end{bmatrix}, \\
Q_2 &= 10^{-8} \times \begin{bmatrix} 0.5583 & 0.1060 \\ 0.1060 & 0.3813 \end{bmatrix}.
\end{aligned} \tag{3.52}$$

Figs. 3.8 and 3.9 show that the system with the designed controller is stable in the nominal case.

Using Lemma 3.1, we obtain

$$\begin{aligned}
A &= 10^{-8} \times \begin{bmatrix} 0.2758 & 0 \\ 0 & 0.3010 \end{bmatrix}, \quad B = 10^{-9} \times \begin{bmatrix} 0 & 0.1059 \\ 0.1032 & 0 \end{bmatrix}, \\
\Gamma &= 10^{-6} \times \begin{bmatrix} 0.0278 & 0.0087 \\ 0.0630 & 0.1265 \end{bmatrix}.
\end{aligned}$$

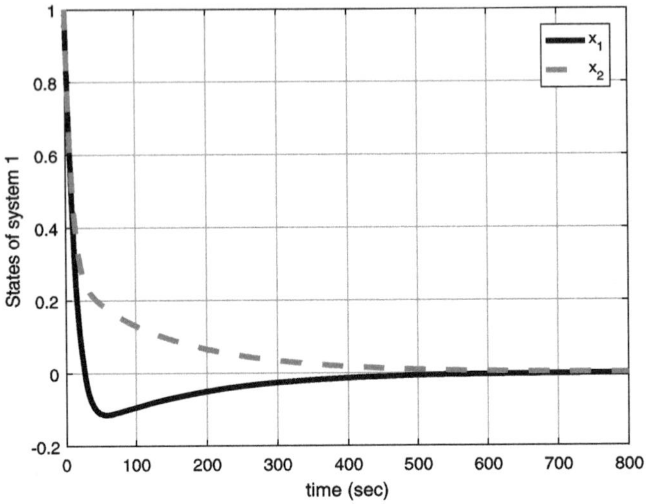

FIGURE 3.8 States of subsystem 1 in the nominal situation.

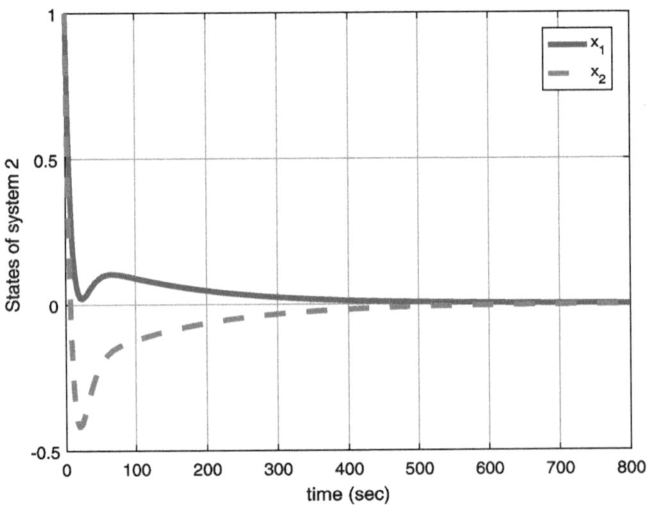

FIGURE 3.9 States of subsystem 2 in the nominal situation.

In addition, the parameters σ_1 and σ_2 are calculated to be (0.1570) and (0.1316), respectively. So σ is selected to be (0.1). Based on Assumption 3.3, we select a round-robin sampling interval $\Delta = 0.01$ s. By applying these parameters we found ω_1 and ω_2 to be $(1.2445(10)^4)$ and (0.2050), respectively.

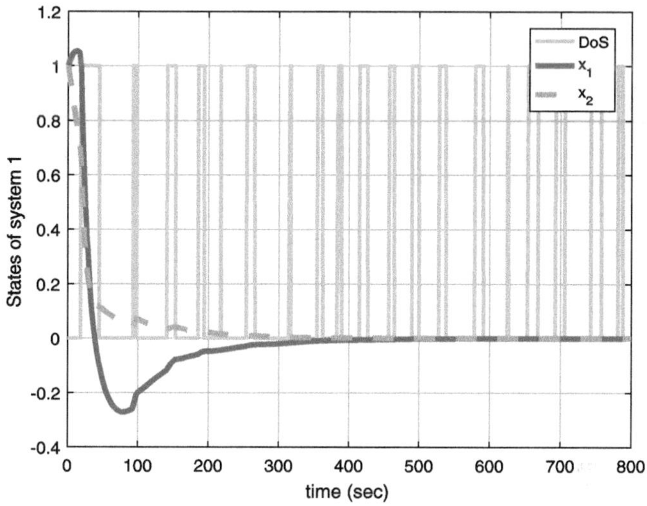

FIGURE 3.10 States of subsystem 1 under a DoS attack.

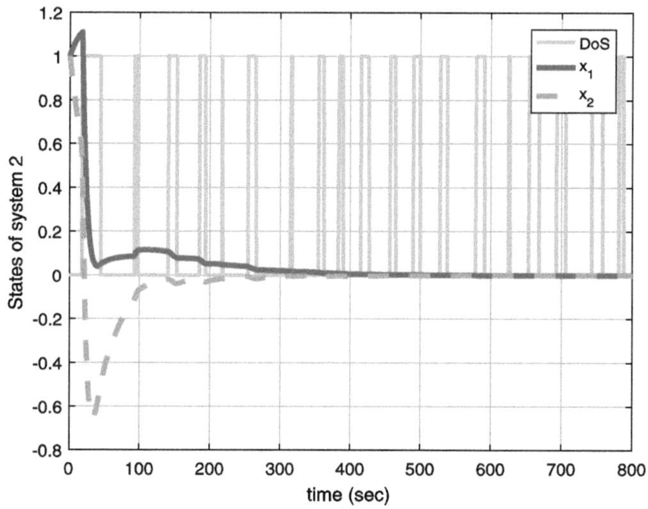

FIGURE 3.11 States of subsystem 2 under a DoS attack.

The characteristics of the DoS attack were designed using Theorem 3.3. As shown in Figs. 3.10 and 3.11, the designed controller maintains the stability of the system in the presence of a DoS attack.

3.6 Notes

Cloud Control Systems (CCSs) are defined as integrations of computation, communication, and control, and are used to achieve the desired performance of physical processes. Security threats have a high possibility of affecting CCSs and can be affected by several cyber attacks without providing any indication of failure. In this chapter, we discussed the stabilization of distributed CPSs affected by denial-of-service (DoS) attack. First, a static output feedback controller was designed to achieve the stability of a nominal distributed system. Then, a simple and typical scenario where the communication sequence is purely round-robin was considered, and a bound of attack frequency and duration was calculated to ensure the stability of the distributed CCS. Finally, a numerical example was provided to demonstrate the feasibility of the proposed system.

Chapter 4

Distributed cloud control systems

Contents

4.1 Introduction and wireless control design challenge

Time-critical and safety-critical automation systems are at the heart of essential infrastructures such as oil refineries, automated factories, logistics, and power generation systems. To meet the reliability requirements, automation systems are traditionally severely constrained along three dimensions, namely

Cloud Control Systems. https://doi.org/10.1016/B978-0-12-818701-2.00012-3

operating resources, scalability of interconnected systems, and flexibility to mode changes.

Oil refineries, for example, are built to operate without interruption for over 25 years and can never be shut down for preventive maintenance or upgrades. They are built with rigid ranges of operating throughput and require a significant overhaul to adapt to changes in crude oil quality and market conditions. This rigidity has resulted in systems with limited scope for reappropriation of resources during faults and retooling to match design changes on demand. For example, automotive assembly lines lose an average of $22,000 per minute of downtime during system faults [173]. This has created a culture where the operating engineer is forced to patch a faulty unit in an ad hoc manner, which often necessitates masking certain sensor inputs to let the operation proceed. This process of indiscriminate alteration to the system exacerbates the problem and makes the assembly line difficult and expensive to operate, maintain, and modify.

Embedded Wireless Sensor-Actuator-Controller (WSAC) networks are emerging as a practical means of monitoring and operating automation systems with lower setup and maintenance costs. While the physical benefits of wireless are apparent, for example in terms of cable replacement, plant owners have increasing interest in the logical benefits. With multi-hop WSAC networks it is possible to build wireless plug-n-play automation systems that can be swapped in and efficiently reconnect hundreds of input/output (I/O) lines. These modular systems can be dynamically assigned to be the primary or the backup on the basis of available resources or the availability of the desired calibration. Modularity allows for incremental expansion of the plant and is a major consideration in emerging economies. WSAC networks allow runtime configuration where resources can be reappropriated on demand, for example when throughput targets change due to lower electricity price during off-peak hours or due to seasonal changes in end-to-end demand. The current generation of embedded wireless systems has largely focused on open-loop sensing and monitoring applications.

To address actuation in closed-loop wireless control systems there is a strong need to rethink the communication architectures and protocols for reliability, coordination, and control [174]. Wireless Networked Control Systems (NCSs), or Networked Cyber-Physical Systems (Networked-CPSs), fundamentally differ from standard distributed systems in that the dynamics of the network (variable channel capacity, probabilistic connectivity, topological changes, node and link failures) can change the operating points and physical dynamics of the closed-loop system [175], [148]. The most important objective of control in a Networked-CPS is to provide stability of the closed-loop system. It is therefore necessary for the network (along with its interfaces between sensors and actuators) to be able to provide some form of guarantee of the control system's stability in the face of the nonidealities of the wireless links and the communication constraints of the wireless swarm network. A secondary goal

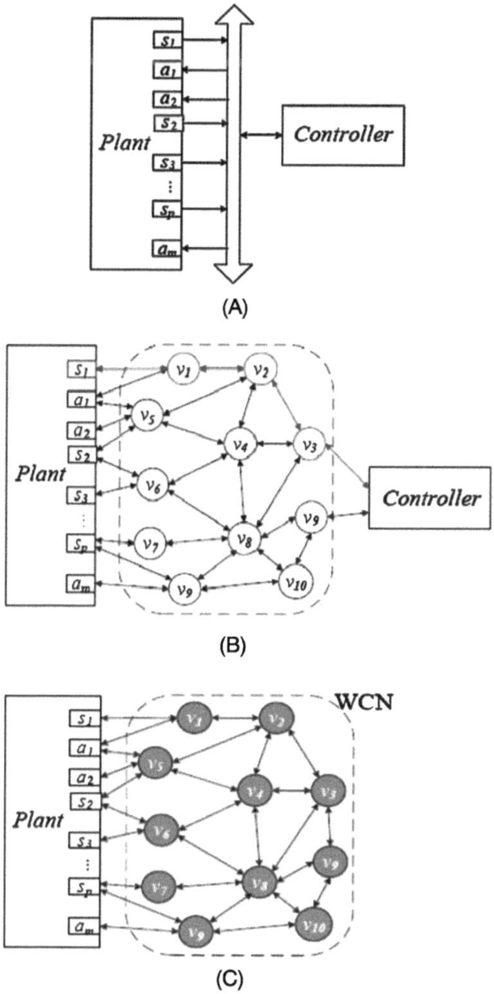

FIGURE 4.1 Standard architectures for networked control systems. (A) Wired network control; (B) Wireless network controlled system; (C) Wireless control network.

in Networked-CPSs is to allow the composition of additional controllers and plants within the same network without requiring reconfiguration of the entire network operation.

Remark 4.1. Note in Fig. 4.1[1] that part (A) is a wired system with a shared bus and dedicated controller; part (B) is a red links/nodes that rout data from the plant's sensors to the controller and blue links/nodes that rout data from the

1. For interpretation of the colors in the figure, the reader is referred to the web version of this article.

controller to the plant's actuators; and part (C) is a multi-hop wireless control network used as a distributed controller.

The most common approach to incorporating a Networked-CPS into the feedback loop is to use it primarily as a communication medium: the nodes in the network simply route information to and from one or more dedicated controllers, which are usually specialized central processing units (CPUs) capable of performing computationally expensive procedures (see Fig. 4.1B). The use of dedicated controllers imposes a routing requirement along one or more fixed paths through the network, which must meet the stability constraints, encapsulated by end-to-end delay requirements [176], [177]. However, this assignment of routes is a static setup that commonly requires global reorganization for changes in the underlying topology, node population, and wireless link capacities.

Routing links the communication, computation, and control problems [178], [179], [180]. Therefore, when a new route is required due to topological changes, the computation and control configurations must also be recalculated. Merely inserting a wireless network controlled system (WNCS) into the standard network architecture of sensor → channel → controller/estimator → channel → actuator requires the addition of significant software support [189], [181] as the cost of completely overhauling the computation and control configurations, due to topological changes or packet drops, is too expensive and does not scale.

Remark 4.2. Note in Fig. 4.2 that part (A) represents a wireless sensor, actuator, and controller network; part (B) is the algorithm assignment to a set of controllers, each mapped to the respective nodes; part (C) depicts three virtual components (VCs), each composed of several network elements; and part (D) displays a decoupled virtual task (VT) and physical nodes with runtime task mapping.

Providing closed-loop stability and performance guarantees for Networked CPS is a challenging problem. On the one hand, the control system community typically abstracts away the system's details and solves the problem for semi-idealized networks with approximated noise distributions and link perturbations [175]. While this approach provides mathematical certainty of the properties of the network, it fails to provide a systematic path to real-world network design. On the other hand, the network system community uses hardware and software approaches to address open-loop issues, but these fail to provide any guarantees to maintaining stability and performance of closed-loop control. We propose a control scheme over wireless networks that provides closed-loop stability and optimality with respect to standard metrics, while maintaining ease of implementation in real-world networks.

While there has been considerable research in the general area of wireless sensor networks, most of the work has been on open-loop and non-real-time applications. As we extend the existing programming paradigm to closed-loop

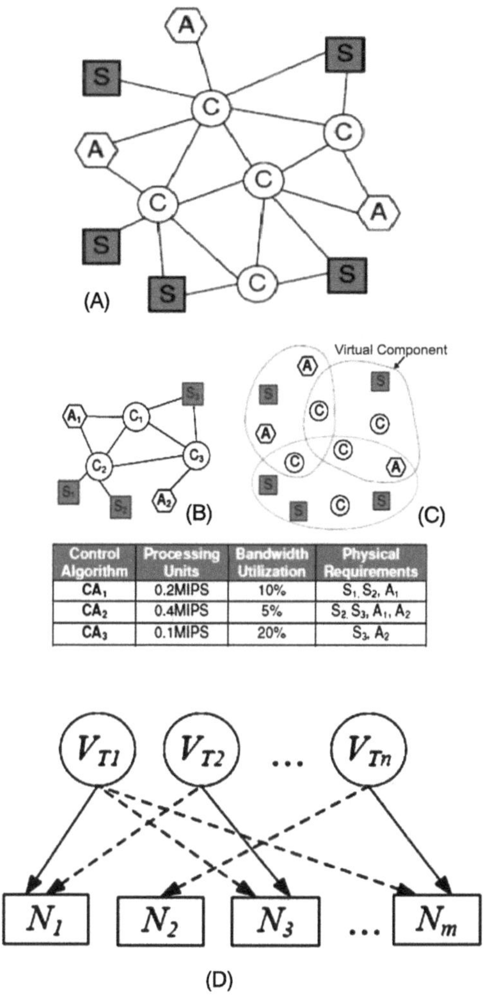

Control Algorithm	Processing Units	Bandwidth Utilization	Physical Requirements
CA_1	0.2MIPS	10%	S_1, S_2, A_1
CA_2	0.4MIPS	5%	S_2, S_3, A_1, A_2
CA_3	0.1MIPS	20%	S_3, A_2

FIGURE 4.2 Embedded virtual machines.

control applications with tight timeliness and safety requirements, we identify four primary challenges with the design, analysis, and deployment of WSAC networks:

1. The current approaches of programming motes in the event-triggered paradigm [182] are tedious for control networks. Time-triggered architectures are required as they naturally integrate communication, computation, and physical aspects of control networks [183], [184].
2. Programming of sensor networks is currently at the physical node-level [185] and is the main reason for the lack of robustness of higher-level control applications.

3. The design of NCSs with flexible topologies is difficult with physical node-level programming as the set of tasks (or responsibilities) is associated with the physical node [186].
4. Fault diagnostics, repair, and recovery are manual and template-driven for the majority of NCSs [187], [188]. Runtime adaptation is necessary to maintain the stability and performance of the higher-level control system.

Furthermore, the networks might be shared among control loops (i.e., a node may be involved in several feedback loops), and new feedback loops may be added at runtime. Adding new communication loops in a standard wireless network control system could affect the performance of the existing loops, and the system must be analyzed as a whole. Although techniques have been developed for compositional analysis of these networks (e.g., [178]), their complexity limits their use. Therefore, it is necessary to derive a composable control scheme, where control loops can be easily added and a simple compositional analysis can be performed at runtime, to ensure that one loop does not affect the performance of other loops.

The applications of interest in this work are industrial process control systems (such as natural gas refineries and paper pulp manufacturing plants) and building automation systems. In general the plant time constants are on the order of several seconds to a few minutes, and the control network is expected to operate at rates of hundreds of milliseconds. While these plants may have as many as 80,000 to 110,000 control loops, they are organized in a hierarchal manner such that networks span 10–20 wireless nodes (per gateway) for low-level control. Therefore, in this work we focus on the networks with up to a few dozen nodes (see Fig. 4.3).

4.2 Embedded virtual machines

While wireless system engineers optimize the physical, link, and network layers to provide an expected packet error rate, this does not translate accurately to the stability of the control problem at the application layer. For example, planned and unplanned changes in the network topology with node or link failures are currently not easily captured or specifiable in the metrics and requirements for control engineers. For a given plant connected to its set of controllers via wireless links (see Fig. 4.1A–B), it is necessary for the controller to process the sensor inputs and to perform actuation within a bounded sampling interval. While one approach is to design specialized wireless control algorithms that are robust to a specified range of packet errors [148], [175], it is nontrivial to design the same for frequent topological changes. Furthermore, it is difficult to extend the current network infrastructure to add or remove nodes and to redistribute control algorithms to suit environmental changes such as battery drain for battery-operated nodes, increased production during off-peak electricity pricing, seasonal production throughput targets, and operation mode changes.

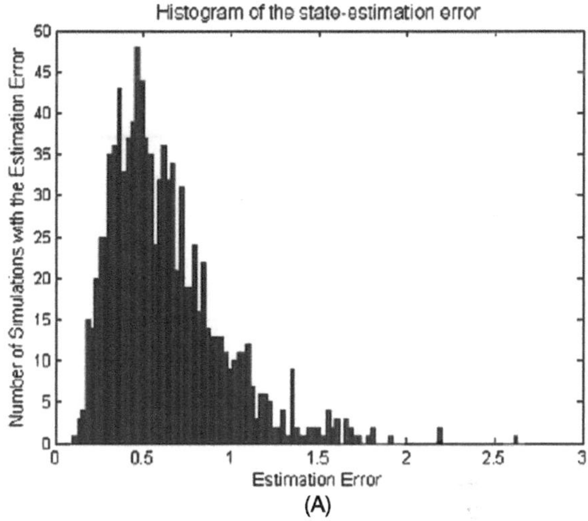

(A)

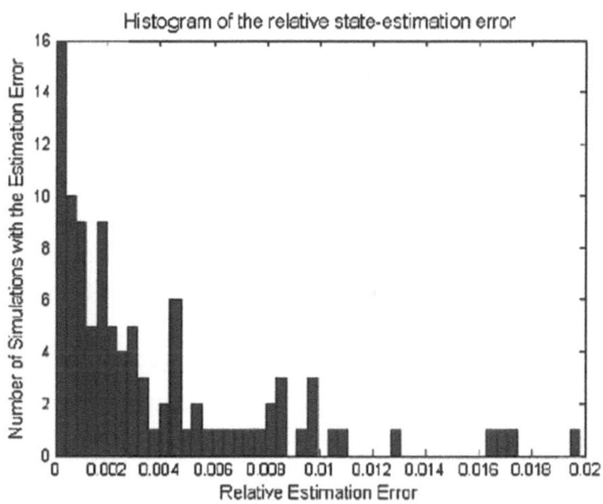

(B)

FIGURE 4.3 Task migration for real-time operation (instructions, stack, data, and & timing/fault tolerance metadata) from one physical node to another. (A) Histogram for a system with error bound 155.98; (B) Histogram of the maximal relative state-estimation error for all 100 system.

The Embedded Virtual Machine (EVM) approach is adopted to allow control engineers to use the same network control algorithms on the wireless network without knowledge of the underlying network protocols, node-specific operating systems, or hardware platforms. The virtual machine executing on each node (within the VC) instructs the VC to adapt and reconfigure to changes while ensuring that the control algorithm is within its stability constraints. This approach is complementary to the body of network control algorithms as it provides a logical abstraction of the underlying physical node topology.

4.2.1 Network CCS related work

There have been several variants of virtual machines, such as Matte [194], Scylla [195], and SwissQM [196], and flexible operating systems, such as TinyOS [182], SOS [197], Contiki [198], Mantis [199], Pixie [200], and LiteOS [201], for wireless sensor networks. The primary differences that set the EVM approach apart from prior work is that it is centered on real-time operation of controllers and actuators. Within the design of the EVM's operating system, link protocol, and programming abstractions and operation, timeliness is the most important and thus all operations are synchronized. The EVM does not have a single-node perspective of mapping operations to one virtualized processor on a particular node; instead, it maintains coordinated operation across a set of controllers within a VC. The Virtual Node Layer [202] provides a programming abstraction where each virtual node is identified with a particular region and it is emulated by one of the physical nodes in its region. On the other hand, EVM uses several physical nodes and allows the user to consider the VC as a single logical entity.

In recent years, several different systems for macroprogramming in wireless sensor network (WSN) have been developed. [185] has defined a set of abstractions representing local communication between nodes in order to expose control over resource consumption along with providing feedback on its performance. An extension of these ideas was used to develop Regiment [203], a high-level language based on the functional reactive programming. The work of [204] allows a programmer to describe code execution for each of the nodes in a network using a centralized approach where details about code generation, remote data access, and management along with node interactions are hidden from the programmer. EVM is not a generic macroprogramming system as it focuses on closed-loop control with native runtime support for task assignment and migration.

The development of control algorithms able to deal with the unreliability of the wireless channel for NCSs is an active area of research in the control system community [148], [175], [205]. Few efforts consider networked control over arbitrary topologies ([186], [206], [207]). In these articles the authors assume the existence of a single actuation point and a single sensing point on the plant. They show that the optimal position of the controller is at the actuation point, but they ignore the wireless channel in the estimation of the plant's state. In a general case the problem of assigning the best location of the controller node is very complex. Finally, Etherware [208] presents challenges in software development for NCSs along with abstractions and architectures used to implement control algorithms for NCSs. The authors describe a middleware for control systems, but do not provide algorithms that might be used to guarantee that designed middleware satisfies requirements for the control algorithms.

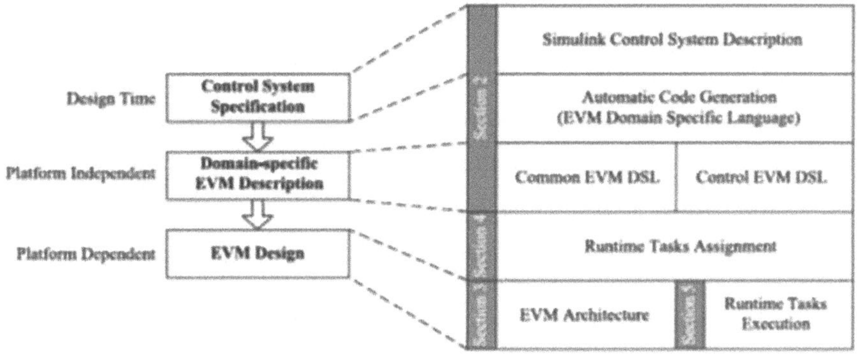

FIGURE 4.4 Design flow of embedded virtual machines (EVMs).

4.2.2 Design flow of embedded virtual machines

Our focus is on the design and implementation of wireless controllers and on providing these controllers with runtime mechanisms for robust operation in the face of spatiotemporal topological changes. We focus exclusively on controllers and not on sensors or actuators, as the latter are largely physical devices with node-bound functionality. A three-layered design process is presented to allow control engineers to design wireless control systems in a manner that is largely independent of platform/protocol/hardware/architecture and also extensible to different domains of control systems (e.g., discrete processes, aviation, medical). This section describes the design flow from a control problem formulation in Simulink, an automatic translation of control models from Simulink to the platform-independent EVM interpreter-based code, and finally to platform-dependent binaries (see Fig. 4.4). These binaries are assigned to physical nodes within a VC using the assignment and scheduling algorithms presented earlier. The binaries are executed as VTs within the platform-dependent architecture described above.

At design time, control systems are usually designed using software tools, such as Matlab® and Simulink, that incorporate both modeling and simulating capabilities. Therefore, to automatize the design flow the EVM is able to automatically generate functional models from the Simulink control system description. These functional models define the processes by which input sampled data is manipulated into output data for feedback and actuation. The models are represented by a generated code and metadata for platform- and node-independent system description. This allows a system designer to exclusively focus on the control problem design. In addition to the functional description in the platform-independent and domain-specific language (DSL), the EVM design flow automatically extracts additional para-functional properties from the Simulink model, like timing and intertask dependencies. These properties, along with the functional descriptions, are used to define a platform-optimized binary for each VT.

4.2.3 Platform-independent domain-specific language

To generate a functional description of the designed system, the EVM programming language is based on FORTH, a structured, stack-based, extensible, interpreter-based programming language [208]. Since the goal of the EVM design is to allow flexibility and designing utilities independent of the chosen programming language, the intermediate programming language is not constrained to the EVM programming language. The interpreter used to execute modules described in the EVM programming language can also execute precompiled binaries. The EVM implementation presented in Section 4.6 executes binaries derived from embedded C code. This enables the execution of code binaries developed in other languages used to describe control system implementation.

- Dictionary consists of predefined blocks/functions that are usually used for control systems description
- PID ($K_P K_I K_D$ *addr_in addr_out* –)
- PI ($K_P K_I$ *addr_in addr_out* –)
- TF (*m n addr_alpha addr_beta addr_in addr_out* –)

$$G(z) = \frac{N(z)}{D(z)} \quad \begin{aligned} N(z) &= \alpha_0 + \alpha_1 z^{-1} + \alpha_2 z^{-2} + \cdots + \alpha_m z^{-m}, \\ D(z) &= \beta_0 + \beta_1 z^{-1} + \beta_2 z^{-2} + \cdots + \beta_n z^{-n}. \end{aligned}$$

- LTI (*p m n addr_A addr_B addr_C a_in a_x a_out* -)

$$x[k+1] = Ax[k] + Bu[k],$$
$$y[k] = Cx[k], A \in R^{n\times n}, B \in R^{n\times n}, C \in R^{p\times n}$$

(A) Control-EVM

- Arithmetic operations (on 16 bit)
- Logical operations
- Comparison and testing
- Controlling programming flow
- Memory manipulation
- Sensor/Actuator handling
  - RDSensG (sensID - n1)
  - RDSensL (sensID - n1)
  - WRActG (value actID -)
  - WRActL (value actID -)
- Networking
  - PktSendG (addr n nodeID -)
- Task handling
  - TaskActivate (addrTCB actID -)

(B) Common-EVM

FIGURE 4.5 A representative control EVM.

Using the EVM intermediate programming language enables domain-specific constructs, where basic programming libraries are related to the type of application that is being developed. For example, for use in embedded wireless networks for industrial control we developed two predefined libraries, Common EVM and Control EVM; a full list of the application programming interfaces (APIs) is provided in [209]. A common EVM (Fig. 4.5B) is based on the standard FORTH library [208]. Except for the: word, used to define new words, all other words can be separated into the following categories: 1) arithmetic operations, 2) logical operations, 3) memory manipulation, 4) sensor and actuator handling, and 5) networking. A control EVM (Fig. 4.5A) contains functionalities widely used to develop control applications. The first three words specified in Fig. 4.5A are used for Singe-Input–Single-Output (SISO) systems. Although these words can be described using the Linear-Time Invariant (LTI) word, which describe LTI systems, their wide use in control systems recommends their specific use.

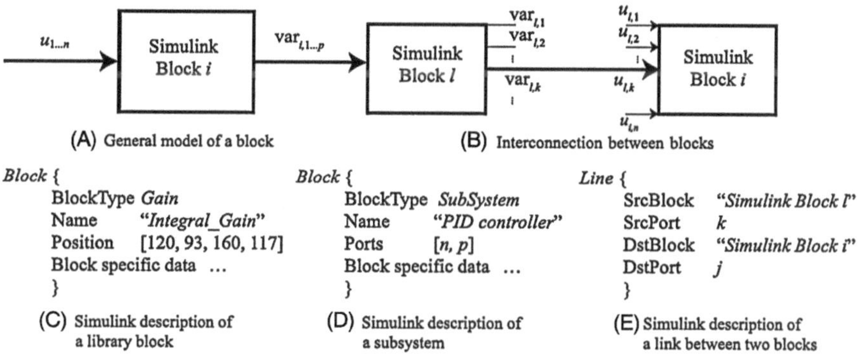

FIGURE 4.6 General relations between Simulink blocks.

The extensibility of the EVM allows the definition of additional domain-specific libraries such as Automotive EVM, Aviation EVM, or Medical EVM libraries, which contain functionalities specific to each of these application fields. Using EVM libraries the code generator creates a system description from the predefined components, thus creating a task description file for each of the VTs.

4.2.4 Control problem synthesis

We now describe the procedure to automatically extract the functional description of a VT from a Simulink design. Within Simulink, each block (and thus the model itself) is represented as a hierarchical composition of other Simulink blocks, either subsystems or library-defined blocks. This organization of Simulink models allows a natural extraction of a structured functional description using predefined words from the platform-independent EVM DSL dictionary. When a new Simulink block is defined as a composition of previously defined blocks, a new word is defined for the EVM functional description using previously defined words. The process is repeated until a level is reached where all words belong to the EVM dictionary.

A VT description is obtained by parsing the Simulink model file. This is done by searching for new block definitions along with the interconnections between blocks. In a Simulink model file (i.e., mdl file) blocks are presented as shown in Fig. 4.6C and Fig. 4.6D where Block Type parameter describes whether the block is a part of the Simulink library or of a subsystem consisting of other Simulink blocks. To extract the VT description we require that the task be implemented in a singular, discrete-time Simulink subsystem, such as the example shown in Fig. 4.7. The synthesis of the platform-independent specification from the model is carried out in three steps:

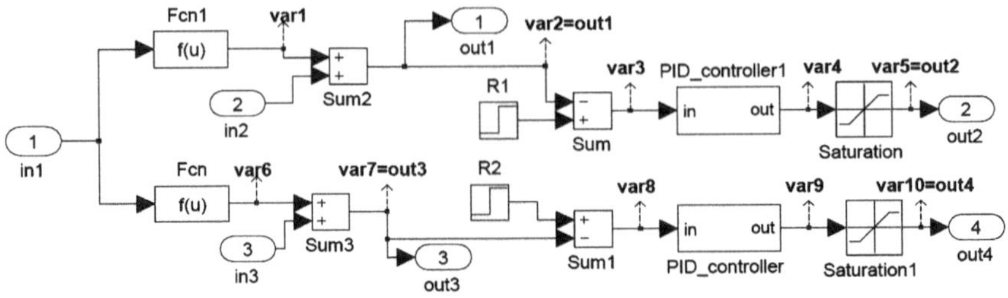

FIGURE 4.7 Simulink model of an extended PID controller.

Step 1: Intermediate words/variables

: W1 in1 ? f var1 @ ;
: W2 var1 ? in2 sum out1 @ ;
: W3 out1 ? NEG R1 sum var3 @ ;
: W4 var3 ? Kp1 Ki1 Kd1 PID var4 @ ;
: W5 var4 ? thr SAT out2 @ ;
: W6 in1 ? f var6 @ ;
: W7 var6 ? in3 ? sum out3 @ ;
: W8 R2 out3 ? NEG sum var8 @ ;
: W9 var8 ? Kp2 Ki2 Kd2 PID var8 @ ;
: W10 var9 ? thr SAT out4 @ ;

Step 2: Composition

: Vctrl W1 W2 W3 W4 W5 W6 W7 W8 W9 W10 out1 ? out2 ? out3 ? out4 ?

Step 3: Optimization

: W1 in1 ? f var1 @ ;
: W2 var1 ? in2 sum out1 @ ;
: W3 out1 ? NEG R1 sum Kp1 Ki1 Kd1 PID thr SAT out2 @ ;
: W7 var1 ? in3 ? sum out3 @ ;
: W8 R2 out3 ? NEG sum Kp2 Ki2 Kd2 PID thr SAT out4 @ ;
: Vctrl W1 W2 W3 W7 W8 out1 ? out2 ? out3 ? out4 ? ;

FIGURE 4.8 EVM functional description extracted from Simulink model shown in Fig. 4.7.

(1) Definition of intermediate words and variables: Each block i is associated with a word W_i from the EVM DSL, where the output of the block is assigned to a variable var_i. To illustrate this, consider the extended proportional-integral-derivative (PID) controller in Fig. 4.7. The outputs of all intermediate blocks are assigned to variables, as shown in Fig. 4.7. For example, the EVM description of block "Sum1" is described with word W_8 and its output with variable var8. As the EVM DSL is stack-based with reversed Polish notation the block is described as

$$: W_8 \; R2 \; out3 \; ? \; NEG \; sum \; var_8 \; @;$$

where ? and @ are read and write operators respectively. In the general case, for a block presented in Fig. 4.6A the parser defines the word

$$: \; W_i \; u_1 \; ? \, u_2 \; ? \; \cdots u_n \; ?coeffs \; BlockWord \; var_{i_1} \} \\ @ \; var_{i_2} \; @ \; \cdots var_{i_p} @ \; ;$$

where **BlockWord**, depending on BlockType, corresponds to either a predefined word (if a library block is used) or a new word that needs to be defined using the same parser algorithm (if the block is a subsystem). Variables presented as **coeffs** are extracted from the Block Specific data in cases when they are contained in the block description (from Fig. 4.6C, D), along with initial values for variables var_i. For example, consider definition of word $\mathbf{W}_4$. Since block PID controller1 contains coefficients for K_p, K_i, and K_d, their values are included in the definition. Finally, in the previous formulation variables $u_{[1..n]}$ are replaced with appropriate system variables with respect to connections between blocks. To illustrate this, consider a connection (i.e., a line) between blocks in Fig. 4.6B. Simulink defines the line as in Fig. 4.6E. Thus, for Simulink Block i, in the definition of word $\mathbf{W}_i$ each variable $u_{i,j}$ is replaced with the appropriate variable $var_{l,k}$.

(2) Composing extracted words: The intermediate words are composed to create functional description of the system (e.g., $\mathbf{VT}_{ctrl}$). The parser is recursively executed for all subsystems until all the words are part of the library. The description for the example in Fig. 4.7 is presented in Step 2, Fig. 4.8. It is worth noting that the intermediate words are executed in the block execution order for the Simulink model. The order is either specified explicitly in the model or determined implicitly based on block connectivity and sample time propagation [210].

(3) DSL code optimization: Intermediate blocks with elementary functions can be pruned in a single word. For the example in Fig. 4.7 the optimized description is shown in Step 3, Fig. 4.8. Words $\mathbf{W_3}$, $\mathbf{W_4}$, $\mathbf{W_5}$ and $\mathbf{W_8}$, $\mathbf{W_9}$, $\mathbf{W_{10}}$ are combined into a single word ($\mathbf{W_3}$ and $\mathbf{W_8}$, respectively). Also, instead of word $\mathbf{W_6}$ and variable var6, $\mathbf{W_1}$ and var1 are used. The code optimization reduces the number of defined words and used variables. Currently the optimization is restricted to a small set of control system configurations. A more general approach is an avenue for future work.

As our intention is to map the control problem to a scheduling problem, timing parameters (i.e., period and worst-case execution time) are also extracted from the model. We consider only discrete-time controllers as potential VTs. For these, Simulink design rules force the designer to define a sampling rate for each (discrete-time) block. Currently we cover cases where the controller is designed in a single clock domain (i.e., all blocks use the same sampling period). In general case, when a controller contains several clock domains, each subdomain is represented by its respective VT, and a set of dependencies between the tasks is extracted. Finally, to extract the worst-case execution time, a simple static analysis is performed using the execution time measurements for library-defined words with respect to the specific platform.

4.3 EVM architecture

We now describe the node-specific architecture which implements the mechanisms for the virtual machine on each node. The Common-EVM and Control-EVM description are scoped within VTs that are mapped at runtime by the Task Assignment procedure presented in the next section. This description is interpreted by the Virtual Component Interpreter running on each node. The

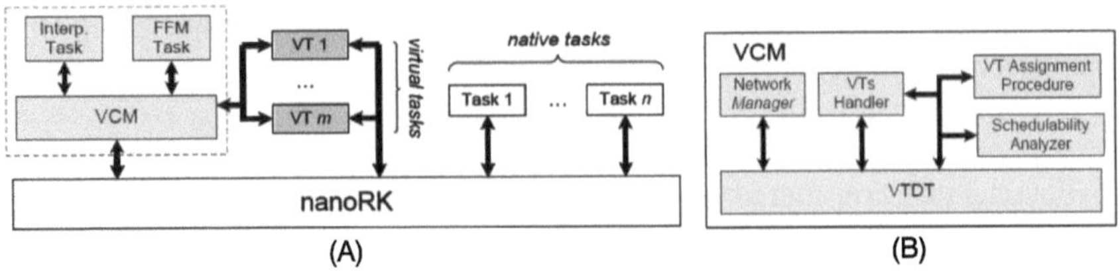

FIGURE 4.9 EVM architecture with the virtual component manager running as a super-task alongside native nano-RK tasks. (A) EVM block-level reference architecture; (B) Structure of the VCM.

EVM runtime system is built as a supertask on top of the nano-resource kernel (RK) real-time operating system (RTOS) [211], allowing node-specific tasks to execute native tasks and VTs (i.e., those that are dynamically coupled with a node) to run within the EVM. The EVM block-level reference architecture is presented in Fig. 4.9A. This allows the EVM to maintain node-specific functionalities and be extensible to runtime task evocation of existing or new VTs.

The interface between nano-RKs and all VTs is realized using the Virtual Component Manager (VCM). The VCM maintains local resource reservations (CPU, network slots, memory, etc.) within nano-RK, the local state of the VTs and global mapping of VTs within the VC. The VCM is responsible for memory and network management for all VTs-to-physical nodes and presents a mapping between local and remote ports that is transparent to all local VTs. It includes a FORTH-like interpreter for generic and domain-specific runtime operations and a Fault/Failure Manager (FFM) for runtime fault-tolerant operations. The VCM is implemented in modular form so the interpreter, FFM, and other specialized modules can be swapped with extensions over time and for domain-specific applications.

4.3.1 Embedded virtual machine extensions to the nano-RK RTOS

Nano-RK is a fully preemptive RTOS with multi-hop networking support that runs on a variety of sensor network platforms (8-bit Atmel-AVR, 16-bit TI-MSP430, Crossbow motes, FireFly) [211]. Nano-RK uses the RT-Link [212], a real-time link protocol. It supports fixed-priority preemptive scheduling to ensure that task deadlines are met, along with support for enforcement of CPU and network bandwidth reservations. Nano-RK was designed as a fully static operating system, configured at design time. Thus, to allow parametric and programmatic runtime code changes nano-RK was redesigned and extended with several new features:

- *Runtime Parametric Control:* Support for dynamic change of the sampling rates, runtime tasks, peripheral activation/deactivation, and runtime modification of the task utilization was added. These facilities are exposed and executed via the Common-EVM programmer interface.

• *Runtime Programmatic Control:* As a part of the EVM design a procedure for dynamic task migration was implemented. This requires runtime schedulability analysis, capability checks to migrate a subset of the task data, instructions, required libraries, and task control blocks. Based on the procedure presented in the previous sections, tasks may be activated or migrated between primary and backup nodes. These facilities are triggered by the primary-backup policy implemented on top of the EVM architecture.

• *Dynamic Memory Management:* Both Best-fit and First-fit memory allocation methods are supported. In addition, a Garbage Collector (GC) has been designed to reclaim all memory segments owned by tasks that had been terminated. The GC is scheduled only when its execution does not influence execution of other tasks.

4.3.2 Virtual component interpreter

The Virtual Component Interpreter provides an interface to define and execute all VTs. Every VT is defined as a word within the VCM library. When a new VT description is received over the network the VCM calls the interpreter, which defines a new word using the description file of the task and existing VC libraries. After a VT is activated each execution of the VT is realized as a scheduled activation of the interpreter with the VT's word provided as an input. To allow preemptivity of the tasks, each call of the interpreter uses a VT-specific stack and dedicated memory segments. In addition, during its execution, each VT is capable of dynamically allocating new memory blocks of fixed size (currently 128B) using the EVM's memory manager. Therefore, the interpreter is designed to use logical addresses in the form (*block index; address in block*).

Each node maintains a local copy of standard Common-EVM and Control-EVM dictionaries. If a new word needs to be included in the existing library, the interpreter first checks the global word identifier and revision number to discard obsolete versions.

4.3.3 Virtual tasks

Each VT is described using the Virtual Task Description Table (VTDT), comprised of global and local descriptions of a VT. Copies of the table are stored on all members of the VC. While this requirement for consistency currently results in an issue of scalability, a large fraction of the higher-speed control in supervisory control and data acquisition (SCADA) systems requires networks with fewer than 20 nodes and is hence within the practical limits of the current approach. Each VT's global description has information about the memory requirements, stack size, and number of fixed-size memory blocks (128B). In addition to the above metadata, network requirements in terms of number of RT-Link transmit and receive slots are specified at design time.

The above descriptors are specified within the VCM's Task Control Block (TCB) for each task, which is an extension to the native nano-RK TCB (for details see [209]).

4.3.4 Virtual component manager

The fundamental difference between the native nano-RK and the VCM is that the scope of nano-RK's activities is local, node-specific, and defined completely at design time, while the scope of the VCM is the VC that may span multiple physical nodes. The VCM subcomponents are presented in Fig. 4.9B. The current set of supported runtime functionalities is as follows:

4.3.4.1 Virtual task handling (controlled by the VT handler)

4.3.4.1.1 VC state

VC state includes the mapping of VTs to physical nodes and quality of links between physical nodes. The VCM in each controller node within the VC maintains the VC state and periodically broadcasts it to keep consistency between all members of the VC. Currently, a centralized consensus protocol is used, while a distributed consensus protocol is needed to scale operations.

4.3.4.1.2 VT migration and activation

VT migration and activation that can be triggered as a result of a fault/failure procedure or by a request from either the VT or the VCM. As a part of a task migration, the task's VTDT is sent along with all memory blocks utilized by the task. If the VT is already defined on a backup node (checked by exchange of hash values), only the task parameters are exchanged. In addition, before migrating a VT to a particular node the Schedulability Analyzer performs network and CPU schedulability analysis for nodes that are potential candidates (details are provided in the next section). If the analysis shows that no node can execute the task correctly, an error message is returned. Finally, after a VT is defined, to activate the task the host node performs a local CPU and network schedulability analysis to ensure that the task will not adversely affect correct execution of previously defined VTs.

4.3.4.1.3 Control of tasks executed on other nodes

For all VTs in the backup mode, the VT Handler shadows the execution of the VT in the primary mode. If a departure from the desired operation is observed (e.g., low battery level, decreased received packet signal strength), backup nodes may be assigned to the primary mode based on the policy.

4.3.4.1.4 VT assignment

VT Assignment procedure is activated to assign execution of the VTs to specific nodes when incremental and local reassignment fail. The procedure determines the best set of physical controller nodes to execute VTs given a snapshot of the current network conditions along with the initial communication and computation schedules for the nodes.

4.3.4.2 Network management (performed by the network manager)

4.3.4.2.1 Transparent radio interface

Using the message header which contains information about message type, the VCM determines tasks that should be informed about the message arrival. Messages containing tasks and their parameter definitions are first processed by the VCM before the VCM activates the interpreter.

4.3.4.2.2 Logical-to-physical address mapping

Communication between VTs is done via the VCM. Since a VT does not have information on which nodes other VTs are deployed, the VCM performs logical-to-physical address mapping. In cases where both tasks are on the same node, the VCM directly passes a message to the receiving task's buffer.

4.4 Virtual task assignment

With the knowledge of the underlying EVM architecture, we now discuss the algorithm used for the VT Assignment procedure. The procedure determines the initial assignment of the VT's executions along with the communication and computation schedules. The criteria for triggering reassignment calculation are described in Section 4.5. We derived a general case problem formulation for the VT's assignment as a binary integer linear optimization problem, which is then solved efficiently using well-known techniques (branch and bound) [213]. In addition, since standard link protocols for wireless factory automation, such as WirelessHART [214], recommend that only one physical node may transmit in each time slot, we were able to obtain an efficient reformulation of the relaxed assignment problem. In this case each control loop (operating across the same physical set of controllers) can be considered separately, which considerably simplifies task assignments, as it allows a compositional system design.

4.4.1 General formulation

To develop an assignment algorithm we considered a multi-hop control network that corresponds to our model of a VC. The network consists of $p \geq 1$ processes ($\mathbb{J} = \{1, \cdots, p\}$ denotes the set of all processes) and a set of nodes (sensors, actuators, and controllers), where all nodes have a radio transceiver along with memory and computing capabilities (see Fig. 4.10A). The nodes communicate using a protocol based on time-division multiple access (TDMA) (i.e., in a time-triggered manner) with frame size F_S. The network is described with a directed graph $G = (V, E)$ that represents radio connectivity in the network. Set $V = \{v_1, v_2, \cdots, v_m\}$ denotes a set of physical nodes in the network,[2] while $E = \{(v_i, v_j)|v_i \text{ and } v_j \text{ are connected}\}$ is a set of all links. In addition, each link e is described with its link quality $LQ(e)$. To extract a problem formulation it is necessary to enumerate all paths in the network that should be used for communication between a node and a sensor (or an actuator).[3,4] Thus, the l-th path between node v_i and sensor/actuator k is denoted as $\psi^l_{i,k}$.

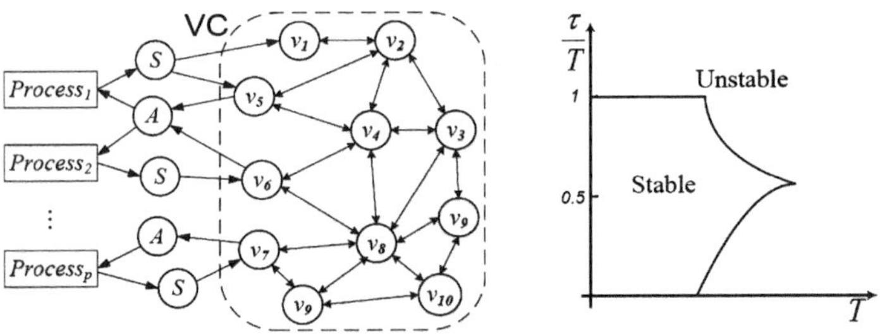

FIGURE 4.10 (A) Reference model of a multi-hop wireless network. (B) An example of a stability region for such a network.

Remark 4.3. Note that in part (A) of Fig. 4.10 the network is used for control p physical plants (processes) and consists of multiple sensors (S), actuators (A) and controllers (v_i). The VC includes multiple physical controller nodes, whereas in part (B) of Fig. 4.10, T is a controller sampling period, while τ is the network-induced delay.

The goal of the assignment procedure is to determine the following:

(1) An assignment of the VTs, i.e., Control Algorithms (CAs), to the set of nodes V, where each VT is assigned to one node in the *Primary* mode and to R nodes in the *Backup* mode;

(2) A communication schedule that determines active links at each time slot;

(3) A computational schedule that determines in which time slot each VT is executed.

2. In the following, V will also denote the set that contains the node indexes $\{1, 2, \cdots, m\}$.
3. A path is represented as a directed path connecting the sender with exactly one receiver.
4. Including all paths could significantly increase the complexity of the optimization problem; therefore, the user might opt to enumerate only selected paths with the best characteristics (e.g., a small number of hops, a high packet delivery ratio).

In addition, to define the problem as an optimization problem, the following assumptions were made:

A.1 For each process j the *Primary* and all *Backup* nodes assigned with the j-th VT are scheduled in the same time slot(s);

A.2 The VTs are mutually independent;

A.3 A process i (for all values of i) will remain stable if its sampling period is less than some predefined value T_i. Therefore, we require that $F_S \leq min(T_1, T_2, \cdots, T_p)$.

Remark 4.4. The first assumption simplifies the problem formulation and paves the way to an easier schedulability analysis scheme. The second assumption is appropriate since a significant class of process controllers execute a large number of simple and independent control loops. The third assumption is justified by using the approach described in [175].

For example, consider a closed-loop control of a plant modeled with continuous-time LTI dynamics:

$$\dot{x}(t) = Ax(t) + Bu(t); \quad y(t) = Cx(t).$$

The controller employs a discrete-time state feedback control with $u(kT) = -Kx(kT)$, where T denotes the plant's sampling period. If the network-induced delay τ_k is less than one sampling period,[5] the control feedback has the following form:

$$u(t^+) = -Kx(kT), \quad t \in [kT + \tau_k, (k+1)T + \tau_{k+1}).$$

Thus, $u(t)$ is a piecewise continuous function that changes values only at time instances $kT + \tau_k$.

The EVM utilizes fully synchronous networks, which allows scheduling the actuators to apply new input values at the same time, after the messages have been delivered to all of them. This guarantees the same delay for all the plant's inputs at each sampling period ($\tau_k = \tau, \forall k$). Using the methods based on simulation, as in [175], the stability region can be determined with respect to sampling period T and the induced delay τ. The region is used to establish the maximum sampling period for which the system maintains stability if a network delay is less than the period ($\frac{\tau}{T} \leq 1$, an example is shown in Fig. 4.10B).

To formulate the problem the following decision variables are used:

- *2mp binary assignment variables,* $x_{i,j}^{st} \in \{0, 1\}$, *where* $i \in V$, $j \in \mathbb{J}$, $st \in \{a, b\}$ and

$$x_{i,j}^{a} = \begin{cases} 1, & v_i \text{ is the } \textit{Primary} \text{ for the } j\textit{-th VT} \\ 0, & \textit{otherwise} \end{cases}$$

5. A similar approach can be used even if the delay is longer than the sampling period.

$$x^b_{i,j} = \begin{cases} 1, & v_i \text{ is the } \textit{Backup} \text{ for the } j\textit{-th VT} \\ 0, & \textit{otherwise} \end{cases}$$

- *Routing binary variables* $y^l_{i,k} \in \{0, 1\}$, *where*

$$y^l_{i,k} = \begin{cases} 1, & \text{the } l\textit{-th} \text{ path between node } v_i \\ & \text{and the sensor/actuator } k \text{ is used} \\ 0, & \textit{otherwise} \end{cases}$$

- *Communication schedule binary variables* $\eta^{l,n}_{i,k} \in \{0, 1\}$ *where* $n \in \{1, \cdots F_s\}$ and:

$$\eta^{l,n}_{i,k} = \begin{cases} 1, & \text{the } l\textit{-th} \text{ path between node } v_i \\ & \text{and the sensor/actuator } k \text{ is active in the } n\textit{-th} \text{ slot} \\ 0, & \textit{otherwise} \end{cases}$$

- *Computation schedule binary variables* $\mu^n_i \in \{0, 1\}$ *where* $n \in \{1, \cdots F_s\}$ and

$$\mu^n_j = \begin{cases} 1, & \text{the } j\textit{-th} \text{ VT is scheduled for execution in the } n\textit{-th} \text{ slot} \\ 0, & \textit{otherwise} \end{cases}$$

Our goal is to describe the assignment problem in the form:

$$min\ f(\mathbf{x}, \mathbf{y}, \eta, \mu), \text{ subject to } \quad \mathbf{x}, \mathbf{y}, \eta, \mu \in \mathbb{S}_{\mathbb{C}},$$

where vectors $\mathbf{x}, \mathbf{y}, \eta, \mu$ contain the above-mentioned decision variables and $\mathbb{S}_{\mathbb{C}}$ describes a set that satisfies all constraints, ensuring the desired system behavior. The constraints take into account the requirements for the control problem along with dependencies between communication and computation schedules. In the following the imposed set of constraints is described.

1) Assignment of the control algorithms: Each VT has to be assigned to exactly one node in the *Primary* mode and R additional *Backup* nodes (different from the *Primary* node for the CA). These constraints are described as

$$\sum_{i=1}^{m} x^a_{i,j} = 1,$$

$$\sum_{i=1}^{m} x^b_{i,j} = R, \quad \text{and}$$

$$x^a_{i,j} + x^b_{i,j} \leq 1, \quad \forall j \in \mathbb{J}, \forall i \in \mathbf{V}.$$

2) Requirements for robust design: Additional sets of constraints are introduced to improve performance of the closed-loop system. *Link reliability* constraints require that only links with quality above a given threshold are considered, which reduces complexity of the problem formulation. Logical pruning of graph G results in a graph $G_T = (V, E_T)$, where $E_T = \{(v_i, v_j) \in E | LQ(v_i, v_j) \geq THR)\}$.

The *routing constraints* describe a means of increasing system robustness to the link failures via the use of different paths for data routing. For example, WirelessHART recommends that each node should use at least two separate paths to route data [184]. Thus, we require that the Primary node for each VT use two different paths to deliver information to all actuators related to the process control. In addition, the *Primary* and all Backup nodes have to be connected with all sensors related to the VT.[6] Denoting as A_j and S_j the sets of actuators and sensors, respectively, related to the j-th process, these constraints are described as

$$\sum_{\forall l} y^l_{i,k_a} = 2x^a_{i,j}, \quad \sum_{\forall l} y^l_{i,k_s} = x^a_{i,j} + x^b_{i,j},$$
$$\forall j \in \mathbb{J},\ k_a \in A_j, k_s \in S_j,\ \forall i \in V.$$

Finally, a set of *monitoring constraints* is imposed, where all *Backup nodes* monitor the execution of a VT on the *Primary* node. Thus, to alleviate the system design and VT migration when the *Primary* node fails, constraints are enforced that all *R Backup nodes* have to be one-hop neighbors of the *Primary* node. Denoting as N_i the set of all neighbors of node v_i, these constraints are described as $\sum_{k \in N_i} x^b_{k,j} \geq R.x^a_{i,j}, \forall j \in \mathbb{J}, \forall i \in V$.

3) Computation schedule constraints: From assumptions A.3 and A.1, we require that computations of each VT on the *Primary* and *Backup* nodes should be scheduled exactly once in a frame. This implies that all VTs have the same sampling rate and could result in a more frequent computation of a VT. In most automation systems the increase in the sampling rate cannot endanger the closed-loop system stability. On the contrary, it can increase the performance of the implemented controller if the optimal discrete-time controller is used [215].[7] Thus, the constraints are expressed as $\sum_{n=1}^{Fs} \mu^n_j = 1, \forall j \in \mathbb{J}$.[8]

4) Communication schedule constraints: From assumption A.3 the closed-loop system stability is guaranteed if the end-to-end communication delay (i.e., delay from the sensors to the assigned controller and from the controller to the actuators) along with the time needed for the controller computation is less than

6. It is worth noting here that a different routing policy could be used. However, even if that is the case, these constraints could be expressed in a similar way.
7. Future extensions of this work will allow CAs to have different sampling periods.
8. In the constraint formulation we assume that each VT can be executed in one time slot. In general this might not be the case. However, it would just require a formulation change where instead of 1, the execution time necessary for execution of the j-th VT (i.e., e_j) is placed. Even more general, if the network contains nodes with different computational power, the previous term should be expressed as $\max_{i=1}^{m}(x^a_{i,j}e^a_j + x^b_{i,j}e^b_j)$. To simplify the notation, we decided to use the above-mentioned assumption.

F_S. Thus, the first requirements for the communication schedule is that only used paths are scheduled and that the number of slots assigned to the used path is exactly equal to the path's length (i.e., the number of hops on the path):

$$\eta_{i,k}^{l,n} \leq y_{i,k}^{l}, \quad \forall n,\ 1 \leq n \leq F_S,$$
$$\sum_{n=1}^{F_S} \eta_{i,k}^{l,n} = y_{i,k}^{l}.d(\psi_{i,k}^{l}), \quad \forall i,\ k \in V,\ \forall l. \tag{4.1}$$

Additionally, the schedule has to be collision free (i.e., two interfering nodes cannot transmit in the same time slot). To express these constraints, for each path $\psi_{i,k}^{l}$ where k is a sensor, all links are enumerated in increasing order starting from the link with origin at sensor k and ending with the link with the destination at node i. Similarly, for each path $\psi_{i,k}^{l}$ where k is an actuator, enumeration starts at node i and ends at actuator k. This is used to create the interference links table for each pair of paths $(\psi_{i_1,k_1}^{l_1}, \psi_{i_2,k_2}^{l_2})$. An element (n_1, n_2) is a member of the $(\psi_{i_1,k_1}^{l_1}, \psi_{i_2,k_2}^{l_2})$ interference table (IT) if transmissions over the n_1^{st} link of the path $\psi_{i_1,k_1}^{l_1}$ interferes with transmissions over the n_2^{nd} link of the path $\psi_{i_2,k_2}^{l_2}$. Constraints for interference-free schedule can be described as follows:

For all n, $1 \leq n \leq F_s$, $\forall i_1, i_2 \in V$,

$$\left| \sum_{n_0=1}^{n} \eta_{i_1,k_1}^{l_1,n_0} - n_1 \right| + \left| \sum_{n_0=1}^{n} \eta_{i_2,k_2}^{l_2,n_0} - n_2 \right| \geq 1,$$
$$\forall k_1, k_2 \in S \cup A, \ (n_1, n_2) \in IT(\psi_{i_1,k_1}^{l_1}, \psi_{i_2,k_2}^{l_2}). \tag{4.2}$$

5) Dependencies between the schedules: Communication and computation schedules must be aligned, meaning that measured data (i.e., data from sensors) is routed to the controller prior to the VT's activation. Also, data designated to the actuators are forwarded after the computation of the VT:

For all n, $1 \leq n \leq F_s$,

$$\eta_{i,k_s}^{l,n} \leq (1 - \sum_{n_0=1}^{n} \mu_j^{n_0}),$$
$$\eta_{i,k_a}^{l,n} \leq (\sum_{n_0=1}^{n} \mu_j^{n_0}), \ \forall j \in \mathbb{J},\ \forall i \in V, \forall l, k_s \in S_j, k_a \in A_j. \tag{4.3}$$

6) Objective function: The goal of the assignment procedure is to minimize the aggregate number of used links while maximizing the aggregate link quality. In addition, we want to maximize the use of disjoint routing. Thus, a cost for sharing links is introduced, both in paths from sensors to controllers and from the *Primary* controller to the actuators. As can be seen, the objective function (i.e., cost) does not depend on the utilized scheduling. Therefore, it is defined as a weighted sum $f(\mathbf{x},\mathbf{y}) = w_1 f_{LN} + w_2 f_{LQ} + w_3 f_{SL}$, where weights w_1, w_2, and w_3 are used to emphasize the impacts of the following cost functions:

1. Aggregate number of used links: $f_{LN}(\mathbf{x},\mathbf{y}) = \sum_{j=1}^{p} \sum_{l,k,i} y_{i,k}^{l} . d(\psi_{i,k}^{l})_j$, where $d(\psi_{i,k}^{l})$ is a distance (i.e., length, number of hops) of path $\psi_{i,k}^{l}$;
2. Negative aggregate link quality: $f_{LQ}(\mathbf{x},\mathbf{y}) = -\sum_{j=1}^{p} \sum_{l,k,i} y_{i,k}^{l} . LQ(\psi_{i,k}^{l})_j$;
3. Cost of the shared links:

$$f_{SL}(\mathbf{x},\mathbf{y}) = \sum_{j=1}^{p} \sum_{\substack{l \le i \le t \le m \\ l_i, l_t,\ k \in S_j \cup A_j}} y_{i,k}^{l_i} . y_{t,k}^{l_t} . SH(\psi_{i,k}^{l_i}, \psi_{t,k}^{l_t})$$

where $SH(\psi_{i,k}^{l_i}, \psi_{t,k}^{l_t})$ is the number of links shared between paths $\psi_{i,k}^{l_i}$ and $\psi_{t,k}^{l_t}$.

Therefore, the assignment problem can be formulated as a binary integer programming optimization problem and solved using some of the well-known techniques (branch and bound) [213]. One caveat is in order. Since the problem formulation has a large number of decision variables, even for a small network it can be computationally expensive to solve the problem. Thus, we translated the problem into the satisfiability problem by transforming each constraint into conjunctive normal form (CNF) (for details see [209]). The satisfiability problem is then solved using zChaff [216], a very efficient satisfiability solver. This allows us to solve the previous problem in real time even for large-scale networks.

4.4.2 Problem relaxation

When only one node in the VC can transmit in each time slot, the number of slots needed to send a message from node v1 to node v2 is equal to the distance between the nodes. This is used for the relaxed problem formulation as it eliminates the need to include communication and computation decision variables used in the general formulation, and therefore significantly reduces the complexity of the optimization problem. In addition, the collision-free communication requirement, which is the most complex set of constraints from the general formulation, becomes redundant. The requirement is inherently fulfilled with the policy that allows a single transmission per time slot for the whole VC.

As the first step for the problem formulation, two maximum node-disjoint paths r_{i,a_c}^{1}, r_{i,a_c}^{2} are determined for each node v_i and each actuator a_c. The existence of two node disjoint paths from a node to all sensors and actuators can be checked using Menger's theorem [217] (for details see [209]). When two node-disjoint paths exist for the node, using a polynomial time algorithm (MIN-SUM 2-paths [218]) paths r^1_{i,a_c}, r^2_{i,a_c} with the minimum total length can be determined. Otherwise, path r^1_{i,a_c} is computed in polynomial time as the shortest path to the actuator. Path r^2_{i,a_c} is calculated as the shortest path to the actuator after removing nodes from path r^1_{i,a_c}, while preserving connectivity. Using a similar approach, for each node v_i and all its neighbors $v_{i_1}, \cdots, v_{i_{n_i}}$ (n_i is a degree of node v_i), a set of $n_i + 1$ paths is created between each sensor s and the nodes. We denote these distances as $(d_{i,s}, d_{i_1,s}, \cdots, d_{i_{n_i},s})$.

To extract the formulation of the relaxed problem we use only $2mp$ binary assignment variables $x^a_{i,j}$ and $x^b_{i,j}$, defined as in the general problem formulation. This allows us to formulate the problem as

$$min\ w_1.f_{LN}(\mathbf{x}) + w_2.f_{LQ}(\mathbf{x}), \text{ with respect to } \mathbf{x} \in \{0,1\}^{2mp},$$

which contains the above-mentioned decision variables. The feasible set is described with the following set of constraints:

$$\sum_{i=1}^{m} x^a_{i,j} = 1, \quad \sum_{i=1}^{m} x^b_{i,j} = R, \quad x^a_{i,j} + x^b_{i,j} \leq 1,$$

$$\sum_{k \in N_i} x^b_{k,j} \geq R.x^a_{i,j}, \ \forall j \in \mathbb{J}, \ \forall i \in V,$$

$$\sum_{\substack{i \in V \\ j \in \{1,\cdots,p\}}} \{\sum_{s \in S_i} (x^a_{i,j}.d_{i,s}) + \sum_{k \in N_i} (x^b_{i,j}.x^a_{i,j}.d_{i_k,s}) +$$

$$\sum_{a \in A_i} x^a_{i,j}.(d(r^1_{i,a}) + d(r^1_{i,a}))\} + 1 \leq F_s.$$

The last constraint requires that all communication be done within one frame, and therefore that it meet the timing requirements necessary for the system's stability. This constraint is the only one that depends on the number of VTs and utilized data routing, thus a suboptimal yet feasible solution can be obtained (if and only if a feasible solution exists) using *compositional analysis*. In this case each control loop operating across the same physical set of controllers is considered separately. Optimizing only for the cost function f_{LN} and for each loop separately provides an optimal assignment for each loop, which uses the minimum number of communication slots (see details in [209]). Note that if $w_1/w_2 >> 1$, the approach provides the optimal solution for the relaxed assignment problem in general. In addition, for a sufficiently high link quality threshold (while deriving graph G_T) the impact of function f_{LQ} is reduced. This enables the use of the compositional design, which significantly simplifies the system analysis and schedule extraction. Since the EVM is focused on networks with less than 20 nodes, we are able to run the optimization algorithm on all nodes in a VC, as the *VT Assignment Procedure*.

4.5 EVM runtime operation

Given the task migration mechanisms and the algorithms to (re)assign tasks, we now describe the relationship between primary and backup nodes for planned and unplanned scenarios. More specifically, we consider the criterion for triggering task migration and the node and network schedulability analysis that must be conducted prior to migration. To completely address the issues in wireless NCSs, we must consider (a) the mechanisms for runtime adaptation, (b) the algorithms for runtime task (re)assignment to physical nodes, and (c) the fault tolerance policy. In this chapter we focus on the first two aspects and apply them to simple network models with non-Byzantine single-node and single-link failures. As the fault tolerance policy is dependent on the control application and the fault/failure model is a function of the specific environment, we do not consider specific policies here. We will address Byzantine errors such as software errors in future work.

4.5.1 Adaptation to planned and unplanned network changes

Planned adjustments occur in situations when a *Primary* node is informed of changes in VC state (e.g., when a node detects that its battery level is below some threshold). To determine which *Backup* node can migrate its task, the *Primary* node has to execute computation and communication schedulability analyses in the neighborhood of $k = 1$ hop and select a Backup node that maximizes the communication slack value while maintaining computation schedulability.

For unplanned changes caused by potential failures we consider the following cases:

- The *Primary* node dies: Computation and communication schedulability analysis in the neighborhood of $k = 1$ hop is initiated. Since state data of the *Primary* node is maintained at *Backup* nodes, a new *Primary* node continues VT execution.
- A *Backup* node dies: The *Primary* node detects the *Backup* has died and selects a new *Backup* from one of its neighbors.
- A forwarding node dies or a link's quality falls below some criterion: The detection of a forwarding node failure is performed by its predecessor or successor on the routing path. Again, a communication schedulability analysis is performed (only for the affected sensor and actuator) to determine a new routing scheme.

To decrease response time for the schedulability analyses, each node uses its idle computation time to calculate in advance the optimal reaction to a set of potential failures. In addition to decreasing the response time, this approach also enables triggering the execution of the *Assignment Procedure* if it is determined that for some failures there is no adjustment that can meet all of the constraints. Also, if the procedure cannot derive a feasible assignment, an alarm is raised notifying system operators to add more nodes in the network to prevent a potential failure.

4.5.2 Communication schedulability analysis

The goal of communication schedulability is to determine whether we can *incrementally* reassign the available communication slots due to the change in the task assignment without executing a global reassignment of communication slots. To accomplish this we determine the current communication slack and evaluate whether it is sufficient for the incremental slot reassignment. When a VT is to be migrated from a node v_i to a node v_j, we define sets S_{VT} and A_{VT} of all sensors and actuators, respectively, related to the VT. Then for each $s \in S_{VT}$ we denote as $v_{i,s}^k$ a node that is k hops away from node v_i on the route from sensor s to node v_i. Similarly, for each $a \in A_{VT}$, $v_{i,a}^k$ denotes a node that is k hops away from node v_i on the route to the actuator a. In addition, we denote as N_u^i the number of unused time slots in the time interval between the first slot in which all nodes $v_{i,s}^k$ were supposed to receive values from sensors in S_{VT} and a first slot in the frame where at least one node $v_{i,a}^k$ was scheduled to receive information from the node v_i. The parameter k determines the set of candidate backup nodes to which the task may be reassigned.

More specifically, the goal of communication schedulability is to determine whether we can reassign (with the respect to the current communication schedule) the available communication slots and slots used to send data in the k hop neighborhood of a node v_i. The reassignment should reroute all sensor and actuator data from these nodes to node v_j. A new feasible communication schedule can be generated if $\Delta \geq 0$, where Δ denotes communication slack value defined as

$$\begin{aligned}\Delta = \sum_{s \in S_{VT}} d(v_i, v_{i,s}^k) + \sum_{a \in A_{VT}} d(v_i, v_{i,a}^k) + N_u^i \\ - \sum_{s \in S_{VT}} d(v_j, v_{j,s}^k) - \sum_{a \in A_{VT}} d(v_j, v_{j,a}^k),\end{aligned}$$

where $d(v_p, v_q)$ is the distance between nodes v_p and v_q. If more than one task is migrated from a node, a similar analysis is performed with the previous equation adjusted to contain sums of all sensors and actuators related to the tasks. In addition, if tasks are migrated from node v_i to separate nodes, the schedulability test is performed on a pairwise basis.

4.5.3 Computation schedulability analysis

For the computation schedulability analysis we use a standard real-time response analysis [219] and the mode-change protocol, presented in [220] and [221], adapted for the EVM. Consider a node v_i that executes a task set

$$\mathbb{T} = \{T_{i_1}, \cdots, T_{i_m}, VT_{i_1}, \cdots, VT_{i_n}\},$$

where tasks T_{i_j} are local node-specific tasks, while tasks VT_{i_j} are VTs assigned to the node (in descending order of priority). We define a set $HP\_VT(T)$ as the set of all VTs with higher priority than local task T and, similarly, a set $HP\_T(VT)$ as a set of all node-specific tasks, with higher priority than task VT. To allow an assignment of a new VT, a schedulability analysis is performed where both active and inactive tasks are considered as active. Although this approach is conservative, it eliminates the need for repeated schedulability analysis prior to task activation. Each node-specific task is denoted as $T_j = (p_{T_j}, e_{T_j})$ and each VT as $VT_j = (p_{VT_j}, e_{VT_j}, \phi_{VT_j}, d_{VT_j})$ (i.e., period, execution time, offset, and deadline respectively). Schedulability of a new task set is performed by checking only the schedulability of each task with a lower priority than the new VT_k, using its time-demand function $w(t)$ [219].

As mentioned in Section 4.4, we currently consider the case where all VTs have the same execution period. Since execution of a VT is triggered by the reception of sensed signals and must be finished before its scheduled communication to actuators, its deadline is significantly lower than its period. Thus, from the activation of a VT until its deadline, all other VTs can be active at most once, so for a task $VT_i, i \geq k$:

$$w_{VT_i}(t) = e_{VT_i}(t) + \sum_{j \in HP\_T(VT_i)} \left\lceil \frac{t}{t_{T_j}} \right\rceil .e_{T_j} + \sum_{j=1}^{i-1} e_{VT_j}.$$

The equation is too conservative as it assumes that all VTs can be activated at the same time. However, VTs are activated when a last radio message containing necessary data is received. In addition, since all VT periods are multiples of TDMA slot duration, when a communication schedule is known, all possible offset combinations of a task activation can be easily calculated. Therefore, for a task VT_i, released at time t_i, for all possible combinations of release times t_j of VTs with higher priority, the time-demand function for $t \geq t_i$ is defined as

$$\begin{aligned} w_{VT_i}^{(t_0, t_1, \cdots, t_{i-1})}(t) &= e_{VT_i} + \sum_{k \in HP\_T(VT_i)} \left\lceil \frac{t}{t_{T_k}} \right\rceil .e_{T_k} \\ &+ \sum_{\substack{j=1 \\ t_j \leq t \leq t_i + d_i}}^{i-1} min(e_{VT_j}, t - t_j) \end{aligned}$$

$$+ \sum_{\substack{j=1 \\ t_i \in [t_j, t_j + d_j]}}^{i-1} min(e_{VT_j}, t_j + d_j - t_i).$$

Here the second term corresponds to the execution of all higher-priority native tasks; the third term corresponds to the demand from higher-priority VTs that are activated after the i-th task's activation, but before its deadline. Finally, the last term describes the demand of the higher priority VTs when the i-th task is activated between the higher-priority task's activation and deadline. For schedulability we are interested in time instances where $w_{VT_i}^{(t_0, t_1, \cdots, t_{i-1})}(t) = t$. These points can be obtained using efficient recurrence procedure described in [219]. The task is schedulable if, for all combinations of activation times, the solution of recurrence procedure is less than the task's deadline (d_{VT_i}).

Although the previous equation seems complicated, in the case when all VTs are executed once per frame there is only one combination of release times $(t_0, t_1, \cdots, t_{i-1})$ (i.e., only one set of task offsets as the TDMA schedule is fixed). Even in the general case there is no need to cover a large number of possible combinations since for most control systems all loops usually have the same sampling period, or all sampling periods are integer multiples of one of the periods. A similar approach is used for schedulability analysis of a node-specific task T_i.

4.6 EVM implementation

To evaluate the EVM's performance in a real setting with multiple coordinated controller operations, we used the factory simulation module shown in Fig. 4.11A. The Fischertechnik model factory consists of 22 sensors and actuators (Fig. 4.11B) that are controlled in a coordinated and timely manner. A block of wood is passed through a conveyor, pushed by a rammer onto a turn table and operated upon by up to three milling/cutting/pneumatic machines. The factory module was initially controlled by wired programmable logic controllers (PLCs). We converted it to use wireless control with FireFly embedded

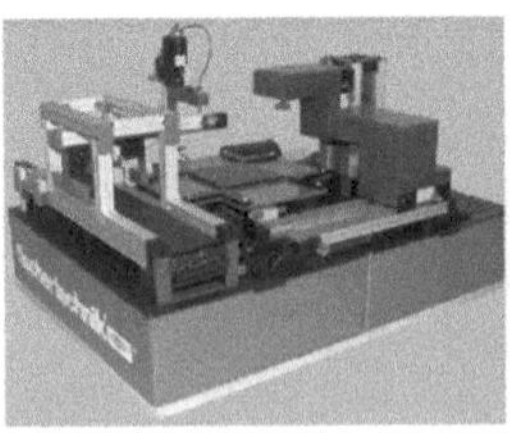

(A)

Component	Sensors			Actuators	
	Proximity Home	Proximity Out	IR Position	Non-reversing Motor	Reversing Motor
Conveyor belt			X		
Ram	X	X		X	X
Horizontal slotting mill					
• Horizontal direction	X	X		X	X
• Vertical direction	X	X			
Vertical Gluing Machine	X	X	X	X	
Table Rotation				X	X
Vertical boring machine	X	X			X

(B)

FIGURE 4.11 Fischertechnik factory module with 22 sensors and actuators. (A) Work-cell module; (B) Module components.

wireless nodes [222] that control all sensors and actuators via a set of electrical relays. FireFly is a low-power platform based on Atmel ATmega1281 8-bit microcontroller with 8KB of RAM and 128KB of ROM along with a Chipcon CC2420 IEEE 802.15.4 standard-compliant radio transceiver. FireFly nodes support tight global hardware-based time synchronization for real-time TDMA-based communication with the RT-Link protocol [212]. The EVM also works on TI MSP430 architectures.

In our experiments we demonstrate the following:

1. Online capacity expansion when a node joins the VC;
2. Redistribution of VTs when adding/removing nodes;
3. Planned VT migration triggered by the user;
4. Unplanned VT migration due to a node or a communication link failure;
5. Multiple coordinated work-flows.

We tested the setup with a batch of 10 input blocks consisting of three different types that require different processing procedures. This is an example of the logical benefits of the EVM as it enables a more agile form of manufacturing. Details of the experiments and the videos can be seen in [223].

4.6.1 EVM case study

As this is an early effort to describe the main functionalities of the EVM, we limit our case study to a simple simulated control network. We simulated the performance of the EVM for the case when a wireless network is used for control in the Shell Problem, a well-known problem in process control theory concerning the control of a heavy oil fractionator [224], [225]. The controlled variables (outputs) are differences of the top product end point ($Y1$) and the bottom reflux temperature ($Y2$) from predefined (reference) values. Fig. 4.12A presents a Simulink framework used for the simulation where the Controller (shown in Fig. 4.7) and the Plant are similar to models from [224]. The major difference is that the Plant's dynamics were sped up to be able to test the system's performance.

The functional description of the VT, shown in Fig. 4.8, was derived earlier. Since all continuous outputs of the Plant have to be sampled before being processed with a discrete-time controller, the sampling period defined in *Sample-and-Hold* blocks in the Simulink model is used to extract the period of each VT.

Fig. 4.12B presents the initial topology of the VC along with the *Primary* and the *Backup* nodes. To be able to address the effects of message drops, we assigned each link in the network a Packet Delivery Ratio (PDR) that is less than 1 (i.e., 100%). A TDMA protocol with 32 slots per frame was used for communication between nodes, where 24 slots were used for transfer of data related to the control problem, while the 8 remaining slots per frame were used to exchange messages about the VC's status. The system response to a series of different step inputs (a new one set to arrive every 60 s) for the initial topology is

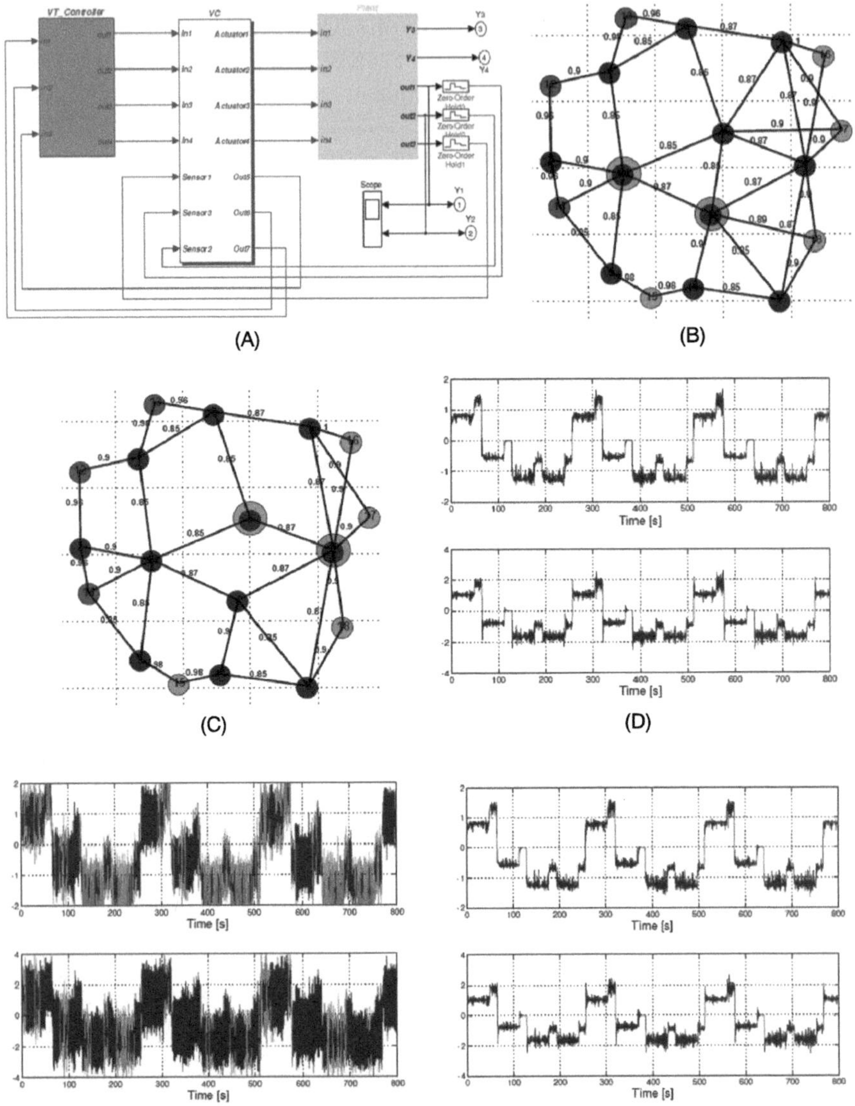

FIGURE 4.12 Simulation of EVM behavior when used for Shell Problem control; Nodes: green – actuators, red – sensors, blue circle – the Primary node, orange circle – the Backup node. (A) Simulink model for Shell problem; (B) Initial network topology; (C) Topology after link failures; (D) System response for initial configuration, showing outputs Y1 (top) and Y2 (bottom); (E) System response when EVM is not used (when only re-routing is used), Y1 (top) and Y2 (bottom); (F) System response when EVM adapts to changes in network conditions, Y1 (top) and Y2 (bottom). (For interpretation of the colors in the figure, the reader is referred to the web version of this article.)

presented in Fig. 4.12D. Also, a scenario was simulated where the initial topology changes after some of the links fail (as shown in Fig. 4.12C). Fig. 4.12E presents the response of the system without the EVM, where only rerouting algorithms are used without changing positions of the *Primary* and *Backup* nodes. This results in a system response that rapidly deteriorates. The system becomes unstable due to the increase in end-to-end communication time from all sensors to the *Primary* node to all actuators.

Fig. 4.12F shows how the EVM's adaptation to unplanned changes in link quality keeps the system's response similar to that in the initial topology. For the case presented in Fig. 4.12F, we simulated the system when at time $t = 60$ s the network topology changes to that presented in Fig. 4.12C. Due to the task reassignment, one execution of the control algorithm is omitted, but as can be seen, without significant influence to the overall system performance. This was expected since, from the perspective of the *Plant*, this case is equivalent to packet drops, which already occurs because PDR is less than 100%.

4.6.2 Limitations of the EVM approach

• **Complexity of consensus:** The complexity of reaching consensus forces our current implementation to maintain an update frequency. This limits the scalability of the current EVM approach to small networks with ≤ 20 nodes. While this is "good enough" for a large number of small embedded wireless control applications such as natural gas processing with slowly varying operating parameters, it is essential to explore distributed algorithms to maintain state across the VC.

• **Centralized approach:** The centralized algorithm has been used to solve the assignment problem. This limitation motivated us to explore a distributed solution for incremental strategies for control-loop implementation. Using the entire node population within a VC as a distributed controller would remove the need for the VT's assignment procedure.

So far, we present an initial stab at a problem that unravels a series of difficulties at the heart of networked CPSs. We investigated several fundamental challenges with the use of wireless networks for time-critical closed-loop control problems. Our approach was to build the networking infrastructure to maintain state across physical node boundaries, allowing tasks to be decoupled from the underlying unreliable physical substrate. We present a modular architecture used for control applications in wireless sensor/actuator/controller networks that allows component integration and system reconfiguration at runtime without any negative effects on the execution of already assigned functionalities. The EVM enables a simple transition from the controller design in widely used simulation tools to the actual, physical "plug-and-play" deployment for wireless networks.

To overcome the shortcomings of EVM, we now present the Wireless Control Network (WCN) approach for distributed network control.

4.7 Wireless control networks

We consider the problem of stabilizing a plant with a multi-hop network of resource-constrained wireless nodes. We present a distributed scheme used for control over a network of wireless nodes. Unlike traditional NCSs where the nodes simply route information to and from a dedicated controller (perhaps performing some encoding along the way), our WCN approach treats the network itself as the controller. In other words, the computation of the control law is done in a fully distributed way inside the network. In the WCN approach, at each time-step each node updates its internal state to be a linear combination of the states of the nodes in its neighborhood. This causes the entire network to behave as a linear dynamical system, with sparsity constraints imposed by the network topology. We demonstrate that with observer style updates, the WCN's robustness to link failures is substantially improved. Furthermore, we show how to design a WCN that can maintain stability even in cases of node failure. We also address the problem of WCN synthesis with guaranteed optimal performance of the plant, with respect to standard cost functions. We extend the synthesis procedure to deal with continuous-time plants and demonstrate how the WCN can be used on a practical, industrial application, using a process-in-the-loop setup with real hardware.

Given the fundamental unreliability of wireless communication, the WCN method handles topological constraints while maintaining MSS for packet drop rates up to 20% for a specific network topology and plant. This bridges the gap between the basic WCN and the theoretical upper bound of robustness to packet drops [191]. We also present a method to synthesize a WCN robust to a certain level of node failure before we extended the synthesis procedures to allow for the use of the WCN for the control of continuous-time plants. Finally, we illustrate the use of the WCN on a real-world industrial case study, for control of a distillation column.

While in the past efforts we considered scenarios where the network topology was already set, in recent efforts [192] we investigate a dual problem: How to synthesize the network so that a stable WCN configuration exists? The topological conditions from [192], along with the results from [190] provide the essential building blocks for an integrated decentralized wireless control network design framework. Early experiments in an industrial process control case study of a distillation column in a process-in-the-loop testbed to demonstrate optimal control of continuous-time physical processes that maintain system stability in the presence of node and link failures.

4.7.1 An intuitive overview

The role of feedback control is to apply inputs to the plant (based on observed outputs) in order to elicit the desired behavior. The exact mapping between observed behavior and applied inputs depends on a mathematical model of the plant that describes how inputs affect the system over time. Here we start with a common discrete-time LTI model of the form[9]

$$\begin{aligned} \mathbf{x}[k+1] &= \mathbf{A}\mathbf{x}[k]+\mathbf{B}\mathbf{u}[k]+\mathbf{B_w}\mathbf{u_w}[k] \\ \mathbf{y}[k] &= \mathbf{C}\mathbf{x}[k], \end{aligned} \tag{4.4}$$

where $\mathbf{x} \in \mathbb{R}^n$ and $\mathbf{y} \in \mathbb{R}^p$ denote the plant's state and output, $\mathbf{u} \in \mathbb{R}^m$ is the plant's (controllable) input, and $\mathbf{u_w} \in \mathbb{R}^{m_w}$ is the disturbance input.[10] Accordingly, the matrices $\mathbf{A}$, $\mathbf{B}$ and $\mathbf{B_w}$, $\mathbf{C}$ have suitable dimensions.

Standard dynamical feedback controllers collect the observed plant outputs $\mathbf{y}[k]$ and generate the control input $\mathbf{u}[k]$ as the output of a linear system of the form

$$\begin{aligned} \mathbf{x_c}[k+1] &= \mathbf{A_c}\mathbf{x_c}[k]+\mathbf{B_c}\mathbf{y}[k] \\ \mathbf{u}[k] &= \mathbf{C_c}\mathbf{x_c}[k]+\mathbf{D_c}\mathbf{y}[k]. \end{aligned} \tag{4.5}$$

The vector $\mathbf{x_c}[k]$ denotes the state of the controller, and the matrices $\mathbf{A_c}$, $\mathbf{B_c}$, and $\mathbf{C_c}$, $\mathbf{D_c}$ are designed using standard tools from control theory to ensure that the control inputs are stabilizing. Depending on the control method used, the state of the controller can often be as large as the state of the system itself.

In this traditional approach to controller design, a wireless network would simply be placed between the controller and the plant to carry information back and forth. The goal of our work is to derive a truly networked and fully distributed control scheme where the collective computation and communication capabilities of the wireless nodes are fully leveraged to compute the control inputs in-network. Intuitively, we propose a simple scheme for each node in the network (using only information from its nearest neighbors at each time-step) that results in the desired network behavior. Essentially, we would like each wireless node to act as a small dynamical controller, with two main differences: (i) the state of the controller at each node will be constrained to be rather small (in order to account for resource and computational constraints) and (ii) in its updates, each node only uses the states of its nearest neighbors (which could include the plant's outputs if the node is within transmission range of the outputs). Note that the second condition precludes the need to route information from the plant to each controller in order for it to perform its update. In the rest of this section, we will make these conditions more mathematically precise.

4.7.2 Model development

To model the WCN we consider the basic WCN setup in Fig. 4.1C where the plant is to be controlled using a multi-hop, fully synchronized wireless network

9. Later on we will show how continuous-time plants can be cast in this framework using discretization.

10. We do not have any control over the disturbances.

with N nodes. In the following we extend the proposed scheme to allow for the design of a WCN that applies inputs in an optimal manner (according to a cost function that we will define later). The plant model is given by (4.4), where the output vector $\mathbf{y}[k]$ contains the plant's output measurements provided by the sensors $s_1, \cdots, s_p$, while the input vector $\mathbf{u}[k]$ corresponds to the signals applied to the plant by actuators $a_1, \cdots, a_m$. The wireless network is described by a graph $\mathcal{G} = \{\mathcal{V}, \mathcal{E}\}$, where $\mathcal{V} = \{v_1, v_2, \cdots, v_N\}$ is the set of N nodes and $\mathcal{E} \subseteq \mathcal{V} \times \mathcal{V}$ represents the radio connectivity (communication topology) in the network (i.e., edge $(v_j, v_i) \in \mathcal{E}$ if node v_i can receive information directly from node v_j).

As mentioned earlier, our scheme views each node v_i as a (small) linear dynamical controller, with (possibly vector) state z_i. Each node updates the state of its controller as a linear combination of the states of its neighbors and its own state. The state update for node v_i can also include a linear combination of the plant outputs from all plant sensors in the v_i neighborhood.

For example, consider the network presented in Fig. 4.13 where at the beginning of a time frame each node has an initial state value z_i (Fig. 4.13A). If each node maintains a scalar state, the size of the state is just **2 bytes**.[11] In the first time slot of a frame (Fig. 4.13B) node v_4 transmits its state, and in the second slot node v_5 transmits the state, etc.

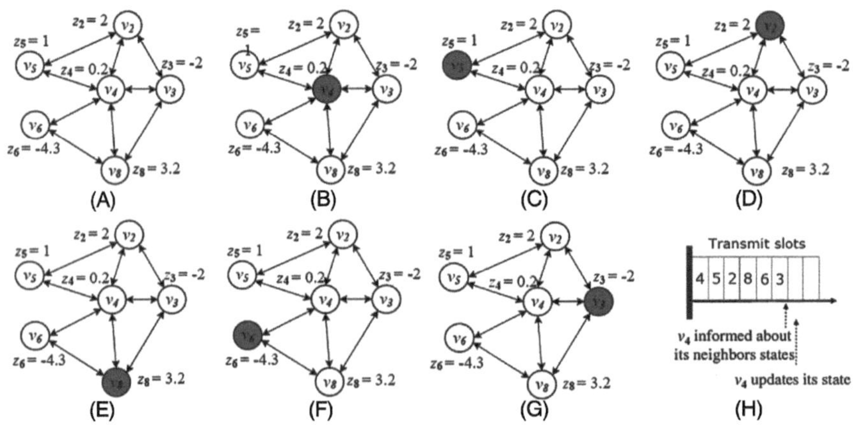

FIGURE 4.13 An illustration of the WCN scheme for a simple network. (A) Initial state; (B) Slot 1: v_4 transmits; (C) Slot 2: v_5 transmits; (D) Slot 3: v_2 transmits; (E) Slot 4: v_8 transmits; (F) Slot 5: v_6 transmits; (G) Slot 7: v_3 transmits; (H) Communication schedule.

Finally, in the 6-th slot node v_3 is the last node in the frame to transmit its state (Fig. 4.13G). This results in a communication schedule as depicted in

11. Given that standard analog-to-digital converters have a precision of 12–16 bits, two bytes are sufficient for scalar values.

Fig. 4.13H. After slot 6, node v_4 is informed about all its neighbors' states, which enables it to update its own state by activating the WCN task. The task has to compute the updated state value before the node is scheduled for transmission in the next frame.

In the general case, if $z_i[k]$ denotes the i-th node's state at time-step (i.e., communication frame) k, the **runtime update** procedure is

$$z_i[k+1] = w_{ii}z_i[k] + \sum_{v_j \in \mathcal{N}_{v_i}} w_{ij}z_j[k] + \sum_{s_j \in \mathcal{N}_{v_i}} h_{ij}y_j[k], \tag{4.6}$$

where the neighborhood of a vertex v is represented as $\mathcal{N}_v$ and $y_j[k]$ is the measurement provided by sensor s_j. We model the resource constraints of each node in the network by limiting the size of the state vector that can be maintained by each node.[12] Note the similarity of update (4.6) to the state update equation for traditional dynamical controllers of the form (4.5); the state $z_i[k]$ plays the role of $x_c[k]$; the weights w_{ii} and w_{ij} play the role of $\mathbf{A_c}$ and the columns of $\mathbf{B_c}$, respectively.

To enable interaction between the network and the plant, each actuator a_i applies input $u_i[k]$, which is computed as a linear combination of states from the nodes in the neighborhood of the actuator:

$$u_i[k] = \sum_{j \in \mathcal{N}_{a_i}} g_{ij}z_j[k]. \tag{4.7}$$

Once again, note the resemblance of this applied input to the input applied by a standard controller of the form (4.5). Therefore, the behavior of each node in the network is determined by values w_{ij}, h_{ij}, and g_{ij}. Aggregating the state values of all nodes at time-step k into the value vector $z[k]$, we see that the above individual controllers at each node collectively cause the entire network to act as a dynamical controller of the form

$$z[k+1] = \underbrace{\begin{bmatrix} w_{11} & w_{12} & \cdots & w_{1N} \\ w_{21} & w_{22} & \cdots & w_{2N} \\ \vdots & \vdots & \ddots & \vdots \\ w_{N1} & w_{N2} & \cdots & w_{NN} \end{bmatrix}}_{\mathbf{W}} \mathbf{z}[k]$$

12. To present the subsequent results, we will focus on the case where each node's state is scalar. The general case, where each heterogeneous node can maintain a vector state with possibly different dimensions, can be treated with a natural extension of our approach (see, e.g., [190]).

$$
\begin{aligned}
&+\underbrace{\begin{bmatrix} h_{11} & h_{12} & \cdots & h_{1p} \\ h_{21} & h_{22} & \cdots & h_{2p} \\ \vdots & \vdots & \ddots & \vdots \\ h_{N1} & h_{N2} & \cdots & h_{Np} \end{bmatrix}}_{\mathbf{H}} \mathbf{y}[k] \\
&= \mathbf{Wz}[k] + \mathbf{Hy}[k] \\
\mathbf{u}[k] &= \underbrace{\begin{bmatrix} g_{11} & g_{12} & \cdots & g_{1N} \\ g_{21} & g_{22} & \cdots & g_{2N} \\ \vdots & \vdots & \ddots & \vdots \\ g_{N1} & g_{N2} & \cdots & g_{NN} \end{bmatrix}}_{\mathbf{G}} \mathbf{z}[k] = \mathbf{Gz}[k]
\end{aligned}
$$

for all $k \in \mathbb{N}$. Since for all $i \in \{1, \cdots, N\}$, $w_{ij} = 0$ if $v_j \notin \mathcal{N}_{v_i}$, $h_{ij} = 0$ if $s_j \notin \mathcal{N}_{v_i}$, and $g_{ij} = 0$ if $v_j \notin \mathcal{N}_{a_i}$ the matrices **W**, **H**, and **G** are *structured*, with sparsity constraints determined by the network topology at design time. Throughout the rest of the chapter, we will define Ψ to be the set of all tuples $(\mathbf{W}, \mathbf{H}, \mathbf{G}) \in \mathbb{R}^{N\times N} \times \mathbb{R}^{N\times p} \times \mathbb{R}^{m\times N}$ satisfying the above-mentioned sparsity constraints. Denoting the overall system state (plant state and states of all nodes in the network) as $\hat{x}[k] = [x[k]^T \;\; z[k]^T]^T$, the closed-loop system evolves as

$$
\begin{aligned}
\hat{x}[k+1] &= \underbrace{\begin{bmatrix} \boldsymbol{A} & \boldsymbol{BG} \\ \boldsymbol{HC} & \boldsymbol{W} \end{bmatrix}}_{\hat{\boldsymbol{A}}} \underbrace{\begin{bmatrix} x[k] \\ z[k] \end{bmatrix}}_{\hat{x}[k]} + \underbrace{\begin{bmatrix} \boldsymbol{B}_w \\ \mathbf{0} \end{bmatrix}}_{\hat{\boldsymbol{B}}} \boldsymbol{u}_w[k] \\
&= \hat{\boldsymbol{A}}\hat{x}[k] + \hat{\boldsymbol{B}}\boldsymbol{u}_w[k].
\end{aligned} \tag{4.8}
$$

To use the WCN runtime scheme it is essential to determine an appropriate set of link weights (w_{ij}, h_{ij}, and g_{ij}) at **design time**, so that the closed-loop system is asymptotically stable.[13] When there are no disturbances (i.e., $\mathbf{u}_w[k] \equiv 0$), an initial procedure is proposed for the basic WCN that guarantees that the closed-loop system is stable, or has MSS, if the communication links are unreliable.[14]

1) Advantages of the WCN: The WCN introduces very low communication and computation overhead. The linear iterative runtime procedure (4.6)

---

13. As a standard result, a linear system $\mathbf{x}[k+1] = \mathbf{Ax}[k]$ is asymptotically stable if for any $\mathbf{x}[0]$, $lim_{k\to\infty}\mathbf{x}[k] = \mathbf{0}$. This is equivalent to saying that all eigenvalues of **A** have a magnitude of less than 1.

14. A switched system described as $\mathbf{x}[k+1] = \mathbf{A}_{\theta(\mathbf{k})}\mathbf{x}[k]$, where subscript $\theta(k)$ describes time variations caused by (probabilistic) drops of communication packets, is mean-square stable if for any initial state $(x[0], \theta(0))$, $lim_{k\to\infty}\mathbb{E}[\|\mathbf{x}[k]\|^2] = 0$, where the expectation is with respect to the probability distribution of the packet drop sequence $\theta(k)$ [226], [227].

is computationally very inexpensive as each node only computes a linear combination of its value and the values of its neighbors. This makes it suitable for resource-constrained, low-power wireless nodes (e.g., Tmote). Furthermore, the communication overhead is also very low as each node needs to transmit only its own state once per frame. If a node maintains a scalar state it transmits only 2 bytes in each message, making it suitable to combine this scheme with periodic message transmissions in existing wireless systems.

Another important benefit is that the WCN can easily handle plants with multiple geographically distributed sensors and actuators, a case that is not easily handled by the sensor → channel → controller/estimator → channel → actuator setup commonly adopted in networked control design. The presence of a centralized controller might impose a requirement that the sampling time of the plant is greater than or equal to the sum of communication delays from sensors to the controller and from the controller to the actuator, along with the time required for the computation of the control algorithm. The WCN does not rely on the presence of centralized controllers, and inherently captures the case of nodes exchanging values with the plant at various points in the network. Therefore, when the WCN is used, the network diameter does not affect the sampling period of the plant.

Finally, the WCN utilizes a simple transmission schedule where each node is active only once during a TDMA cycle and the control loop does not impose end-to-end delay requirements. This allows the network operator to decouple the computation schedule from the communication schedule, which significantly simplifies closed-loop system design and enables compositional design and analysis. As long as each node can send additional states in a single transmission packet, and schedule the computation of additional linear procedures, adding a new control loop will not affect the performance of the existing control loops. For example, consider IEEE 802.14.5 networks that have a maximum packet size of 128 bytes. If each plant is controlled using the WCN scheme where all nodes maintain a scalar 16-bit state value, then up to 64 plants can be controlled in parallel.

In this section, we provide an enhanced WCN scheme that maintains all of these desirable properties, and further incorporates optimality and robustness metrics into the basic scheme.

2) Synchronization requirements: For the network sizes considered here it is necessary to use either the hardware MANGHARAM based on out-of-band synchronization or some of the built-in synchronization protocols that guarantee low synchronization error between neighboring nodes (e.g., the approach described in [228] guarantees that the maximum synchronization error between neighboring nodes is less than 1 μs). Even for a 10 μs synchronization error between neighboring nodes, for large-scale networks with a network diameter of less than 100 nodes, the maximum synchronization error between nodes is less than 1 ms, which is significantly smaller than standard sampling rates of the plant when WCN is used. For example, if communication frames that consist of 16 slots are used, where each slot is 10 ms wide, the sampling period of the plant is equal to 160 ms. In this case the synchronization errors would take less than 1% of the sampling period. We employ a synchronized network and use the RT-Link [212] time-synchronized protocol in our evaluation. Time-synchronized network protocols are the norm in the control automation industry, and two recent standards, WirelessHART [229] and ISA 100.11a [230], utilize a time division multiplexing link protocol.

4.8 Synthesis of an optimal wireless control network

In this section we present a design-time method to determine a WCN configuration (i.e., link weights for a network with predefined topology) that minimizes the effects of the disturbances acting on the system. More specifically, consider the model of the closed-loop system from (4.8), and assume that we want to minimize the influence of the disturbance input $\mathbf{u}_w$ on the vector $\hat{\mathbf{y}} = \hat{\mathbf{C}}\hat{\mathbf{x}}[k]$, for some matrix $\hat{\mathbf{C}}$. For example, if we want to focus on minimizing the effects on the plant's state $\hat{\mathbf{x}}$, we define $\hat{\mathbf{C}} = [\mathbf{I} \;\; \mathbf{0}]$. Thus, we can consider the vector $\hat{\mathbf{y}}$ as the output of the system:

$$\begin{aligned} \hat{\mathbf{x}}[k+1] &= \hat{\mathbf{A}}\hat{\mathbf{x}}[k] + \hat{\mathbf{B}}\mathbf{u}_w[k] \\ \hat{\mathbf{y}} &= \hat{\mathbf{C}}\hat{\mathbf{x}}[k] \end{aligned} \quad (4.9)$$

To determine the effect of the disturbance on the system's outputs, it is necessary to define a unit of measure to capture the size of the discrete-time signals. We use the norms $\|\mathbf{v}\|_{\ell_2} \triangleq \left(\sum_{k=0}^{\infty} \|\mathbf{v}[k]\|^2\right)^{1/2}$ and $\|\mathbf{v}\|_{\ell_\infty} \triangleq sup_{k\geq 0} \|\mathbf{v}[k]\|$. Furthermore, the notion of a system gain is introduced to classify the worst-case system response to limited energy input disturbances.

Definition 4.1. ([231]) System gains for the discrete-time system (4.9) are defined as

- **Energy-to-Peak Gain:** $\gamma_{ep} = sup_{\|\mathbf{u}_w\|_{\ell_2}\leq 1} \left\|\hat{\mathbf{y}}\right\|_{\ell_\infty}$;
- **Energy-to-Energy Gain:** $\gamma_{ee} = sup_{\|\mathbf{u}_w\|_{\ell_2}\leq 1} \left\|\hat{\mathbf{y}}\right\|_{\ell_2}$.

We will require the following result from [232].

Theorem 4.1. *Suppose that system (4.9) is asymptotically stable, and consider any nonnegative $\gamma \in \mathbb{R}$.*

(a) $\gamma_{ep} < \gamma$ *if and only if there exist matrices* $\mathcal{X} \succ 0$, $\Upsilon \succeq 0$, *and* $\mathcal{Z}$ *such that* $\Upsilon \prec \gamma \mathbf{I}$ *and*

$$\mathcal{R}(\mathcal{X}, \mathcal{Z}, \Upsilon, \mathcal{X}^{-1}) = \begin{bmatrix} \mathcal{X} & \mathcal{Z} & \hat{\mathbf{A}} & \hat{\mathbf{B}} \\ \mathcal{Z}^T & \Upsilon & \hat{\mathbf{C}} & \mathbf{0} \\ \hat{\mathbf{A}}^T & \hat{\mathbf{C}}^T & \mathcal{X}^{-1} & \mathbf{0} \\ \hat{\mathbf{B}}^T & \mathbf{0} & \mathbf{0} & \mathbf{I} \end{bmatrix} \succ 0 \tag{4.10}$$

(b) $\gamma_{ee} < \gamma$ *if and only if there exist matrices* $\mathcal{X} \succ 0$, $\Upsilon \succeq 0$ *such that* $\Upsilon \prec \gamma^2 \mathbf{I}$ *and (4.10) hold for* $\mathcal{Z} = 0$.

Only the matrix $\hat{\mathbf{A}}$ contains the WCN parameters, aggregated in the structured matrices $\mathbf{W}, \mathbf{G}, \mathbf{H}$ (from (4.8)). Our goal is to determine matrices $\mathbf{W}, \mathbf{G}, \mathbf{H}$ that satisfy the imposed structural constraints, along with matrices $\mathcal{X}, \mathcal{Z}, \Upsilon$, for which the value γ is minimized.

The constraint (4.10) is linear with respect to all variables, except the matrix $\mathcal{X}$ (due to the presence of the term $\mathcal{X}^{-1}$). This term causes the problem of solving the matrix inequality to be nonconvex. To help fix this issue and efficiently solve the optimization problem, we linearize the $\mathcal{X}^{-1}$ term. As shown in [232], the Taylor series expansion of $\mathcal{X}^{-1}$ "around" any matrix $\mathcal{X}_k$ is

$$LIN(\mathcal{X}^{-1}, \mathcal{X}_k) = \mathcal{X}_k^{-1} - \mathcal{X}_k^{-1}(\mathcal{X} - \mathcal{X}_k)\mathcal{X}_k^{-1} \tag{4.11}$$

With the above linearization we obtain a linear matrix inequality (LMI) for the constraint 10. As in [232] and [233], we can now define an iterative algorithm to minimize, while ensuring that the constraint from (4.10) is satisfied. This is achieved by replacing the term $\mathcal{X}^{-1}$ with $LIN(\mathcal{X}^{-1}, \mathcal{X}_k)$ in each iteration, which results in Algorithm 1. Note that $\hat{\mathbf{A}}(\mathbf{W}, \mathbf{G}, \mathbf{H})$ denotes the matrix $\hat{\mathbf{A}}$ obtained from matrices $\mathbf{W}, \mathbf{G}, \mathbf{H}$ as defined in (4.8). Finally, for γ obtained from Algorithm 1, $\sqrt{\gamma}$ should be used if we optimized for γ_{ee}.

Consider the sequence $\{\gamma_k\}_{k \geq 0}$ obtained from Algorithm 1. As shown in [232], the linearization from (4.11) guarantees that for each $k \geq 0$, in step $k + 1$, there exists a feasible matrix in an open neighborhood of the point $\mathcal{X}_k$ for which there exists a value of γ, such that $\gamma \leq \gamma_k$. Since γ_{k+1} is the minimum in that iteration, it follows that $\gamma_{k+1} \leq \gamma$. Thus, the sequence $\{\gamma_k\}_{k \geq 0}$ is nonincreasing and bounded ($\gamma_k \geq 0$), meaning that it will always converge. Since we are optimizing a convex function over a nonconvex set, by linearizing the constraints we might obtain a suboptimal WCN configuration. The final result and the convergence rate depend on the initial point (from Step 1 of the algorithm). Finally, the smallest ϵ for which we can find an optimal controller can be obtained using a bisection on the parameter ϵ.

Algorithm 1 Design-time procedure used to extract optimal WCN configuration

1. Set $\epsilon > 0$, $k = 0$. Find a feasible point $\mathcal{X}_0, \mathcal{Y}_0, \Upsilon_0 \succ 0$, $\hat{\mathbf{A}}(\mathbf{W}_0, \mathbf{H}_0, \mathbf{G}_0)$, such that $\mathcal{R}(\mathcal{X}_0, \mathcal{Z}, \Upsilon_0, \mathcal{Y}_0) \in \Psi$. If a feasible point does not exist, it is not possible to stabilize the system with this network topology.

2. At iteration $k(k \geq 0)$, from $\mathcal{X}_k$ obtain the matrix $\mathcal{X}_{k+1}$ and scalar γ_{k+1} by solving the LMI problem:

$$\mathcal{X}_{k+1} = arg \min_{\mathcal{X},\mathcal{Z},\Upsilon,\mathbf{W},\mathbf{H},\mathbf{G},\gamma_{k+1}} \gamma_{k+1} \tag{4.12}$$

$$\mathcal{R}(\mathcal{X}, \mathcal{Z}, \Upsilon, LIN(\mathcal{X}^{-1}, \mathcal{X}_k)) \succ 0 \tag{4.13}$$

$$\Upsilon \prec \gamma_{k+1}\mathbf{0} \tag{4.14}$$

$$(\mathbf{W}, \mathbf{H}, \mathbf{G}) \in \Psi, \ \ \mathcal{X} \succ 0, \Upsilon \succeq 0; \tag{4.15}$$

if γ_{ee} is being optimized, add the constraint $\mathcal{Z} = \mathbf{0}$.

3. If $\gamma_{k+1} < \epsilon$ stop the algorithm. Otherwise, set $k = k + 1$ and go to step 2.

4.8.1 Robustness to link failures

We now describe the main limitation of the basic WCN, and extend the WCN scheme to improve its robustness to link failures.

The unreliability of wireless communication links is one of the main drawbacks when wireless networks are used for control. When communication links in the feedback loop fail according to a given probability distribution, the notion of asymptotic stability is typically relaxed to settle for mean-square stability MSS, where the expected value of the norm of the state stays bounded. For the basic WCN we propose a design-time procedure that can be used to extract a stabilizing configuration that guarantees MSS despite unreliable communication links [190]. For example, consider the system in Fig. 4.14 with a scalar plant, where $\alpha = 2$ (the plant is unstable), and assume that the link between node v_2 and the actuator is reliable (i.e., never drops packets). The basic scheme, where each node maintains a scalar state, guarantees that the closed-loop system is MSS for probabilities of packet drops $\leq 1.18\%$.

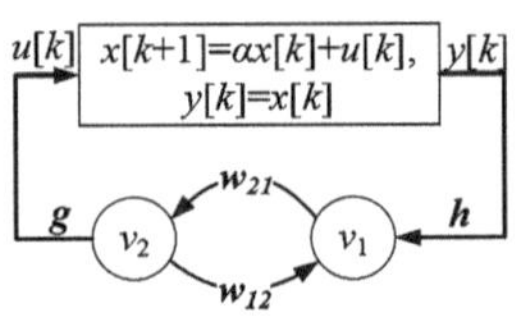

FIGURE 4.14 An example of the WCN: A plant with a scalar state controlled by a WCN.

To place this result in context, it is worth comparing it with the theoretical limit of robustness in lossy networks from [191]. The work in [191] considers a system with a plant controlled by a centralized controller, which is connected to the plant using a single wireless link between a sensor and the controller. In addition, the controller is connected to the actuators with a set of wired connections. It was shown that for this setup the system cannot be stabilized with a linear controller for probability of message drops p greater than $\frac{1}{\|\lambda_{max}\|^2}$, where λ_{max} denotes the maximum norm of the plant's eigenvalues (i.e., eigenvalues of $\mathbf{A}$ from (4.4)). For the plant in Fig. 4.14 this would mean that a centralized controller in the above-mentioned setup cannot provide MSS of the plant if the probability of message drops is higher than 25% (since $\alpha = 2$). This value is significantly higher than the 1.18% value obtained when the basic WCN scheme is used. We now show how the basic WCN formulation presented in (4.6) and (4.7) can be modified to significantly improve tolerance to packet drops.

4.8.2 Wireless control networks with observer style updates

To improve WCN robustness to independent link failures, we now allow each node in the network to use different weights in each time-step that depend on which neighbors' transmissions were successfully received. Thus, we define the update procedure as[15]

$$z_j[k+1] = \tilde{w}_{jj} z_j[k] + \sum_{i \in \mathcal{N}_{v_j}} \tilde{w}_{ji} z_i[k], \tag{4.16}$$

where $\tilde{w}_{ji} = 0$ if the message from the node v_i is not received, or w_{ji} otherwise.[16] More importantly, $\tilde{w}_{jj}$ depends on a newly introduced set of link weights $(q_{ji}) : \tilde{w}_{jj} = w_{jj} - \sum_{i \in \mathcal{N}_{v_j}} \tilde{q}_{ji}$. Here $\tilde{q}_{ji} = 0$ if the message from the node v_i is not received, and q_{ji} (a free parameter that will be carefully designed) otherwise.

To model the WCN that employs the above scheme, we need to model the links in the network. We utilize the approach proposed in [227] where each unreliable link $\xi_{ji} = (v_i, v_j)$ (i.e., $v_i \rightarrow v_j$) can be modeled as a memoryless, discrete, and independent and identically distributed (i.i.d.) random process ξ_{ji}. Here i.i.d. implies that the random variables $\{\xi_{ji}[k]\}_{k \geq 0}$ are i.i.d.[17] For each link, these random processes map each transmitted value t_{ji} into a received value $\xi_{ji}[k] t_{ji}$ (see Fig. 4.15).

15. A similar update is introduced for nodes that receive sensor values. This part has been omitted for ease of exposition.

16. Although these weights are technically time varying (i.e., they depend on k), we use this notation for simplicity.

17. We will address these assumptions later in this section.

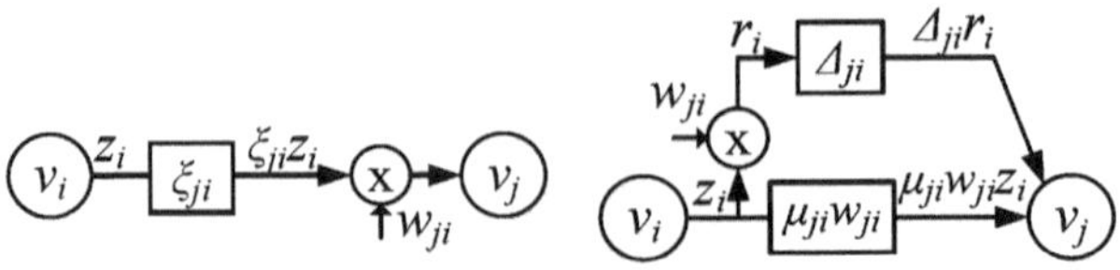

FIGURE 4.15 Communication over a nondeterministic channel. (left) A link between nodes v_i and v_j; (right) Link transformation into a robust control form.

With this link model, (4.16) can be described as

$$z_j[k+1] = (w_{jj} - \sum_{i \in \mathcal{N}_{v_j}} \xi_{ji} q_{ji})\, z_i[k] + \sum_{i \in \mathcal{N}_{v_j}} \xi_{ji} w_{ji} z_i[k].$$

Remark 4.5. If we consider the case with reliable communication links, the update procedure for each node v_j in the network can be described as

$$z_j[k+1] = w_{jj} z_j[k] + \sum_{i \in \mathcal{N}_{v_j}} (w_{ji} z_i[k] - q_{ji} z_i[k]). \tag{4.17}$$

Since Eq. (4.17) has the standard observer structure [234], we refer to this scheme as the WCN with observer style updates (as in [206]).

Following the approach from [227], each link described with a random process ξ_{ji} can be specified with a fixed gain, corresponding to the mean value of the random variable and to the zero-mean random part: $\xi_{ji} = \mu_{ji} + \Delta_{ji}$. For example, if each link (i.e., random process ξ_{ji}) is described as a Bernoulli process with probability $p_{ji} \leq 1$ (i.e., the link delivers the transmitted message with probability p_{ji}), then $\mu_{ji} = p_{ji}$ and Δ_{ji} can have values $-p_{ji}$ and $1 - p_{ji}$, with probabilities $1 - p_{ji}$ and p_{ji}, respectively. Therefore, the above procedure becomes

$$\begin{aligned} z_j[k+1] &= (w_{jj} - \sum_{i \in \mathcal{N}_{v_j}} \mu_{ji} q_{ji})\, z_i[k] + \sum_{i \in \mathcal{N}_{v_j}} \mu_{ji} w_{ji} z_i[k] \\ &\quad + \sum_{i \in \mathcal{N}_{v_j}} \Delta(w_{ji} z_i[k] - q_{ji} z_i[k]). \end{aligned}$$

We define $r_t[k] := (w_{ji} z_i[k] - q_{ji} z_i)$, for each link $t = (v_i, v_j)$. Also, for each link $t = (s_i, v_j)$ we denote $r_t[k] := (h_{ji} y_i[k] - q_{ji} z_j[k])$. After aggregating all of the $r_t[k]$ values in a vector $r[k]$ of length N_l (where N_l is the number of links), we obtain

$$\mathbf{r}[k] = \mathbf{J}^{or} \begin{bmatrix} \mathbf{y}[k] \\ \mathbf{z}[k] \end{bmatrix} = \underbrace{\mathbf{J}^{or} \begin{bmatrix} \mathbf{C} & \mathbf{0} \\ \mathbf{0} & \mathbf{I}_N \end{bmatrix}}_{\hat{\mathbf{J}}^{or}} \hat{\mathbf{x}}[k]. \tag{4.18}$$

Each row of the matrix $\mathbf{J}^{or} \in \mathbb{R}^{N_l \times (N+p)}$ contains up to two nonzero elements, equal to a gain w_t, h_t, g_t or $-q_t$.

This allows us to model the behavior of the closed-loop system with unreliable communication. Specifically, the update equation for each node v_j is

$$\begin{aligned} z_j[k+1] &= (w_{jj} - \sum_{i\in\mathcal{N}_{v_j}} \mu_{ji}q_{ji})z_j[k] + \sum_{t=(v_i,v_j)} \mu_t w_t z_i[k] \\ &\quad + \sum_{t=(s_i,v_j)} \mu_t h_t y_i[k] + \sum_{t=(v_i,v_j)} \Delta_t[k]r_t[k] \\ &\quad + \sum_{t=(s_i,v_j)} \Delta_t[k]r_t[k]. \end{aligned}$$

Similarly, the input value applied by each actuator at time k is

$$u_j[k] = \sum_{t=(v_i,a_j)} \mu_t g_t z_i[k] + \sum_{t=(v_i,a_j)} \Delta_t[k]r_t[k].$$

Finally, denoting $\Delta[k] = diag(\{\Delta_t[k]\}_{t=1}^{N_l})$, the above expressions can be written in vector form as

$$\mathbf{z}[k+1] = \mathbf{W}_\mu \mathbf{z}[k] + \mathbf{H}_\mu \mathbf{y}[k] + \mathbf{J}_v^{dst}\Delta[k]\mathbf{r}[k], \tag{4.19}$$

$$\mathbf{u}[k] = \mathbf{G}_\mu z[k] + \mathbf{J}_u^{dst}\Delta[k]\mathbf{r}[k], \tag{4.20}$$

where all elements of matrices $\mathbf{W}_\mu$, $\mathbf{H}_\mu$, and $\mathbf{G}_\mu$ (except the diagonal entries of $\mathbf{W}_\mu$) are of the form $\mu_{ji}w_{ji}$, $\mu_{ji}h_{ji}$, and $\mu_{ji}g_{ji}$, respectively. The diagonal entries of $\mathbf{W}_\mu$ are of the form $w_{jj} - \sum_{i\in\mathcal{N}_{v_j}} \mu_{ji}q_{ji}$. The binary matrices $\mathbf{J}_v^{dst}$ and $\mathbf{J}_u^{dst}$ are designed in a way that each row of the matrices selects elements of the vector $\Delta[k]r[k]$ that are added to the linear combinations calculated by the nodes and the actuators. If we denote $\mathbf{J}^{dst} = [\mathbf{J}_u^{dst} \ \ \mathbf{J}_v^{dst}]^T$ the overall system with unreliable links can be modeled as

$$\begin{aligned} \hat{\mathbf{x}}[k+1] &= \underbrace{\begin{bmatrix} \mathbf{A} & \mathbf{B}\mathbf{G}_\mu \\ \mathbf{H}_\mu\mathbf{G} & \mathbf{W}_\mu \end{bmatrix}}_{\hat{\mathbf{A}}_\mu} \hat{\mathbf{x}}[k] \\ &\quad + \underbrace{\begin{bmatrix} \mathbf{B} & \mathbf{0} \\ \mathbf{0} & \mathbf{I}_N \end{bmatrix} \mathbf{J}^{dst}}_{\hat{\mathbf{J}}^{dst}} \Delta[k]\mathbf{r}[k] \end{aligned} \tag{4.21}$$

with $\mathbf{r}[k]$ given by (4.18). Using the same approach as in [227] and [190], the following theorem can now be proven.

Theorem 4.2. *The system from (4.21) is mean-square stable if and only if there exist matrices* $\mathcal{X}, \mathcal{Y} \succ 0$ *and scalars* $\alpha_1, \cdots, \alpha_{N_l}$ *such that*

$$\begin{bmatrix} \mathcal{X} - \hat{\mathbf{J}}^{dst} diag\{\alpha\}(\hat{\mathbf{J}}^{dst})^T & \hat{\mathbf{A}}_\mu \\ \hat{\mathbf{A}}_\mu^T & \mathcal{Y} \end{bmatrix} \succ 0 \tag{4.22}$$

$$\mathcal{Y} = \mathcal{X}^{-1} \tag{4.23}$$

$$\alpha_i \geq \sigma_i^2 (\hat{\mathbf{J}}^{or})_i \mathcal{Y}^{-1} (\hat{\mathbf{J}}^{or})_i^T, \quad \forall i \in \{1, \cdots, N_l\} \tag{4.24}$$

where $(\hat{\mathbf{J}}^{or})_i$ *denotes the i-th row of the matrix* $\hat{\mathbf{J}}^{or}$.

A procedure based on LMIs, with the same structure as Algorithm 1, can be used in this case to compute a WCN configuration that guarantees MSS of the closed-loop system with error-prone links. The difference with Algorithm 1 is that in Step 2 the following problem should be solved,

$$\mathcal{X}_{k+1} = arg \min_{\mathcal{X}, \mathcal{Y}, \Upsilon, \mathbf{W}, \mathbf{H}, \mathbf{G}} tr(\Upsilon),$$
$$\mathcal{Y} - LIN(\mathcal{X}^{-1}, \mathcal{X}_k) \prec \Upsilon, \quad \mathcal{X} \succeq \mathcal{Y}^{-1},$$

such that the constraints from (4.23), (4.24), (4.15) are valid, where $tr(\mathbf{A})$ denotes the trace of the matrix $\mathbf{A}$. Note that the above algorithm adds only one additional LMI constraint for each link in the network.

1) Validity of the Assumptions: While developing the model of the WCN from (4.20), we have assumed that all links in the network are memoryless and independent. Memoryless channels can be obtained if channel hopping is used at the network layer [235]. However, the physical placement of the nodes might introduce correlation between some of the network links.

If these i.i.d. assumptions are not valid (or are too simplistic), we must model correlation between links along with more complex link failures (such as those induced by a Markov process). In these cases an approach similar to [226] can be used, which would result in an exponential number of additional constraints introduced to deal with link failures (compared to the linear number of additional constraints introduced under the i.i.d. assumption of independent and memoryless channels). Except for very large-scale systems, the observer style update procedure is practical as the computation of WCN configurations $\mathbf{W}, \mathbf{H}, \mathbf{G}$ is only required at design time.

2) Evaluation: We evaluated the performance of the proposed scheme by modeling all links as independent Bernoulli processes. To analyze the robustness of the WCN with observer style updates, we first analyzed the performance of WCNs with $N \geq 2$ nodes that create a complete graph. The WCN is used for control of the single-state plant shown in Fig. 4.14 (with $\alpha > 1$). Node v_1 receives the plant output $y[k] = x[k]$ at each time-step k, and the input to the plant is derived as a scaled version of the transmission of the node v_2 (i.e., $u[k] = gz_2[k]$ for a scalar g). Using the bisection method from [226], we extracted the

	WCN (scalar state)	WCN ($\mathbb{R}^2$ state)	$oWCN$ (scalar state)	$oWCN$ ($\mathbb{R}^2$ state)
$N = 2$	$p_m = 0.69\%$	$p_m = 0.72\%$	$p_m = 1.64\%$	$p_m = 1.82\%$
$N = 3$	$p_m = 0.74\%$	$p_m = 0.77\%$	$p_m = 1.66\%$	$p_m = 1.88\%$
$N = 4$	$p_m = 0.77\%$	$p_m = 0.79\%$	$p_m = 1.66\%$	$p_m = 1.88\%$

(A)

	WCN (scalar state)	WCN ($\mathbb{R}^2$ state)	$oWCN$ (scalar state)	$oWCN$ ($\mathbb{R}^2$ state)
$N = 2$	$p_m = 1.18\%$	$p_m = 1.30\%$	$p_m = 10.46\%$	$p_m = 17.82\%$
$N = 3$	$p_m = 1.32\%$	$p_m = 1.46\%$	$p_m = 11.24\%$	$p_m = 17.88\%$
$N = 4$	$p_m = 1.41\%$	$p_m = 1.54\%$	$p_m = 11.46\%$	$p_m = 17.88\%$
		$oWCN$ ($\mathbb{R}^3$ state)	$oWCN$ ($\mathbb{R}^4$ state)	$oWCN$ ($\mathbb{R}^5$ state)
$N = 2$		$p_m = 20.40\%$	$p_m = 20.48\%$	$p_m = 20.64\%$

(B)

FIGURE 4.16 Maximum probabilities of link failure for which the closed-loop system in Fig. 4.14 ($\alpha = 2$) has MSS, when controlled with observer style updates (oWCN) and without (WCN). (A) With all links being unreliable; (B) With a reliable link between the node v_2 and actuator.

maximum probabilities of message drops (p_m) for which there exits a stabilizing configuration that ensures MSS.

We considered two scenarios. In the first scenario, we compared the performance of the basic WCN with that of the WCN with observer style updates (denoted WCN). We analyzed networks where all the links are unreliable, described with the same probability of packet drops p (including the links between the plant and the network nodes). The results are presented in Fig. 4.16A. In addition, we investigated the case where the link between node v_2 and the plant's actuator is reliable (without any packet drops). The results are shown in Fig. 4.16B. As can be observed, the proposed scheme **significantly** improves system robustness to link failures. *For example, the WCN with observer style updates guarantees MSS for the system in Fig. 4.14 even when the probability of link failure is more than 20% (compared to 1.5% for the basic WCN).* Similarly, going back to the discussion at the beginning of the section, we show in this simple example that the WCN performance is much closer to that of the optimal centralized controllers used for control over wireless links (guaranteeing MSS with up to 25% packet drops).

Using the observer style updates, similar significantly improved results were obtained for the more complex examples from [190], including larger plants with multiple inputs and outputs, controlled by a mesh network with nine nodes.

4.9 Robustness to node failure

The stability of the closed-loop system, described by (4.8), can be affected by node crash failures (i.e., nodes that stop working and drop out of the network). So far we have considered two approaches to deal with the node failures. One obvious way to deal with up to k node failures is to precompute at the design-time a set of $N_k = \sum_{j=0}^{k} \binom{N}{j}$ different stabilizing configurations $(\mathbf{W}, \mathbf{H}, \mathbf{G})$ that correspond to all possible choices of k or fewer failed nodes. In this case each node would need to maintain N_k different sets of link weights for all its incoming links. For example, if each node in the WCN maintains a scalar state, a node with d neighbors would have to maintain on the order of $d.N_k$ different scalar weights. The switching between the precomputed stabilizing configurations can be done either by implementing the detection algorithm from [193] or by having the neighbors of failed nodes broadcast the news of the failures throughout the network, which will prompt all nodes to switch to the appropriate choice of $(\mathbf{W}, \mathbf{H}, \mathbf{G})$.

A more sophisticated method for dealing with node failure would be to design the WCN in such a way that even if some of the nodes fail the closed-loop system remains stable. For simplicity, consider a WCN that can deal with a single node failure. Let us denote with $\hat{\mathbf{A}}_i$ the matrix $\hat{\mathbf{A}}$ from (4.8) in the case when node i dies. This is equivalent to setting to zero the i-th row of matrices $\mathbf{W}$ and $\mathbf{H}$, along with the i-th column of $\mathbf{W}$ and $\mathbf{G}$:

$$\hat{\mathbf{A}}_i \triangleq \begin{bmatrix} \mathbf{A} & \mathbf{B}\mathbf{G}\mathbf{I}_N^i \\ \mathbf{I}_N^i\mathbf{H}\mathbf{C} & \mathbf{I}_N^i\mathbf{W}\mathbf{I}_N^i \end{bmatrix}, \quad i = 1, \cdots, N. \tag{4.25}$$

Here, $\mathbf{I}_N^i$ denotes $N \times N$ diagonal matrix, with all ones on the diagonal except at the i-th position. A sufficient condition for system stability in this case is that there exists a positive definite matrix $\mathcal{X}$ (and thus a common Lyapunov function $V(\hat{\mathbf{x}}) = \hat{\mathbf{x}}^T \mathcal{X} \hat{\mathbf{x}}$ such that $\mathcal{X} - \hat{\mathbf{A}}^T \mathcal{X} \hat{\mathbf{A}} \succ \mathbf{0}$ and

$$\mathcal{X} - \hat{\mathbf{A}}_i^T \mathcal{X} \hat{\mathbf{A}}_i \succ \mathbf{0} \;\; i = 1, 2, \cdots, N. \tag{4.26}$$

Therefore, the procedure from the previous section with additional N LMI constraints can be used to extract a stabilizing configuration that can deal with a single node failure. However, in this case it is necessary to design the network in a way that guarantees that such a stabilizing configuration exists. Initial results on these topological conditions have been presented in [192].

4.10 Control of continuous-time plants

Optimal and stabilizing WCN configurations can be obtained using algorithms developed from the closed-loop system model (4.8) that contains a discrete-time model of the plant (4.4). However, a similar framework can be used for control of continuous-time plants by discretizing the controlled plant while taking into account a subtle delay introduced by the communication schedule. To illustrate this, consider a standard continuous-time plant model,

$$\begin{aligned} \dot{\mathbf{x}}(t) &= \mathbf{A}_c\mathbf{x}(t) + \mathbf{B}_c\mathbf{u}(t), \\ \mathbf{y}(t) &= \mathbf{C}_c\mathbf{x}(t), \end{aligned} \tag{4.27}$$

with input $\mathbf{x}(t) \in \mathbb{R}^n$, output $\mathbf{y}(t) \in \mathbb{R}^p$, $\mathbf{u}(t) \in \mathbb{R}^m$, and matrices $\mathbf{A}_c$, $\mathbf{B}_c$, $\mathbf{C}_c$ of the appropriate dimensions.[18] We denote the sampling period of the plant as T, and we assume that all sensors sample the plant outputs at the beginning of the zero-th slot (as shown in Fig. 4.17A). We also assume that all actuators are scheduled to apply their newly calculated inputs at the beginning of the h-th time slot. Note that $h > 0$, because from (4.7) each actuator has to first receive state values from all of its neighbors before calculating its next plant input. Similarly, from (4.7) $h \geq max(d_{a_i})$, where d_{a_i} denotes the number of neighbors of the actuator a_i.

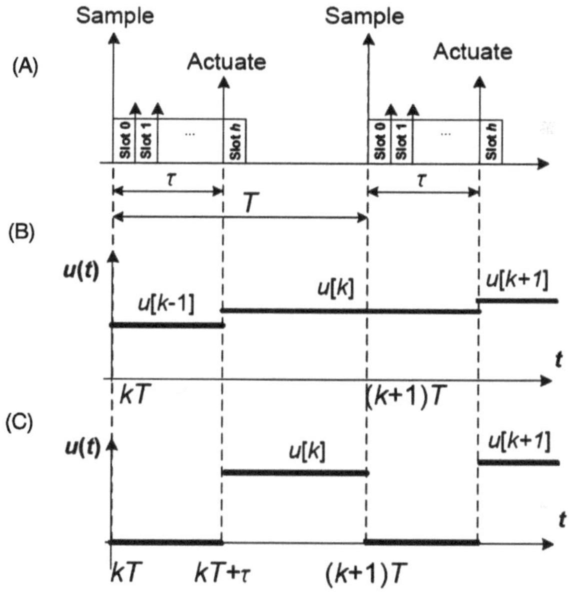

FIGURE 4.17 (A) Scheduling sampling/actuation at the start of the slots; (B) Timing diagram for the first type of plant inputs; (C) Plant inputs when actuators reset the inputs at the beginning of the frames.

Therefore, the new inputs will be applied to the plant with the delay $\tau = hT_{sl}$, where T_{sl} is the size of communication slots. This results in the input signal with the form shown in Fig. 4.17B. Denoting the number of slots in a communication frame by F, we can write $T = FT_{sl}$. Using the approach from [148] and [175], we describe the system,

$$\begin{aligned}\dot{\mathbf{x}}(t) &= \mathbf{A}_c\mathbf{x}(t) + \mathbf{B}_c\mathbf{u}(t),\\ \mathbf{y}(t) &= \mathbf{C}_c\mathbf{x}(t), \;\; t \in [kT + \tau, (k+1)T + \tau),\\ \mathbf{u}(t^+) &= \mathbf{G}\mathbf{z}[k], \;\; t \in \{kT + \tau, k = 0, 1, 2, \cdots\},\end{aligned} \tag{4.28}$$

18. For simplicity we do not model disturbance inputs to the plant. However, the approach presented in this section can readily handle that scenario.

where $\mathbf{u}(t^+)$ is a piecewise continuous function and only changes values at time instances $kT+\tau, k=0,1,2,\cdots$. From the above equations the discretized model of the system with the sampling period T can be represented as [234]

$$\begin{aligned}\mathbf{x}[k+1] &= \mathbf{A}\mathbf{x}[k]+\mathbf{B}\mathbf{G}\mathbf{z}[k]+\mathbf{B}^{-}\mathbf{G}\mathbf{z}[k-1],\\ \mathbf{y}[k] &= \mathbf{C}\mathbf{x}[k],\end{aligned} \tag{4.29}$$

where $\mathbf{x}[k]=\mathbf{x}[k](kT)$, $k\geq 0$, and

$$\begin{aligned}\mathbf{A} &= e^{\mathbf{A}_c T}\ \ \mathbf{B}=\int_0^{T-\tau} e^{\mathbf{A}_c\delta}\mathbf{B}_c d\delta,\\ \mathbf{B}^{-} &= \int_{T-\tau}^{T} e^{\mathbf{A}_c\delta}\mathbf{B}_c d\delta.\end{aligned} \tag{4.30}$$

When the communication schedule is extracted and the network is configured, the matrices $\mathbf{A}$, $\mathbf{B}$ and $\mathbf{B}^{-}$ obtain fixed values that depend on the continuous-time plant dynamics, communication frame size T (i.e., the sampling period of the plant), and the utilized communication schedule (as it determines the value for h).

If each actuator applies its current input only until the end of the corresponding frame and then forces its input to zero until the next actuation slot (i.e., the h-th slot), the input signals would have the form shown in Fig. 4.17C (instead of the form in Fig. 4.17B). In this case the discretized system could be specified as in (4.29) and (4.30), with the difference that $\mathbf{B}^{-}=\mathbf{0}$. Therefore, the discrete-time system takes the form from (4.4), and stabilizing and optimal configurations can be obtained using the procedures described in the previous sections. However, due to the delay τ, the resulting discrete-time system could be uncontrollable, which in the general case would mean that there is no stabilizing configuration for the closed-loop system.

In situations where $\mathbf{A}$, $\mathbf{B}$ is not controllable it is necessary for all actuators to apply their old inputs until new inputs are available (as shown in Fig. 4.17B). This results in a discrete-time plant that does not have the form from (4.4), and the previous algorithms cannot be directly employed. However, by defining a new vector $\tilde{\mathbf{x}}[k]\triangleq\begin{bmatrix}\mathbf{x}[k]^T & \mathbf{u}[k-1]^T\end{bmatrix}$ the discrete-time system can be described as

$$\begin{aligned}\tilde{\mathbf{x}}[k+1] &= \begin{bmatrix}\mathbf{A} & \mathbf{B}^{-}\\ \mathbf{0} & \mathbf{0}\end{bmatrix}\tilde{\mathbf{x}}[k]+\begin{bmatrix}\mathbf{B}\\ \mathbf{I}\end{bmatrix}\mathbf{u}[k]\\ \tilde{\mathbf{x}}[k+1] &= \tilde{\mathbf{A}}\tilde{\mathbf{x}}[k]+\tilde{\mathbf{B}}\mathbf{u}[k]\\ \mathbf{y}[k] &= \begin{bmatrix}\mathbf{C} & \mathbf{0}\end{bmatrix}\tilde{\mathbf{x}}[k]=\tilde{\mathbf{C}}\tilde{\mathbf{x}}[k]\end{aligned}$$

The above system has the same form as (4.4), and therefore we can use the above-mentioned algorithms to obtain a stabilizing or optimal configuration of the WCN.

4.11 Process control application

The WCN was deployed on a process-in-the-loop testbed with a plant running in Simulink and the plant's sensors and actuators connected to analog interfaces (see Fig. 4.19A). We first describe the plant's model, then the closed-loop wireless control testbed, and finally demonstrate the WCN use for control of the plant.

4.11.1 Case description

To illustrate the use of the WCN, we consider the distillation column control (Fig. 4.18A), a well-known process control problem described in [236]. Four input flows (in [mols/s]) are available for the column control: reflux (L), boilup (V), distillate (D), and bottom flow (B). The goal is to control four outputs: x_D – top composition, x_B – bottom composition, M_D – liquid levels in condenser, and M_B – liquid levels in the reboiler (in [mol]). Finally, the column has two disturbances, feed flow-rate F and feed composition z_F. The columns are described using the continuous-time LTI model from [236], where the state-space contains eight states.

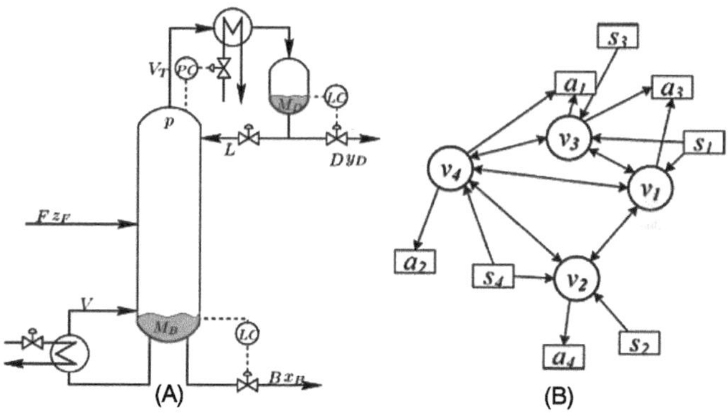

FIGURE 4.18 (left) Structure of the distillation column [236]; (right) The network topology of the WCN corresponding to the sensor and actuator positions.

4.11.2 Wireless control network experimental platform

We implemented the WCN scheme on FireFly embedded wireless nodes [222] and TI MSP430F5438 Experimenter Boards, both equipped with IEEE 802.15.4 standard-compliant radio transceivers. FireFly is a low-cost, low-power platform based on Atmel ATmega1281 8-bit microcontroller, while the experimenter board uses a 16-bit MSP430 microcontroller. Both platforms can be used for TDMA-based communication with the RT-Link protocol [212], and support in-band synchronization provided as a part of the protocol.

The WCN procedure on each wireless node was implemented as a simple task executed on top of the nano-RK, an RTOS [211]. The WCN task had a 140.64 ms period, equal to the RT-Link frame size (RTLink was configured to use 16 slots of size 8.79 ms). Since the WCN requires a TTA,[19] nano-RK has been modified to enable scheduling of sensing and actuation at the start of the desired slots. This guarantees synchronized actions at all sensors and all actuators.

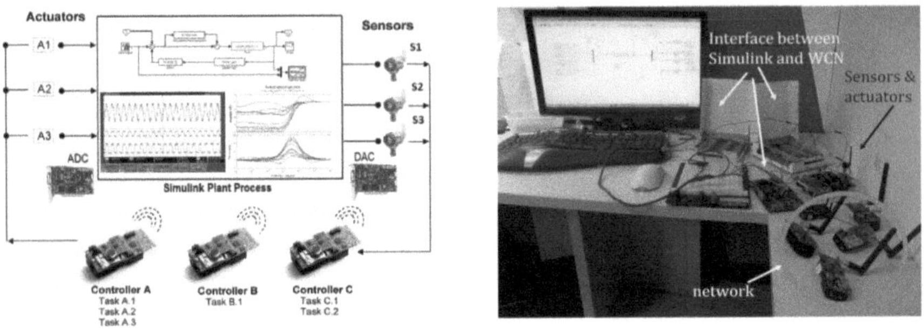

FIGURE 4.19 Process-in-the-loop simulation of the distillation column control. (left) The plant model is simulated in Simulink, while the WCN is implemented on FireFly nodes; (right) Experimental setup used for the WCN validation.

The column, modeled as a continuous-time LTI system along with disturbances and measurement noise, was run in Simulink in real-time using the Real-Time Windows Target [237]. The interface between the model and the real hardware were two National Instruments (NI) PCI-6229 boards, which provided analog outputs that correspond to the Simulink model's outputs (see Fig. 4.19A). The output signals were saturated between −4V and 4V, due to the limitations of the NI boards. Also, to provide inputs to the Simulink model the boards sampled the analog input signals in the range [−4V, 4V] at a 1 kHz rate. Finally, Simulink's input and output signals were monitored and controlled with four sensors and four actuators positioned according to the distillation column structure (Fig. 4.18A). In addition, four real wireless controller nodes ($v_1 - v_4$) were added, resulting in the topology shown in Fig. 4.18B.

4.11.3 Wireless control networks results

From the communication and computation schedules, we obtained the discrete-time plant model using the discretization procedure from Section 4.10 (Eqs. (4.29), (4.30)), with sampling rate T = 140:64 ms (RT-Link frame size).

We first investigated the problem of providing MSS of the closed-loop system with uncorrelated random link failures and single node failures. Assigning

19. TTA is the turn-around time corresponding to the time between the end of the RTS packet and the beginning of the CTS packet.

each node to maintain a scalar state, using the procedures from Sections 4.11.1 and 4.11.2 we derived a stabilizing WCN configuration for the topology presented in Fig. 4.18B and the discretized LTI plant model. To solve the convex optimization problems we used CVX, a package for specifying and solving convex programs [238].

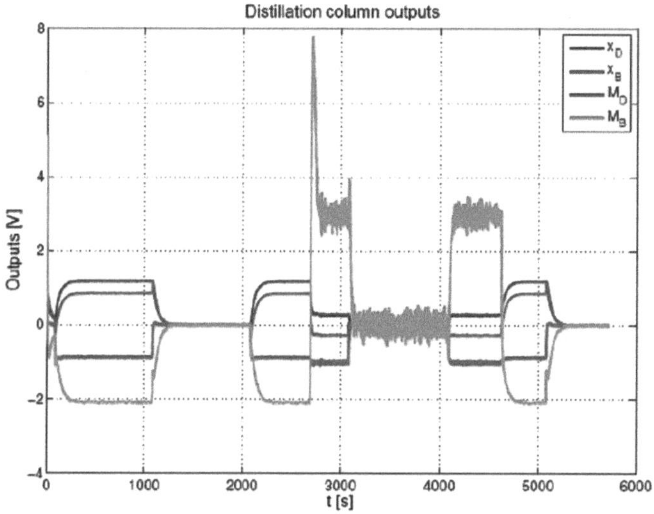

FIGURE 4.20 Plant outputs for a stabilizing WCN configuration.

Note in Fig. 4.20 that v_1 has been turned off at time $t = 1680$ s and turned back on at $t = 4560$ s.

We were able to obtain only WCN configurations that maintain stability if one of the nodes $v_1 - v_3$ fails, meaning that the constraint from (4.26) for the node v_4 was violated (without v_4 the topology violates the conditions from [192] for the presence of a stabilizing configuration). Fig. 4.20 shows the obtained measurements where the disturbance inputs F, z_F were set to zero, while we provided periodical pulses to the input L. Although the output of the plant degrades when the node v_1 is turned off, the WCN maintains system stability. However, if the node v_4 is turned off, the system becomes unstable (shown in Fig. 4.21; after the node is turned back on the system slowly returns to stability, due to the output saturation). Finally, we showed that if a node is added, connected to actuator a_2, sensor s_4, and nodes v_2, v_4, we can maintain stability if one of the node fails.

We also considered optimal WCN design that minimizes effects of disturbance inputs F, z_F. Using Algorithm 1 we computed an optimal WCN configuration for energy-to-peak minimization. The obtained measurements for a setup with periodical F impulses are shown in Fig. 4.22. Figs. 4.22B and 4.22A present the plant outputs for the optimal and stable WCN configurations. As shown in Fig. 4.22C, the norm of the output controlled with the optimal

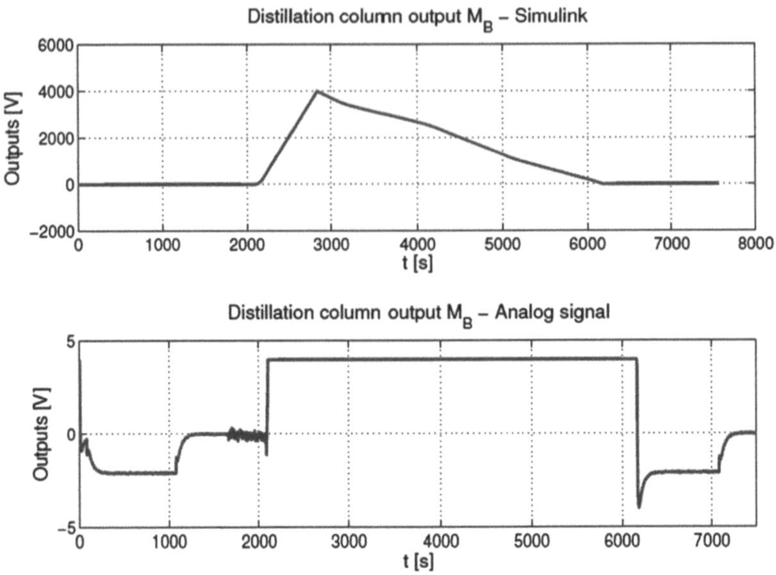

FIGURE 4.21 Distillation column output M_B. Node v_4 is turned off at $t = 2140$ s and back on at $t = 2860$ s. Top – Simulink signal; bottom – analog signal, saturated at 4V.

configuration is almost five times smaller than the norm with the stabilizing WCN.

4.12 Notes

A brief overview of NCSs was given and new trends in NCSs were also pointed out, namely with the development of cloud computing and the processing techniques of big data the dawn of cloud control systems will emerge soon. A preliminary structure and algorithm were proposed. Some new results on cloud control systems will be found in our future publications. We believe that there will be more interesting and important results produced in this new research area.

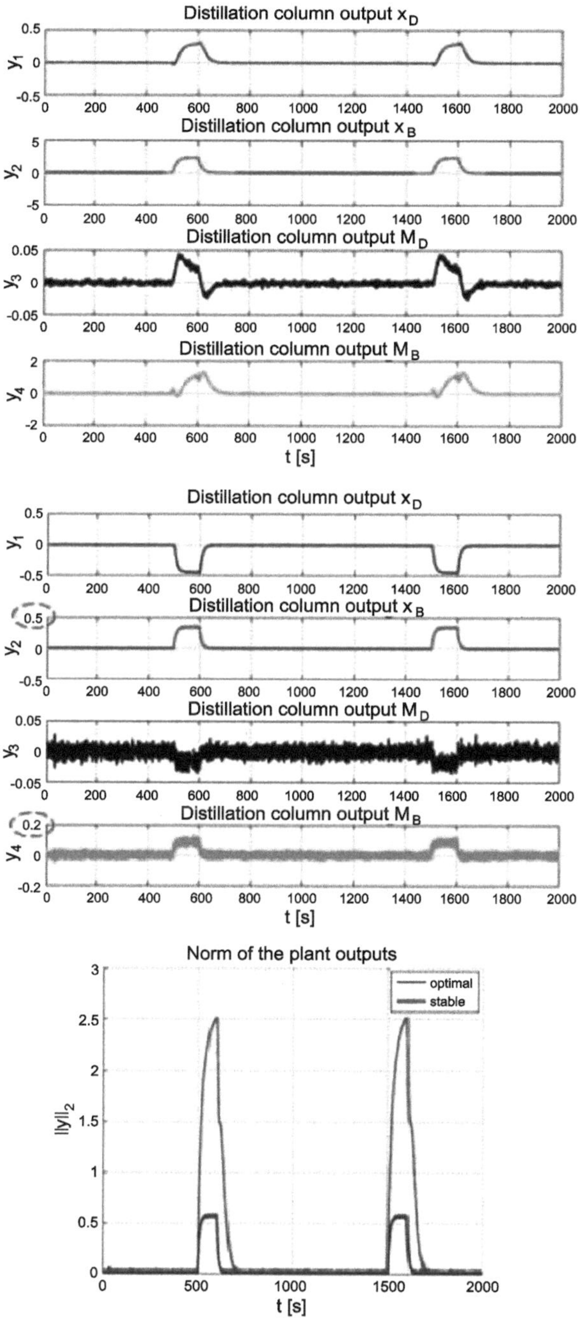

FIGURE 4.22 Distillation column outputs: (Top) For a stable WCN configuration; (Middle) For an optimal WCN configuration (note the axis scales); (Bottom) Comparison of the output vector norms for the stable and the optimal WCN configurations.

Chapter 5

Secure stabilization of distributed systems

Contents

5.1 Introduction

Cloud control systems (CCSs) are increasingly appealing for industry nowadays thanks to the development of computation and communication infrastructures. The application of CCSs ranges from local control systems to large-scale systems, examples being house temperature control systems and regional grid control systems. Owing to the advances in economic growth and possibly reliability, systems tend to be large-scale, interconnected, and spatially distributed, and communications are operated via wireless network [239]. This pushes attention towards networked control of large-scale interconnected systems, which are possibly safety-critical and potentially exposed to malicious attacks [240].

The concept of cyber-physical security (CPS) mostly concerns security against intelligent attacks. These attacks are usually classified as either deceptive attacks or Denial-of-Service (DoS) attacks. Deceptive attacks affect the trustworthiness of transmitted data [241], [242]. Instead, DoS compromises the timeliness of information exchange; for example, in the presence of DoS, communication is not possible [243], [90].

This chapter investigates DoS attacks. We consider a large-scale system composed of interconnected subsystems, which are possibly spatially distributed. The information exchange between distributed systems and controllers takes place over a shared communication channel, which implies that all the communication attempts can be denied in the presence of DoS.

Cloud Control Systems. https://doi.org/10.1016/B978-0-12-818701-2.00013-5

The literature on distributed/decentralized networked control [244–250] and centralized system under DoS attacks [90,251–253,116,254,103,89,255,256,104,111,102] is large and diversified. In [249], based on a small gain approach, the authors propose a parsimonious event-triggered design, which is able to prevent Zeno behavior and stabilize nonlinear distributed systems asymptotically. In [245], [247], event-triggered approaches are discussed within large-scale interconnected systems. By introducing a constant in the triggering condition, the authors prove that the system converges to a region around equilibrium without the occurrence of Zeno behavior. In [251] the authors consider a scenario where malicious attacks and genuine packet losses coexist, where the effect of malicious attacks and random packet losses are merged and characterized by an overall packet drop ratio. In [111] the authors formulate a two-player zero-sum stochastic game framework to consider a remote secure estimation problem where the signals are transmitted over a multichannel network under DoS attacks. A problem similar to zero-sum games between controllers and strategic jammers is considered in [253]. In [116] the authors investigate DoS from the attacker's viewpoint, and the objective is to consume limited energy and maximize the effect induced by DoS attacks. The paper [102] considers a stabilization problem where transmissions are event-based and the network is corrupted by periodic DoS attacks. In [103] and [89] a framework is introduced where DoS attacks are characterized by frequency and duration. The contribution is an explicit characterization of DoS frequency and duration under which stability can be preserved through state-feedback control. Extensions have been considered dealing with dynamic controllers [255], [256] and nonlinear systems [104].

In this chapter we consider networked distributed systems under DoS attacks, which has not yet been investigated under the class of DoS attacks introduced in [103] and [89]. Previously in [103,89,255,256,104] the authors analyzed the behavior of systems in a centralized-system manner, where the major characteristic is that all the states are assumed to be collected and sent in one transmission attempt. In this chapter, we analyze the problem from the distributed system point of view, where the interconnected subsystems share one communication channel and transmission attempts of the subsystems take place asynchronously. The contribution of this chapter is twofold. First, we consider a simple but typical scenario where the communication sequence is purely round-robin, and we explicitly compute a bound on attack frequency and duration under which the large-scale system is asymptotically stable. Second, trading off between system resilience and communication load, we design a hybrid transmission strategy. Specifically, in the absence of DoS attacks, we design a distributed event-triggered control using small gain argument, which guarantees the practical stability of the closed-loop system while preventing the occurrence of Zeno behavior. During DoS active periods, communication switches to round-robin, aiming to quickly restore communications. This hybrid communication strategy surprisingly ends up with the same bound as pure round-robin transmission, but promotes the possibility of saving communication resources.

5.2 Networked distributed system

Consider a large-scale system consisting of N interacting subsystems, whose dynamics satisfy

$$\dot{x}_i(t) = A_i x_i(t) + B_i u_i(t) + \sum_{j\in N_i} H_{ij} x_j(t), \tag{5.1}$$

where A_i, B_i, and H_{ij} are matrices with appropriate dimensions; $t \in \mathbb{R}_{>0}$; and $x_i(t)$ and $u_i(t)$ are state and control inputs of subsystem i, respectively. Here we assume that all the subsystems are full state output. N_i denotes the set of neighbors of subsystem i. Subsystem i physically interacts through $\sum_{j\in N_i} H_{ij}x_j(t)$ with its neighbor subsystem(s) $j \in N_i$. Here we consider bidirectional edges (i.e., $j \in N_i$ when $i \in N_j$).

The distributed systems are controlled via a shared networked channel, through which distributed plants broadcast the measurements and controllers send control inputs. The computation of control inputs is based on the transmitted measurements. The received measurements are in sample-and-hold fashion such as $x_i(t_k^i)$, where t_k^i represents the sequence of transmission instants of subsystem i. We assume that there exists a feedback matrix K_i such that $\phi_i = A_i + B_i K_i$ is Hurwitz. Therefore, the control input applied to subsystem i is given by

$$u_i(t) = K_i x_i(t_k^i) + \sum_{j\in N_i} L_{ij} x_j(t_k^j), \tag{5.2}$$

where L_{ij} is the coupling gain in the controller. Here we assume that the channel is noiseless and there is no quantization. Moreover, we assume that the network transmission delay and the computation time of control inputs are zero.

5.2.1 Denial-of-service attacks–frequency and duration

We refer to DoS as the phenomenon for which transmission attempts may fail. In this chapter we do not distinguish between transmission failures due to channel unavailability and transmission failures because of DoS induced packet corruption. Since the network is shared, DoS simultaneously affects the communication attempts of all the subsystems.

Clearly the problem in question does not have a solution if the DoS amount is allowed to be arbitrary. Following [89], we consider a general DoS model that constrains the attacker action in time by only posing limitations on the frequency of DoS attacks and their duration. Let $\{h_n\} n \in N_0$, $h_0 \geq 0$, denote the sequence of DoS *off/on* transitions, in other words the time instants at which DoS exhibits a transition from zero (transmissions are possible) to one (transmissions are not possible). Hence,

$$H_n := \{h_n\} \cup [h_n, h_n + \tau_n[\tag{5.3}$$

represents the n-th DoS time-interval, of a length $\tau_n \in \mathbb{R}_{\geq 0}$, over which the network is in DoS status. If $\tau_n = 0$, then H_n takes the form of a single pulse at h_n. If $\tau_n \neq 0$, $[h_n, h_n + \tau_n[$ represents an interval from the instant h_n (including h_n) to $(h_n + \tau_n)^-$ (arbitrarily close to but excluding $h_n + \tau_n$). Similarly, $[\tau, t[$ represents an interval from τ to t^-. Given $\tau, t \in \mathbb{R}_{\geq 0}$ with $t \geq \tau$, let $n(\tau, t)$ denote the number of DoS off/on transitions over $[\tau, t[$, and let

$$\Xi(\tau, t) := \bigcup_{n \in \mathbb{N}_0} H_n \bigcap [\tau, t] \tag{5.4}$$

denote the subset of $[\tau, t]$ where the network is in DoS status. The subset of time where DoS is absent is denoted by

$$\Theta(\tau, t) := [\tau, t] \setminus \Xi(\tau, t). \tag{5.5}$$

We make the following assumptions.

Assumption 5.1 (DoS frequency). There exist constants

$$\Xi(\tau, t) := \bigcup_{n \in \mathbb{N}_0} H_n \bigcap [\tau, t], \quad \eta \in \mathbb{R}_{\geq 0}, \quad \tau_D \in \mathbb{R}_{>0}$$

such that

$$n(\tau, t) \leq \eta + \frac{t - \tau}{\tau_D} \tag{5.6}$$

for all $\tau, t \in \mathbb{R}_{\geq 0}$ with $t \geq \tau$.

Assumption 5.2 (DoS duration). There exist constants $\kappa \in \mathbb{R}_{\geq 0}$ and $T \in \mathbb{R}_{>1}$ such that

$$|\Xi(\tau, t)| \leq \kappa + \frac{t - \tau}{T} \tag{5.7}$$

for all $\tau, t \in \mathbb{R}_{\geq 0}$ with $t \geq \tau$.

Remark 5.1. Assumptions 5.1 and 5.2 only constrain a given DoS signal in terms of its average frequency and duration. Actually, τ_D can be defined as the average dwell time between consecutive DoS off/on transitions, while η is the chattering bound. Assumption 5.2 expresses a similar requirement with respect to the duration of DoS. It expresses the property that, on average, the total duration of interrupted communication does not exceed a certain fraction of time, as specified by $1/T$. Like η, the constant κ plays the role of a regularization term. It is needed because during a DoS interval, we have $|\Xi(h_n, h_n + \tau_n)| = \tau_n > \tau_n / T$. Thus κ serves to make (5.7) consistent. Conditions $\tau_D > 0$ and $T > 1$ imply that DoS cannot occur at an infinitely fast rate or always be active.

5.3 Analytical results

In this section, our objective is to find stability conditions for the networked distributed systems under DoS attacks. We first study the stabilization problem of large-scale systems under a digital communication channel in the absence of DoS.

5.3.1 A small-gain approach

For each subsystem i, we denote as $e_i(t)$ the error between the value of the state transmitted to its neighbors and the current state:

$$e_i(t) = x_i(t_k^i) - x_i(t), \quad i = 1, 2, \cdots, N. \tag{5.8}$$

Then combining (5.1), (5.2), and (5.8), the dynamics of subsystem i can be written as

$$\begin{aligned} \dot{x}_i(t) \quad = \quad & \Phi_i x_i(t) + B_i K_i e_i(t) \\ & + \sum_{j \in N_i} (B_i L_{ij} + H_{ij}) x_j(t) \\ & + B_i \sum_{j \in N_i} L_{ij} e_j(t) \end{aligned} \tag{5.9}$$

from which it can be seen that the dynamics of subsystem i depends on the interconnected neighbors $x_j(t)$ and on as $e_i(t)$, $e_j(t)$, and the coupling parameters. Intuitively, if the couplings are weak and e remains small, then stability can be achieved. Here the notion of "smallness" of e can be characterized by the x-dependent bound $||e_i(t)|| \leq \sigma_i ||x_i(t)||$, where σ_i is a suitable design parameter. Note that this is not the network update rule.

We implement a periodic sampling protocol (e.g., round-robin) as our update law. In this respect, we make the following hypothesis.

Assumption 5.3 (Intersampling of round-robin). In the absence of DoS attacks, there exists an intersampling interval Δ such that

$$||e_i(t)|| \leq \sigma_i ||x_i(t)|| \tag{5.10}$$

holds, where σ_i is a suitable design parameter.

For centralized settings, values of Δ satisfying a bound like (5.10) can be explicitly determined. On the other hand, in [248] and [257] the authors compute and apply a lower bound of time elapsed between two events to prevent Zeno behavior, where the distributed or decentralized systems are asymptotically stable. The problem of obtaining Δ is left for future research.

As mentioned in the above argument, σ_i should be designed carefully. Otherwise, even if there exists a Δ for which (5.10) holds, in the event of an inappropriate σ_i, stability can be lost as well.

Given any symmetric positive definite matrix Q_i, let P_i be the unique solution of the Lyapunov equation $\Phi_i^T P_i + P_i \Phi_i + Q_i = 0$. For each i, consider the Lyapunov function $V_i = x_i^T P_i x_i$, which satisfies

$$\lambda_{min}(P_i)||x_i(t)||^2 \leq V_i(x_i(t)) \leq \lambda_{max}(P_i)||x_i(t)||^2, \tag{5.11}$$

where $\lambda_{min}(P_i)$ and $\lambda_{max}(P_i)$ represent the smallest and largest eigenvalue of P_i, respectively. The following lemma presents the design of σ_i guaranteeing stability.

Lemma 5.1. *Consider a distributed system as in (5.1) along with a control input as in (5.2). Suppose that the spectral radius $r(A^{-1}B) < 1$. The distributed system is asymptotically stable if σ_i satisfies*

$$\sigma_i < \sqrt{\frac{l_i}{j_i}}, \tag{5.12}$$

where l_i is the i-th entry of row vector $L := \mu^T(A - B) = [l_1, l_2, \cdots, l_N]$ and j_i is the j-th entry of row vector $J := \mu^T \Gamma [j_1, j_2, \cdots, j_N]$. $\mu \in \mathbb{R}_+^N$ is an arbitrary column vector satisfying $\mu^T(-A + B) < 0$. The matrices A, B, and Γ are given by

$$A = \begin{bmatrix} \alpha_1 & & \\ & \ddots & \\ & & \alpha_N \end{bmatrix}, \tag{5.13}$$

$$B = \begin{bmatrix} 0 & \beta_{12} & \cdots & \beta_{1N} \\ \beta_{21} & 0 & \cdots & \beta_{2N} \\ \vdots & \vdots & \ddots & \vdots \\ \beta_{N1} & \beta_{2N} & \cdots & 0 \end{bmatrix}, \tag{5.14}$$

$$\Gamma = \begin{bmatrix} 0 & \gamma_{12} & \cdots & \gamma_{1N} \\ \gamma_{21} & 0 & \cdots & \gamma_{2N} \\ \vdots & \vdots & \ddots & \vdots \\ \gamma_{N1} & \gamma_{2N} & \cdots & 0 \end{bmatrix}, \tag{5.15}$$

with

$$\alpha_i = \lambda_{min}(Q_i) - \delta, \tag{5.16}$$

$$\beta_{ij} = \frac{||P_i||^2||B_i L_{ij} + H_{ij}||^2}{\delta}, \tag{5.17}$$

$$\gamma_{ii} = \frac{||P_i||^2||B_i K_i||^2}{\delta}, \tag{5.18}$$

$$\gamma_{ij} = \frac{||P_i||^2||B_iL_{ij}||^2}{\delta}, \tag{5.19}$$

where δ is positive and real, such that $\alpha_i > 0$ and $\lambda_{min}(Q_i)$ is the smallest eigenvalue of Q_i for $i = 1, 2, \cdots, N$.

Proof. Recalling that $V_i = x_i^T P_i x_i$, the derivative of V_i along the solution to (5.9) satisfies

$$\begin{aligned} \dot{V}_i(x_i(t)) \leq\ & -\lambda_{min}(Q_i)||x_i(t)||^2 \\ & +||2P_iB_iK_i||\ ||x_i(t)||\ ||e_i(t)|| \\ & +\sum_{j\in N_i} ||2P_i(B_iL_{ij}+H_{ij})||\ ||x_i(t)||\ ||x_j(t)|| \\ & +\sum_{j\in N_i} ||2P_iB_iL_{ij}||\ ||x_i(t)||\ ||e_j(t)||. \end{aligned} \tag{5.20}$$

Observe that for any positive real δ the Young inequalities yield

$$\begin{aligned} & ||2P_iB_iK_i||\ ||x_i(t)||\ ||e_i(t)|| \\ \leq\ & \delta||x_i(t)||^2 + \frac{||P_i||^2||B_iK_i||^2}{\delta}||e_i(t)||^2 \end{aligned} \tag{5.21}$$

$$\begin{aligned} & ||2P_i(B_iL_{ij}+H_{ij})||\ ||x_i(t)||\ ||x_j(t)|| \\ \leq\ & \delta||x_i(t)||^2 + \frac{||P_i||^2||B_iL_{ij}+H_{ij}||^2}{\delta}||x_j(t)||^2 \end{aligned} \tag{5.22}$$

$$\begin{aligned} & ||2P_iB_iL_{ij}||\ ||x_i(t)||\ ||e_j(t)|| \\ \leq\ & \delta||x_i(t)||^2 + \frac{||P_i||^2||B_iL_{ij}||^2}{\delta}||e_j(t)||^2. \end{aligned} \tag{5.23}$$

Hence, the derivative of V_i along the solution to (5.9) satisfies

$$\begin{aligned} \dot{V}_i(x(t)) \leq\ & -\alpha_i||x_i(t)||^2 + \sum_{j\in N_i} \beta_{ij}||x_j(t)||^2 \\ & +\gamma_{ii}||e_i(t)||^2 + \sum_{j\in N_i} \gamma_{ij}||e_j(t)||^2, \end{aligned} \tag{5.24}$$

where α_i, β_{ij}, γ_{ii}, and γ_{ij} are as in Lemma 5.1. Note that we can always find a δ such that $\alpha_i > 0$ for $i = 1, 2, \cdots, N$.

By defining vectors

$$\begin{aligned} V_{vec}(x_i(t)) &:= [V_1(x_1(t)), V_2(x_2(t)), \cdots, V_N(x_N(t))]^T, \\ ||x(t)||_{vec} &:= [||x_1(t)||^2, ||x_2(t)||^2, \cdots, ||x_N(t)||^2]^T, \\ ||e(t)||_{vec} &:= [||e_1(t)||^2, ||e_2(t)||^2, \cdots, ||e_N(t)||^2]^T, \end{aligned}$$

the inequality (5.24) can be compactly written as

$$\dot{V}_{vec}(x_i(t)) \leq (-A+B)||x(t)||_{vec} + \Gamma||e(t)||_{vec} \tag{5.25}$$

with A, B, and Γ being as in Lemma 5.1.

If the spectral radius satisfies $r(A^{-1}B) < 1$, there exists a positive vector $\mu \in \mathbb{R}^n_+$, such that $\mu^T(-A+B) < 0$. We refer readers to [258] for more details. We select the Lyapunov function $V(x(t)) := \mu^T V_{vec}(x_i(t))$. Then the derivative of V yields

$$\begin{aligned} \dot{V}(x(t)) &= \mu^T \dot{V}_{vec}(x_i(t)) \\ &\leq \mu^T(-A+B)||x(t)||_{vec} + \mu^T\Gamma||e(t)||_{vec}. \end{aligned} \tag{5.26}$$

By noticing that $\mu^T(-A+B) < 0$, we have

$$\dot{V}(x(t)) \leq -L||x(t)||_{vec} + J||e(t)||_{vec}, \tag{5.27}$$

where $L := \mu^T(A-B)$ and $J := \mu^T\Gamma$ are row vectors. We denote l_i and j_i as the entries of L and J, respectively. Then (5.27) yields

$$\begin{aligned} \dot{V}(x(t)) &\leq -\sum_{i\in N} l_i||x(t)||^2 + \sum_{i\in N} j_i||e(t)||^2 \\ &= -\sum_{i\in N}\left(l_i||x(t)||^2 - j_i||e(t)||^2\right), \end{aligned} \tag{5.28}$$

which implies asymptotic stability with $\sigma_i < \sqrt{\frac{l_i}{j_i}}$. □

Remark 5.2. Lemma 5.1 can only deal with the case where $j_i > 0$. The case $j_i = 0$ is only possible whenever every entry in the column i of Γ is zero. In fact, $j_i = 0$ implies that the error $||e_i(t)||$ never contributes to the system dynamics via (5.28), which in turn implies that $||e_i(t)||$ does not affect stability at all. Therefore, in the case $j_i = 0$, no constraint on $||e_i(t)||$ is imposed.

5.3.2 Stabilization under denial of service

In the previous analysis, we introduced the design of a suitable σ_i and hence error bound under which the system is asymptotically stable in the absence of DoS. By hypothesis, we also assumed the existence of a round-robin transmission that satisfies such an error bound. In the presence of DoS, (5.10) is possibly violated even though the sampling strategy is still round-robin. Under these circumstances stability can be lost. Hence, we are interested in the stabilization problem when the round-robin network is under DoS attacks.

Theorem 5.1. *Consider a distributed system as in (5.1) along with a control input as in (5.2). The plant-controller information exchange takes place over a* *shared network in which the communication protocol is round-robin with sampling interval Δ, as in Assumption 5.3. The large-scale system is asymptotically stable for any DoS sequence satisfying Assumptions 5.1 and 5.2 with arbitrary η and κ, and with τ_D and T if*

$$\frac{1}{T} + \frac{\Delta_*}{\tau_D} < \frac{\omega_1}{\omega_1 + \omega_2}, \tag{5.29}$$

in which $\Delta_* = N\Delta$, $\omega_1 := min\{\frac{l_i - \sigma_i^2 j_i}{\lambda_{max}(P_i)\mu_i}\}$, and $\omega_2 := \frac{4max\{j_i\}}{min\{\mu_i \lambda_{min}(P_i)\}}$. l_i, j_i, μ_i and σ_i are as in Lemma 5.1.

Proof. The proof is divided into three steps:

Step 1. Lyapunov function in DoS-free periods. In DoS-free periods, by hypothesis of Assumption 5.3, (5.10) holds true with σ_i as in Lemma 5.1, and (5.28) is negative. Therefore, the derivative of the Lyapunov function satisfies

$$\begin{aligned} \dot{V}(x(t)) &\leq -\sum_{i \in N}(l_i - j_i\sigma_i^2)||x_i(t)||^2 \\ &\leq -\sum_{i \in N}\frac{l_i - \sigma_i^2 j_i}{\lambda_{max}(P_i)\mu_i}\mu_i V_i \\ &= -\omega_1 V, \end{aligned} \tag{5.30}$$

where $\omega_1 := min\{\frac{l_i - \sigma_i^2 j_i}{\lambda_{max}(P_i)\mu_i}\}$. Thus for $t \in [h_n + \tau_n, h_n + 1[$ (DoS-free time), the Lyapunov function yields

$$V(x(t)) \leq e^{-\omega_1(t - h_n - \tau_n)} V(x(h_n + \tau_n)). \tag{5.31}$$

Step 2. Lyapunov function in DoS active periods. Here we let z_m^i denote the last successful sampling instant before the occurrence of DoS. Recalling the definition of $e_i(t)$, we obtain that

$$e_i(t) = x_i(z_m^i) - x_i(t) = x_i(h_n) - x_i(t) \tag{5.32}$$

and

$$||e_i(t)||^2 \leq ||x_i(h_n)||^2 + 2||x_i(t)||\,||x_i(h_n)|| + ||x_i(t)||^2 \tag{5.33}$$

for $t \in H_n$. By summing up $||e_i(t)||^2$ for $i \in N$, we obtain

$$\begin{aligned} \sum_{i \in N}||e_i(t)||^2 &\leq \sum_{i \in N}||x_i(h_n)||^2 + \sum_{i \in N}||x_i(t)||^2 \\ &\quad + \sum_{i \in N}\left(||x_i(h_n)||^2 + ||x_i(t)||^2\right) \\ &= 2\sum_{i \in N}||x_i(h_n)||^2 + 2\sum_{i \in N}||x_i(t)||^2. \end{aligned} \tag{5.34}$$

If

$$\sum_{i\in N} ||x_i(h_n)||^2 \leq \sum_{i\in N} ||x_i(t)||^2,$$

we have

$$\sum_{i\in N} ||e_i(t)||^2 \leq 4\sum_{i\in N} ||x_i(t)||^2.$$

Otherwise, we have

$$\sum_{i\in N} ||e_i(t)||^2 \leq 4\sum_{i\in N} ||x_i(h_n)||^2.$$

Recalling (5.28), it is simple to see that

$$\dot{V}(x(t)) \leq \sum_{i\in N} j_i ||e_i(t)||^2. \tag{5.35}$$

Thus, for all $t \in H_n$ (DoS active time) in the case that $\sum_{i\in N} ||x_i(h_n)||^2 \leq \sum_{i\in N} ||x_i(t)||^2$, the derivative of the Lyapunov function yields

$$\begin{aligned} \dot{V}(x(t)) &\leq max\{j_i\}\sum_{i\in N} ||e_i(t)||^2 \\ &\leq 4max\{j_i\}\sum_{i\in N} ||x_i(t)||^2 \\ &\leq \frac{4max\{j_i\}}{min\{\mu_i\lambda_{min}(P_i)\}}\sum_{i\in N} \mu_i V(x_i(t)) \\ &= \omega_2 V(x(t)) \end{aligned} \tag{5.36}$$

with $\omega_2 := \frac{4max\{j_i\}}{min\{\mu_i\lambda_{min}(P_i)\}}$. On the other hand, for all $t \in H_n$ such that

$$\sum_{i\in N} ||x_i(h_n)||^2 > \sum_{i\in N} ||x_i(t)||^2$$

we have

$$\dot{V}(x(t)) \leq \omega_2 V(x(h_n)). \tag{5.37}$$

Thus, (5.36) and (5.37) imply the Lyapunov function during H_n satisfies

$$V(x(t)) \leq e^{\omega_2(t-h_n)} V(x(h_n)). \tag{5.38}$$

Step 3. Switching between stable and unstable modes. Consider a DoS attack with period τ_n, at the end of which the overall system has to wait an additional period with length $N\Delta$ to have a full round of communications. Hence, the period when at least one subsystem transmission is not successful can be upper bounded by $\tau_n + N\Delta$. For all $\tau, t \in \mathbb{R}_{\geq 0}$ with $t \geq \tau$, the total length when communication is not possible over $[\tau, t[$, say $|\bar{\Xi}(\tau, t)|$, can be upper bounded by

$$\begin{aligned} |\bar{\Xi}(\tau, t)| &\leq |\Xi(\tau, t)| + (1 + n(\tau, t))\Delta_* \\ &\leq \kappa_* + \frac{t-\tau}{T_*}, \end{aligned} \tag{5.39}$$

where $\Delta_* = N\Delta$, $\kappa_* := \kappa + (1+\eta)\Delta_*$, and $T_* := \frac{\tau_D T}{\tau_D T + T\Delta_*}$. Considering the additional waiting time due to round-robin, the Lyapunov function in (5.31) yields $V(x(t)) \leq e^{-\omega_1(t-h_n-\tau_n-N\Delta)} V(h_n + \tau_n + N\Delta)$ for $t \in [h_n + \tau_n + N\Delta, h_{n+1}[$ and $V(x(t)) \leq e^{\omega_2(t-h_n)} V(h_n)$ for $t \in [h_n, h_n + \tau_n + N\Delta[$.

Thus, the overall behavior of the closed-loop system can be regarded as a switching system with two modes. Applying simple iterations to the Lyapunov functions in and out of DoS status, we have

$$\begin{aligned} V(x(t)) &\leq e^{-\omega_1|\bar{\Theta}(0,t)|} e^{2|\bar{\Xi}(0,t)|} V(x(0)) \\ &\leq e^{\kappa_*(\omega_1+\omega_2)} e^{-\beta_* t} V(x(0)), \end{aligned} \tag{5.40}$$

where $\beta_* := \omega_1 - (\omega_1 + \omega_2)(\frac{\Delta_*}{\tau D} + \frac{1}{T})$. By constraining $\beta_* < 0$, we obtain the desired result in (5.29). Hence, stability is implied at once. □

Remark 5.3. The resilience of the distributed systems depends on the largeness of ω_1 and the smallness of ω_2. To achieve this, we can try to find K_i and L_{ij} such that $||B_i K_i||$ and $||B_i L_{ij}||$ are small. On the other hand, the sampling interval of round-robin also affects stability in the sense that it determines how quickly the overall system can restore the communication. We can always apply smaller round-robin intersampling time to reduce the left-hand side of (5.29) at the expense of higher communication load.

5.4 Approximation of resilience with reduced communication

In the above argument (cf. Remark 5.3), we show that system resilience depends on the round-robin sampling. The faster the round-robin sampling, the more quickly the overall system restores communication. On the other hand, in DoS-free periods we are interested in the possibility of reducing the communication load while maintaining the comparable robustness attained earlier. To realize this we propose a hybrid transmission strategy: in the absence of DoS the communications of the distributed systems are event-based; if DoS occurs the communications switch to round-robin until the moment when every subsystem has one successful update.

The advantage of event-triggered control is that it saves communication resources. However, the effectiveness of prolonging transmission intervals, in turn, appears to be a disadvantage in the presence of DoS. The main shortcoming is that event-triggered control could potentially prolong DoS status. For example, consider that the sampling strategy is purely event-based. After a DoS attack there is a short period where communications are possible, during which the error bounds as in (5.10) are not violated so that systems do not update. If the DoS appears quickly, this is equivalent to the scenario that systems face a longer DoS attack. This indicates that a better strategy is to save communications in the absence of DoS and restore communications as soon as possible when DoS is over, which indeed leads to a hybrid communication strategy.

5.4.1 Zeno-free event-triggered control

Abusing the notation, in this section we denote as $\{t_k^i\}$ the triggering time sequence of subsystem i under an event-triggered control scheme. For a given initial condition $x_i(0)$, if t_k^i converges to a finite t^{i*}, we say that the event-triggered control induces Zeno behavior [248], [249]. Hence, Zeno-freeness implies an event-triggered control scheme preventing the occurrence of Zeno behavior. The following lemma addresses the Zeno-free event-triggered control.

Lemma 5.2. *Consider a distributed system as in (5.1) along with a control input as in (5.2). Suppose that the spectral radius $r(A^{-1}B) < 1$. In the absence of DoS the distributed system is practically stable and Zeno-free if the event-triggered law satisfies*

$$||e_i(t)|| \leq max\{\sigma_i ||x_i(t)||, c_i\}, \tag{5.41}$$

in which c_i is positive, finite, and real, and

$$\sigma_i < min\{\sqrt{\frac{l_i}{j_i}}, 1\}, \tag{5.42}$$

where l_i and j_i are the same as in Lemma 5.1.

Proof. According to Lemma 5.1, if the spectral radius $r(A^{-1}B) < 1$, (5.28) holds true. Then we observe that the event-triggered control law (5.41) would lead (5.28) to

$$\begin{aligned}
\dot{V}(x(t)) &\leq -\sum_{i \in N} \Big(l_i ||x_i(t)||^2 - j_i max\{\sigma^2 ||x_i(t)||^2, c_i^2\} \Big) \\
&\leq max\{-\sum_{i \in N}(l_i - j_i\sigma_i^2)||x_i(t)||^2, \\
&\quad -\sum_{i \in N} l_i ||x_i(t)||^2 + \sum_{i \in N} j_i c_i^2\} \\
&\leq -\sum_{i \in N}(l_i - j_i\sigma_i^2)||x_i(t)||^2 + \sum_{i \in N} j_i c_i^2,
\end{aligned} \tag{5.43}$$

which implies practical stability with $\sigma_i < min\{\sqrt{\frac{l_i}{j_i}}, 1\}$ and finite c_i.

Then we introduce the analysis about Zeno-freeness of this distributed event-triggered control law. Since $\dot{e}_i(t) = -\dot{x}_i(t)$, then the dynamics of e_i satisfy

$$\begin{aligned}\dot{e}_i(t) &= A_i e_i(t) - \Phi_i x_i(t_k^i) - \sum_{i\in N}(B_i L_{ij} + H_{ij})x_j(t_k^j) \\ &+\sum_{i\in N} H_{ij} e_j(t). \end{aligned} \tag{5.44}$$

From the triggering law (5.41), we can obtain

$$||x_i(t_k^i) - x_i(t)||^2 \leq max\{\sigma_i ||x_i(t)||, c_i\},$$

and further calculations yield $||x_i(t)|| - ||x_i(t_k^i)|| \leq \sigma_i ||x_i(t)|| + c_i$. Thus, it is simple to verify that $||e_i(t)|| \leq \bar{\sigma}_i ||x_i(t_k^i)|| + \bar{\sigma}_i c_i$, where $\bar{\sigma}_i := \frac{\sigma_i}{1-\sigma_i}$.

For each i, at the instant t_{k+1}^i, $||e_i(t)||$ satisfies

$$\begin{aligned}||e_i(t_{k+1}^i)|| \leq\ & f_i ||\Phi_i||\ ||x_i(t_k^i)|| \\ &+ f_i \sum_{j\in N_i} ||B_i L_{ij} + H_{ij}||m \\ &+ f_i \sum_{j\in N_i} ||H_{ij}||\bar{\sigma}_j (m + c_j), \end{aligned} \tag{5.45}$$

where $f_i := \int_{t_k^i}^{t_{k+1}^i} e^{A(t_{k+1}^i - \tau)} d\tau$, $m = max\{||x_j(t_p^j)||\}$ for $t_k^i \leq t_p^j < t_{k+1}^i$ and $j \in N_i$. Meanwhile the triggering law in (5.41) implies that $||e_i(t_{k+1}^i)|| \geq c_i$. Then we immediately see that

$$\begin{aligned} t_{k+1}^i - t_k^i &\geq z_i, && \text{if } \mu_{A_i} \leq 0, \\ t_{k+1}^i - t_k^i &\geq \frac{1}{\mu_{A_i}} log(z_i \mu_{A_i} + 1), && \text{if } \mu_{A_i} > 0, \end{aligned} \tag{5.46}$$

in which

$$z_i := \frac{c_i}{||\Phi_i||\ ||x_i(t_k^i)|| + m\sum_{j\in N_i} \zeta_{ij} + \sum_{j\in N_i} ||H_{ij}||\bar{\sigma}_j c_j},$$

where $\zeta_{ij} := ||B_i L_{ij} + H_{ij}|| + ||H_{ij}||\bar{\sigma}_j$ and μ_{A_i} is the logarithmic norm of A_i. Note that the system is practically stable, so that $||x_i(t_k^i)||$ and m are bounded. This implies that $z_i > 0$ and hence $t_{k+1}^i - t_k^i > 0$. □

5.4.2 Hybrid transmission strategy under DoS

As a counterpart of Assumption 5.3, here we assume that there exists a round-robin sampling interval Δ satisfying (5.41). Now we are ready to present the following result.

Theorem 5.2. *Consider a distributed system as in (5.1) along with a control input as in (5.2). The plant-controller information exchange takes place over a shared network implementing the event-triggered control law (5.41) in the absence of DoS. Suppose that there exists a round-robin sampling interval Δ such that (5.41) holds. The network is subject to DoS attacks regulated by Assumptions 5.1 and 5.2, during which the communication switches to round-robin until every subsystem updates successfully. Then the distributed system is practically stable if (5.29) holds true.*

Proof. Similar to the proof of Theorem 5.1, considering the additional waiting time $N\Delta$ due to round-robin for the restoring of communications, in DoS-free periods the Lyapunov function satisfies

$$V(x(t)) \leq e - \omega_1(t - h_n - \tau_n - N\Delta)V(x(h_n + \tau_n + N\Delta)) + \frac{c}{\omega_1} \tag{5.47}$$

for $t \in [h_n + \tau_n + N\Delta, h_{n+1}[$, where ω_1 is as in Theorem 5.1 and $c := \sum_{i=1}^{N} j_i c_i^2$. On the other hand, (5.38) still holds for $t \in [h_n, h_n + \tau_n + N\Delta[$.

Applying the very similar calculation to that in Step 3 in the proof of Theorem 5.1, we obtain

$$\begin{aligned} V(x(t)) &\leq e^{-\omega_1|\bar{\Theta}(0,t)|}e^{\omega_2|\bar{\Xi}(0,t)|}V(x(0)) \\ &\quad + \sum_{n=0}^{q} e^{-\omega_1|\bar{\Theta}(h_n,t)|}e^{\omega_2|\bar{\Xi}(h_n,t)|}\frac{c}{\omega_1} + \frac{c}{\omega_1} \\ &\leq e^{\kappa_*(\omega_1+\omega_2)}e^{-\beta_* t}V(x(0)) \\ &\quad + e^{\kappa_*(\omega_1+\omega_2)}\sum_{n=0}^{q} e^{-\beta_*(t-h_n)}\frac{c}{\omega_1} + \frac{c}{\omega_1}, \end{aligned} \tag{5.48}$$

where $n \in \mathbb{N}_0$, $q := sup\{q \in \mathbb{N}_0 | h_q \leq t\}$ and β_* is as in the proof of Theorem 5.1. Note that $t - h_n \geq \tau_D n(h_n, t) - \tau_D \eta$ by exploiting Assumption 5.1. Then the Lyapunov function yields

$$\begin{aligned} V(x(t)) &\leq e^{\kappa_*(\omega_1+\omega_2)}e^{-\beta_* t}V(x(0)) \\ &\quad + e^{\kappa_*(\omega_1+\omega_2)+\beta*\tau_D\eta}\sum_{n=0}^{q} e^{-\beta_*\tau_D n(h_n,t)}\frac{c}{\omega_1} + \frac{c}{\omega_1}. \end{aligned} \tag{5.49}$$

Recalling the definition of Assumption 5.1, we have that $n(h_n, t) - n(h_{n+1}, t) \geq 1$ for $t \geq h_{n+1}$. This implies that

$$\sum_{n=0}^{q} e^{-\beta_*\tau_D n(h_n,t)} \leq \frac{1}{1 - e^{-\beta_*\tau_D}}. \tag{5.50}$$

Finally, (5.49) can be written as

$$\begin{aligned} V(x(t)) &\leq e^{\kappa_*(\omega 1+\omega 2)} e^{-\beta_* t} V(x(0)) \\ &+ \frac{e^{\kappa_*(\omega 1+\omega 2)+\beta_* \tau_D \eta}}{1-e^{-\beta_* \tau_D}} \frac{c}{\omega_1} + \frac{c}{\omega_1}. \end{aligned} \tag{5.51}$$

If (5.29) holds, it is simple to verify that $\beta_* < 0$, which implies practical stability. □

5.5 Simulation results

5.5.1 Simulation example 1

The numerical example is taken from [259]. The systems are open-loop unstable such that

$$\dot{x}_1(t) = x_1(t) + u_1(t) + x_2(t)$$
$$\dot{x}_2(t) = x_2(t) + u_2(t)$$

under distributed control inputs such that

$$u_1(t) = -4.5x_1(t_k^1) - 1.4x_2(t_k^2) u_2(t) = -6x_2(t_k^2) - x_1(t_k^1).$$

Solutions of the Lyapunov equation $\Phi_i^T P_i + P_i \Phi_i + Q_i = 0$ with $Q_i = 1$ $(i = 1, 2)$ yield $P_1 = 0.1429$ and $P_2 = 0.1$. The matrices are $A = [0.7\ \ 0;\ 0\ \ 0.9]$, $B = [0\ \ 0.0327;\ 0.1\ \ 0]$, and $\Gamma = [4.1327\ \ 0.4;\ 0.1\ \ 3.6]$ according to Lemma 5.1. From these parameters we obtain that the spectral radius $r(A^{-1}B) = 0.072$, $\sigma_1 < 0.3765$, and $\sigma_2 < 0.4657$. We let $\sigma_1 = \sigma_2 = 0.2$. Based on Assumption 5.3, we choose a round-robin sampling interval $\Delta = 0.01$ s.

With these parameters, we obtain the bound $\frac{\omega_1}{\omega_1+\omega_2} \approx 0.0175$ with $\omega_1 \approx 3.0149$ and $\omega_2 \approx 169.3061$. This implies that a maximum duty cycle of 1.75% of a sustained DoS would not destabilize our systems in the example. Actually, this bound is conservative. The systems under inspection can endure more DoS without losing stability. As shown in Fig. 5.1, lines represent states and gray stripes represent the presence of DoS. Over a simulation horizon of 20 s the DoS corresponds to parameters of $\tau_D \approx 1.8182$ and $T \approx 2.5$, and $\sim 40\%$ of transmission failures. According to (5.29), we obtain $\frac{\Delta_*}{\tau_D} + \frac{1}{T} = 0.411$, which violates the theoretical bound, but the system is still stable.

Meanwhile, the hybrid transmission strategy is able to reduce communications effectively. As shown in Fig. 5.1 the transmissions with the hybrid transmission strategy are only 10% of the transmissions with the pure round-robin strategy.

5.5.2 Simulation example 2

In this example, we consider a physical system in [260]. The system is composed of N inverted pendulums interconnected as a line by springs, whose states are

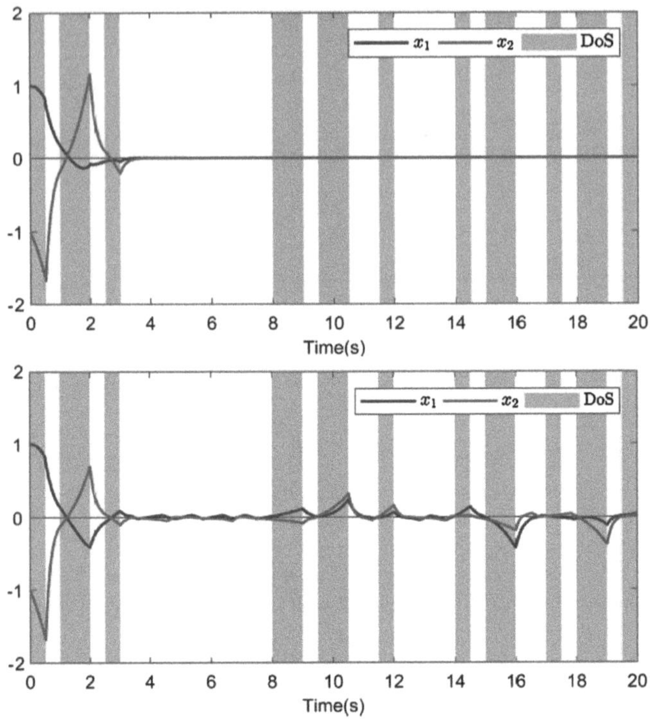

FIGURE 5.1 Example 1: Top: States under pure round-robin communication where there are 1200 transmissions in total; Bottom: States under hybrid communication strategy where there are 112 transmissions.

$x_i = [\bar{x}_i, \tilde{x}_i]^T$ for $i = 1, 2, \cdots, N$. Here we consider a simple case where $N = 3$. The parameters of the pendulums are

$$A_1 = A_3 = \begin{bmatrix} 0 & 1 \\ -3.75 & 0 \end{bmatrix}, \quad A_2 = \begin{bmatrix} 0 & 1 \\ -2.5 & 0 \end{bmatrix},$$

$$B_1 = B_2 = B_3 = \begin{bmatrix} 0 \\ 0.25 \end{bmatrix},$$

$$H_{12} = H_{21} = H_{23} = H_{32} = \begin{bmatrix} 0 & 0 \\ 1.25 & 0 \end{bmatrix}.$$

The parameter of designed controllers are given by

$$K_1 = K_3 = [-23 \quad -12], \quad K_2 = [-18 \quad -12],$$
$$L_{12} = L_{32} = [-5 \quad 0.25], \quad L_{21} = L_{23} = [-4.75 \quad -0.25].$$

With the solutions of the Lyapunov function $\Phi_i^T P_i + P_i \Phi_i + Q_i = 0$, where $Q_i = I$ and $i = 1, 2, 3$, we obtain

$$A = \begin{bmatrix} 0.67 & 0 & 0 \\ 0 & 0.45 & 0 \\ 0 & 0 & 0.67 \end{bmatrix},$$

$$B = \begin{bmatrix} 0 & 0.0608 & 0 \\ 0.1217 & 0 & 0.1217 \\ 0 & 0.0608 & 0 \end{bmatrix},$$

$$\Gamma = \begin{bmatrix} 47.7983 & 24.4007 & 0 \\ 22.0276 & 33.2386 & 22.0276 \\ 0 & 24.4007 & 47.7983 \end{bmatrix}.$$

With A, B, and Γ we obtain that $r(A^{-1}B) = 0.2216$, $\sigma_1 < 0.0646$, $\sigma_2 < 0.0844$, and $\sigma_3 < 0.0646$. We select $\sigma_1 = \sigma_2 = \sigma_3 = 0.01$. The round-robin sampling interval is chosen as $\Delta = 0.001$ s according to Assumption 5.3. Following the same procedure as in Example 1, we obtain $\frac{\omega_1}{\omega_1+\omega_2} \approx 0.00012$, which is consid-

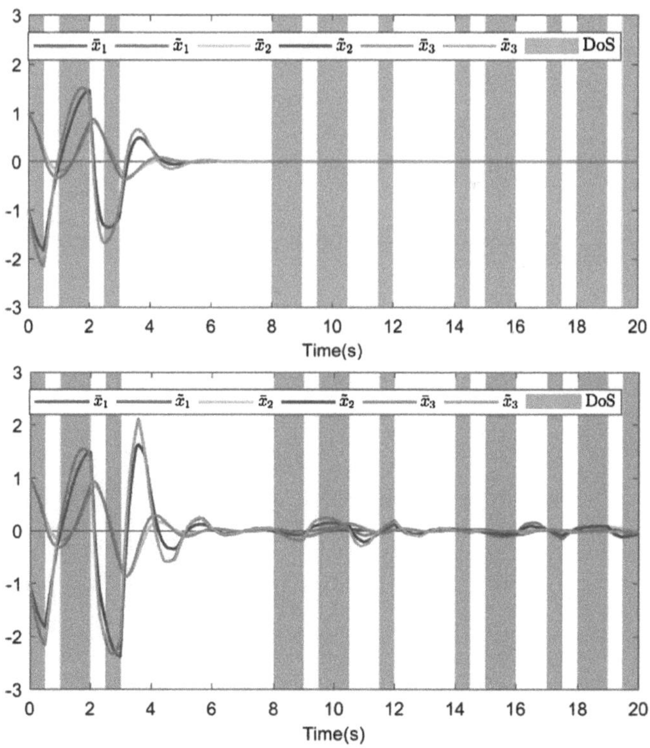

FIGURE 5.2 Example 2: Top: states under pure round-robin communication during which there are 11997 transmissions. Bottom: states under hybrid communication strategy where there are 254 transmissions.

erably conservative. In fact, if the systems are under the same DoS attacks as in Example 1 they are still stable, which can be seen from Fig. 5.2. The conservativeness is due to the unstable dynamics of the inverted pendulums, the feedback gain K_i, and the coupling parameter L_{ij} in the controllers. It is worth investigating how to design suitable K_i and L_{ij} to mitigate this effect (*cf. Remark 5.3*).

5.6 Notes

In this work, we investigated the problem of stabilizing distributed systems under Denial of Service, characterizing the DoS frequency and duration under which stability can be preserved. In order to save communication resources, we also considered a hybrid communication strategy. It turns out that the hybrid transmission strategy can reduce the communication load effectively and can prevent Zeno behavior while preserving the same robustness as a pure round-robin protocol. An interesting research direction is the stabilization problem of networked distributed systems, where only a fraction of subsystems, possibly time-varying, are under DoS. It is also interesting to investigate the problem where DoS attacks imposing on systems are asynchronous with different frequencies and durations. Finally, in the hybrid transmission strategy the effect of event-triggered control with communication collision can be an interesting direction from a practical viewpoint.

Chapter 6

False data injection attacks

Contents

6.1 Related work

Networked control systems (NCSs) are control systems in which the controller and the plant are connected via communication networks; there are many merits to this organization, such as simple installation and maintenance, reduced weight and power requirement, and high flexibility and reliability. However, the introduction of networks into the control loop inevitably causes some adverse effects such as network-induced delay and packet dropout, which may deteriorate the system performance or even destabilize the closed-loop system. Therefore, NCSs have become an active research topic in the past decade [149, 150,261–263].

Today NCSs have found numerous applications in various fields such as process control, intelligent transportation, and the measurement and control of critical infrastructures (e.g., electricity, water, and gas distribution). In these systems measurement data and control commands travel through the open and unprotected network, which are susceptible to be corrupted by attackers [264,122,265, 266]. For example, the typical malware such as Stuxnet and Duqu have been reported to disrupt the control systems of critical infrastructures [10]. These attacks may significantly hamper the economy and environment, and even endanger human lives. Therefore, the security of NCSs is of paramount importance for various applications.

Network attacks can be classified into two categories: 1) denial-of-service (DoS) attacks and 2) deception attacks [267,135,242]. The DoS attacks aim to obstruct the transmission of data. To handle them, some secure control schemes

Cloud Control Systems. https://doi.org/10.1016/B978-0-12-818701-2.00014-7

have been proposed in [90,268,269]. Deception attacks are implemented to compromise the integrity of data, which are usually more subtle and stealthy than DoS attacks. Typical deception attacks include data replay attacks and false data injection (FDI) attacks. Mo and Sinopoli [142] and Mo et al. [71] analyzed the performance of the control system under replay attacks and provided model-based countermeasures to improve the probability of attack detection.

The FDI attacks against the measurement data and control commands are to a certain degree similar to sensor faults and actuator faults, respectively. However, the faults are usually assumed to be random, independent events with a fixed failure-rate probability. On the contrary, FDI attacks can be carefully designed by smart attackers so as to cause the greatest possible damage without being detected, which may thus result in more serious consequences. In this case, such smart attacks would be difficult to detect by existing fault-detection techniques [270–272].

During the past five years, FDI attacks have received increasing attention. Mo and Sinopoli [273] proposed a simple FDI attack model to compromise the sensors of a linear control system. Manandhar et al. [274] showed that the FDI attack in [273] could be detected by the proposed Euclidean-based detector. Niu and Huie [275] analyzed the impact of the sensor FDI attack on the performance of the Kalman filter for linear dynamic systems. Teixeira et al. [276] studied the cyber security of state estimators in supervisory control and data acquisition systems, and showed that undetectable FDI attacks could be designed even when an attacker had limited resources. Kwon et al. [277] gave the conditions under which the FDI attacks on the sensors and/or actuators could fail the state estimators while successfully bypassing the monitoring system.

As can be seen, studies on stealthy FDI attacks are only in their embryonic stage. Furthermore, in the above-mentioned works [273–277], there exist some common drawbacks:

1. Not all of them are concerned with the network-induced constraints, although they are inevitable in practical NCSs;
2. In [273–276] only the FDI attacks on the measurement data are considered, and in [277], although the FDI attacks on both sensors and actuators are considered, only the case of open-loop control is investigated;
3. The theoretical results in [273–277] are tested only by numerical simulation.

The foregoing three factors motivated the present study.

Notation. The notations used here are fairly standard: $\Delta x(k)$ is defined as $\Delta x(k) = x(k) - x(k-1)$; $x(k+i|k)$ refers to the ith-step-ahead predictive value of $x(k)$ based on the data up to time k; $\mathbf{E}(.)$ denotes the mathematical expectation operation.

6.2 Kalman filter-based systems

A networked predictive output tracking control (NPOTC) system, as depicted in Fig. 6.1, consists of five parts: 1) a physical plant; 2) a data buffer in the sensor; 3) a communication network; 4) a control prediction generator in the controller; and 5) a network delay compensator in the actuator. Each part will be described in the following sections. It is assumed that the sensor and actuator are time-driven and synchronous, while the controller is event-driven.

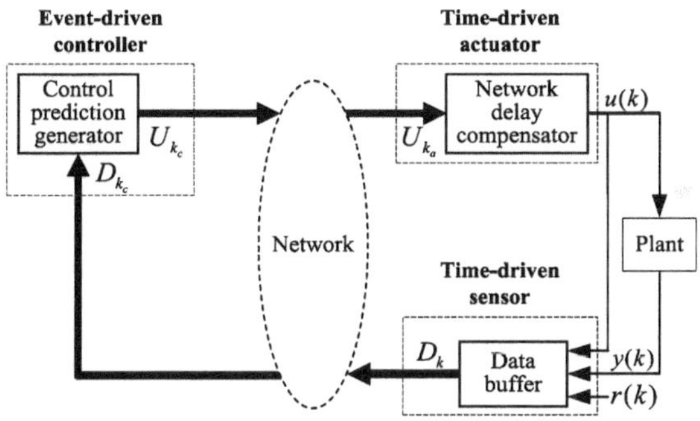

FIGURE 6.1 NPOTC systems.

6.2.1 Physical plant

Suppose that the physical plant in Fig. 6.1 is described by the linear system

$$\begin{aligned} x(k+1) &= Ax(k) + Bu(k) + \omega(k), \\ y(k) &= Cx(k) + \upsilon(k), \end{aligned} \tag{6.1}$$

where $x(k) \in \Re^n$ is the system state, $u(k) \in \Re^m$ is the control input, $y(k) \in \Re^q$ is the measurement output, $\omega(k) \in \Re^n$ is the system noise, and $\upsilon(k) \in \Re^q$ is the measurement noise. A, B, and C are system matrices with appropriate dimensions. $\omega(k)$ and $\upsilon(k)$ are the uncorrelated Gaussian white noise with

$$\omega(k) \sim \mathcal{N}(0, Q) \text{ and } \upsilon(k) \sim \mathcal{N}(0, R),$$

where Q and R are the covariance matrices. It is assumed that (A, C) is observable, (A, B) is controllable, and the matrix $\begin{bmatrix} A - I_n & B \\ C & 0_{q\times m} \end{bmatrix}$ has full row rank.

The incremental form of (6.1) is

$$\begin{aligned} \Delta x(k+1) &= A\Delta x(k) + B\Delta u(k) + \Delta\omega(k), \\ \Delta y(k) &= C\Delta x(k) + \Delta\upsilon(k). \end{aligned} \tag{6.2}$$

The output tracking error is defined as

$$e(k) = r(k) - y(k), \tag{6.3}$$

where $r(k) \in \Re^q$ is the reference input. It is obtained from (6.2) and (6.3):

$$\begin{aligned} e(k+1) &= e(k) - CA\Delta x(k) - CB\Delta u(k) + \Delta r(k+1) \\ &\quad -C\Delta\omega(k) - \Delta\upsilon(k+1). \end{aligned} \tag{6.4}$$

From (6.2) and (6.4), we obtain the augmented system

$$\begin{aligned} x_e(k+1) &= A_e x_e(k) + B_e\Delta u(k) + E_e\Delta r(k+1) \\ &\quad +W_e\Delta\omega(k) + V_e\Delta\upsilon(k+1), \\ \Delta y(k) &= C_e x_e(k) + \Delta\upsilon(k), \end{aligned} \tag{6.5}$$

where

$$\begin{aligned} x_e(k) &= \begin{bmatrix} \Delta x(k) \\ e(k) \end{bmatrix} \in \mathbb{R}^{\bar{n}}, \quad A_e = \begin{bmatrix} A & 0_{n\times q} \\ -CA & I_q \end{bmatrix}, \\ B_e &= \begin{bmatrix} B \\ -CB \end{bmatrix}, \quad E_e = \begin{bmatrix} 0_{n\times q} \\ I_q \end{bmatrix}, \quad W_e = \begin{bmatrix} I_n \\ -C \end{bmatrix}, \\ V_e &= \begin{bmatrix} 0_{n\times q} \\ -I_q \end{bmatrix}, \quad C_e = \begin{bmatrix} C & 0_{q\times q} \end{bmatrix}, \quad \bar{n} = n + q. \end{aligned}$$

Thus, the output tracking problem of system (6.1) can be solved by the feedback control of the augmented state $x_e(k)$.

6.2.2 Data buffer

In general, the full state of the plant is not directly measurable. To obtain the estimation of the state $x(k)$ in the controller, at each sampling instant k, the following data

$$D_k = \begin{bmatrix} y(k)^T & u(k-1)^T & R(k)^T \end{bmatrix}^T \tag{6.6}$$

are transmitted to the controller together with the timestamp k, where $R(k) = [r(k)^T \quad r(k+1)^T \quad \cdots \quad r(k+\bar{\tau})^T]^T$.

6.2.3 Communication network

The Ethernet-like network is considered in this chapter. The packets travel through the network from the sensor to the controller and then from the controller to the actuator. As a result, network-induced delays are inevitable during the packet transmission, which are generally random with unknown distribution. In this chapter, it is assumed that the round-trip time (RTT) delay τ_k is bounded by $\bar{\tau}$.

6.2.4 Control prediction generator

To obtain the state estimation $\hat{x}(k_c)$ the following Kalman filter is usually used [71]

$$\begin{cases} P_{k_c|k_c-1} = AP_{k_c-1}A^T + Q \\ K_{k_c} = P_{k_c|k_c-1}C^T(CP_{k_c|k_c-1}C^T + R)^{-1} \\ P_{k_c} = (I - K_{k_c}C)P_{k_c|k_c-1} \\ \hat{x}(k_c|k_c-1) = A\hat{x}(k_c-1) + Bu(k_c-1) \\ \hat{x}(k_c) = \hat{x}(k_c|k_c-1) + K_{k_c}\left(y(k_c) - C\hat{x}(k_c|k_c-1)\right) \end{cases} \tag{6.7}$$

with the initial conditions

$$\hat{x}(0) = \mathbf{E}(x(0)),\ P_0 = \mathbf{E}\left((x(0) - \hat{x}(0))(x(0) - \hat{x}(0))^T\right),$$

where $k_c \leq k$ is the timestamp of the following feedback data available in the controller:

$$D_{k_c} = \left[y(k_c)^T u(k_c-1)^T R(k_c)^T\right]^T. \tag{6.8}$$

Although the filter gain K_{k_c} in (6.7) is time-varying, it usually converges in a few steps [71]. Hence, K can be defined as

$$K \triangleq PC^T(CPC^T + R)^{-1}, \tag{6.9}$$

where $P \triangleq \lim_{k_c\to\infty} P_{k_c|k_c-1}$, and thus the Kalman filter in (6.7) is reduced to the following estimator with a fixed gain:

$$\begin{cases} \hat{x}(k_c|k_c-1) = A\hat{x}(k_c-1) + Bu(k_c-1) \\ \hat{x}(k_c) = \hat{x}(k_c|k_c-1) + K\left(y(k_c) - C\hat{x}(k_c|k_c-1)\right). \end{cases} \tag{6.10}$$

A state feedback control law is designed,

$$\Delta\hat{u}(k_c|k_c) = -L\hat{x}_e(k_c), \tag{6.11}$$

where $\hat{x}_e(k_c) = [\Delta\hat{x}(k_c)^T\ e(k_c)^T]^T$, and $L \in \Re^{m\times\bar{n}}$ is the gain matrix. Then the predicted augmented states and control increments from k_c+1 to $k_c+\bar{\tau}$ are obtained as

$$\hat{x}_e(k_c+i|k_c) \quad = \quad A_e\hat{x}_e(k_c+i-1|k_c)$$

$$
\begin{aligned}
& +B_e \Delta \hat{u}(k_c + i - 1|k_c) & \\
& +E_e \Delta r(k_c + i) & (6.12) \\
\Delta \hat{u}(k_c + i|k_c) \quad = \quad & -L\hat{x}_e(k_c + i|k_c) & (6.13)
\end{aligned}
$$

for $i = 1, 2, \cdots, \bar{\tau}$, where $\hat{x}_e(k_c + i|k_c) = [\Delta\hat{x}(k_c + i|k_c)^T e(k_c + i|k_c)^T]^T$, and $\hat{x}_e(k_c|k_c) = \hat{x}_e(k_c)$. Thus, we obtain the following i-step control predictions

$$\hat{u}(k_c + i|k_c) = \hat{u}(k_c + i - 1|k_c) + \Delta\hat{u}(k_c + i|k_c) \tag{6.14}$$

for $i = 0, 1, 2, \cdots, \bar{\tau}$, where $\hat{u}(k_c - 1|k_c) = u(k_c - 1)$. Clearly, (6.14) yields the control prediction sequence

$$U_{k_c} = [\hat{u}(k_c|k_c)^T \hat{u}(k_c + 1|k_c)^T \cdots \hat{u}(k_c + \bar{\tau}|k_c)^T]^T, \tag{6.15}$$

which is sent to the actuator together with the timestamp k_c.

6.2.5 Network delay compensator

In the actuator the network delay compensator is designed to store the latest control prediction sequence and then use it to control the plant. Without loss of generality the latest control prediction sequence at time k is expressed as

$$U_{k_a} = [\hat{u}(k_a|k_a)^T \hat{u}(k_a + 1|k_a)^T \cdots \hat{u}(k_a + \bar{\tau}|k_a)^T]^T, \tag{6.16}$$

where $k_a \leq k_c$ is the timestamp of U_{k_a}. Its RTT delay is

$$\tau_k = k - k_a. \tag{6.17}$$

To compensate for the RTT delay the following control signal is chosen to control the plant at time k:

$$u(k) = \hat{u}(k_a + \tau_k|k_a) = \hat{u}(k|k - \tau_k). \tag{6.18}$$

6.3 FDI attacks

It is assumed that the attacker is able to

1) read the data transmitted through the feedback and forward channels and modify them arbitrarily;
2) know the system parameters, i.e., A, B, C, Q, and R.

The objective now is to design stealthy FDI attacks on the feedback data and the control data (see Fig. 13.3), i.e., D_{k_c} in (6.8) and U_{k_a} in (6.16), such that the resulting NPOTC system becomes unstable while the two-channel FDI attacks fail to be detected.

As shown in Fig. 6.2, under FDI attacks the feedback data arriving at the controller are assumed to be modified as

$$D^a_{k_c} = \left[y_a(k_c)^T u(k_c - 1)^T R(k_c)^T \right]^T \tag{6.19}$$

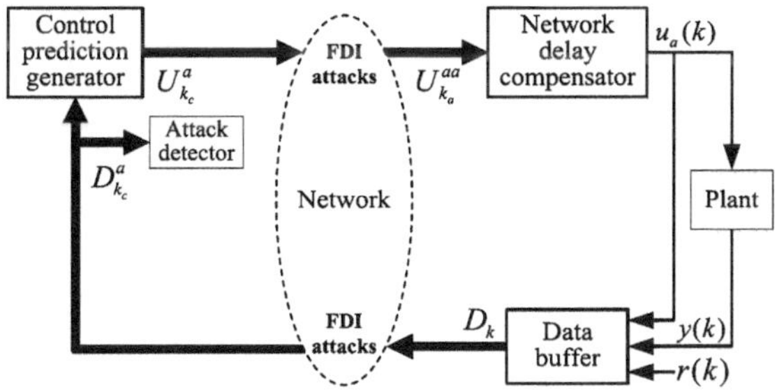

FIGURE 6.2 NPOTC systems under two-channel FDI attacks.

with

$$y_a(k_c) = y(k_c) + \alpha(k_c), \tag{6.20}$$

where $y_a(k_c)$ is the attacked output and $\alpha(k_c)$ is the feedback channel attack. Similarly, the control data arriving at the actuator are falsified by the attacker as

$$U^{aa}_{k_a} = [\hat{u}_a(k_a|k_a)^T \hat{u}_a(k_a+1|k_a)^T \cdots \hat{u}_a(k_a+\bar{\tau}|k_a)^T]^T \tag{6.21}$$

with

$$\hat{u}_a(k_a+i|k_a) = \hat{u}(k_a+i|k_a) + \beta(k_a+i) \tag{6.22}$$

for $i = 0, 1, 2, \cdots, \bar{\tau}$, where $\hat{u}_a(k_a+i|k_a)$ is the attacked control prediction and $\beta(k_a+i)$ is the forward channel attack.

Remark 6.1. It should be noted that the FDI attacks in (6.20) and (6.22) are related to the timestamps of the packets transmitted respectively through the feedback and forward channels (i.e., k_c and k_a). In the NPOTC system the packet transmitted through networks is with a timestamp. As a consequence, although the measurement data and control data are randomly delayed in their transmission due to the presence of random network-induced delays, with the help of the timestamps the FDI attacks in (6.20) and (6.22) can still be easily designed.

To detect these FDI attacks a general strategy is to deploy a detector in the controller, as shown in Fig. 13.3. Here an attack detector is designed using the Kalman filter in (6.10) and the feedback data $D^a_{k_c}$ in (6.19). Due to the presence of FDI attacks the Kalman filter in (6.10) becomes

$$\begin{cases} \hat{x}_a(k_c|k_c-1) = A\hat{x}_a(k_c-1) + Bu(k_c-1) \\ \hat{x}_a(k_c) = \hat{x}_a(k_c|k_c-1) + K\left(y_a(k_c) - C\hat{x}_a(k_c|k_c-1)\right) \end{cases}, \tag{6.23}$$

where $\hat{x}_a(k_c)$ is the state estimation under attack. Then the residual $z_a(k_c)$ is defined as

$$\begin{aligned} z_a(k_c) &= y_a(k_c) - \hat{y}_a(k_c) \\ &= y_a(k_c) - C\left(A\hat{x}_a(k_c-1) + Bu(k_c-1)\right), \end{aligned} \tag{6.24}$$

where $\hat{y}_a(k_c)$ is the output estimation under attack. If some rough FDI attacks are performed in the feedback and forward channels, they usually lead to a large value of $||z_a(k_c)||$, which thus induces the detector to trigger an alarm.

If no attacks are injected into the NPOTC system the residual is

$$\begin{aligned} z(k_c) &= y(k_c) - \hat{y}(k_c) \\ &= y(k_c) - C\left(A\hat{x}(k_c-1) + Bu(k_c-1)\right). \end{aligned} \tag{6.25}$$

Lemma 6.1. *[71]: The residual $z(k_c)$ in (6.25) is Gaussian independent and identically distributed (i.i.d.) with zero mean and covariance $S = CPC^T + R$, i.e.,*

$$z(k_c) \sim \mathcal{N}(0, S). \tag{6.26}$$

Under the FDI attacks in (6.20) and (6.22) the physical plant is expressed as

$$\begin{aligned} x_a(k+1) &= Ax_a(k) + B\left(u(k) + \beta(k)\right) + \omega(k), \\ y(k) &= Cx_a(k) + \upsilon(k), \end{aligned} \tag{6.27}$$

where $x_a(k) \in \Re^n$ is the system state under attack. Eqs. (6.11)–(6.13) also become

$$\Delta\hat{u}(k_c|k_c) = -L\hat{x}_{ea}(k_c) \tag{6.28}$$

$$\begin{aligned} \hat{x}_{ea}(k_c+i|k_c) &= A_e\hat{x}_{ea}(k_c+i-1|k_c) \\ &\quad + B_e\Delta\hat{u}(k_c+i-1|k_c) \\ &\quad + E_e\Delta r(k_c+i) \end{aligned} \tag{6.29}$$

$$\Delta\hat{u}(k_c+i|k_c) = -L\hat{x}_{ea}(k_c+i|k_c) \tag{6.30}$$

for $i = 1, 2, \cdots, \bar{\tau}$, where $\hat{x}_{ea}(k_c+i|k_c) = [\Delta\hat{x}_a(k_c+i|k_c)^T \, e_a(k_c+i|k_c)^T]^T$ and $\hat{x}_{ea}(k_c|k_c) = \hat{x}_{ea}(k_c) = [\Delta\hat{x}_a(k_c)^T \, e_a(k_c)^T]^T$ with

$$e_a(k_c) = r(k_c) - y_a(k_c). \tag{6.31}$$

6.3.1 Design results

In this section we investigate the design of stealthy FDI attacks in the feedback and forward channels for the NPOTC system. The following definition is first given.

Definition 6.1. The NPOTC system is successfully attacked if the residual $z_a(k)$ follows the same distribution as $z(k)$, i.e.,

$$z_a(k) \sim \mathcal{N}(0, S) \tag{6.32}$$

and for the constant reference input

$$\lim_{k\to\infty} \mathbf{E}\left(e_a(k)\right) = 0, \tag{6.33}$$

$$\lim_{k\to\infty} \mathbf{E}\left(e(k)\right) = \infty. \tag{6.34}$$

In this chapter the feedback channel attack is designed as

$$\alpha(k) = -y(k) + CA\hat{x}_a(k-1) + CBu(k-1) + \xi(k), \tag{6.35}$$

where $\xi(k) \sim \mathcal{N}(0, S)$ is the Gaussian white noise. It is clear from (6.24)–(6.26) that the feedback channel attack in (6.35) always satisfies the condition in (6.32), i.e.,

$$z_a(k) = \xi(k) \sim \mathcal{N}(0, S). \tag{6.36}$$

Next the attacked output tracking error $e_a(k)$ in (6.33) is analyzed.

Theorem 6.1. *Under the feedback channel FDI attack in (6.35),* $\lim_{k\to\infty} \mathbf{E}\left(e_a(k)\right) = 0$ *for the constant reference input if and only if the eigenvalues of matrix $A_e - B_e L$ are within the unit circle.*

Proof. According to the network delay compensation strategy in (6.18), it is obtained from (6.28) and (6.30) that

$$\Delta u(k) = \Delta\hat{u}(k|k-\tau_k) = -L\hat{x}_{ea}(k|k-\tau_k), \tag{6.37}$$

where

$$\begin{aligned} \hat{x}_{ea}(k|k-\tau_k) &= A_e\hat{x}_{ea}(k-1|k-\tau_k) \\ &\quad + B_e\Delta\hat{u}(k-1|k-\tau_k). \end{aligned} \tag{6.38}$$

Under the feedback channel attack in (6.35), from (6.20) we have

$$y_a(k) = CA\hat{x}_a(k-1) + CBu(k-1) + \xi(k), \tag{6.39}$$

and then from (6.23) and (6.31) we obtain

$$\Delta\hat{x}_a(k+1) = A\Delta\hat{x}_a(k) + B\Delta u(k) + K\Delta\xi(k+1), \tag{6.40}$$

$$\begin{aligned} e_a(k+1) &= e_a(k) - \Delta y_a(k+1) \\ &= e_a(k) - CA\Delta\hat{x}_a(k) - CB\Delta u(k) \\ &\quad - \Delta\xi(k+1). \end{aligned} \tag{6.41}$$

The combination of (6.40) and (6.41) yields

$$\mathbf{E}\left(\hat{x}_{ea}(k+1)\right) = A_e\mathbf{E}\left(\hat{x}_{ea}(k)\right) + B_e\Delta u(k), \tag{6.42}$$

where $\hat{x}_{ea}(k) = [\Delta\hat{x}_a(k)^T \;\; e_a(k)^T]^T$. With (6.37), subtracting (6.29) from (6.42) leads to

$$\begin{aligned}\mathbf{E}(\hat{x}_{ea}(k \;\; + \;\; 1)) - \hat{x}_{ea}(k+1|k-\tau_k) \\ &= A_e\left(\mathbf{E}(\hat{x}_{ea}(k)) - \hat{x}_{ea}(k|k-\tau_k)\right) \\ &= A_e^{\tau_k+1}\left(\mathbf{E}(\hat{x}_{ea}(k-\tau_k)) - \hat{x}_{ea}(k-\tau_k|k-\tau_k)\right) \\ &= 0 \end{aligned} \tag{6.43}$$

since $\hat{x}_{ea}(k-\tau_k|k-\tau_k) = \hat{x}_{ea}(k-\tau_k)$. That is,

$$\hat{x}_{ea}(k|k-\tau_k) = \mathbf{E}\left(\hat{x}_{ea}(k)\right). \tag{6.44}$$

Then (6.37) can be rewritten as

$$\Delta u(k) = -L\mathbf{E}(\hat{x}_{ea}(k)). \tag{6.45}$$

Thus, substituting (6.45) into (6.42) gives rise to

$$\mathbf{E}\left(\hat{x}_{ea}(k+1)\right) = (A_e - B_e L)\mathbf{E}\left(\hat{x}_{ea}(k)\right). \tag{6.46}$$

It is obvious from (6.46) that $\lim_{k\to\infty}\mathbf{E}\left(e_a(k)\right) = 0$ if and only if the eigenvalues of $A_e - B_e L$ are within the unit circle. □

Finally, we analyze the real output tracking error $e(k)$ in (6.34) for the constant reference input. Under the feedback channel attack in (6.35) it is obtained from (6.23) that

$$\hat{x}_a(k+1) = A\hat{x}_a(k) + Bu(k) + K\xi(k+1). \tag{6.47}$$

Then from (6.27) and (6.47) we have

$$\begin{aligned}\tilde{x}_a(k+1) &= x_a(k+1) - \hat{x}_a(k+1) \\ &= A\tilde{x}_a(k) + B\beta(k) + \omega(k) - K\xi(k+1). \end{aligned} \tag{6.48}$$

In this chapter the forward channel attack is designed as

$$\beta(k+1) = F\beta(k), \tag{6.49}$$

where $F \in \Re^{m\times m}$ is the attack matrix. Combining (6.48) and (6.49) yields

$$X_\beta(k+1) = \Lambda X_\beta(k), \tag{6.50}$$

where

$$X_\beta(k) = \begin{bmatrix}\mathbf{E}(\tilde{x}_a(k)) \\ \beta(k)\end{bmatrix}, \quad \Lambda = \begin{bmatrix}A & B \\ 0 & F\end{bmatrix}.$$

Then from (6.3), (6.27), (6.31), (6.39), and (6.47), we have

$$\mathbf{E}(e(k)) - \mathbf{E}(e_a(k)) = -C\mathbf{E}(\tilde{x}_a(k)). \tag{6.51}$$

Thus, we obtain

$$\lim_{k\to\infty} \mathbf{E}(e(k)) = -C \lim_{k\to\infty} \mathbf{E}(\tilde{x}_a(k)) \tag{6.52}$$

since $\lim_{k\to\infty} \mathbf{E}(e_a(k)) = 0$ if the matrix $A_e - B_e L$ is stable.

Obviously the matrix Λ in (6.50) is a block upper triangular matrix. It is well known that a block upper triangular linear system is stable if and only if each block diagonal subsystem is stable. Thus, it can be concluded from (6.50) and (6.52) that, with the stable matrix $A_e - B_e L$, if A is stable and F is unstable, or if A is unstable, we have $\lim_{k\to\infty} \mathbf{E}(e_a(k)) = \infty$. Therefore, we can obtain the following main results.

Theorem 6.2. *Under the feedback channel attack in (6.35) and the forward channel attack in (6.49) the closed-loop NPOTC system is stable, and further* $\mathbf{E}(e_a(\infty)) = 0$, *if and only if the matrices* $A_e - B_e L$, A, *and* F *are stable.*

Theorem 6.3. *If the system matrix* A *is stable, the NPOTC system can be attacked successfully without being detected by injecting the feedback channel attack in (6.35) and the forward channel attack in (6.49) with an unstable matrix* F.

Theorem 6.4. *If the system matrix* A *is unstable, the NPOTC system can be attacked successfully without being detected by injecting the feedback channel attack in (6.35) and any arbitrary attack in the forward channel.*

Remark 6.2. It is easy to observe from Theorems 6.3 and 6.4 that whether the system matrix A is unstable or stable, the control system can be attacked successfully without being detected. Instead, in [273] and [277] it is required that matrix A be unstable, where only the FDI attack on the sensor data is considered. Therefore, in this chapter, by performing two-channel FDI attacks simultaneously, the derived results are more general.

6.4 Simulation results

In this section, numerical simulations are carried out for three cases: 1) A and F are stable; 2) A is stable and F is unstable; and 3) A is unstable and F is stable. The network-induced delays in two channels are considered, which lead to the RTT delays of 0–4 steps shown in Fig. 6.3.

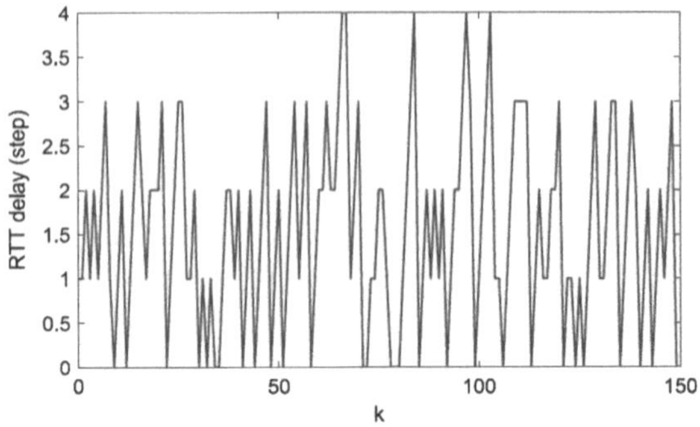

FIGURE 6.3 RTT delays.

6.4.1 Case 1: A and F are stable

Consider a stable system with matrices

$$A = \begin{bmatrix} 0.2071 & 0.3705 & 0.0439 \\ 0.6072 & 0.5751 & 0.0272 \\ 0.6299 & 0.4514 & 0.3127 \end{bmatrix}, \quad B = \begin{bmatrix} 0.1730 & 0.2523 \\ 0.9797 & 0.8757 \\ 0.2714 & 0.7373 \end{bmatrix},$$

$$C = \begin{bmatrix} 0.1365 & 0.8939 & 0.2987 \\ 0.0118 & 0.1991 & 0.6614 \end{bmatrix}. \tag{6.53}$$

With $Q = 0.0001$ and $R = 0.01$, by using the Kalman filter in (6.7) we obtain the filter gain

$$K = \begin{bmatrix} 0.0241 & 0.0171 \\ 0.0495 & 0.0323 \\ 0.0514 & 0.0434 \end{bmatrix}. \tag{6.54}$$

By using the assignment of the closed-loop poles $[0.6 \pm 0.3j \;\; 0.2 \;\; 0.1]$, the controller gain L is designed to be

$$L = \begin{bmatrix} 0.0244 & -0.7056 & -0.4480 & -0.0840 & 1.5769 \\ 0.7696 & 1.1064 & 0.4683 & -0.2284 & -1.3080 \end{bmatrix}. \tag{6.55}$$

The feedback channel attack in (6.35) and the forward channel attack in (6.49) with $\beta(50) = [10 \;\; 10]^T$ and the stable matrix

$$F = \begin{bmatrix} 0.0579 & 0.8132 \\ 0.3529 & 0.0099 \end{bmatrix} \tag{6.56}$$

are injected into the NPOTC system at time $k = 50$. The simulation result is shown in Fig. 6.4. It can be seen that the attacked output $y_a(k)$ tracks the refer-

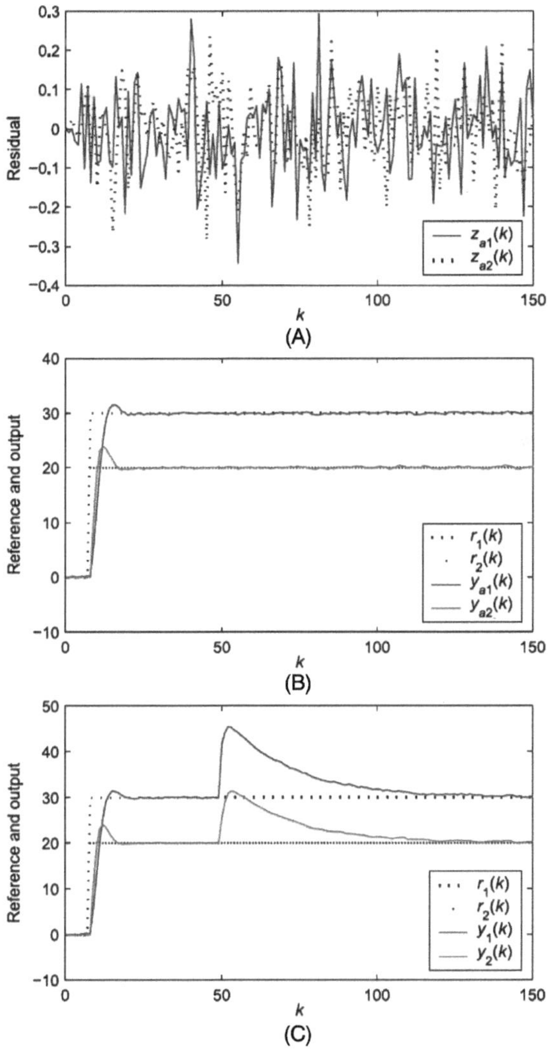

FIGURE 6.4 Simulation results of Case 1. (A) Residual $z_a(k)$. (B) Attacked output response. (C) Real output response.

ence signal $r(k)$ well, and the real output $y(k)$ ultimately converges to $r(k)$ with the diminishment of the forward channel attack $\beta(k)$, which coincides with the result of Theorem 6.2. Furthermore, the two-channel attacks cannot be detected by using the residual $z_a(k)$ in Fig. 6.4A.

6.4.2 Case 2: A is stable and F is unstable

Consider the same stable system as Case 1 with the same K and L. The feedback channel attack in (6.35) as well as the forward channel attack in (6.49) with

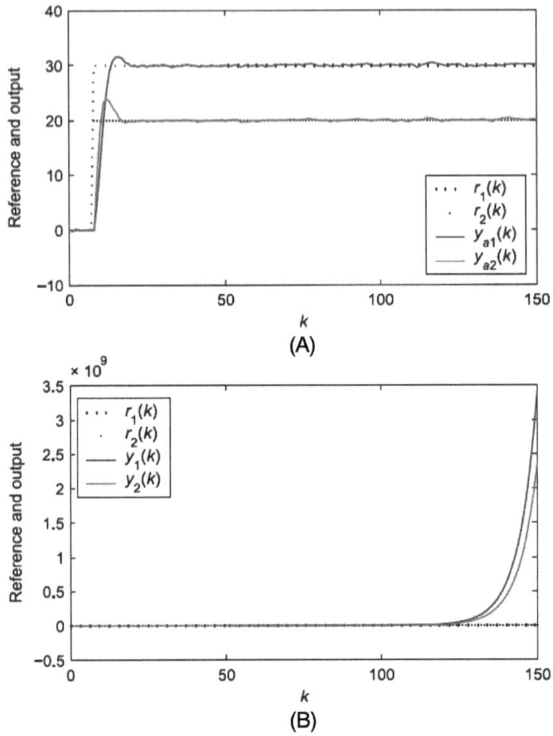

FIGURE 6.5 Simulation results of Case 2. (A) Attacked output response. (B) Real output response.

$\beta(50) = [10 \;\; 10]^T$ and the unstable matrix

$$F = \begin{bmatrix} 0.9501 & 0.6068 \\ 0.2311 & 0.4860 \end{bmatrix} \tag{6.57}$$

are injected into the NPOTC system at time $k = 50$. The simulation result is shown in Fig. 6.5. It can be seen that the attacked output $y_a(k)$ still tracks the reference signal $r(k)$ well, but the real output $y(k)$ ultimately diverges from $r(k)$, which coincides with the result of Theorem 6.3.

6.4.3 Case 3: A is unstable and F is stable

Consider an unstable system with matrix

$$A = \begin{bmatrix} 0.2312 & 0.6724 & 0.5630 \\ 0.4161 & 0.9383 & 0.1189 \\ 0.2988 & 0.3431 & 0.1690 \end{bmatrix} \tag{6.58}$$

and matrices B and C in (6.53). Using the same design procedure as Case 1, the matrices K and L are obtained as

$$K = \begin{bmatrix} 0.2740 & 0.1274 \\ 0.3267 & 0.1502 \\ 0.1627 & 0.0813 \end{bmatrix} \tag{6.59}$$

$$L = \begin{bmatrix} 0.0506 & 0.8601 & 0.3087 & -0.5182 & 2.0070 \\ 0.4128 & 0.0181 & -0.2153 & 0.1520 & -1.5753 \end{bmatrix}. \tag{6.60}$$

The feedback channel attack in (6.35) as well as the forward channel attack in (6.49) with $\beta(50) = [10\ \ 10]^T$ and the stable matrix in (6.56) are injected into the NPOTC system at time $k = 50$. The simulation result is shown in Fig. 6.6. It can be seen that the attacked output $y_a(k)$ still tracks the reference signal $r(k)$ well. Although the injected forward channel attack $\beta(k)$ is convergent, the closed-loop system still becomes unstable, which coincides with the result of Theorem 6.4.

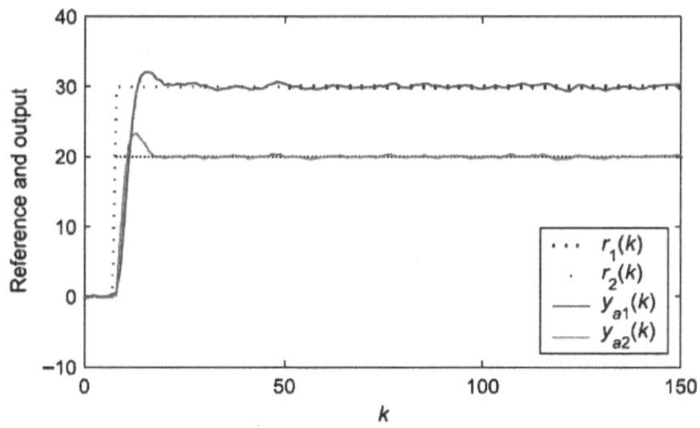

(A)

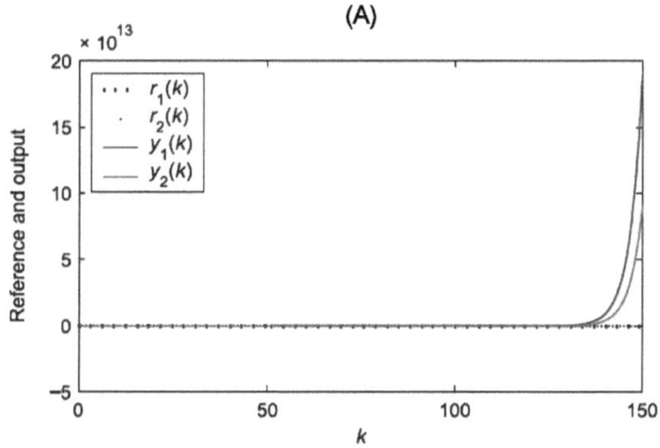

(B)

FIGURE 6.6 Simulation results of Case 3. (A) Attacked output response. (B) Real output response.

6.5 Experimental results

To further test the proposed method an Internet-based servomotor system test rig was constructed (Fig. 6.7). It consists mainly of a servomotor system, a networked controller, a local control board, and the Internet from Tsinghua University, Beijing, China, to the University of South Wales, Pontypridd, UK. The RTT delays of the Internet vary randomly from 3 to 8 steps. For details of the experimental setup, refer to [278].

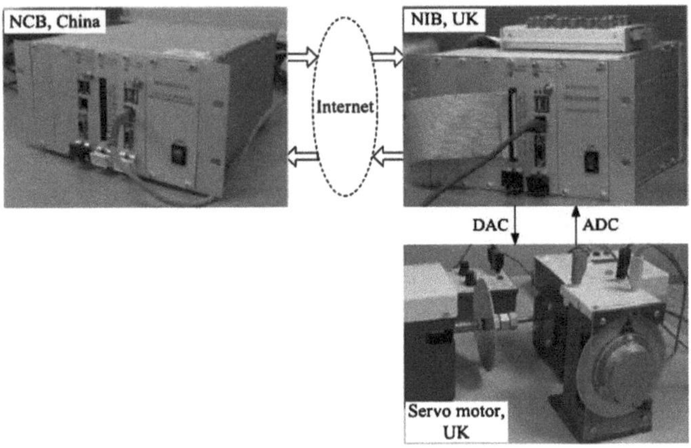

FIGURE 6.7 Internet-based servomotor system.

Our objective is to control the position of the servomotor system. With the sampling period of 0.04 s the model of the servomotor system is identified as

$$A = \begin{bmatrix} 1.2998 & -0.4341 & 0.1343 \\ 1 & 0 & 0 \\ 0 & 1 & 0 \end{bmatrix}, \quad B = \begin{bmatrix} 1 \\ 0 \\ 0 \end{bmatrix},$$

$$C = \begin{bmatrix} 3.5629 & 2.7739 & 1.0121 \end{bmatrix}, \tag{6.61}$$

whose input and output are the control voltage (-10 to 10 V) and the angle position ($-120°$, $-120°$), respectively. The filter gain K and the controller gain L are chosen as

$$K = \begin{bmatrix} 0.1070 & 0.0877 & 0.0178 \end{bmatrix}^T \tag{6.62}$$

$$L = \begin{bmatrix} 0.7125 & -0.2593 & 0.1253 & -0.0245 \end{bmatrix}. \tag{6.63}$$

From (6.61) we know that the servomotor system is open-loop critically stable. In the following, practical experiments are performed for two cases: 1) F is stable and 2) F is unstable.

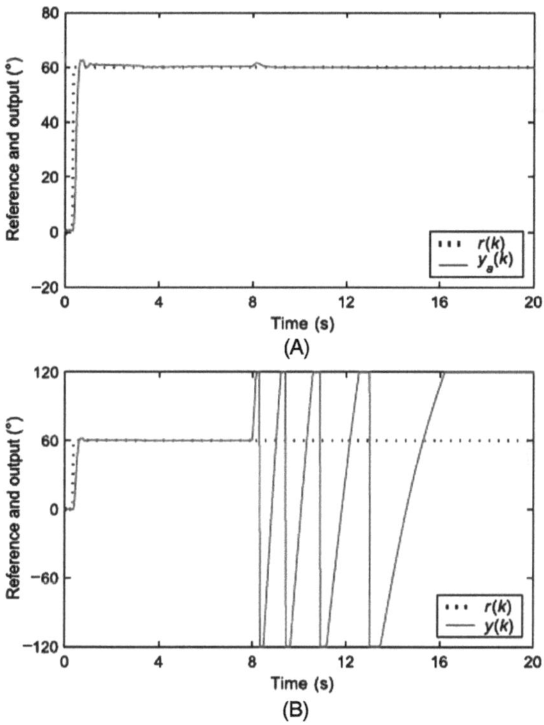

FIGURE 6.8 Experimental results of Case 1. (A) Attacked output response. (B) Real output response.

6.5.1 Case 1: F is stable

The feedback channel attack in (6.35) as well as the forward channel attack in (6.49) with the initial value $\beta(199) = 2$ and the stable matrix $F = 0.99$ are injected into the NPOTC system at time $t = 7.96$ s. The experimental result is shown in Fig. 6.8, which indicates that the attacked output $y_a(k)$ tracks the reference signal $r(k)$ well, but the real position of the servomotor deviates from the reference signal $r(k)$. With the disappearance of the forward channel attack $\beta(k)$, the servomotor finally stops at a certain position rather than the reference signal $r(k)$.

6.5.2 Case 2: F is unstable

The feedback channel attack in (6.35) as well as the forward channel attack in (6.49) with the initial value $\beta(180) = 0.1$ and the unstable matrix $F = 1.02$ are injected into the NPOTC system at time $t = 7.20$ s. The experimental result is shown in Fig. 6.9. It is clear that the two-channel attacks lead to the instability of the closed-loop control system.

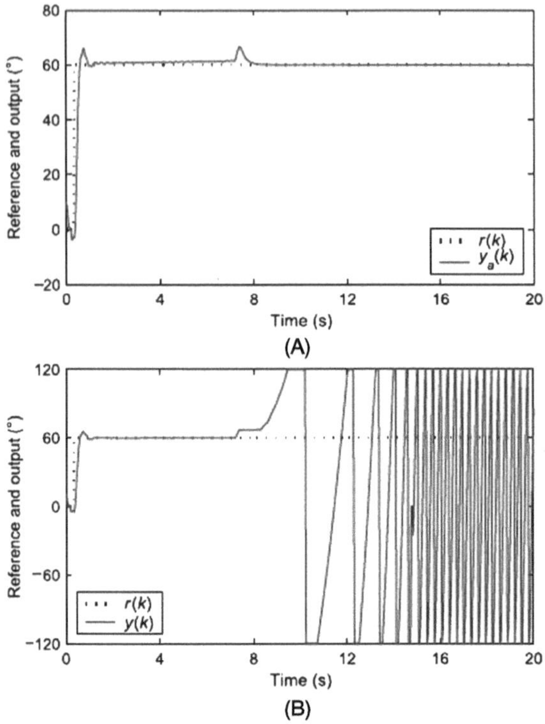

FIGURE 6.9 Experimental results of Case 2. (A) Attacked output response. (B) Real output response.

It should be pointed out that in Figs. 6.8A and 6.9A, when the two-channel attacks are added, slight fluctuations occur in the attacked output $y_a(k)$, which do not appear in the above-mentioned numerical simulations. This phenomenon results from the mismatch between the model in (6.61) and the practical servomotor system.

6.6 Notes

This chapter investigated the design problem of FDI attacks against the output tracking control of networked systems. To compensate for two-channel network-induced delays, a Kalman filter-based NPOTC method was proposed for stochastic linear systems. Then from an attacker's viewpoint, stealthy FDI attacks were designed for the measurement data in the feedback channel and the control data in the forward channel that can avoid being detected by a Kalman filter-based detector. Both simulation and experimental results illustrated the effectiveness of the proposed method.

It is worth mentioning that, in general, the research on FDI attacks includes three aspects: 1) attack design, 2) attack detection, and 3) secure control design. This chapter mainly focused on the first aspect, the design of stealthy FDI attacks. The other two aspects are more important and interesting, and thus deserve further investigation in our future research.

Chapter 7

Stabilization schemes for secure control

Contents

7.1 Introduction and objectives

Recent years have witnessed a growing interest towards cyber-physical systems (CPSs), systems with a close connection between computational and physical resources. Their field of application is immense, ranging from autonomous vehicles and supply chains to power and transportation networks. Many of these applications are safety-critical. This has generated considerable attention to networked systems in the presence of attacks, bringing the question of cyber-security into filtering and control theories [267], [279].

As argued in [267] and [279], security in CPSs drastically differs from security in general-purpose computing systems. In CPSs, attacks can in fact cause disruptions that transcend the cyber realm and affect the physical world. For instance, if a critical process is open-loop unstable, failures in the plant-controller

Cloud Control Systems. https://doi.org/10.1016/B978-0-12-818701-2.00015-9

communication network can result in environmental damage. Control theory, on the other hand, is typically concerned with well-defined uncertainties or faults. In fact, most of networked control approaches assume that communication failures follow a given class of probability distributions [280], [148], which is hardly justified in the case of a malicious adversary.

In a networked control system (NCS), attacks to the communication links can be classified as either deception attacks or denial-of-service (DoS) attacks. The former affect the trustworthiness of data by manipulating the packets transmitted over the network (see [281,282,241,77,283] and the references therein). DoS attacks are instead primarily intended to affect the timeliness of the information exchange, for example to cause packet losses (see, e.g., [284], [285] for an introduction to the topic).

Focusing on DoS attacks, we consider a sampled-data control system in which the plant-controller communication is networked; the attacker's objective is to induce instability in the control system by denying communication on measurement (sensor-to-controller) and control (controller-to-actuator) channels. Under DoS attacks the process evolves in an open-loop according to the last transmitted control sample. The problem of interest is that of finding the conditions under which closed-loop stability, in some suitably defined sense, can be preserved.

A basic question for this problem is concerned with the modeling of the DoS attacks. As previously noted, it is hard to justify the incentive for an attacker to follow a probabilistic packet drop model. In this chapter, no assumption is made regarding the DoS attack underlying strategy. We consider a general attack model that only constrains the attacker action in time by posing limitations on the frequency of DoS attacks and their duration. This makes it possible to capture many different types of DoS attacks, including trivial, periodic, random, and protocol-aware jamming attacks [285,243,286,287].

An explicit characterization of the frequency and duration of DoS attacks under which closed-loop stability can be preserved are established. The result is intuitive as it relates stability with the ratio between the on and off periods of jamming. The analysis taken here is reminiscent of stability problems for switching systems [288], a modeling tool which has already proved effective in networked systems [289–291]. In this paper, however, the peculiarity of the problem under consideration leads to specific design solutions.

The design of the transmission times turns out to be key. To get stability the transmission times are selected in such a way that whenever communication is possible the closed-loop trajectories satisfy a suitable norm bound. This choice has two main advantages: i) it can ensure *global exponential input-to-state stability* (ISS) with respect to process disturbances even in the presence of DoS; and ii) it is flexible enough to allow the designer to choose from several implementation options that can be used to trade off between performance and communication resources. The design of the network transmission times has interesting and perhaps surprising connections with the *event-based* sampling approach of [292], though substantial modifications are needed to account for the presence of DoS and disturbances. More specifically, the adoption of sampling rules that suitably constrain the closed-loop trajectories is crucial for achieving a simple Lyapunov-based analysis of the ISS property during the *on/off* periods of DoS.

In the control literature, contributions to this research topic have been reported in [293,294,90,252,102,254]. In [293] and [294] the authors consider the problem of finding optimal control policies when DoS attacks either evolve according to a Bernoulli process or follow a hidden Markov process model; however, as noted, this problem is closer to classical NCS literature. A scenario more similar to the present one is considered in [90] and [252], where the problem is to find optimal control and attack strategies assuming a maximum number of jamming actions over a prescribed (finite) control horizon. There are two main differences with respect to our framework. First, in [90] and [252] the authors consider a pure discrete-time setting, while here we deal with sampled-data networked systems and the performance analysis is concerned with the continuous-time process state. Second, we do not formulate the problem as an optimal control design problem. The controller can be designed according to any suitable design method, robustness against DoS attacks being achieved thanks to the design of the network transmission times.

Perhaps the closest references to our work are [102] and [254]. In these papers the authors consider DoS attacks in the form of pulse-width modulated signals. The goal is to identify the salient features of the DoS signal such as maximum *on/off* cycle in order to suitably schedule the transmission times. For the case of *periodic jamming* (of unknown period and duration) an identification algorithm is derived that makes it possible to desynchronize the transmission times from the on periods of DoS. This framework should be therefore looked at as complementary rather than an alternative to the present one in order to deal with cases where the jamming signal is "well-structured" so that desynchronization from attacks can be achieved. Such a feature is conceptually impossible to achieve in scenarios such as the one considered in this section where the jamming strategy is not prefixed (the attacker can modify the attack strategy online).

7.1.1 Process dynamics and ideal control action

The framework of interest is schematically represented in Fig. 7.1. The process to be controlled is described by the differential equation

$$\frac{d}{dt}x(t) = Ax(t) + Bu(t) + w(t), \tag{7.1}$$

where $t \in \mathbb{R}_{\geq 0}$; $x \in \mathbb{R}^n$ is the state and $u \in \mathbb{R}^m$ is the control input; A and B are matrices of appropriate size; $w \in \mathbb{R}^n$ is an unknown disturbance: it accounts for process input disturbances and the noise on control (controller-to-actuator) and measurement (sensor-to-controller) channels.

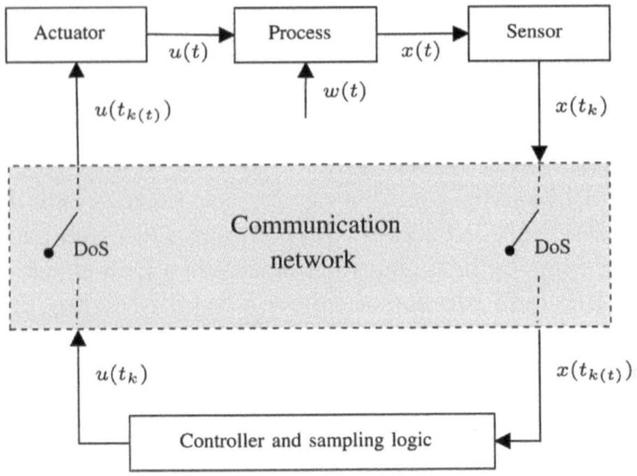

FIGURE 7.1 Block diagram of the closed-loop system.

The control action is implemented over a sensor–actuator network. We assume that (A, B) is stabilizable and that a state-feedback matrix K has been designed in such a way that all the eigenvalues of $A + BK$ have a negative real part. The control signal is sampled using a *sample-and-hold device*. Let $\{t_k\}_{k \in \mathbb{N}_0}$ represent the sequence of time instants at which it is desired to update the control action, where by convention $t_0 := 0$. Accordingly, whatever the logic generating the sequence $\{t_k\}_{k \in \mathbb{N}_0}$, in the ideal situation where data can be sent and received at any desired instant of time, the control input applied to the process is given by

$$u_{\text{ideal}}(t) = Kx(t_k) \tag{7.2}$$

for all $t \in \mathcal{I}_k := [t_k, t_{k+1}[$.

7.1.2 DoS and actual control action

We refer to DoS as the phenomenon that may prevent (7.2) from being executed at each desired time. In principle, this phenomenon can affect measurement and control channels separately. In this chapter we consider the case of DoS simultaneously affecting both measurement and control channels. This amounts to assuming that, in the presence of DoS, data can be *neither sent nor received*. Specifically, let $\{h_n\}_{n \in \mathbb{N}_0}$ denote the sequence of DoS off/on transitions, the time instants at which DoS exhibits a transition from zero (communication is possible) to one (communication is interrupted), where $h_0 \geq 0$. Then

$$H_n := \{h_n\} \cup [h_n, h_n + \tau_n[\tag{7.3}$$

represents the n-th DoS time-interval, of a length $\tau_n \in \mathbb{R}_{\geq 0}$, over which communication is not possible. If $\tau_n = 0$, the n-th DoS takes the form of a single pulse at time h_n.

In the presence of DoS the actuator generates an input that is based on the *most recently received* control signal. Given $\tau,\ t \in \mathbb{R}_{\geq 0}$ with $t \geq \tau$, let

$$\Xi(\tau, t) := \bigcup_{n \in \mathbb{N}_0} H_n \bigcap [\tau, t], \tag{7.4}$$

$$\Theta(\tau, t) := [\tau, t] \setminus \Xi(\tau, t). \tag{7.5}$$

In other words, for each interval $[\tau, t]$, $\Xi(\tau, t)$ and $\Theta(\tau, t)$, represent the sets of time instants where communication is denied and allowed, respectively. The reason for considering generic intervals $[\tau, t]$ rather than simply $[0, t]$ will become clear in Section IV. Accordingly, for each $t \in \mathbb{R}_{\geq 0}$ the control input applied to the process can be expressed as

$$u(t) = Kx(t_{k(t)}), \tag{7.6}$$

where

$$k(t) := \begin{cases} -1, & \text{if } \Theta(0, t) = \emptyset \\ \sup\{k \in \mathbb{N}_0 \mid t_k \in \Theta(0, t)\}, & \text{otherwise.} \end{cases} \tag{7.7}$$

In other words, for each $t \in \mathbb{R}_{\geq 0}$, $k(t)$ represents the last successful control update. Note that $h_0 = 0$ implies $k(0) = -1$, which raises the question of assigning a value to the control input when communication is not possible at the process startup. In this respect, we assume that when $h_0 = 0$, then $u(0) = 0$, and we let $x(t_{-1}) := 0$ for notational consistency.

7.1.3 Control objectives

The problem of interest is that of finding sampling logic that achieves robustness against DoS while ensuring that the control inter-execution times are bounded away from zero. While robustness is concerned with stability and performance of the closed-loop system, positive inter-execution times are required for the control scheme to be physically implementable over a network.

The following definitions reflect the stated goals.

Definition 7.1. [295] Let Σ be the control system resulting from (7.1) under a control signal, as in (7.6). System Σ is said to be input-to-state stable if there exists a $\mathcal{KL}$ function β and a $\mathcal{K}_\infty$ function γ such that for each $w \in \mathcal{L}_\infty(\mathbb{R}_{\geq 0})$ and $x(0) \in \mathbb{R}^n$,

$$\|x(t)\| \leq \beta(\|x(0)\|, t) + \gamma(\|w_t\|_\infty) \tag{7.8}$$

for all $t \in \mathbb{R}_{\geq 0}$. If (7.8) holds when $w \equiv 0$, then Σ is said to be globally asymptotically stable (GAS).

Definition 7.2. A control update sequence $\{t_k\}_{k\in\mathbb{N}_0}$ is said to have a finite sampling rate property if there exists $\underline{\Delta} \in \mathbb{R}_{>0}$ such that

$$\Delta_k := t_{k+1} - t_k \geq \underline{\Delta} \tag{7.9}$$

for all $k \in \mathbb{N}_0$.

In the following it is assumed that the network can send information at the sampling rate induced by $\underline{\Delta}$.

7.1.4 Stabilizing control update policies

We first introduce a class of control update policies ensuring ISS in the absence of DoS. The results will serve as a basis for the subsequent developments.

Consider the closed-loop system resulting from (7.1) under a control signal as in (7.6). As a first step, we rewrite it in a form that is better suited for analysis purposes. Let

$$e(t) := x(t_{k(t)}) - x(t) \tag{7.10}$$

represent the error between the value of the process state at the last successful control update and the value of the process state at the current time, where $t \in \mathbb{R}_{\geq 0}$. The closed-loop system can be therefore rewritten as

$$\frac{d}{dt}x(t) = \Phi x(t) + BKe(t) + w(t), \tag{7.11}$$

where $\Phi := A + BK$.

The closed-loop system now depends on the control update rule through e, which enters the dynamics as an additional disturbance term. It is then intuitively clear that the stability will not be destroyed if we adopt control update rules that keep e small in a suitable sense. The notion of "smallness" considered here, which characterizes the control update rules of interest, is expressed in terms of the following boundedness inequality:

$$\|e(t)\| \leq \sigma \|x(t)\| + \sigma \|w_t\|_\infty, \tag{7.12}$$

where $\sigma \in \mathbb{R}_{>0}$ is a suitable design parameter. We anticipate that (7.12) is not the control update rule we are going to implement because of its dependence on the supremum norm of the disturbance w, which is in general unknown. We instead adopt different update rules that guarantee that (7.12) is always satisfied. These different update rules, motivated by Lemma 7.1 below, are discussed in detail later on.

As the next result shows, provided that σ is suitably chosen, any control update rule that restricts e to satisfy (7.12) is stabilizing. This can be proved by resorting to standard Lyapunov arguments. Given any positive definite matrix $Q = Q^\top \in \mathbb{R}^{n\times n}$, let P be the unique solution of the Lyapunov equation

$$\Phi^\top P + P\Phi + Q = 0. \tag{7.13}$$

Then by taking $V(x) = x^\top Px$ as a Lyapunov function, and computing it along the solution of (7.11), it is simple to verify that

$$\alpha_1\|x(t)\|^2 \le V(x(t)) \le \alpha_2\|x(t)\|^2 \tag{7.14a}$$

$$\begin{aligned}\frac{d}{dt}V(x(t)) &\le -\gamma_1\|x(t)\|^2 + \gamma_2\|x(t)\|\,\|e(t)\| \\ &\quad +\gamma_3\|x(t)\|\,\|w(t)\|\end{aligned} \tag{7.14b}$$

hold for all $t \in \mathbb{R}_{\ge 0}$, with α_1 and α_2 equal to the smallest and largest eigenvalue of P, respectively; γ_1 equals to the smallest eigenvalue of Q; and $\gamma_2 := \|2PBK\|$ and $\gamma_3 := \|2P\|$. It is then immediate to see that under (7.12), the second term of (7.14) always satisfies a dissipation-like inequality whenever σ is chosen small enough.

Theorem 7.1. *Consider the control system Σ composed of (7.1) and control input (7.6), where K is such that all the eigenvalues of $\Phi = A + BK$ have a negative real part. Given any positive symmetric definite matrix $Q \in \mathbb{R}^{n\times n}$, let P be the unique solution of the Lyapunov equation $\Phi^\top P + P\Phi + Q = 0$. Let $V(x) = x^\top Px$. Consider any control update sequence occurring at a finite sampling rate and satisfying (7.12) for all $t \in \mathbb{R}_{\ge 0}$, with σ such that*

$$\gamma_1 - \sigma\gamma_2 > 0, \tag{7.15}$$

where γ_1 and γ_2 are as in (7.14b). Then, Σ is ISS.

Proof. Substituting (7.12) into (7.14b) yields

$$\frac{d}{dt}V(x(t)) \le -\gamma_4\|x(t)\|^2 + \gamma_5\|x(t)\|v(t), \tag{7.16}$$

where $v(t) := \sup\{\|w(t)\|, \|w_t\|_\infty\}$, $\gamma_4 := (\gamma_1 - \sigma\gamma_2)$, and $\gamma_5 := (\gamma_3 + \sigma\gamma_2)$. Here, we recall that γ_1 is the minimal eigenvalue of Q, $\gamma_2 = \|2PBK\|$, $\gamma_3 = \|2P\|$, and $\sigma < \gamma_1/\gamma_2$. Observe that for any positive real δ, Young's inequality (see, e.g., [305]) yields

$$2\,\|x(t)\|\,v(t) \le \frac{1}{\delta}\,\|x(t)\|^2 + \delta v^2(t). \tag{7.17}$$

By letting $\delta := \gamma_5/\gamma_4$, we get

$$\begin{aligned}\frac{d}{dt}V(x(t)) &\le -\frac{\gamma_4}{2}\|x(t)\|^2 + \gamma_6 v(t)^2 \\ &\le -\omega_1 V(x(t)) + \gamma_6 v(t)^2,\end{aligned} \tag{7.18}$$

where $\omega_1 := \gamma_4/(2\alpha_2)$ and $\gamma_6 := \gamma_5^2/(2\gamma_4)$. Note now that $\|v_t\|_\infty = \|w_t\|_\infty$ for any $t \in \mathbb{R}_{>0}$. Thus, standard comparison results for differential inequalities yield

$$V(x(t)) \leq e^{-\omega_1 t} V(x(0)) + \gamma_7 \|w_t\|_\infty^2, \tag{7.19}$$

where $\gamma_7 := \gamma_6/\omega_1$. Using (7.14a), we get

$$\|x(t)\|^2 \leq \frac{\alpha_2}{\alpha_1} e^{-\omega_1 t} \|x(0)\|^2 + \frac{\gamma_7}{\alpha_1} \|w_t\|_\infty^2. \tag{7.20}$$

Since $a^2 + b^2 \leq (a+b)^2$ for any pair of positive real a and b, we finally get

$$\|x(t)\| \leq \sqrt{\frac{\alpha_2}{\alpha_1}} e^{-(\frac{\omega_1}{2})t} \|x(0)\| + \sqrt{\frac{\gamma_7}{\alpha_1}} \|w_t\|_\infty, \tag{7.21}$$

which yields the desired result. □

Inequality (7.15) can always be satisfied by selecting σ sufficiently small in that $\gamma_1 > 0$. Given any σ satisfying (7.15), the only question that arises regards the possibility of designing a sampling logic that guarantees (7.12) with a finite sampling rate. As the next result shows, in the absence of DoS this is always possible. Given a matrix $M \in \mathbb{R}^{n \times n}$, let

$$\mu_M := \max \left\{ \lambda \mid \lambda \in \text{spectrum} \left\{ \frac{(M + M^\top)}{2} \right\} \right\} \tag{7.22}$$

denote the logarithmic norm of M [296].

Lemma 7.1. *Consider the same notation as in Theorem 7.1. Then, in the absence of DoS, any control update rule with inter-sampling times smaller than or equal to*

$$\bar{\Delta}_\sigma := \left(\frac{\sigma}{1+\sigma} \right) \frac{1}{\max\{\|\Phi\|, 1\}} \tag{7.23}$$

when $\mu_A \leq 0$, and

$$\bar{\Delta}_\sigma := \frac{1}{\mu_A} \log \left[\left(\frac{\sigma}{1+\sigma} \right) \frac{1}{\max\{\|\Phi\|, 1\}} \mu_A + 1 \right] \tag{7.24}$$

when $\mu_A > 0$, satisfies (7.12) for all $t \in \mathbb{R}_{\geq 0}$.

Proof. In the absence of DoS any control update attempt is successful. Thus, in accordance with (7.10), the dynamics of e satisfy

$$\begin{aligned} \frac{d}{dt} e(t) &= -Ax(t) - BKx(t_k) - w(t) \\ &= Ae(t) - \Phi x(t_k) - w(t) \end{aligned} \tag{7.25}$$

for all $t \in \mathcal{I}_k$ and for all $k \in \mathbb{N}_0$, where $e(t_k) = 0$. Recall now that $\|e^{At}\| \le e^{\mu_A t}$ for all $t \in \mathbb{R}_{\ge 0}$. Using this property we then have

$$\|e(t)\| \le \kappa_1 \int_{t_k}^{t} e^{\mu_A (t-s)} \left[\|x(t_k)\| + \|w(s)\|\right] ds \tag{7.26}$$

for all $t \in \mathcal{I}_k$ and all $k \in \mathbb{N}_0$, where $\kappa_1 := \max\{\|\Phi\|, 1\}$. Let

$$f(t - t_k) := \int_{t_k}^{t} e^{\mu_A (t-s)} ds.$$

Given that $x(t_k) = e(t) + x(t)$, we obtain

$$\begin{aligned} \|e(t)\| \quad &\le \quad \kappa_1 f(t - t_k)\|e(t)\| \\ &\quad\quad + \kappa_1 f(t - t_k)\left(\|x(t)\| + \|w_t\|_\infty\right). \end{aligned} \tag{7.27}$$

Observe now that $f(0) = 0$ and $f(t - t_k)$ is monotonically increasing with t. Accordingly, for any positive real Δ such that

$$f(\Delta) \le \frac{1}{\kappa_1} \frac{\sigma}{(1 + \sigma)}, \tag{7.28}$$

then any control update rule such that $\Delta_k \le \Delta$ will satisfy (7.12) for all $t \in \mathbb{R}_{\ge 0}$. To conclude the proof, we derive an explicit expression for Δ. Let first $\mu_A \le 0$. In this case $f(\Delta) \le \Delta$, so that (7.23) yields the desired result. If instead $\mu_A > 0$, we have

$$f(\Delta) = \frac{1}{\mu_A}\left(e^{\mu_A \Delta} - 1\right) \tag{7.29}$$

and (7.24) yields the desired result. □

Remark 7.1. A boundedness inequality similar to (7.12) was considered by [292] in the context of event-based control for disturbance-free processes. The difference here lies in the bound imposed on e, which is dictated by the need to take into account w. With Lemma 7.1, we explicitly determine inter-sampling times that ensure (7.12) without prior knowledge of an upper bound on w. A detailed discussion on sampling logic based on Lemma 7.1 is given later on.

7.2 Input-to-state stability under denial of service

The above analysis relies on the possibility of satisfying condition (7.12) *for all* $t \in \mathbb{R}_{\ge 0}$. According to Lemma 7.1, in the absence of DoS this is always possible. In the presence of DoS the analysis becomes more involved since certain control update attempts need not be successful, no matter how we sample. Failing to reset e may cause (7.12) to be violated and stability can be lost in that (7.14b) need no longer satisfy a dissipation-like inequality. Thus, a natural question arises on how the conclusions of Theorem 7.1 can be extended so as to account for the presence of DoS. The remainder of this section addresses this question. In the following, we introduce and discuss the class of DoS signals under consideration. An interesting simplification arising in the disturbance-free case is discussed later on.

7.2.1 Assumptions of time-constrained denial of service

The first question to be addressed is that of determining the amount of DoS that a system can tolerate before undergoing instability. In this respect it is simple to see that this amount is not arbitrary, and that suitable conditions must be imposed on both DoS *frequency* and *duration*.

1) DoS Frequency: Consider first the frequency at which DoS can occur, and let $\Lambda_n := h_{n+1} - h_n$, $n \in \mathbb{N}_0$ denote the time elapsing between any two successive DoS triggers. We immediately see that if $\Lambda_n \leq \underline{\Delta}$ for all $n \in \mathbb{N}_0$ (DoS can occur at the same rate as the minimum possible sampling rate $\underline{\Delta}$), then stability can be lost regardless of the adopted control update policy. It is intuitively clear that in order to get stability the frequency at which DoS can occur must be sufficiently small compared to the minimum sampling rate. As discussed in what follows, a natural way to express this requirement is via the concept of *average dwell time*, as introduced by [297]. Given $\tau, t \in \mathbb{R}_{\geq 0}$ with $t \geq \tau$, let $n(\tau, t)$ denote the number of DoS off/on transitions occurring on the interval $[\tau, t[$.

Assumption 7.1. There exists $\eta \in \mathbb{R}_{\geq 0}$ and $\tau_D \in \mathbb{R}_{>\underline{\Delta}}$ such that

$$n(\tau, t) \leq \eta + \frac{t - \tau}{\tau_D} \tag{7.30}$$

for all $\tau, t \in \mathbb{R}_{\geq 0}$ with $t \geq \tau$.

2) DoS Duration: In addition to the DoS frequency, we also need to constrain the DoS *duration*, namely the length of the interval over which communication is interrupted. To see this, consider for example a DoS sequence consisting of the singleton $\{h_0\}$. Assumption 7.1 is clearly satisfied with $\eta \geq 1$. However, if $H_0 = \mathbb{R}_{\geq 0}$ (communication is never possible), then stability is lost regardless of the adopted control update policy. Recalling the definition of $\Xi(\tau, t)$ in (7.4), the assumption that follows provides a quite natural counterpart of Assumption 7.1 with respect to the DoS duration.

Assumption 7.2 (DoS duration). There exists $\kappa \in \mathbb{R}_{\geq 0}$ and $T \in \mathbb{R}_{>1}$ such that

$$|\Xi(\tau, t)| \leq \kappa + \frac{t - \tau}{T} \tag{7.31}$$

for all $\tau, t \in \mathbb{R}_{\geq 0}$ with $t \geq \tau$.

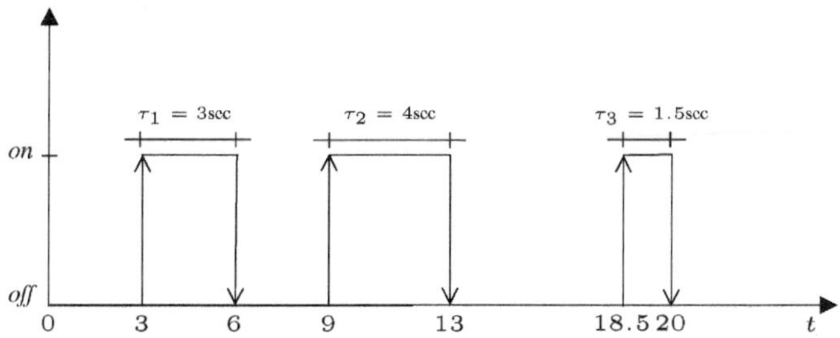

FIGURE 7.2 Example of DoS signal. Off/on transitions are represented as $\uparrow$, while on/off transitions are represented as $\downarrow$. Off/on transitions occur at 3 s, 9 s, and 18.5 s and the corresponding DoS intervals have a duration of 3 s, 4 s, and 1.5 s, respectively. This yields for instance $n(0,1) = 0$, $n(1,10) = 2$ and $n(10,20) = 1$, while $\Xi(0,1) = \emptyset$, $\Xi(1,10) = [3,6[\cup[9,10[$, and $\Xi(10,20) = [10,13[\cup[18.5,20[$.

Fig. 7.2 exemplifies values of $n(\tau,t)$ and $\Xi(\tau,t)$ for a given DoS pattern.

Remark 7.2. Assumptions 7.1 and 7.2 specify the class of DoS signals that will be considered throughout the remainder of this chapter. It is worth noting that no assumption is made on the information available to the attacker about the process dynamics, state-feedback matrix, and sampling logic. The assumptions only constrain DoS in terms of its frequency and duration. In addition to rendering the control problem meaningful, limiting the DoS frequency and duration also has a practical motivation. There are several steps that can be taken in order to mitigate DoS attacks, including spreading techniques and high-pass filtering (see, e.g., [284], [286], [103]). These provisions decrease the chance that a DoS attack will be successful, and limit in practice the frequency and duration of the time intervals over which communication is effectively denied.

7.2.2 Input-to-state stability under denial of service

We are now in position to derive the main result of this section, which can be expressed as follows: *Any control update rule attaining the conditions of Lemma 7.1 preserves ISS for any DoS signal that satisfies Assumptions 7.1 and 7.2 with τ_D and T sufficiently large.* Although the proof of this result is rather involved, the underlying approach is very intuitive. We decompose the time axis into intervals where it possible to satisfy (7.12) and intervals where, due to the occurrence of DoS, (7.12) need not hold. We then analyze the closed-loop dynamics as a system switching between stable and unstable modes, and determine values of τ_D and T under which the stable behavior is predominant with respect to the unstable one.

Consider a sequence $\{t_k\}_{k\in\mathbb{N}_0}$ of sampling times, along with a DoS sequence $\{h_n\}_{n\in\mathbb{N}_0}$. Let

$$\mathscr{S} := \left\{ k \in \mathbb{N}_0 \mid t_k \in \bigcup_{n \in \mathbb{N}_0} H_n \right\} \tag{7.32}$$

denote the set of integers related to a control update attempt occurring under DoS.

The following result holds.

Theorem 7.2. *Consider the control system Σ composed of (7.1) and control input (7.6), where K is such that all the eigenvalues of $\Phi = A + BK$ have a negative real part. Given any positive symmetric definite matrix $Q \in \mathbb{R}^{n \times n}$, let P be the unique solution of the Lyapunov equation $\Phi^\top P + P\Phi + Q = 0$. Let $V(x) = x^\top Px$. Consider any control update sequence occurring at a finite sampling rate and with inter-sampling times smaller than or equal to $\bar{\Delta}_\sigma$ as in Lemma 7.1, with σ satisfying (7.15). Consider any DoS sequence satisfying Assumptions 7.1 and 7.2 with arbitrary η and κ, and with τ_D and T such that*

$$\frac{\Delta_*}{\tau_D} + \frac{1}{T} < \frac{\omega_1}{\omega_1 + \omega_2}, \tag{7.33}$$

where Δ_* is a nonnegative constant satisfying

$$\sup_{k \in \mathscr{S}} \Delta_k \leq \Delta_*, \tag{7.34}$$

Δ_k is as in (7.9), $\omega_1 := (\gamma_1 - \gamma_2\sigma)/2\alpha_2$ and $\omega_2 := 2\gamma_2/\alpha_1$, where α_1, α_2, γ_1, and γ_2 are as in (7.14). Then Σ is ISS.

The idea is to decompose the time axis into intervals where it possible to satisfy (7.12) and intervals where, due to the occurrence of DoS, (7.12) need not hold. We then analyze the closed-loop dynamics as a system that switches between stable and unstable modes. For clarity of exposition the proof is divided into three steps.

Step I. Modeling of the Intervals Related to Stable and Unstable Dynamics: In this step we characterize the intervals of time where (7.12) holds and those where it need not hold. During these intervals the closed-loop system evolves obeying to stable and possibly unstable dynamics, respectively. The characterization of these intervals is essential for the Lyapunov-based analysis we carry out in the next steps and can be formalized as follows:

Lemma 7.2. *For any $\tau, t \in \mathbb{R}_{\geq 0}$, with $0 \leq \tau \leq t$, the interval $[\tau, t]$ is the disjoint union of $\bar{\Theta}(\tau, t)$ and $\bar{\Xi}(\tau, t)$, where $\bar{\Theta}(\tau, t)$ (respectively, $\bar{\Xi}(\tau, t)$) is the union of subintervals of $[\tau, t]$ over which (7.12) holds (respectively, need not hold). Specifically, there exist two sequences of nonnegative and positive real numbers $\{\zeta_m\}_{m \in \mathbb{N}_0}$, $\{v_m\}_{m \in \mathbb{N}_0}$ such that*

$$\bar{\Xi}(\tau, t) := \bigcup_{m \in \mathbb{N}_0} Z_m \cap [\tau, t], \tag{7.35}$$

$$\bar{\Theta}(\tau,t) \quad := \quad \bigcup_{m\in\mathbb{N}_0} W_{m-1} \cap [\tau,t], \tag{7.36}$$

where

$$Z_m \quad := \quad \{\zeta_m\} \cup [\zeta_m, \zeta_m + \upsilon_m[, \tag{7.37}$$
$$W_m \quad := \quad \{\zeta_m + \upsilon_m\} \cup [\zeta_m + \upsilon_m, \zeta_{m+1}[, \tag{7.38}$$

and where $\zeta_{-1} = \upsilon_{-1} := 0$.

Proof. Let $\mathscr{S}_n := \{k \in \mathbb{N}_0 | t_k \in H_n\}$ denote the set of integers related to a control update attempt occurring over $H_n, n \in \mathbb{N}_0$. Define

$$\lambda_n \quad := \quad \begin{cases} \tau_n, & \text{if } \mathscr{S}_n = \emptyset \\ t_{\sup\{k\in\mathbb{N}_0 : k\in\mathscr{S}_n\}} - h_n, & \text{otherwise} \end{cases} \tag{7.39}$$

$$\Lambda_n \quad := \quad \begin{cases} 0, & \text{if } \mathscr{S}_n = \emptyset \\ \Delta_{\sup\{k\in\mathbb{N}_0 : k\in\mathscr{S}_n\}}, & \text{otherwise.} \end{cases} \tag{7.40}$$

Thus,

$$\bar{H}_n := \{h_n\} \cup [h_n, h_n + \lambda_n + \Lambda_n[\tag{7.41}$$

specifies the n-th time interval where (7.12) need not hold, which consists of H_n plus the corresponding DoS-induced actuation delay. Note that $\lambda_n + \Lambda_n \geq \tau_n$ for all $n \in \mathbb{N}_0$. Note now that the intervals $\bar{H}_n$ and $\bar{H}_{n+1}$ may overlap each other in that h_{n+1} may belong to $\bar{H}_n$. For analysis purposes, it is convenient to regard these overlapping intervals as a single interval of the form (7.37). This can be done by defining an auxiliary sequence $\{\zeta_m\}_{m\in\mathbb{N}_0}$, which is recursively defined from $\{h_n\}_{n\in\mathbb{N}_0}$ as follows:

$$\zeta_0 \quad := \quad h_0 \tag{7.42}$$
$$\zeta_{m+1} \quad := \quad \inf\{\, h_n > \zeta_m \mid h_n > h_{n-1} + \lambda_{n-1} + \Lambda_{n-1}\} \tag{7.43}$$

for all $m \in \mathbb{N}$, and letting

$$\upsilon_m := \sum_{\substack{n\in\mathbb{N}_0; \\ \zeta_m \leq h_n < \zeta_{m+1}}} |\bar{H}_n \backslash \bar{H}_{n+1}| \tag{7.44}$$

for all $m \in \mathbb{N}_0$. Given $\tau, t \in \mathbb{R}_{\geq 0}$, define $\bar{\Xi}(\tau,t)$ as in (7.35). Hence, by construction, $\bar{\Xi}(\tau,t)$ represents the union of subintervals of $[\tau,t]$ where, due to the occurrence of DoS, (7.12) need not hold.

Let W_m be as in (7.38). Given $\tau, t \in \mathbb{R}_{\geq 0}$, let $\bar{\Theta}(\tau,t)$ as in (7.36). Note that over each $[\tau,t]$ the sets $\bar{\Xi}(\tau,t)$ and $\bar{\Theta}(\tau,t)$ are complementary since their union equals $[\tau, t]$ and their intersection is the empty set. Also note that $\bar{\Theta}(\tau, t)$ represents the union of subintervals of $[\tau, t]$ where (7.12) holds true. By construction, for each $m \in \mathbb{N}_0$, a successful control update necessarily occurs at $\zeta_m + v_m$ and there is no DoS over W_m. □

Step II. Lyapunov Function Analysis: Consider the intervals $W_m, m \in \mathbb{N}_0$, where (7.12) holds true by construction. From (7.19) we get

$$V(x(t)) \leq e^{-\omega_1(t-\zeta_m-v_m)} V(x(\zeta_m + v_m)) + \gamma_7 \|w_t\|_\infty^2 \tag{7.45}$$

for all $t \in W_m$ and all $m \in \mathbb{N}_0$. Here and in the following, all the relevant constants are as in the proof of Theorem 7.1. In particular we recall that $\gamma_7 = \alpha_2(\gamma_3 + \sigma\gamma_2)^2/(\gamma_1 - \sigma\gamma_2)^2$ with $\sigma < \gamma_1/\gamma_2$.

Consider next the intervals $Z_m, m \in \mathbb{N}_0$, over which the inequality (7.12) does not necessarily hold. To derive a bound on the growth of $V(x(t))$ in this case, some intermediate steps are needed. We first prove that for each $m \in \mathbb{N}_0$,

$$\|e(t)\| \leq (1+\sigma)\|x(\zeta_m)\| + \|x(t)\| + \sigma\|w_t\|_\infty \tag{7.46}$$

for all $t \in Z_m$. Recall that

$$e(t) = x\left(t_{k(\zeta_m)}\right) - x(t) \tag{7.47}$$

for all $t \in Z_m$, where $x(t_{k(\zeta_m)})$ represents the value of the process state at the last successful control update up to ζ_m. If $\zeta_0 = 0$, then $x(t_{k(\zeta_0)}) = 0$ by definition so that (7.46) is valid. Consider next the case $\zeta_0 > 0$. By construction, (7.12) holds true for all $t \in W_m$. Hence, by continuity of x we have

$$\|e(\zeta_m)\| \leq \sigma\|x(\zeta_m)\| + \sigma\|w_{\zeta_m}\|_\infty \tag{7.48}$$

for all $m \in \mathbb{N}_0$. Hence,

$$\|x(t_{k(\zeta_m)}) - x(\zeta_m)\| \leq \sigma\|x(\zeta_m)\| + \sigma\|w_{\zeta_m}\|_\infty \tag{7.49}$$

and (7.46) follows by applying the triangular inequality.

Substituting (7.46) into (7.14b) yields

$$\begin{aligned}\frac{d}{dt}V(x(t)) \quad &\leq \quad (\gamma_2 - \gamma_1)\|x(t)\|^2 + \gamma_2(1+\sigma)\|x(t)\|\,\|x(\zeta_m)\| \\ &\quad + (\gamma_3 + \sigma\gamma_2)\|x(t)\|v(t),\end{aligned} \tag{7.50}$$

where $v(t) := \sup\{\|w(t)\|, \|w_t\|_\infty\}$. We then proceed as in the proof of Theorem 7.1. Using (7.17) with $\delta = (\gamma_3 + \gamma_2\sigma)/(\gamma_1 - \sigma\gamma_2)$, simple calculations yield

$$\frac{d}{dt}V(x(t)) \le \gamma_2(1-\sigma)\|x(t)\|^2 \\ +\gamma_2(1+\sigma)\|x(t)\|\,\|x(\zeta_m)\| + \gamma_6 v^2(t), \tag{7.51}$$

where we recall that $\gamma_6 = (\gamma_3 + \gamma_2\sigma)^2/(2(\gamma_1 - \sigma\gamma_2))$. Note that

$$\frac{d}{dt}V(x(t)) \le \omega_2 \max\{V(x(t)), V(x(\zeta_m))\} + \gamma_6 v^2(t), \tag{7.52}$$

where $\omega_2 := 2\gamma_2/\alpha_1$. Since $\|v_t\|_\infty = \|w_t\|_\infty$ for any $t \in \mathbb{R}_{\ge 0}$, we then have

$$V(x(t)) \le e^{\omega_2(t-\zeta_m)} V(x(\zeta_m)) + \gamma_8 e^{\omega_2(t-\zeta_m)} \|w_t\|_\infty^2 \tag{7.53}$$

for all $t \in Z_m$, where $\gamma_8 := \gamma_6/\omega_2$.

Combining (7.45) and (7.53), we can prove the following result.

Lemma 7.3. *For all $t \in \mathbb{R}_{\ge 0}$, the Lyapunov function satisfies*

$$V(x(t)) \le e^{-\omega_1|\bar{\Theta}(0,t)|} e^{\omega_2|\bar{\Xi}(0,t)|} V(x(0)) \\ +\gamma_* \left[1 + 2 \sum_{\substack{m\in\mathbb{N}_0;\\ \zeta_m \le t}} e^{-\omega_1|\bar{\Theta}(\zeta_m+v_m,t)|} e^{\omega_2|\bar{\Xi}(\zeta_m,t)|} \right] \|w_t\|_\infty^2, \tag{7.54}$$

where $\gamma_* := \max\{\gamma_7, \gamma_8\}$.

Hereafter, in accordance with (7.36), it is understood that $|\bar{\Theta}(\zeta_m + v_m, t)| = 0$ whenever $t < \zeta_m + v_m$.

Proof of Lemma 7.3. We use an induction argument. First we show that the inequality holds true over $W_{-1} = [0, \zeta_0]$. If $\zeta_0 = 0$, the claim trivially holds. Suppose $\zeta_0 > 0$. Over W_{-1} the Lyapunov function obeys (7.45); thus, (7.54) follows by noting that $|\bar{\Theta}(0,t)| = t$ and $|\bar{\Xi}(0,t)| = 0$ for all $t \in W_{-1}$ and the sum term in (7.54) is zero.

Assume next that (7.54) holds true over the interval $[0, \zeta_p]$, where $p \in \mathbb{N}_0$. By hypothesis, and since $V(x)$ is continuous, we have

$$V(x(\zeta_p)) \le e^{-\omega_1|\bar{\Theta}(0,\zeta_p)|} e^{\omega_2|\bar{\Xi}(0,\zeta_p)|} V(x(0)) \\ +\gamma_* \left[1 + 2 \sum_{\substack{m\in\mathbb{N}_0;\\ \zeta_m < \zeta_p}} e^{-\omega_1|\bar{\Theta}(\zeta_m+v_m,\zeta_p)|} e^{\omega_2|\bar{\Xi}(\zeta_m,\zeta_p)|} \right] \|w_{\zeta_p}\|_\infty^2. \tag{7.55}$$

Then, consider first the interval Z_p. Over Z_p, the Lyapunov function obeys (7.53) with

$$V(x(t)) \le e^{\omega_2(t-\zeta_p)} V\left(x(\zeta_p)\right) + \gamma_* e^{\omega_2(t-\zeta_p)} \|w_t\|_\infty^2. \tag{7.56}$$

Note that $t - \zeta_p = |\bar{\Xi}(\zeta_p, t)|$ for all $t \in Z_p$. Hence,

$$|\bar{\Xi}(0,t)| = t - \zeta_p + |\bar{\Xi}(0,\zeta_p)| \tag{7.57}$$

$$|\bar{\Theta}(0,t)| = |\bar{\Theta}(0,\zeta_p)| \tag{7.58}$$

for all $t \in Z_p$. Thus, multiplying $e^{\omega_2(t-\zeta_p)}$ by the first term on the right-hand side (RHS) of (7.55) we get the first term on the RHS of (7.54). In addition, multiplying $e^{\omega_2(t-\zeta_p)}$ by the sum term in (7.55) we obtain

$$\begin{aligned} & e^{\omega_2(t-\zeta_p)} \sum_{\substack{m\in\mathbb{N}_0;\\ \zeta_m<\zeta_p}} e^{-\omega_1|\bar{\Theta}(\zeta_m+\upsilon_m,\zeta_p)|} e^{\omega_2|\bar{\Xi}(\zeta_m,\zeta_p)|} \\ & = \sum_{\substack{m\in\mathbb{N}_0;\\ \zeta_m<\zeta_p}} e^{-\omega_1|\bar{\Theta}(\zeta_m+\upsilon_m,t)|} e^{\omega_2|\bar{\Xi}(\zeta_m,t)|} \end{aligned} \tag{7.59}$$

for all $t \in Z_p$ because $|\bar{\Theta}(\zeta_m + \upsilon_m, \zeta_p)| = |\bar{\Theta}(\zeta_m + \upsilon_m, t)|$ and $t - \zeta_p + |\bar{\Xi}(\zeta_m, \zeta_p)| = |\bar{\Xi}(\zeta_m, t)|$ for all $t \in Z_p$. Overall, the factor multiplying the disturbance term can therefore be rewritten as

$$\begin{aligned} & 2\gamma_* e^{\omega_2(t-\zeta_p)} + 2\gamma_* \sum_{\substack{m\in\mathbb{N}_0;\\ \zeta_m<\zeta_p}} e^{-\omega_1|\bar{\Theta}(\zeta_m+\upsilon_m,t)|} e^{\omega_2|\bar{\Xi}(\zeta_m,t)|} \\ & = 2\gamma_* \sum_{\substack{m\in\mathbb{N}_0;\\ \zeta_m\le\zeta_p}} e^{-\omega_1|\bar{\Theta}(\zeta_m+\upsilon_m,t)|} e^{\omega_2|\bar{\Xi}(\zeta_m,t)|} \\ & = 2\gamma_* \sum_{\substack{m\in\mathbb{N}_0;\\ \zeta_m\le t}} e^{-\omega_1|\bar{\Theta}(\zeta_m+\upsilon_m,t)|} e^{\omega_2|\bar{\Xi}(\zeta_m,t)|}, \end{aligned} \tag{7.60}$$

where the first equality follows from $|\bar{\Theta}(\zeta_p + \upsilon_p, t)| = 0$ and $t - \zeta_p = |\bar{\Xi}(\zeta_p, t)|$ for all $t \in Z_p$, while the second equality holds because the condition $\zeta_m \le \zeta_p$ is equivalent to $\zeta_m \le t$ for all $t \in Z_p$. Hence, (7.54) holds true for all $t \in [0, \zeta_p + \upsilon_p]$.

Consider next the interval W_p. Over W_p the Lyapunov function obeys (7.45). In particular,

$$\begin{aligned} V(x(t)) \le\ & e^{-\omega_1(t-\zeta_p-\upsilon_p)} e^{\omega_2\upsilon_p} V(x(\zeta_p)) \\ & + \gamma_* \left[1 + e^{-\omega_1(t-\zeta_p-\upsilon_p)} e^{\omega_2\upsilon_p}\right] \|w_t\|_\infty^2 \end{aligned} \tag{7.61}$$

for all $t \in W_p$, with $V(x(\zeta_p))$ as in (7.55). To see that this implies again (7.54), note that $t - \zeta_p - \upsilon_p = |\bar{\Theta}(\zeta_p + \upsilon_p, t)|$ and $|\bar{\Theta}(0, \zeta_p)| = |\bar{\Theta}(0, \zeta_p + \upsilon_p)|$. Hence,

$$|\bar{\Theta}(0,t)| = t - \zeta_p - v_p + |\bar{\Theta}(0,\zeta_p)| \tag{7.62}$$

for all $t \in W_p$. Moreover, $v_p = |\bar{\Xi}(\zeta_p, \zeta_p + v_p)|$ and $|\bar{\Xi}(\zeta_p + v_p, t)| = 0$ for all $t \in W_p$. Hence,

$$|\bar{\Xi}(0,t)| = v_p + |\bar{\Xi}(0,\zeta_p)| \tag{7.63}$$

for all $t \in W_p$. Thus, multiplying $e^{-\omega_1(t-\zeta_p-v_p)}e^{\omega_2 v_p}$ by the first term on the RHS of (7.55) we get the first term on the RHS of (7.54). Moreover, multiplying $e^{-\omega_1(t-\zeta_p-v_p)}e^{\omega_2 v_p}$ by the sum term in (7.55), we obtain

$$\begin{aligned} & e^{-\omega_1(t-\zeta_p-v_p)}e^{\omega_2 v_p} \sum_{\substack{m\in\mathbb{N}_0;\\ \zeta_m<\zeta_p}} e^{-\omega_1|\bar{\Theta}(\zeta_m+v_m,\zeta_p)|}e^{\omega_2|\bar{\Xi}(\zeta_m,\zeta_p)|} \\ & = \sum_{\substack{m\in\mathbb{N}_0;\\ \zeta_m<\zeta_p}} e^{-\omega_1|\bar{\Theta}(\zeta_m+v_m,t)|}e^{\omega_2|\bar{\Xi}(\zeta_m,t)|}. \end{aligned} \tag{7.64}$$

Overall the factor multiplying the disturbance term can be therefore rewritten as

$$\begin{aligned} & \gamma_* \left[1 + 2e^{-\omega_1(t-\zeta_p-v_p)}e^{\omega_2 v_p}\right] \\ & \quad +2\gamma_* \sum_{\substack{m\in\mathbb{N}_0;\\ \zeta_m<\zeta_p}} e^{-\omega_1|\bar{\Theta}(\zeta_m+v_m,t)|}e^{\omega_2|\bar{\Xi}(\zeta_m,t)|} \\ & = 2\gamma_* \sum_{\substack{m\in\mathbb{N}_0;\\ \zeta_m\le\zeta_p}} e^{-\omega_1|\bar{\Theta}(\zeta_m+v_m,t)|}e^{\omega_2|\bar{\Xi}(\zeta_m,t)|} \\ & = 2\gamma_* \sum_{\substack{m\in\mathbb{N}_0;\\ \zeta_m\le t}} e^{-\omega_1|\bar{\Theta}(\zeta_m+v_m,t)|}e^{\omega_2|\bar{\Xi}(\zeta_m,t)|}. \end{aligned} \tag{7.65}$$

As before, the second equality holds because, for all $t \in W_p$, the condition $\zeta_m \le \zeta_p$ is equivalent to $\zeta_m \le t$. Thus, (7.54) holds true for all $t \in [0, \zeta_{p+1}]$, which concludes the proof. □

Step III. Bounds on DoS Frequency and Duration: In order to conclude the proof, we need to bound the sum term in (7.54). The following can be proven:

Lemma 7.4. *Under Assumptions 7.1 and 7.2 and condition (7.33), the sum*

$$\sum_{\substack{m\in\mathbb{N}_0;\\ \zeta_m\le t}} e^{-\omega_1|\bar{\Theta}(\zeta_m+v_m,t)|}e^{\omega_2|\bar{\Xi}(\zeta_m,t)|} \tag{7.66}$$

is bounded from above by

$$e^{(\omega_1+\omega_2)\kappa_*}\frac{e^{\beta_*\tau_D\eta}}{1-e^{-\beta_*\tau_D}}, \tag{7.67}$$

where the constants ω_1, ω_2, τ_D, κ_*, and β_* are defined in Theorem 7.2, (7.85), and (7.86).

Proof. First note that

$$|\bar{\Xi}(\tau,t)| \leq |\Xi(\tau,t)| + (1+n(\tau,t))\Delta_* \tag{7.68}$$

for all $\tau, t \in \mathbb{R}_{\geq 0}$ with $t \geq \tau$. In other words, $|\bar{\Xi}(\tau,t)|$ can be upper bounded by the total length of DoS over $[\tau,t]$ plus the maximum actuation delay Δ_*, which may occur once at the beginning of the interval $[\tau,t]$ (as a result of a previous DoS) plus $n(\tau,t)$ times, where $n(\tau,t)$ represents the number of off/on transitions of DoS occurring over $[\tau,t[$. Using Assumptions 7.1 and 7.2 we have

$$\begin{aligned} |\bar{\Xi}(\tau,t)| &\leq \kappa + \frac{t-\tau}{T} + \left(1+\eta+\frac{t-\tau}{T}\right)\Delta_* \\ &= :\kappa_* + \frac{t-\tau}{T_*}, \end{aligned} \tag{7.69}$$

where $\kappa_* := \kappa + (1+\eta)\Delta_*$ and $T_* := \tau_D T/(\tau_D + T\Delta_*)$. With this in mind, we can now analyze the sum term in (7.66). From the above inequality we have

$$|\bar{\Xi}(\zeta_m,t)| \leq \kappa_* + \frac{t-\zeta_m}{T_*} \tag{7.70}$$

for all $t \in \mathbb{R}_{\geq \zeta_m}$, $m \in \mathbb{N}_0$. Consider next $|\bar{\Theta}(\zeta_m + v_m, t)|$. We have

$$|\bar{\Theta}(\zeta_m + v_m, t)| = t - \zeta_m - |\bar{\Xi}(\zeta_m,t)| \tag{7.71}$$

for all $t \in \mathbb{R}_{\geq \zeta_m}$, $m \in \mathbb{N}_0$. To see this, consider first the interval Z_m. For all $t \in Z_m$, we have $|\bar{\Xi}(\zeta_m,t)| = t - \zeta_m$ so that (7.71) holds true since, by definition, $|\bar{\Theta}(\zeta_m + v_m, t)| = 0$ whenever $t < \zeta_m + v_m$. For all $t \in \mathbb{R}_{\geq \zeta_m + v_m}$ we have

$$\begin{aligned} |\bar{\Theta}(\zeta_m + v_m, t)| &= t - \zeta_m - v_m - |\bar{\Xi}(\zeta_m + v_m, t)| \\ &= t - \zeta_m - |\bar{\Xi}(\zeta_m,t)|, \end{aligned} \tag{7.72}$$

where the second equality follows because

$$\begin{aligned} |\bar{\Xi}(\zeta_m + v_m, t)| &= |\bar{\Xi}(\zeta_m,t)| - |\bar{\Xi}(\zeta_m, \zeta_m + v_m)| \\ &= |\bar{\Xi}(\zeta_m,t)| - v_m \end{aligned} \tag{7.73}$$

for all $t \in \mathbb{R}_{\geq \zeta_m + v_m}$, $m \in \mathbb{N}_0$. Hence,

$$\sum_{\substack{m\in\mathbb{N}_0;\\ \zeta_m \leq t}} e^{-\omega_1|\bar{\Theta}(\zeta_m+v_m,t)|} e^{\omega_2|\bar{\Xi}(\zeta_m,t)|}$$

$$\leq e^{(\omega_1+\omega_2)\kappa_*} \sum_{\substack{m\in\mathbb{N}_0;\\ \zeta_m\leq t}} e^{-[\omega_1-(\omega_1+\omega_2)/T_*](t-\zeta_m)}$$
$$= e^{(\omega_1+\omega_2)\kappa_*} \sum_{\substack{m\in\mathbb{N}_0;\\ \zeta_m\leq t}} e^{-\beta_*(t-\zeta_m)}, \tag{7.74}$$

where $\beta_* = \omega_1 - (\omega_1+\omega_2)/T_*$. Under condition (7.33), we have $\beta_* > 0$. Moreover, Assumption 7.1 yields

$$t - \zeta_m \geq \tau_D n(\zeta_m, t) - \tau_D \eta. \tag{7.75}$$

This implies

$$\sum_{\substack{m\in\mathbb{N}_0;\\ \zeta_m\leq t}} e^{-\beta_*(t-\zeta_m)} \leq e^{\beta_*\tau_D\eta} \sum_{\substack{m\in\mathbb{N}_0;\\ \zeta_m\leq t}} e^{-\beta_*\tau_D n(\zeta_m,t)}. \tag{7.76}$$

Let

$$m(t) := \begin{cases} -1, & \text{if } t < \zeta_0 \\ \sup\{m \in \mathbb{N}_0 | \zeta_m \leq t\}, & \text{otherwise.} \end{cases} \tag{7.77}$$

We then have

$$\sum_{\substack{m\in\mathbb{N}_0;\\ \zeta_m\leq t}} e^{-\beta_*\tau_D n(\zeta_m,t)} = \sum_{m=0}^{m(t)} e^{-\beta_*\tau_D n(\zeta_m,t)}. \tag{7.78}$$

Since

$$n(\zeta_m, t) \geq m(t) - m \tag{7.79}$$

then

$$\sum_{m=0}^{m(t)} e^{-\beta_*\tau_D n(\zeta_m,t)} = \sum_{m=0}^{m(t)} e^{-\beta_*\tau_D (m(t)-m)}$$
$$= \sum_{m=0}^{m(t)} e^{-\beta_*\tau_D m} \leq \frac{1}{1-e^{-\beta_*\tau_D}}. \tag{7.80}$$

The relations (7.74), (7.76), (7.78), and (7.80) yield the desired bound (7.67).

We can now finalize the proof of Theorem (7.2). First observe that the first term on the RHS of (7.54) is bounded by

$$e^{\kappa_*(\omega_1+\omega_2)} e^{-(\omega_1 - \frac{\omega_1+\omega_2}{T_*})t} V(x(0)), \tag{7.81}$$

where we have exploited the equality $|\bar{\Theta}(0,t)| = t - |\bar{\Xi}(0,t)|$ and the inequality (7.69). Moreover, $\beta_* = \omega_1 - (\omega_1 + \omega_2)/T_*$. Hence, from (7.54) and (7.67), we see that the Lyapunov function $V(x)$ computed along any trajectory $x(t)$ satisfies

$$\begin{aligned} V(x(t)) \;&\le\; e^{\kappa_*(\omega_1+\omega_2)} e^{-\beta_* t} V(x(0)) \\ &\quad + \gamma_* \left[1 + 2e^{\kappa_*(\omega_1+\omega_2)} \frac{e^{\beta_* \tau_D \eta}}{1 - e^{-\beta_* \tau_D}} \right] \| w_t \|_\infty^2. \end{aligned} \tag{7.82}$$

Letting $\delta_* := \gamma_*/\alpha_1$, we readily get

$$\begin{aligned} \|x(t)\| \;&\le\; \sqrt{\frac{\alpha_2}{\alpha_1} \frac{e^{\kappa_*(\omega_1+\omega_2)} 2^{-(\beta_*/2)t}}{e}} \; \|x(0)\| \\ &\quad + \sqrt{\delta_*} \left[1 + 2e^{\kappa_*(\omega_1+\omega_2)} \frac{e^{\beta_* \tau_D \eta}}{1 - e^{-\beta_* \tau_D}} \right]^{\frac{1}{2}} \| w_t \|_\infty. \end{aligned} \tag{7.83}$$

Note that $\delta_* = ((\gamma_3 + \gamma_2\sigma)^2/(\gamma_1 - \gamma_2\sigma)) \max\{(\alpha_2/(\alpha_1(\gamma_1 - \gamma_2\sigma))), (1/4\gamma_2)\}$ is a positive constant (i.e., independent of the process initial condition and the disturbance w). Since Δ_* is a positive constant, then also $\kappa_* = \kappa + (1+\eta)\Delta_*$ and $\beta_* = \omega_1 - (\omega_1 + \omega_2)((\Delta_*/\tau_D) + (1/T))$ are independent of the process initial condition and w. Thus ISS follows at once. □

In connection with (7.34) and in accordance with the adopted notation, we note that $\sup_{k \in \mathscr{S}} \Delta_k = 0$ when $\mathscr{S} = \emptyset$.

Remark 7.3. (Performance Bounds): The considered class of sampling logic provides very strong stability properties, namely *exponential* ISS with linear bounds on the disturbance-to-state map. In particular, from the proof of Theorem 7.2 we have

$$\begin{aligned} \|x(t)\| \;&\le\; \sqrt{\frac{\alpha_2}{\alpha_1}} e^{\frac{\kappa_*(\omega_1+\omega_2)}{2}} e^{-(\frac{\beta_*}{2})t} \|x(0)\| \\ &\quad + \sqrt{\delta_*} \left[1 + 2e^{\kappa_*(\omega_1+\omega_2)} \frac{e^{\beta_* \tau_D \eta}}{1 - e^{-\beta_* \tau_D}} \right]^{\frac{1}{2}} \| w_t \|_\infty, \end{aligned} \tag{7.84}$$

where

$$\kappa_* \;:=\; \kappa + (1+\eta)\Delta_* \tag{7.85}$$

$$\beta_* \;:=\; \omega_1 - (\omega_1 + \omega_2)\left(\frac{\Delta_*}{\tau_D} + \frac{1}{T} \right) \tag{7.86}$$

$$\delta_* \;:=\; \frac{(\gamma_3 + \gamma_2\sigma)^2}{(\gamma_1 - \gamma_2\sigma)} \max\left\{ \frac{\alpha_2}{\alpha_1(\gamma_1 - \gamma_2\sigma)}, \frac{1}{4\gamma_2} \right\} \tag{7.87}$$

with α_1, α_2, γ_1, γ_2, and γ_3 as in (7.14). We see that the convergence rate of the closed-loop dynamics and the amount of DoS that we can tolerate can be increased by decreasing the sampling rate upon the DoS occurrence (i.e., by decreasing Δ_*).

Remark 7.4. (DoS-Induced Actuation Delay): Decreasing Δ_* can also by viewed as reducing the DoS-induced actuation delay. Assume for instance that the k-th sampling instant belongs to $\mathscr{S}$. Since $t_k \in \mathscr{S}$, the process will evolve in open-loop under out-of-date control up to time t_{k+1}, even if DoS ceases before t_{k+1}. Thus, decreasing Δ_* has the effect of reducing the delay in the actuation of the new control sample. In general, due to the finite sampling rate, this delay is nonzero. In fact, $\Delta_* = 0$ only in two cases. The first case is when $\mathscr{S} = \emptyset$. This case is not interesting because if $\mathscr{S} = \emptyset$, then DoS is ineffective and ISS follows directly from Theorem 7.1. The second case is when $\mathscr{S} \neq \emptyset$ and, for each DoS interval H_n containing a t_k, the control update is scheduled exactly at $h_n + \tau_n$. This case is, however, unrealistic since h_n and τ_n are unknown. The case $\mathscr{S} \neq \emptyset$ and $\Delta_* = 0$ becomes of interest when considering control under DoS in a purely continuous-time framework. If the control action is continuous, then $\Delta_* = \bar{\Delta}_\sigma = 0$. Under this condition, (7.33) reduces to $1/T < \omega_1/(\omega_1 + \omega_2)$, which is independent of τ_D. In a continuous-time setting, slow-on-the-average DoS in the form of pulses has no effect on closed-loop stability.

7.2.3 Disturbance-free case

An interesting simplification arises in the disturbance-free case. Specifically, when $w \equiv 0$ then Assumptions 7.1 and 7.2 can be relaxed as follows:

Assumption 7.3 (DoS frequency–disturbance-free case). There exists $\eta \in \mathbb{R}_{\geq 0}$ and $\tau_D \in \mathbb{R}_{>\underline{\Delta}}$ such that

$$n(0,t) \leq \eta + \frac{t}{\tau_D} \tag{7.88}$$

for all $t \in \mathbb{R}_{\geq 0}$.

Assumption 7.4 (DoS duration–disturbance-free case). There exists $\kappa \in \mathbb{R}_{\geq 0}$ and $T \in \mathbb{R}_{>1}$ such that

$$|\Xi(0,t)| \leq \kappa + \frac{t}{T} \tag{7.89}$$

for all $t \in \mathbb{R}_{\geq 0}$.

The following result holds.

Corollary 7.1. *Consider the control system Σ composed of (7.1) and control input (7.6), where K is such that all the eigenvalues of $\Phi = A + BK$ have a negative real part. Given any positive symmetric definite matrix $Q \in \mathbb{R}^{n \times n}$, let P be the unique solution of the Lyapunov equation $\Phi^\top P + P\Phi + Q = 0$. Let $V(x) = x^\top P x$. Consider any control update sequence occurring at a finite* *sampling rate and with inter-sampling times smaller than or equal to $\bar{\Delta}_\sigma$, as in Lemma 7.1, with σ as in (7.15). Consider any DoS sequence satisfying Assumptions 7.3 and 7.4 with arbitrary η and κ, and with τ_D and T satisfying (7.33). Then, Σ is GAS.*

Proof. It follows directly from the proof of Theorem 7.2 i) given that the rightmost term in (7.54) is zero and ii) exploiting the equality $|\bar{\Theta}(0,t)| = t - |\bar{\Xi}(0,t)|$ and the inequality (7.69) with $\tau = 0$, where the sets $\bar{\Xi}(\tau,t)$ and $\bar{\Theta}(\tau,t)$ are defined in (7.35) and (7.36), respectively. □

Following the same lines as in Remark 7.3, it is easy to see that the convergence rate of the closed-loop trajectories is exponential.

7.2.4 Resilient control logic

The considered framework is flexible enough to allow the designer to choose from several implementation options that can be used to trade off between performance and communication resources. Although the solutions described in the following originate from a fundamentally different logic, they exhibit the common feature of *resilience*, by which we mean the ability to adapt the sampling rate to the occurrence of DoS and, sometimes, to closed-loop behavior. Here only a few examples of sampling logic will be discussed; combinations and variants of the examples can be easily envisaged.

7.2.5 Periodic sampling logic

The simplest architecture we can think of for implementing the control action is through periodic sampling. In accordance with Theorem 7.2, the sampling rate affects the amount of DoS that can be tolerated via (7.33). Since (7.33) only depends on the sampling rate that is adopted during DoS, a convenient strategy consists in making use of a two-mode switching rule, where upon DoS the sampling rate is increased so as to reduce the DoS-induced delay.

Proposition 7.1. *Let δ_1 and δ_2 be positive constants such that $\delta_1 \leq \delta_2 \leq \bar{\Delta}_\sigma$, where $\bar{\Delta}_\sigma$ is as in Lemma 7.1. For each $k \in \mathbb{N}_0$, let the sampling times be given by*

$$t_{k+1} = \begin{cases} t_k + \delta_1, & \text{if } k \in \mathcal{S} \\ t_k + \delta_2, & \text{otherwise} \end{cases}, \tag{7.90}$$

where $\mathcal{S}$ is as in (7.32). If the conditions of Theorem 7.2 hold true with $\Delta_* = \delta_1$, then the control system Σ composed of (7.1) and (7.6) is ISS. Moreover, the inter-sampling times are bounded from below by $\underline{\Delta} = \delta_1$.

Proof. In accordance with (7.90), the control update occurs at a finite time with an inter-sampling time not greater than $\delta_2 \leq \bar{\Delta}_\sigma$. Hence, ISS follows immediately from Theorem 7.2. The bounds on Δ_* and $\underline{\Delta}$ hold by construction. □

7.3 Event-based periodic sampling logic

As discussed in [292], [298], and [299], event-based sampling is a very effective solution for saving communication resources. The basic idea is that periodic sampling can be relaxed by triggering the control updates only when certain conditions (events) are met. In its basic formulation this is achieved by measuring continuously the state x and triggering a control update whenever

$$\|x(t_k) - x(t)\| = \sigma \|x(t)\|, \tag{7.91}$$

where σ is as in (7.15). In the disturbance-free case this allows for less frequent sampling, while preserving the existence of a lower bound $\underline{\Delta}$ on the inter-execution times, which can be computed as the time satisfying $\phi(\underline{\Delta}) = \sigma$, where $\phi(t)$ is the unique solution of the generalized scalar Riccati equation,

$$\frac{d}{dt}\phi(t) = \|\Phi\| + (\|\Phi\| + \|BK\|)\phi(t) + \|BK\|\phi^2(t), \tag{7.92}$$

with initialization $\phi(0) = 0$. A similar reasoning, though with some noticeable differences, can also be adopted in the present context. Specifically, a suitable strategy still consists of a two-mode switching rule: in the absence of DoS, sampling is based on (7.91); upon occurrence of DoS the sampling rate is increased so as to reduce the DoS-induced delay. Note that in this case there is no need to enforce the upper bound $\bar{\Delta}_\sigma$ on the sampling rate since (7.91) automatically implies (7.12). However, since (7.91) in general guarantees nonzero inter-execution times only in the absence of disturbances, a lower bound on the sampling rate must be imposed a priori.

Let

$$\rho_k := \inf\left\{t \in \mathbb{R}_{>t_k} : \|x(t_k) - x(t)\| = \sigma \|x(t)\|\right\} \tag{7.93}$$

for all $k \in \mathbb{N}_0$. We have at once the following result:

Proposition 7.2. *Let δ_1 and δ_2 be positive constants such that $\delta_1 \le \delta_2 \le \bar{\Delta}_\sigma$, where $\bar{\Delta}_\sigma$ is as in Lemma 7.1. For each $k \in \mathbb{N}_0$, let the sampling times be given by*

$$t_{k+1} = \begin{cases} t_k + \delta_1, & \text{if } k \in \mathcal{S} \\ t_k + \delta_2, & \text{if } k \notin \mathcal{S} \vee \rho_k < t_k + \delta_2\,, \\ \rho_k, & \text{otherwise} \end{cases} \tag{7.94}$$

where $\mathcal{S}$ is as in (7.32). If the conditions of Theorem 2 hold true with $\Delta_* = \delta_1$, then the control system Σ composed of (7.1) and (7.6) is ISS. Moreover, the inter-sampling times are bounded from below by $\underline{\Delta} = \delta_1$.

Proof. The inter-sampling times associated with (7.94) are equal to δ_1, δ_2, or $\rho_k - t_k$. According to (7.94), the sampling time t_{k+1} equals ρ_k only if $k \notin \mathcal{S}$.

Under these circumstances, since t_k does not belong to any DoS interval, $e(t) = x(t_k) - x(t)$ for all $t \in [t_k, t_{k+1}[$, with $e(t_k) = 0$. Hence, if $t_{k+1} = \rho_k$ we have from (7.93) that $\|e(t)\| \leq \sigma \|x(t)\|$ for all $t \in [t_k, t_{k+1}[$ since e is continuous on $[t_k, t_{k+1}[$ and $e(t_k) = 0$. Thus, (7.94) guarantees that either the inter-sampling times equal δ_1 or δ_2, in which case they are not larger than $\bar{\Delta}_\sigma$, or they are equal to ρ_k, in which case (7.12) is satisfied. In view of this, a minor variation of Theorem 7.2 shows the thesis. The bounds on Δ_* and $\underline{\Delta}$ hold by construction. □

Interestingly, logic similar to that used in (7.94) was recently considered in [300] to achieve a finite sampling rate in the context of nonlinear output feedback for event-based controllers.

7.3.1 Self-triggering sampling logic

Event-based sampling has the positive feature of potentially saving communication resources, but requires the continuous monitoring of the process state, and thus hardware specifically available for this purpose. An alternative way to relax periodic implementations consists in selecting Δ_k based on the value of the process state at the measurement instants. Logic of this kind is typically referred to as self-triggering, in that the next update instant is computed directly by the control unit (see, e.g., [299], [301]).

Given $t, \tau \in \mathbb{R}_{\geq 0}$ with $t \geq \tau$, let

$$\chi(t, \tau) := \left[e^{A(t-\tau)} + \int_\tau^t e^{A(t-s)} BK \, ds \right] x(\tau). \tag{7.95}$$

Thus, $\chi(t_k, t_{k(t_k)})$ provides a prediction of $x(t_k)$ based on the last successful measurement of the process state. Note that since the disturbance is not available for measurements, this prediction cannot take into account the influence of the disturbance and is an approximate estimate of the actual state. Define

$$z(t_k) = \begin{cases} \left\| \chi(t_k, t_{k(t_k)}) \right\|, & \text{if } k \in \mathscr{S} \\ \|x(t_k)\|, & \text{otherwise.} \end{cases} \tag{7.96}$$

The idea is then to schedule the next control update based on $z(t_k)$, where $\chi(t_k, t_{k(t_k)})$ replaces $x(t_k)$ when the latter is not available: the larger $z(t_k)$, the smaller Δ_k, and vice versa, which corresponds to increasing the sampling rate as the distance of the process state from the origin gets larger.[1]

Proposition 7.3. *Let δ_1 and δ_2 be positive constants such that $\delta_1 \leq \delta_2 \leq \bar{\Delta}_\sigma$, where $\bar{\Delta}_\sigma$ is as in Lemma 7.1. For each $k \in \mathbb{N}_0$, let the sampling times be given*

1. We implicitly assume that the sampling logic is located on the controller side. In the case that the sampling logic is located on the process side, then (7.97) simplifies. Under these circumstances x can always be assumed to be available for measurements so that $z(t_k)$ reduces to $x(t_k)$.

by

$$t_{k+1} = t_k + \delta_2 - (\delta_2 - \delta_1)\varphi(z(t_k)), \tag{7.97}$$

where $\varphi : \mathbb{R}_{\geq 0} \mapsto [0, 1]$ *is an arbitrary class* $\mathcal{K}$ *function. If the conditions of Theorem 7.2 hold true with* $\Delta_* = \delta_2$*, then the control system* Σ *composed of (7.1) and (7.6) is ISS. Moreover, the inter-sampling times are bounded from below by* $\underline{\Delta} = \delta_1$.

Proof. By definition, φ takes value in the interval $[0, 1]$ and therefore the inter-sampling time induced by the control update rule (7.97) ranges in the interval $[\delta_1, \delta_2]$. This implies that the bounds on Δ_* and $\underline{\Delta}$ hold true. Thus, ISS follows directly from Theorem 7.2. □

Compared with (7.90), (7.97) achieves larger bounds on Δ_*, hence larger DoS-induced actuation delays. However, it potentially decreases the number of control updates since upon DoS the sampling rate is not set to δ_1, but is chosen depending on the inferred process state magnitude.

7.3.2 Simulation examples and discussions

In this section we discuss in greater detail the considered assumptions and provide examples in order to substantiate the analysis.

Slow-on-the-Average DoS As noted earlier, Assumptions 7.1 and 7.2 both pose constraints on DoS that are reminiscent of average dwell-time conditions [297]. In the present context the rationale behind Assumption 7.1 is that occasionally DoS can occur at the same rate as or even faster than $\underline{\Delta}$ (i.e., $\Lambda_n = h_{n+1} - h_n \leq \underline{\Delta}$ for some $n \in \mathbb{N}_0$), but the average interval between consecutive DoS triggering is greater than $\underline{\Delta}$. According to (7.30), we may in fact have intervals where $\Lambda_n \leq \underline{\Delta}$, hence intervals where $n(\tau, t)$ is greater than or equal to the maximum number $\lceil (t - \tau)/\underline{\Delta} \rceil$ of control updates that can be executed over $[\tau, t[$. However, over large time windows (i.e., when the term $(t - \tau)/\tau_D$ is predominant compared to η) the number of DoS triggering events is at most on the order of $(t - \tau)/\tau_D$, hence less than $\lceil (t - \tau)/\underline{\Delta} \rceil$.

Assumption 7.2 expresses a similar requirement with respect to the DoS duration. It expresses the property that, on average, the time instants over which communication is interrupted do not exceed a certain fraction of time, as specified by the constant $T \in \mathbb{R}_{>1}$. Similarly to η, the constant $\kappa \in \mathbb{R}_{\geq 0}$ plays the role of a regularization term. It is needed because during a DoS interval we have $|\Xi(h_n, h_n + \tau_n)| = \tau_n > \tau_n/T$ since $T > 1$. Hence, κ serves to make (7.31) consistent.

The considered assumptions are general enough to capture several different situations, as exemplified in the following. More complex scenarios can be easily envisaged.

Example 4.3 In analogy with [252], consider the situation where, on every interval that contains N communication attempts, a number $M < N$ of these attempts can be denied. A simple way to account for this situation is to regard the DoS as a train of pulses, M of which are superimposed to the sampling times. This implies that Assumption 7.2 holds true with $\kappa = 0$ and $T = \infty$. As for Assumption 7.1, assume a lower bound $\underline{\Delta}$ on the control executions. We sees that

$$n(\tau, t) \leq \left\lceil \frac{t-\tau}{\underline{\Delta} N} \right\rceil M \tag{7.98}$$

for all $\tau, t \in \mathbb{R}_{\geq 0}$ with $t \geq \tau$. Then, for each N, (7.30) holds true with $\eta = M$ and $\tau_D = \underline{\Delta} N/M$. In connection with Theorem 7.2, this means that stability is preserved whenever $(\underline{\Delta} N)/(\Delta_* M) > (\omega_1 + \omega_2)/\omega_1$. Using logic that samples at rate $\underline{\Delta}$ upon DoS, then stability is preserved whenever $N/M > (\omega_1 + \omega_2)/\omega_1$. For instance, this means that if $(\omega_1 + \omega_2)/\omega_1 = 5$, then up to 20% of the communication attempts can be denied without destroying stability. The value of N affects both closed-loop performance and robustness against DoS: the larger N is, the larger the number of consecutive communication attempts that can be denied without destroying stability. However, this potentially results in larger overshoots since $\eta = M$.

Example 4.4 Another interesting scenario is when DoS is sustained. We can account for this situation by modeling the DoS signal as a rectangular wave of a (possibly) variable period and duty cycle [286]. Let P_n and $D_n = \tau_n / P_n$ denote the period and duty cycle of the n-th DoS attack, respectively. Also let $P_{\min} := \inf_{n \in \mathbb{N}_0} P_n$, $D_{\max} := \sup_{n \in \mathbb{N}_0} D_n$ and $\tau_{\max} := \sup_{n \in \mathbb{N}_0} \tau_n$. Suppose that $P_{\min} > \underline{\Delta}$, $D_{\max} < 1$, and $\tau_{\max} < \infty$. Since the maximum number of *off/on* transitions of DoS during the interval $[\tau, t[$ can be upper bounded as $n(\tau, t) \leq \lceil (t - \tau)/P_{\min} \rceil$, Assumption 7.1 holds true with $\eta = 1$ and $\tau_D = P_{\min}$. Now let

$$n(t) := \begin{cases} -1, & \text{if } t < h_0 \\ \sup\{\, n \in \mathbb{N}_0 | h_n \leq t \,\}, & \text{otherwise} \end{cases}, \tag{7.99}$$

where $\tau_{-1} := 0$. For any $\tau, t \in \mathbb{R}_{\geq 0}$ with $t \geq \tau$, it is possible to write

$$\begin{aligned} |\Xi(\tau, t)| \quad \leq \quad & \max\{0, h_{n(\tau)} + \tau_{n(\tau)} - \tau\} \\ & + \min\{t - h_{n(t)}, \tau_{n(t)}\} + D_{\max} \sum_{\substack{n \in \mathbb{N}_0; \\ h_{n(\tau)} < h_n; \\ h_n < h_{n(t)}}} P_n . \end{aligned} \tag{7.100}$$

Hence, we see that Assumption 7.2 holds true with $\kappa = 2\tau_{\max}$ and $T = D_{\max}^{-1}$. This example includes the special case known as periodic jamming, where both period and duty cycle are constant.

Example 4.5 In connection with Example 4.4, we pointed out earlier that the requirements $P_{\min} > \underline{\Delta}$ and $D_{\max} < 1$ are necessary in a wide sense for closed-loop stability unless other conditions are imposed on DoS (as for $\tau_{\max} < \infty$, cf). These conditions, however, rule out the possibility of DoS signals with off/on transitions that are occasionally faster than the maximum transmission rate. Consider first Assumption 7.1 and suppose that $P_{\min} > 0$. Let $f(\tau, t) := (t - \tau)/n(\tau, t)$ represent the average dwell time of DoS off/on transitions on $[\tau, t[$. Assume that for some $\tau_D \in \mathbb{R}_{>\underline{\Delta}}$, there exists a $\gamma \in \mathbb{R}_{>0}$ such that

$$f(\tau, \tau + \gamma) \geq \tau_D \tag{7.101}$$

for all $\tau \in \mathbb{R}_{\geq 0}$, which implies that the DoS off/on transitions are slower than $\underline{\Delta}$ on every sufficiently large time interval. It is then easy to verify that Assumption 7.1 holds true with $\eta = \lceil \gamma / P_{\min} \rceil$. If $t - \tau \leq \gamma$, then $n(\tau, t) \leq n(\tau, \tau + \gamma)$, where $n(\tau, \tau + \gamma) \leq \lceil \gamma / P_{\min} \rceil$. If instead $t - \tau > \gamma$, let m denote the largest integer m such that $m\gamma < t - \tau$. Then

$$\begin{aligned} n(\tau, t) &= \sum_{k=0}^{m-1} n(\tau + k\gamma, \tau + (k+1)\gamma) \\ &\quad + n(\tau + m\gamma, t) \\ &= \sum_{k=0}^{m-1} \gamma / f(\tau + k\gamma, \tau + (k+1)\gamma) \\ &\quad + n(\tau + m\gamma, t) \leq m\gamma/\tau_D + \lceil \gamma / P_{\min} \rceil \end{aligned} \tag{7.102}$$

as follows from (7.101) and the definition of m. Then the claim follows by recalling that $m\gamma < t - \tau$. As for Assumption 7.2, let $D(\tau, t) := |\Xi(\tau, t)|/(t - \tau)$, which can be thought of as the average DoS duty cycle over $[\tau, t]$. Assume that for some $T \in \mathbb{R}_{>1}$, there exists a $\delta \in \mathbb{R}_{>0}$ such that

$$D(\tau, \tau + \delta) \leq 1/T \tag{7.103}$$

for all $\tau \in \mathbb{R}_{\geq 0}$. Reasoning as before, it is simple to verify that Assumption 7.2 holds true with $\kappa = \delta$. In connection with Example 4.3, the conditions stated above allow us to consider more general DoS classes. For instance, let

$$\tau_{ave} := \lim_{m \to \infty} \frac{1}{m+1} \sum_{n=0}^{m} (h_{n+1} - h_n) \tag{7.104}$$

$$D_{ave} \quad := \quad \lim_{m\to\infty} \frac{1}{m+1} \sum_{n=0}^{m} D_n \tag{7.105}$$

denote the average dwell time of DoS off/on transitions and the average DoS duty cycle, respectively. Then (7.101) and (7.103) with $\tau_D = \tau_{ave}$ and $T = D_{ave}^{-1}$ are sufficient to conclude that the DoS signal is also slow-on-the-average in the sense of Assumptions 7.1 and 7.2 with respect to τ_{ave} and D_{ave}.

7.3.3 Numerical example

For the sake of clarity a numerical example illustrating the theory and the above discussions is presented here. Consider the following open-loop unstable system [259]:

$$\frac{d}{dt}x(t) = \begin{bmatrix} 1 & 1 \\ 0 & 1 \end{bmatrix} x(t) + u(t) + w(t) \tag{7.106}$$

under linear quadratic regulator (LQR) gain

$$K = \begin{bmatrix} -2.1961 & -0.7545 \\ -0.7545 & -2.7146 \end{bmatrix}. \tag{7.107}$$

The solution of the Lyapunov equation $\Phi^\top P + P\Phi + Q = 0$ with $Q = I_2$ yields $\alpha_1 = 0.2779$, $\alpha_2 = 0.4497$, $\gamma_1 = 1$, and $\gamma_2 = 2.1080$. From this we deduce that we must select σ such that $\sigma < 0.4744$. Picking for instance $\sigma = 0.26$, Lemma 7.1 yields $\bar{\Delta}_\sigma = 0.1005$, where $\|\Phi\| = 1.9021$ and $\mu_A = 1.5$. $\bar{\Delta}_\sigma$ specifies the inter-sampling time of maximum length that guarantees ISS. Furthermore, $\omega_1/(\omega_1 + \omega_2) = 0.0321$. This value determines the DoS signals that are admissible in accordance with the present analysis. In connection with Example 4.3, this means a maximum of $\sim 3\%$ of communication denials on the average. As for Example 4.4 (Example 4.5), this implies a maximum (average) duty cycle of $\sim 3\%$ in case of a sustained DoS attack. The value obtained for $\omega_1/(\omega_1 + \omega_2)$ is conservative: as shown in Fig. 7.3, the bounds can in practice be much smaller than the theoretical one. This has also been confirmed by extensive simulations.

The conservativeness of the bound comes from two main sources: i) the bounds on the growth of the Lyapunov function under DoS (cf. (7.46)–(7.53)). In this respect the approach in [103], which does not rely on Lyapunov functions (albeit restricted to the disturbance-free case), can provide a possible alternative to the present analysis and ii) the generality of the considered scenario. Tighter bounds are likely to be obtained when more "structure" is assumed for the DoS. In this respect interesting results in the case of periodic jamming have been recently reported in [102], [254].

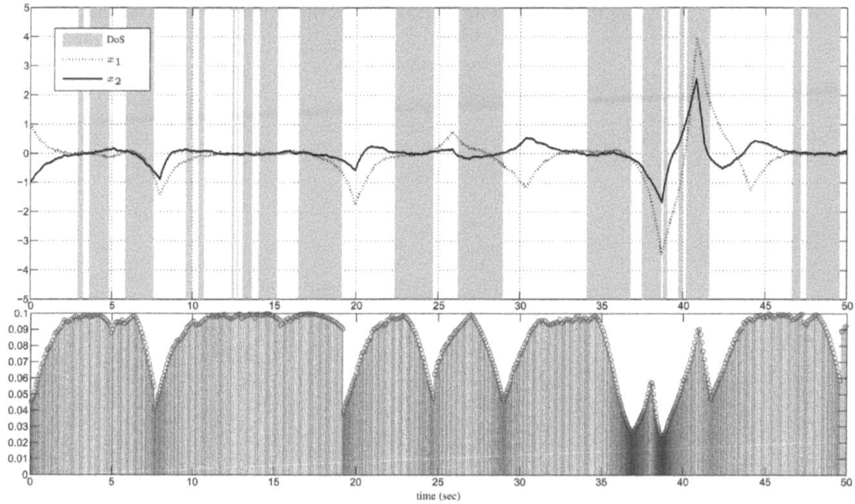

FIGURE 7.3 Simulation example for system (7.106) under state feedback (7.108) and sampling logic (7.97) with $\delta_1 = 0.01$, $\delta_2 = 0.1$, and $\varphi(\cdot) = 2/\pi \arctan(\cdot)$. Top: Closed-loop state response under initial conditions $x = [1 \; -1]^\top$, disturbance w uniformly distributed in $[-1, 1]$, and sustained DoS attack with variable period and duty cycle, generated randomly, with $P_{\min} = 0.01$ s and $P_{\max} = 10$ s. The resulting DoS signal has an average duty cycle of $\sim 42\%$. The vertical gray stripes represent the time intervals over which DoS is active. Bottom: Inter-sampling times determined by the logic (7.97). In terms of regulation performance, very similar results are obtained with the other logic described previously.

It is interesting to observe that the value of $\omega_1/(\omega_1 + \omega_2)$ also depends on a number of design parameters. It depends on the Lyapunov equation $\Phi^\top P + P\Phi + Q = 0$, and thus on Q and the state-feedback matrix K. For instance, a choice of

$$K = \begin{bmatrix} -4.5 & -1 \\ 0 & -6 \end{bmatrix} \tag{7.108}$$

achieves a bound $\omega_1/(\omega_1 + \omega_2) = 0.0971$, thus allowing an average duty cycle of $\sim 10\%$ in the case of a sustained attack. This suggests an investigation of analytic or numeric methods to find the Q and K that could maximize robustness against DoS. In practice, another possibility for increasing $\omega_1/(\omega_1 + \omega_2)$ is to reduce the value of σ; however, this has to be traded off against the inter-sampling times. For instance, using the LQR gain and letting $Q = I_2$, the choice $\sigma = 0.1$ is sufficient to increase $\omega_1/(\omega_1 + \omega_2)$ to 0.0547. As an offset, $\bar{\Delta}_\sigma$ drops to 0.0462. This phenomenon is illustrated in Fig. 7.4.

7.3.4 Slow-on-the-average DoS: disturbance-free case

In the disturbance-free case, Assumptions 7.1 and 7.2 can both be relaxed. While Assumptions 7.3 and 7.4 are similar in concept, they pose constraints

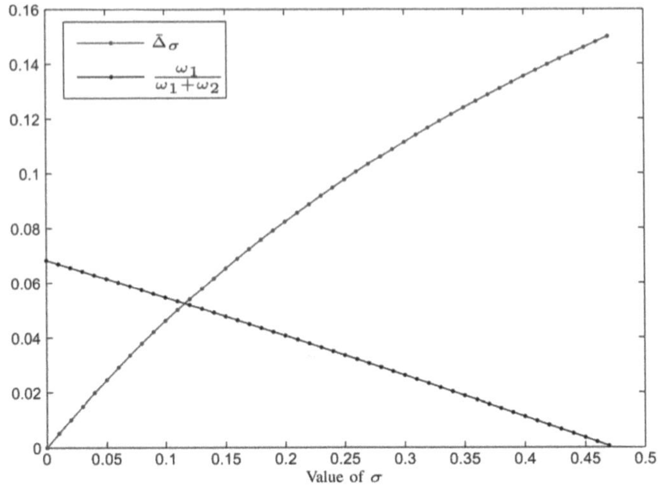

FIGURE 7.4 Maximum inter-sampling time and value of $\omega_1/(\omega_1+\omega_2)$ vs. choice of σ using the LQR gain and $Q = I_2$.

on DoS frequency and duration that must hold on $[0, t)$ only, rather than on each subinterval $[\tau, t)$ of $[0, t)$. This makes it possible to face more general DoS classes, including DoS signals that deny communication for *unbounded* periods of time.

Consider for instance the example in connection with a DoS signal given by

$$\begin{aligned} h_n &= (n+1) + \frac{1}{2}n(n+1) - \frac{1}{\alpha}(n+1), \\ \tau_n &= \frac{1}{\alpha}(n+1), \end{aligned} \tag{7.109}$$

where $n \in \mathbb{N}_0$ and $\alpha \in \mathbb{R}_{>1}$. It is straightforward to verify that the resulting DoS signal satisfies Assumptions 7.3 and 7.4 with $\eta = \tau_D = 1$, $\kappa = 0$, and $T = \alpha$. Picking $\bar{\Delta}_\sigma = 0.01$ and recalling that $\omega_1/(\omega_1+\omega_2) = 0.0321$, we see that (7.33) holds true for $\alpha \geq 50$. Since the conditions of Corollary 7.1 are satisfied, then the closed-loop system is GAS even though the length of the DoS intervals grows unbounded with n. This is possible because the closer the state to the origin, the less the effect of DoS.

In the presence of disturbances the situation just described is no longer true since w may always cause the state to deviate from its nominal trajectory. It is easy to see that for open-loop unstable systems no sampling logic exists that can achieve ISS in the presence of unbounded DoS. As for Theorem 7.2, boundedness of the DoS intervals is implicit in Assumption 7.2. In fact, (7.4) with $\tau = h_n$ and $t = h_n + \tau_n$ implies $\sup_{n\in\mathbb{N}_0} \tau_n \leq \kappa T/(T-1)$.

7.4 Observer-based secure control

Security issues increase the challenges in control of CPSs because these systems have a high possibility of being affected by several cyber attacks without providing any notification of failure. These attacks can lead to a disruption of the physical system, such as the disarrangement of coordination packets in medium-access control layers could be a result of malware injected by an adversary. Moreover, in order to destroy the normal operation, an attacker can illegally obtain access to the supervision centers while obtaining the encryption key. This means that the system dynamics can be disturbed arbitrarily by the attacker, and when there is a lack of security protection in hardware or software strategies the attacker has the capability of inducing perturbations [58].

The communication between the items of control systems (i.e., sensors, actuators, and controllers) occurs through a common network medium. This network needs to be secured to prohibit the vulnerability of attack by adversaries during data transmission. These attacks could lead the system to instability or could drive the plant to undesired operations, as mentioned before. Thus, considering security issues is very important when designing the controllers for such a system. From a control security viewpoint, cyber attacks can be classified into two main types: 1) DoS attacks, which are often used to occupy the communication resources in order to prohibit the transmission of measurement or control signals, and 2) deception attacks, also called false data injection (FDI) attacks, where modification of the data integrity for the transmitted packets occurs in some cyber parts of the CPS.

Control of CPSs under cyber attack is one of the main issues in control engineering, and as such it has attracted a great deal of research. Most of the literature considers one kind of attack, such as [99], [100], [102], [103], and [104] for the case of DoS attacks and [122], [136], and [121] for the case of deception attacks. Some of the literature considers two kind of attacks; in [98] randomly occurring DoS and deception attacks were both considered for the design of an event-based security control system. The optimal control problem was investigated for a class of NCSs subject to DoS, deception, and physical attacks using a delta operator approach and by applying the ϵ-Nash equilibrium [91]. A resilient linear quadratic Gaussian control strategy was designed for NCSs subject to zero dynamic attacks [137]. Dynamic programming was applied for the control strategy, and value iteration methods were applied for the design of power transmission strategy for a class of CPSs subject to DoS attack [317]. An H_∞ observer-based periodic event-triggered control (PETC) framework was used for the design of a resilient control strategy for CPS subject to DoS attacks [318]. In [312], a H_∞ MiniMAX controller was applied in the physical layer by using a delta operator approach to solve a resilient control problem for wireless NCSs subject to DoS attack via a hierarchical game approach.

7.4.1 Problem formulation

A CPS composed of actuator, plant, sensor, and controller is considered in this chapter, where the communication network is used to connect controller and actuator, as shown in Fig. 7.5. The considered system could be affected by both physical and cyber attacks. The physical attack affecting the plant is represented by (A_1) in Fig. 7.5. A reliable network is used for data transmission between the sensor and the controller, while the channel used for communication between controller and actuator is unreliable, so it could be affected by a cyber attack, which could be either a DoS attack or a deception attack, labeled (A_2) and (A_3), respectively, in Fig. 7.5.

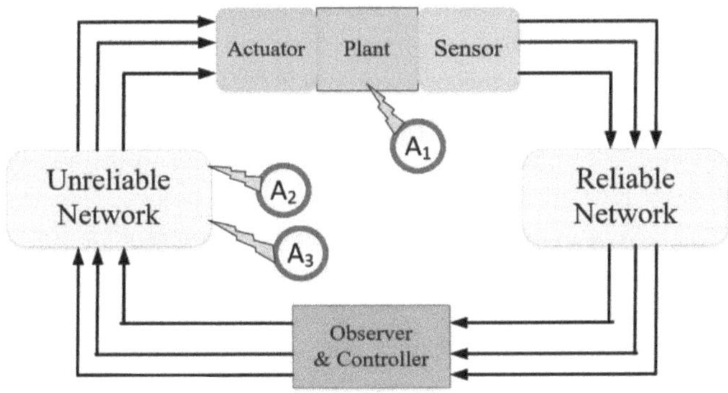

FIGURE 7.5 Model.

The plant is described by the formula

$$\begin{aligned} x(k+1) &= Ax(k) + Bu_p(k) + \eta(k)f(k), \\ y(k) &= Cx(k), \end{aligned} \tag{7.110}$$

where $x(k) \in \mathbb{R}^{n_x}$, $u_p(k) \in \mathbb{R}^{n_u}$, $y_p(k) \in \mathbb{R}^{n_y}$, and $f(k) \in \mathbb{R}^{n_f}$ are the system state, the control signals received by the actuators, the system output, and physical attack signal injected by the attackers, respectively. A, B, and C are known matrices with proper dimensions, and B is partitioned as

$$B = \begin{bmatrix} B_1 & B_2 & \cdots & B_r \end{bmatrix}. \tag{7.111}$$

In addition, the control input received by the actuator $u_p(k)$ is partitioned as

$$u_p(k) = \Gamma(k)\begin{bmatrix} u_1^T & u_2^T & \cdots & u_r^T \end{bmatrix}^T, \tag{7.112}$$

where $\Gamma(k)$ describes the occurrence of the DoS attack as

$$\Gamma(k) = \{\beta_1 \quad \beta_2 \quad \cdots \quad \beta_r\} \tag{7.113}$$

with indicator $\beta_i(k), i \in \mathbf{R} := \{1, \ldots, r\}$ being the Bernoulli distributed white sequence. The physical attack is considered to be source limited and satisfies $\| f(k)\|^2 < \delta_1^2$, where δ_1 is a known constant.

When the full state information is not available, it is desirable to design the following observer-based controller:

$$
\begin{aligned}
& \textit{Observer:} \\
\hat{x}(k+1) &= A\hat{x}(k) + Bu_c(k) + L(y(k) - \hat{y}(k)), \\
\hat{y}(k) &= C\hat{x}(k), \qquad (7.114) \\
& \textit{Controller:} \\
u_c(k) &= K\hat{x}(k). \qquad (7.115)
\end{aligned}
$$

Here $\hat{x}(k) \in \mathbb{R}^{n_x}$ is the estimate of the system states (7.110), $\hat{y}(k) \in \mathbb{R}^{n_y}$ is the observer output, and $L \in \mathbb{R}^{n_x \times n_y}$ and $K \in \mathbb{R}^{n_u \times n_x}$ are respectively the observer and controller gains.

The control signal is subjected to both DoS attacks and deception attacks, so it will be received by the actuator as

$$
u_{p_i}(k) = K_i\hat{x}(k) + \alpha_i(k)\zeta_i(k), \qquad (7.116)
$$

where $\zeta_i(k)$ is the deception attack signal affecting actuator i and $\|\zeta(k)\|^2 < \delta_2^2$, where δ_2 is a known constant. The indicator $\alpha_i(k)$ is a Bernoulli distributed white sequence. Furthermore, the indicators $\eta(k)$, $\alpha_i(k)$ and $\beta_i(k)$, $i \in \mathbf{R}$ are uncorrelated with each other and with the stochastic properties listed in Fig. 7.6.

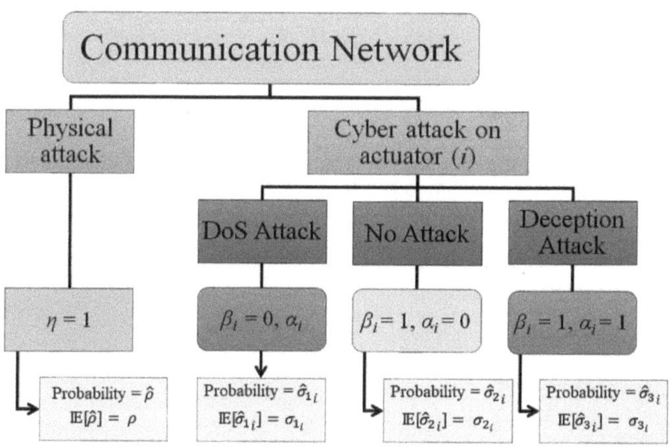

FIGURE 7.6 Attack types.

Defining the estimation error as $e(k) = x(k) - \hat{x}(k)$, the closed-loop system and estimation error are formulated using Eqs. (7.111)–(7.116) as follows:

$$\begin{aligned} x(k+1) &= Ax(k) + \sum_{i=1}^{r} \beta_i(k) B_i K_i x(k) \\ &- \sum_{i=1}^{r} \beta_i(k) B_i K_i e(k) \\ &+ \sum_{i=1}^{r} \beta_i(k)\alpha_i(k) B_i \zeta_i(k) + \eta(k) f(k) \end{aligned} \tag{7.117}$$

$$\begin{aligned} e(k+1) &= (A - LC)e(k) + \sum_{i=1}^{r} (1 - \beta_i(k)) B_i K_i e(k) \\ &- \sum_{i=1}^{r} (1 - \beta_i(k)) B_i K_i x(k) \\ &+ \sum_{i=1}^{r} \beta_i(k)\alpha_i(k) B_i \zeta_i(k) + \eta(k) f(k). \end{aligned} \tag{7.118}$$

In terms of $\xi(k) = [x^T(k) \quad e^T(k)]^T$, systems (7.117) and (7.118) can be cast in the form

$$\xi(k+1) = \bar{A}\xi(k) + \bar{B}\zeta(k) + \bar{C} f(k), \tag{7.119}$$

where $\zeta(k) = [\zeta_1(k), \zeta_2(k), \ldots, \zeta_r(k)]^T$, and

$$\begin{aligned} \bar{A} &= \begin{bmatrix} A + \sum_{i=1}^{r} \beta_i(k) B_i K_i & -\sum_{i=1}^{r} \beta_i(k) B_i K_i \\ -\sum_{i=1}^{r} (1 - \beta_i(k)) B_i K_i & \bar{A}_{22} \end{bmatrix} \\ \bar{A}_{22} &= A - LC + \sum_{i=1}^{r} (1 - \beta_i(k)) B_i K_i \\ \bar{B} &= \begin{bmatrix} \bar{B}_1 & \bar{B}_2 & \cdots & \bar{B}_r \\ \bar{B}_1 & \bar{B}_2 & \cdots & \bar{B}_r \end{bmatrix} \\ \bar{B}_i &= \beta_i(k)\alpha_i(k) B_i, \quad i = 1, 2, \ldots, r \\ \bar{C} &= \begin{bmatrix} \eta(k) I & \eta(k) I \end{bmatrix}^T. \end{aligned} \tag{7.120}$$

Remark 7.5. As noted from (7.117), there are three scenarios for cyber attacks on each channel i:

1) DoS attack, when $\beta_i(k) = 0$ and regardless of the value of $\alpha_i(k)$;

2) Deception attack, when $\beta_i(k) = 1$ and $\alpha_i(k) = 1$;
3) No cyber attack, when $\beta_i(k) = 1$ and $\alpha_i(k) = 0$.

These scenarios and the physical attack are summarized in Fig. 7.6.

Definition 7.3. Given the positive constant scalars $\delta_1, \delta_2, \delta_3$, the observer-based controllers (7.114) and (7.115) are said to be $\delta_1, \delta_2, \delta_3$ secure (when $\mathbb{E}\,\|f(k)\|^2 < \delta_1^2$) if $\|\zeta(k)\|^2 \le \delta_2^2$, then $\mathbb{E}\|e(k)\|^2 \le \delta_3^2$ for all k.

7.4.2 Design results

Theorem 7.3. *Given the positive scalars $\delta_1, \delta_2, \delta_3$ and the control and estimator gains $(K_1, K_2, \ldots, K_r)$ and L, the observer-based controllers (7.114) and (7.115) are δ_1, δ_2 secure if there exist positive definite matrix P and positive scalars ε_1 and ε_2 satisfying the inequalities*

$$\begin{cases} \hat{\Omega} \le 0 \\ \frac{\phi^2 s_0^2}{\lambda_{min}(P)(s_0 - 1)} \le \delta_3^2 \end{cases}, \tag{7.121}$$

where

$$\hat{\Omega} = \begin{bmatrix} \bar{A}^T P \bar{A} & \bar{A}^T P \bar{B} & \bar{A}^T P \bar{C} \\ * & \bar{B}^T P \bar{B} - \varepsilon_2 I & \bar{B}^T P \bar{C} \\ * & * & \bar{C}^T P \bar{C} - \varepsilon_1 I \end{bmatrix}, \tag{7.122}$$

and where $\phi^2 = \varepsilon_1\delta_1 + \varepsilon_2\delta_2$ and $\bar{A}, \bar{B}, \bar{C}$ are defined in (7.120).

Proof. To establish the main theorem, the following Lyapunov function is constructed:

$$V(k) = x^T(k) P x(k). \tag{7.123}$$

Evaluating the difference of $V(k)$, we have

$$\begin{aligned} &\mathbb{E}[\Delta V(k)] = \mathbb{E}\big[V(k+1) - V(k)\big] < \\ &\mathbb{E}\Big[\xi^T(k)\bar{A}^T P \bar{A}\xi(k) + 2\xi^T(k)\bar{A}^T P \bar{B}\zeta(k) \\ +\ & 2\xi^T(k)\bar{A}^T P \bar{C} f(k) + \zeta^T(k)\bar{B}^T P \bar{B}\zeta(k) \\ +\ & 2\zeta^T(k)\bar{B}^T P \bar{C} f(k) + f^T(k)\bar{C}^T P \bar{C} f(k) - \xi^T(k) P \xi(k) \\ +\ & \varepsilon_1(\delta_1 - f^T(k) f(k)) + \varepsilon_2(\delta_2 - \zeta^T(k)\zeta(k))\Big], \end{aligned} \tag{7.124}$$

which can be rewritten as

$$\mathbb{E}\big[\Delta V(k)\big] \le \mathbb{E}\big[\Xi^T(k)\hat{\Omega}\Xi(k) + \phi^2\big], \tag{7.125}$$

where

$$\Xi^T(k) = \begin{bmatrix} \xi(k) & \zeta(k) & f(k) \end{bmatrix}. \tag{7.126}$$

From (7.125), it is known that

$$\mathbb{E}[\Delta V(k)] \leq -\lambda_{min}(-\hat{\Omega})\mathbb{E}\big[||\xi(k)||^2\big] + \phi^2. \tag{7.127}$$

In addition, by referring to the definition of the energy-like functional $V(k)$, it is seen that

$$V(k) \leq \lambda_{max}(P)\mathbb{E}\big[||\xi(k)||^2\big]. \tag{7.128}$$

In addition, a scalar $s > 1$ is introduced, and from (7.127) and (7.128) it follows that

$$\begin{aligned}
&\mathbb{E}[s^{k+1}V(k+1)] - \mathbb{E}[s^k V(k)] \\
&= s^{k+1}\mathbb{E}[\Delta V(k)] + s^{k+1}\mathbb{E}[V(k)] - s^k\mathbb{E}[V(k)] \\
&\leq s^{k+1}\Big[- \lambda_{min}(-\hat{\Omega})\mathbb{E}\big[||\xi(k)||^2\big] + \phi^2\Big] \\
&\quad + s^k(s-1)\mathbb{E}[V(k)] \\
&\leq a(s)s^k\mathbb{E}\big[||\xi(k)||^2\big] + s^{k+1}\phi^2,
\end{aligned} \tag{7.129}$$

where $a(s) = -\lambda_{min}(-\hat{\Omega})s + (s-1)\lambda_{max}(P)$.

For any integer T, summing up both sides of (7.129) from 0 to $T-1$ with respect to k yields

$$\begin{aligned}
&\mathbb{E}[s^T V(T)] - \mathbb{E}[V(0)] \\
&\leq a(s)\sum_{k=0}^{T-1} s^k\mathbb{E}\big[||\xi(k)||^2\big] + \frac{s(1-s^T)}{1-s}\phi^2.
\end{aligned} \tag{7.130}$$

Since $a(1) = -\lambda_{min}(-\hat{\Omega}) < 0$ and $\lim_{s\to\infty} = +\infty$, there exists a scalar $s_0 > 1$ such that $a(s_0) = 0$. Thus, a scalar $s_0 > 1$ can be found such that

$$\mathbb{E}[s_0^T V(T)] - \mathbb{E}[V(0)] \leq \frac{s_0(1-s_0^T)}{1-s_0}\phi^2. \tag{7.131}$$

Noting that

$$\begin{aligned}
\mathbb{E}[s_0^T V(T)] &\geq \lambda_{min}(P)s_0^T\mathbb{E}\big[||\xi(T)||^2\big] \\
&\geq \lambda_{min}(P)s_0^T\mathbb{E}\big[||e(T)||^2\big],
\end{aligned} \tag{7.132}$$

we have

$$\mathbb{E}\big[||e(T)||^2\big] \leq \frac{(s_0^T - 1)\phi^2}{s_0^{T-1}(s_0 - 1)\lambda_{min}(P)}. \tag{7.133}$$

Referring to (7.124), it can be shown that $\mathbb{E}\|e(T)\|^2 \leq \delta_2^2$, which from Definition 7.3 implies that the estimation error system (7.119) is $\delta_1, \delta_2, \delta_3$ secure, and so the proof of Theorem 7.7 is complete. □

Theorem 7.4. *Given the positive scalars $\delta_1, \delta_2, \delta_3$ and the controller and estimator gains $(K_1, K_2, \ldots, K_r)$ and L, the observer-based controllers (7.114) and (7.115) are $\delta_1, \delta_2, \delta_3$ secure if there exists a positive definite matrix P and positive scalars ε_1 and ε_2 satisfying the inequalities*

$$\begin{cases} \Omega \leq 0 \\ \frac{\phi^2 s_0^2}{\lambda_{min}(P)(s_0 - 1)} \leq \delta_3^2 \end{cases}, \tag{7.134}$$

where

$$\begin{bmatrix} \Omega_{11} & \Omega_{12} \\ * & -\bar{X} \end{bmatrix} \tag{7.135}$$

with

$$\Omega_{11} = \begin{bmatrix} -\bar{X} & 0 & 0 \\ * & -\varepsilon_2 I & 0 \\ * & * & -\varepsilon_1 I \end{bmatrix}, \quad \Omega_{12} = \begin{bmatrix} \Psi \\ \bar{B}^T \\ \bar{C}^T \end{bmatrix}, \tag{7.136}$$

where

$$\begin{aligned} \Psi &= \begin{bmatrix} \Psi_1 & \Psi_2 \\ \Psi_3 & \Psi_4 \end{bmatrix}, \\ \Psi_1 &= XA^T + \sum_{i=1}^{r} \beta_i(k) Y_i^T B_i^T, \\ \Psi_2 &= -\sum_{i=1}^{r} (1 - \beta_i(k)) Y_i^T B_i^T, \\ \Psi_3 &= XA^T + \sum_{i=1}^{r} \beta_i(k) Y_i^T B_i^T, \\ \Psi_4 &= XA^T - Z^T + \sum_{i=1}^{r} (1 - \beta_i(k)) Y_i^T B_i^T, \end{aligned}$$

and $K_i = Y_i X^{-1}$, $i = 1, \cdots, r$, and $L = Z X^{-1} C^{\dagger}$.

Proof. $\hat{\Omega}$ in Eq. (7.121) can be rewritten as

$$\hat{\Omega} = \hat{\Omega}_{11} + \hat{\Omega}_{12} \hat{\Omega}_{22} \hat{\Omega}_{12}^T \tag{7.137}$$

with

$$\hat{\Omega}_{11} = \begin{bmatrix} -P & 0 & 0 \\ * & -\varepsilon_2 I & 0 \\ * & * & -\varepsilon_1 I \end{bmatrix}, \quad \hat{\Omega}_{12} = \begin{bmatrix} \bar{A}^T \\ \bar{B}^T \\ \bar{C}^T \end{bmatrix},$$
$$\hat{\Omega}_{22} = P. \tag{7.138}$$

Thus, Eq. (7.121) is formulated using Schur complements as

$$\begin{bmatrix} \hat{\Omega}_{11} & \hat{\Omega}_{12} \\ * & -\hat{\Omega}_{22}^{-1} \end{bmatrix}. \tag{7.139}$$

Now we define $\bar{X} = P^{-1}$, and then multiply Eq. (7.139) on the right and left by

$$diag\{\bar{X}, I, I, I\},$$

and by selecting

$$\bar{X} = \begin{bmatrix} X & 0 \\ 0 & X \end{bmatrix},$$
$$Y_i^T = XK_i^T, \quad i = 1, 2, \ldots, r, \quad Z^T = XC^T L^T$$

we obtain (7.135). □

7.4.3 Illustrative example I

The effectiveness of the proposed method presented in this chapter is shown by solving the following numerical examples. The parameters of system (7.110) are as follows:

$$A = \begin{bmatrix} 0.1 & 1 & 2 & 3 & 4 & 5 & 5 \\ 0 & -0.2 & 3 & 4 & 5 & 6 & 7 \\ 0 & 0 & -0.4 & 2 & 4 & -3 & 2 \\ 0 & 0 & 0 & -0.7 & 2 & 4 & -3 \\ 0 & 0 & 0 & 0 & 0.8 & -3 & 2 \\ 0 & 0 & 0 & 0 & 0 & 0.5 & 2 \\ 0 & 0 & 0 & 0 & 0 & 0 & -0.2 \end{bmatrix}.$$

B is a diagonal matrix of $\{B_{11}, B_{22}, B_{33}\}$, such that $B_{11} = [2; 10]$; $B_{22} = [2.5; 5; 10]$; $B_{33} = [0.25; -11]$. Thus, B can be partitioned as $B = [B_1\ B_2\ B_3]$

with

$$B_1 = \begin{bmatrix} B_{11} \\ 0 \\ 0 \end{bmatrix}, \quad B_2 = \begin{bmatrix} 0 \\ B_{22} \\ 0 \end{bmatrix}, \quad B_3 = \begin{bmatrix} 0 \\ 0 \\ B_{33} \end{bmatrix}.$$

$$C = \begin{bmatrix} 1 & 0 & 0 & 0 & 0 & 0 & 0 \\ 0 & 1 & 0 & 0 & 0 & 0 & 0 \\ 0 & 0 & 1 & 0 & 0 & 0 & 0 \end{bmatrix}.$$

Using YALMIP the gains of the controller and estimator (7.114) and (7.115) were obtained as

$$\begin{aligned}
K_1 &= 10^{-3} \times \big[-1.5799 \quad 3.6214 \quad -166.7601 \quad -253.3419 \\
&\qquad -429.7391 \quad -553.8924 \quad -279.0259\big], \\
K_2 &= 10^{-3} \times \big[-19.9563 \quad -106.2964 \quad 1.2131 \quad 2.9130 \\
&\qquad -55.8918 \quad 147.590 \quad -97.0861\big], \\
K_3 &= 10^{-3} \times \left[\begin{matrix} 81.5008e-3 & 471.3945e-3 & 8.4428 \\ -35.3270e-3 & 228.9993e-3 & 9.4184 \end{matrix}\right. \\
&\qquad \left.\begin{matrix} -16.0644 & -97.3650 & -594.5690 & -3.4557 \\ -6.7051 & -125.0653 & -559.9038 & -3258.1 \end{matrix}\right], \\
L &= 10^{-3} \times \\
&\begin{bmatrix} 230.1872 & 475.3367 & 1237.4 \\ 351.9685 & -401.0426 & 946.5461 \\ 84.9112 & 144.7963 & -240.0106 \\ -1.1219 & 18.0030 & 85.0714 \\ 2.6965 & 7.1696 & 28.7968 \\ -235.4232e-3 & 982.4951e-3 & 5.5035 \\ -76.5793e-3 & -93.6182e-3 & 426.8155e-3 \end{bmatrix}.
\end{aligned} \tag{7.140}$$

Three scenarios of attacks were considered, and the states and error in estimation for them were obtained using MATLAB/Simulink as follows:

1. System without attack, Figs. 7.7–7.9;
2. System under DoS and physical attacks, Figs. 7.10–7.12;
3. System under deception and physical attacks, Figs. 7.13–7.15.

As shown in Figs. 7.7–7.15, the designed observer-based controller shows stability in the states and a small error in estimating the states under all possibilities of attack.

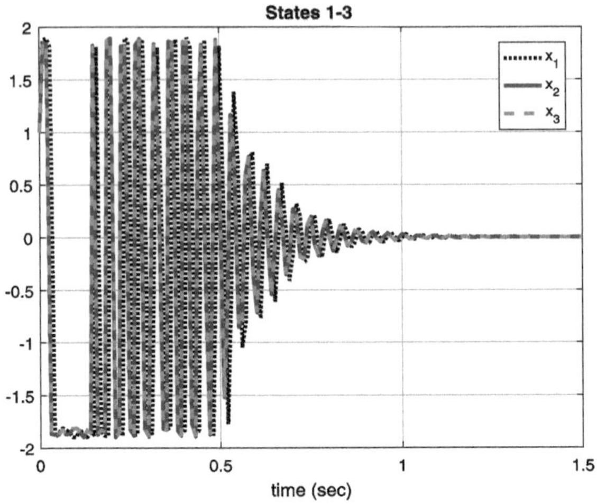

FIGURE 7.7 States 1–3 with no attack.

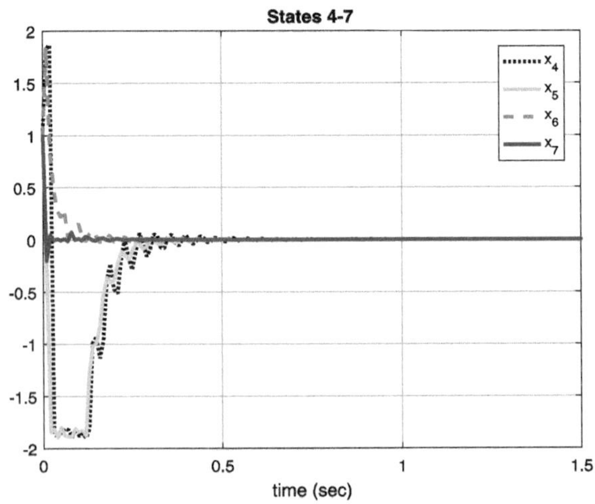

FIGURE 7.8 States 4–7 with no attack.

7.5 Stabilization of discrete-time systems under DoS attack

Security issues increase the challenges in cloud control systems (CCSs) due to the high possibility that it will be affected by several cyber attacks without providing any notification of failure. These attacks can lead to a disruption to the physical system; for example, the disarrangement of coordination packets in medium-access control layers could be a result of malware introduced by an adversary. Moreover, in order to destroy the nominal operation an attacker can illegally obtain access to the supervision centers while obtaining the encryption

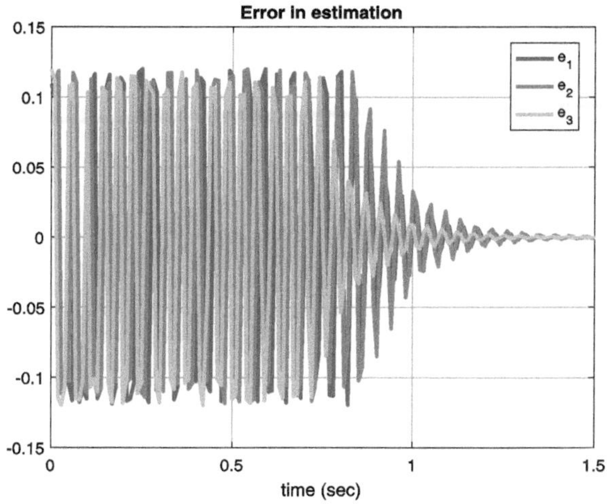

FIGURE 7.9 Error on the estimation of outputs (states 1–3) with no attack.

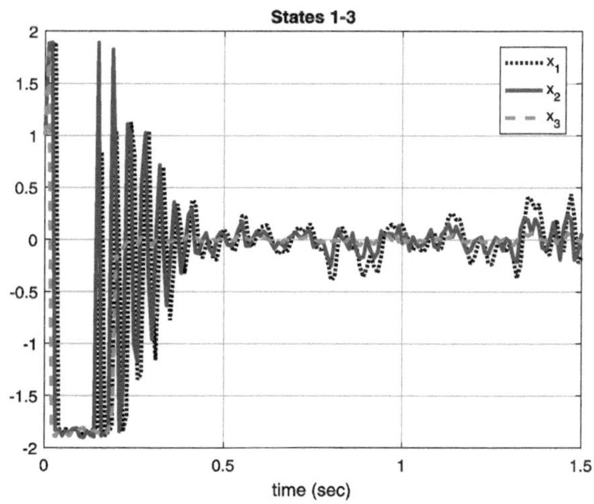

FIGURE 7.10 States 1–3 with DoS and physical attacks.

key. This means that the system dynamics can be disturbed arbitrarily by the attacker, and when there is a lack of security protection either in hardware or software strategies the attacker has the capability of inducing perturbations [58].

The communication between the components of the control systems (i.e., sensors, actuators, and controllers) occurs through a common network medium. This network needs to be secured to prohibit the vulnerability of attack by adversaries during data transmission. These attacks could lead the system to instability or drive the plant to undesired operations as mentioned before. Thus,

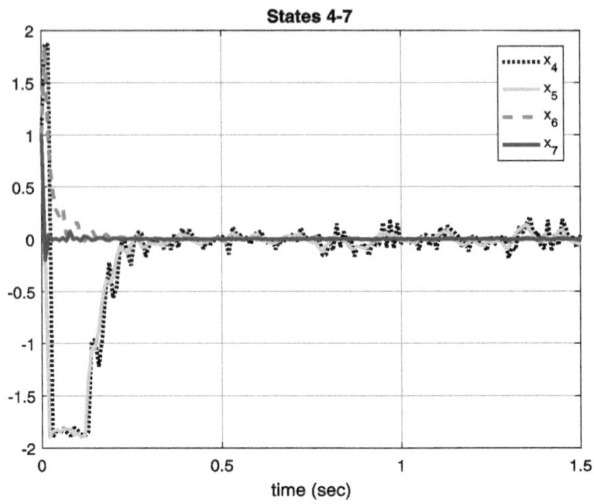

FIGURE 7.11 States 4–7 with DoS and physical attacks.

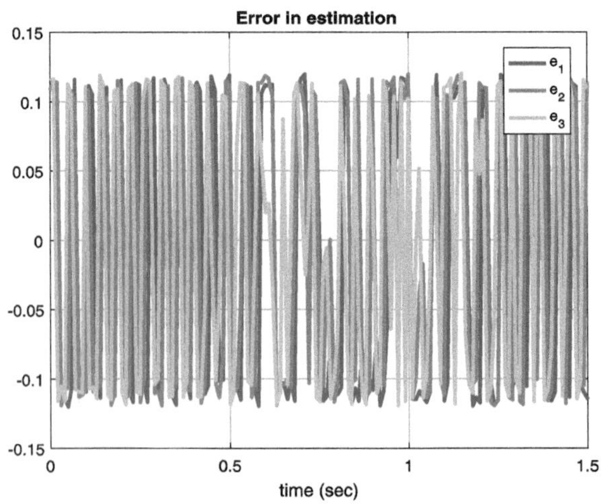

FIGURE 7.12 Error on the estimation of outputs (states 1–3) with DoS and physical attacks.

considering security issues is very important in the design of controllers for such a system [306].

Denial-of-service attacks are strategies that are often used to occupy the communication resources in order to block the transmission of measurement and/or control signals and that cause the maximum possible deterioration of the system performance. Several approaches for controlling systems affected by DoS attacks are applied in the literature, due to its high importance in CCSs [306].

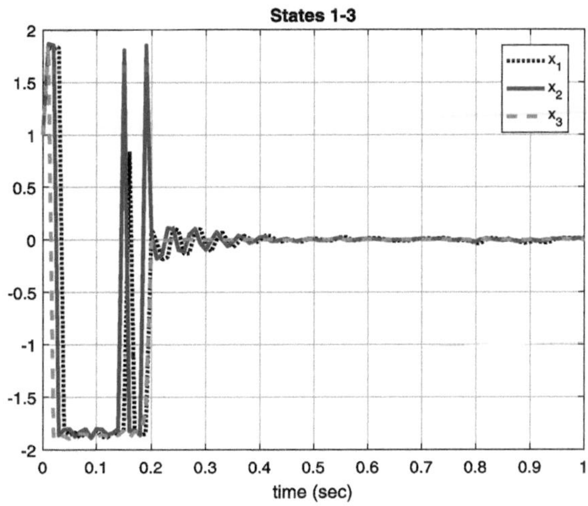

FIGURE 7.13 States 1–3 with deception and physical attacks.

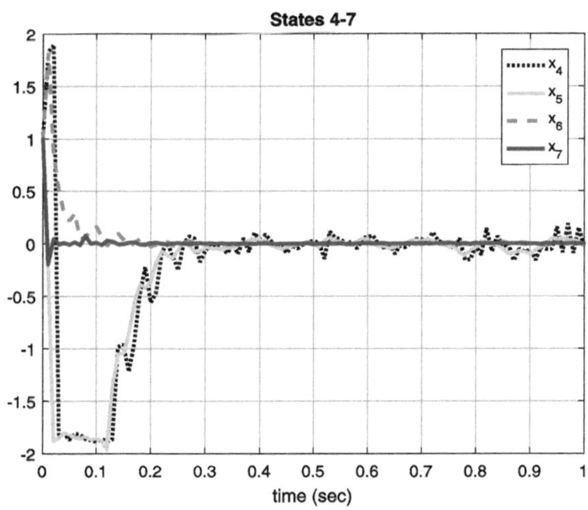

FIGURE 7.14 States 4–7 with deception and physical attacks.

7.5.1 Preliminaries

The cyclic small-gain theorem was implemented to design an output-feedback controller for large-scale nonlinear systems subject to nonsmooth sensor noise [250]. A distributed output feedback control of linear-time invariant (LTI) systems in the presence of unreliable communication was designed by solving an optimization control problem [307]. The problem of lossy sensors and cyber attacks in discrete-time multiagent systems was discussed, and a distributed observer-based consensus controller was proposed using an event-triggering

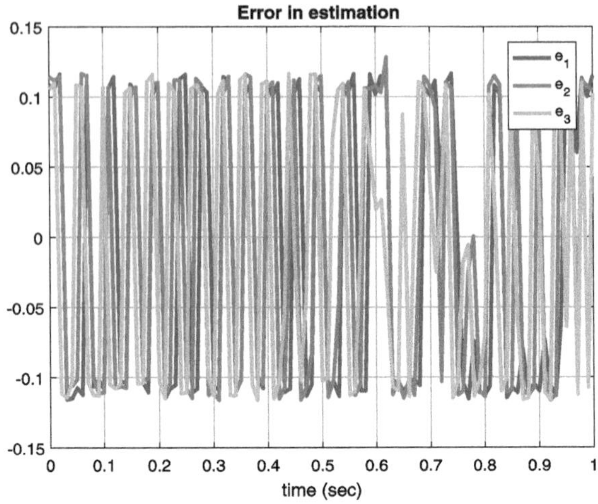

FIGURE 7.15 Error on the estimation of outputs (states 1–3) with deception and physical attacks.

method [121]. The backstepping adaptive approach was implemented for large-scale stochastic nonlinear time-delay systems in the presence of constrained outputs and saturation of actuators [308].

An observer-based controller was proposed for linear systems affected by process disturbances and false data injection attacks by implementing a controller gain scheme and a supervisory switching strategy [309]. A secure distributed controller for power systems subject to time-varying data injection attacks were proposed using the model predictive control approach [11]. A distributed controller was designed for NCSs undergoing stochastic cyber attacks using an event-triggered approach [310]. Another implementation of an event-triggered approach was presented with the help of H_∞ optimization to achieve the stability of neural networks affected by cyber attacks and considering a constrained bandwidth of the network [311].

The small-gain approach was widely applied by researchers to solve the stabilization problem of distributed systems. An event-triggered sampling scheme was presented to ensure the stability of large-scale systems by distributed controllers in the presence of limited communication medium [285], [286]. A hierarchical game method was presented to solve the control problem of a wireless NCS subject to DoS attack [312]. A robust pinning synchronization control problem was proposed to ensure that the initial state would be restored for a complex CPS subject to mixed attacks affecting independent transmitting channels [313].

The analysis problem of distributed systems subject to cyber attacks has attracted many researchers. The duration and frequency of the DoS attacks for a CPS with multiple transmission channels were characterized to ensure the stability of a switched system [314]. The bound of DoS attack frequency and duration was discussed for distributed systems in the presence of pure round-robin communication [101]. The major drawback of these techniques is that they consider the availability of the full state for all subsystems, which is not true for most real practical CCSs. In this chapter we examine the stabilization problem of discrete time distributed systems subject to DoS attacks while considering that partial information of the states are available through the output of each system. The following factors are examined:

- The robustness problem in distributed CCS is discussed by examining the stabilization of these systems in face of DoS attack and elaborating on the published work from various respects;
- A static output feedback control problem of a "nominal discrete-time" distributed system is considered, and an appropriate control law is designed using the linear matrix inequality (LMI) technique to achieve closed-loop stability;
- A bound of attack frequency and duration is derived to ensure the stability of the distributed CPS with partial information by means of a simple and typical scenario where the communication sequence is purely round-robin;
- The feasibility of the proposed system is demonstrated through numerical simulation.

7.5.2 Discrete-time distributed system

Let us consider the following discrete-time distributed system consisting of N interacting subsystems

$$\begin{aligned} x_i(k+1) &= A_i x_i(k) + B_i u_i(k) + \sum_{j \in N_i} H_{ij} x_j(k), \\ y_i(k) &= C_i x_i(k), \end{aligned} \tag{7.141}$$

where A_i, B_i, H_{ij}, and C_i are system matrices with appropriate dimensions; $x_i(k)$, $u_i(k)$, and $y_i(k)$ are state, control input, and output of each subsystem i, respectively; and N_i denotes for the set of neighbors of subsystem i.

The distributed systems are controlled through a communication network that is used by each subsystem to send the output of the sensors to controllers. The controllers use these data to calculate the input signal and send it to the actuators of the systems. The output arrives in sample-and-hold fashion such as $y_i(k_i^l)$, where k_i^l represents the sequence of transmission instants of subsystem i.

Remark 7.6. We assume a feedback gain K_i such that matrix $\bar{A}_i = A_i + B_i K_i C_i$ is Hurwitz. Thus, each control input u_i affecting subsystem i is written as

$$u_i(k) = K_i C_i x_i(k_i^l) + \sum_{j \in N_i} L_{ij} C_j x_j(k_i^l). \tag{7.142}$$

7.5.3 Characteristics of the DoS attacks

In this chapter the effect of the DoS attacks is considered as a failure in the transmission of signals. In addition, this effect is accumulated with the failure caused by channel unavailability. The communication attempts of all the subsystems is simultaneously affected by the DoS attacks because the network is shared.

Similar to [89], the model of the DoS attacks is considered to have a limited frequency and duration. Let $\{H_n\}_{n\in\mathbb{N}_0}, h_0 \geq 0$ refer to the sequence of DoS off/on transitions associated with time instants at which DoS exhibits a transition from possible to impossible transmissions (or zero to one). So, the n-th DoS time interval of a length $\tau_n \in \mathbb{R}_{\geq 0}$ is given by

$$H_n := \{h_n\} \cup [h_n, h_n + \tau_{n-1}]. \tag{7.143}$$

If $\tau_n = 0$, then H_n takes the form of a single pulse at h_n. If $\tau_n \neq 0$, $[h_n, h_n + \tau_{n-1}]$ represents an interval from the instant h_n to $(h_n + \tau_{n-1})$. Similarly, $[\tau, k-1]$ represents an interval from τ to $k-1$. Given $\tau, k \in \mathbb{R} \geq 0$ with $k \geq \tau$, let $n(\tau, k)$ refer to the number of DoS off/on transitions over $[\tau, k-1]$ and let $\Xi(\tau, k)$ refer to the subset of $[\tau, k]$ during which the network affected by the DoS attack such that:

$$\Xi(\tau, k) := \bigcup_{n\in\mathbb{N}_0} H_n \cap [\tau, k]. \tag{7.144}$$

In addition, $\Theta(\tau, k)$ refers to the interval where the attack does not exist as such and is represented by

$$\Theta(\tau, k) := [\tau, k] \backslash \Xi(\tau, k). \tag{7.145}$$

Assumption 7.5 (Frequency of the DoS attack). There exist constants $\eta \in \mathbb{R} \geq 0$ and $\tau_D \in \mathbb{R} > 0$ such that

$$n(\tau, k) \leq \eta + \frac{k - \tau}{\tau_D} \tag{7.146}$$

for all $\tau, k \in \mathbb{R} \geq 0$ with $k \geq \tau$.

Assumption 7.6 (Duration of the DoS attack). There exist constants $\kappa \in \mathbb{R} \geq 0$ and $T \in \mathbb{R} > 1$ such that

$$|\Xi(\tau, k)| \leq \kappa + \frac{k - \tau}{T} \tag{7.147}$$

for all $\tau, k \in \mathbb{R} \geq 0$ with $k \geq \tau$.

Remark 7.7. Assumptions 7.5 and 7.6 constrain the average frequency and duration of the DoS attack signals. τ_D and η in Assumption 7.5 are the average dwell time between consecutive DoS off/on transitions and the chattering bound, respectively. Assumption 7.6 constrains the duration of the DoS attack such that it is limited by a certain fraction of time $1/T$. The constant κ is used for regulation [101].

7.5.4 Design results

The objective in this section is to find stability conditions for the distributed CPS affected by DoS attacks. We first present an output feedback controller to ensure the stability of the system in the nominal case (absence of DoS). We also discuss the stabilization problem of distributed CPS under a digital communication channel in the nominal scenario.

The error between the value of the current state and the transmitted state is defined as $e_i(k)$, where i refers to the subsystem such that

$$e_i(k) = x_i(k_i') - x_i(k), \quad i = 1, 2, \ldots, N. \tag{7.148}$$

The dynamics of each subsystem i can be described by combining (7.141), (7.142) and (7.148) as

$$x_i(k+1) = \bar{A}_i x_i(k) + B_i K_i C_i e_i(k) + \sum_{j \in N_i} \bar{A}_j x_j(k) + B_i \sum_{j \in N_i} L_{ij} C_j e_j(k), \tag{7.149}$$

where $\bar{A}_j = B_i L_{ij} C_j + H_{ij}$. It should be noted that the interconnected neighbors $x_j(k)$ affect the dynamics of subsystem i in addition to $e_i(k)$ and $e_j(k)$.

Remark 7.8. It is clear from (7.149) that the stability can be accomplished in the case of small error e and weak couplings. Moreover, the "smallness" of e can be explained by the x-dependent bound $\|e_i(k)\| \leq \sigma_i \|x_i(k)\|$, with a suitable design parameter σ_i.

Our objective in this section is to design a static output feedback in the form of (7.142) to achieve the asymptotic stability for nominal distributed systems (7.141).

Theorem 7.5. *Let the controller gains K_i and L_{ij} of (7.142) be given. System (7.149) is asymptotically stable if there exist positive matrices P_i satisfying the following inequalities:*

$$\Lambda_i = \begin{bmatrix} \Lambda_{1i} & \Lambda_{2i} & \Lambda_{3i} & \Lambda_{4i} \\ * & \Lambda_{5i} & \Lambda_{6i} & \Lambda_{7i} \\ * & * & \Lambda_{8i} & \Lambda_{9i} \\ * & * & * & \Lambda_{10i} \end{bmatrix} < 0, \tag{7.150}$$

where

$$
\begin{aligned}
\Lambda_{1i} &= \bar{A}_i^T P_i \bar{A}_i - P_i, & \Lambda_{2i} &= \bar{A}_i^T P_i B_i K_i C_i, \\
\Lambda_{3i} &= \bar{A}_i^T P_i \sum_{j\in N_i} \bar{A}_j, & \Lambda_{4i} &= \bar{A}_i^T P_i B_i \sum_{j\in N_i} L_{ij} C_j, \\
\Lambda_{5i} &= C_i^T K_i^T B_i^T P_i B_i K_i C_i, & \Lambda_{6i} &= C_i^T K_i^T B_i^T P_i \sum_{j\in N_i} \bar{A}_j, \\
\Lambda_{7i} &= C_i^T K_i^T B_i^T P_i B_i \sum_{j\in N_i} L_{ij} C_j, & \Lambda_{8i} &= \sum_{j\in N_i} \bar{A}_j^T P_i \bar{A}_j, \\
\Lambda_{9i} &= \sum_{j\in N_i} \bar{A}_j^T P_i B_i L_{ij} C_j, & \Lambda_{10i} &= \sum_{j\in N_i} C_j^T L_{ij}^T B_i^T P_i^T B_i L_{ij} C_j.
\end{aligned}
\tag{7.151}
$$

Proof. To establish the main theorem the following Lyapunov function is constructed:

$$
V_i(k) = x_i^T(k) P_i x_i(k). \tag{7.152}
$$

Evaluating the difference of $V_i(k)$, we have

$$
\begin{aligned}
\Delta V_i(k) &= V_i(k+1) - V_i(k) < 0 \\
\Delta V_i(k) &= x_i^T(k)\left(\bar{A}_i^T P_i \bar{A}_i - P_i\right) x_i(k) + 2x_i^T(k)\bar{A}_i^T P_i B_i K_i C_i e_i(k) \\
&\quad +2x_i^T(k)\bar{A}_i^T P_i \sum_{j\in N_i} \bar{A}_j x_j(k) \\
&\quad +2x_i^T(k)\bar{A}_i^T P_i \sum_{j\in N_i} B_i L_{ij} C_j e_j(k) \\
&\quad +e_i^T(k) C_i^T K_i^T B_i^T P_i B_i K_i C_i e_i(k) \\
&\quad +2e_i^T(k) C_i^T K_i^T B_i^T P_i \sum_{j\in N_i} \bar{A}_j x_j(k) \\
&\quad +2e_i^T(k) C_i^T K_i^T B_i^T P_i \sum_{j\in N_i} B_i L_{ij} C_j e_j(k) \\
&\quad +\sum_{j\in N_i} \left(x_j^T(k)\bar{A}_j^T\right) P_i \sum_{j\in N_i} \left(\bar{A}_j x_j(k)\right) \\
&\quad +2\sum_{j\in N_i} \left(x_j^T(k)\bar{A}_j^T\right) P_i \sum_{j\in N_i} \left(B_i L_{ij} C_j e_j(k)\right) \\
&\quad +\sum_{j\in N_i} \left(e_j^T(k) C_j^T L_{ij}^T B_i^T\right) P_i \sum_{j\in N_i} \left(B_i L_{ij} C_j e_j(k)\right) \\
&< 0.
\end{aligned}
\tag{7.153}
$$

So, it can be rewritten as

$$\Delta V_i(k) = \Omega_i^T(k)\Lambda_i\Omega_i(k) < 0, \tag{7.154}$$

where

$$\Omega_i^T(k) = \begin{bmatrix} x_i(k) & e_i(k) & \sum_{j\in N_i} x_j(k) & \sum_{j\in N_i} e_j(k) \end{bmatrix}^T \tag{7.155}$$

and Λ_i as described in (7.150). □

Theorem 7.6. *If there exist positive definite matrices X_i and matrices Y_i and Y_{ij} such that the following inequality is satisfied*

$$\bar{\Lambda}_i = \begin{bmatrix} -X_i & 0 & 0 & 0 & Y_i^T B_i^T + X_i A_i^T \\ 0 & \epsilon_i I & 0 & 0 & Y_i^T B_i^T \\ 0 & 0 & \epsilon_i I & 0 & \sum_{j\in N_i} Y_{ij}^T B_i^T + X_i H_{ij}^T \\ 0 & 0 & 0 & \epsilon_i I & \sum_{j\in N_i} Y_{ij}^T B_i^T \\ * & * & * & * & -X_i \end{bmatrix} < 0, \tag{7.156}$$

then system (7.149) is asymptotically stable and the controller gains are given by

$$K_i = Y_i X_i^{-1}(C_i)^\dagger, \quad L_{ij} = Y_{ij} X_i^{-1}(C_j)^\dagger. \tag{7.157}$$

Proof. Λ_i in LMI (7.150) can be rewritten as

$$\Lambda_i = \bar{\Lambda}_{i11} + \bar{\Lambda}_{i12} P_i \bar{\Lambda}_{i12}^T < 0 \tag{7.158}$$

with $\bar{\Lambda}_{i11} = \text{diag}\{-P_i, 0, 0, 0\}$ and

$$\bar{\Lambda}_{i12}^T = \begin{bmatrix} \bar{A}_i & B_i K_i C_i & \sum_{j\in N_i} \bar{A}_j & \sum_{j\in N_i} B_i L_{ij} C_j \end{bmatrix}^T.$$

So, LMI (7.158) is formulated using Schur complements as

$$\begin{bmatrix} \bar{\Lambda}_{i11} & \bar{\Lambda}_{i12} \\ \bar{\Lambda}_{i12}^T & -P_i^{-1} \end{bmatrix} < 0. \tag{7.159}$$

Now we define $X_i = P_i^{-1}$, then multiply (7.159) on the right and left by $\text{diag}\{X_i, X_i, X_i, X_i, I\}$. Then by recalling the values of $\bar{A}_i$ and $\bar{A}_j$ and by selecting $Y_i^T := X_i C_i^T K_i^T$ and $Y_{ij}^T := X_i C_j^T L_{ij}^T$, the LMI (7.156) can be obtained. □

7.5.5 The small-gain approach

In this section a periodic sampling protocol, the round-robin protocol, is implemented to update the law.

Assumption 7.7 (Inter-sampling of round-robin). There exists an inter-sampling interval Δ in the absence of DoS attacks satisfying

$$\|e_i(k)\| \leq \sigma_i \|x_i(k)\|, \tag{7.160}$$

where σ_i is a suitable design parameter.

Although there is a Δ satisfying (7.160), σ_i should be selected appropriately to avoid losing stability [101].

Here $Q_i = Q_i^T > 0$, and P_i is the unique solution of the Lyapunov equation $\bar{A}_i^T P_i \bar{A}_i - P_i + Q_i = 0$. We consider the Lyapunov function $V_i(k) = x_i^T(k) P_i x_i(k)$ for each subsystem i satisfying

$$\lambda_{\min}(P_i)\|x_i(k)\|^2 \leq V_i(x_i(k)) \leq \lambda_{\max}(P_i)\|x_i(k)\|^2, \tag{7.161}$$

where $\lambda_{\min}(P_i)$ and $\lambda_{\max}(P_i)$ refer to the smallest and largest eigenvalue of P_i, respectively. The selection of σ_i to ensure the stability of the system is presented by the following lemma.

Lemma 7.5. *For a distributed CPS described by (7.141) controlled by inputs described by (7.142), we assume that the spectral radius $r(A^{-1}B) < 1$. The distributed CPS is asymptotically stable if there is σ_i such that*

$$\sigma_i < \sqrt{\frac{l_i}{j_i}}, \tag{7.162}$$

where l_i is the i-th entry of the row vector $L := \mu^T(A - B) = [l_1, l_2, \ldots, l_N]$ and j_i is the j-th entry of the row vector $J := \mu^T \Gamma = [j_1, j_2, \ldots, j_N]$. $\mu \in \mathbb{R}_+^N$ is an arbitrary column vector satisfying $\mu^T(-A + B) < 0$. The matrices A, B, and Γ are given by

$$A = \begin{bmatrix} \alpha_1 & & \\ & \ddots & \\ & & \alpha_N \end{bmatrix} \tag{7.163}$$

$$B = \begin{bmatrix} 0 & \beta_{12} & \cdots & \beta_{1N} \\ \beta_{21} & 0 & \cdots & \beta_{2N} \\ \vdots & \vdots & \vdots & \vdots \\ \beta_{N1} & \beta_{N2} & \cdots & 0 \end{bmatrix} \tag{7.164}$$

$$\Gamma = \begin{bmatrix} \gamma_{11} & \gamma_{12} & \cdots & \gamma_{1N} \\ \gamma_{21} & \gamma_{22} & \cdots & \gamma_{2N} \\ \vdots & \vdots & \vdots & \vdots \\ \gamma_{N1} & \gamma_{N2} & \cdots & \gamma_{NN} \end{bmatrix} \quad (7.165)$$

with

$$\alpha_i = \lambda_{\min}(Q_i) - \delta, \quad (7.166)$$

$$\beta_{ij} = \delta + \frac{\|\bar{A}_j^T P_i \bar{A}_j\|^2}{\delta} + \|\bar{A}_j^T P_i \bar{A}_j\| + \|C_i^T K_i^T B_i^T P_i \bar{A}_j\|^2, \quad (7.167)$$

$$\gamma_{ii} = \frac{\|\bar{A}_i^T P_i B_i K_i C_i\|^2}{\delta} + \|C_i^T K_i^T B_i^T P_i B_i K_i C_i\|, \quad (7.168)$$

$$\gamma_{ij} = \frac{\|\bar{A}_i^T P_i B_i L_{ij} C_j\|^2}{\delta} + \frac{\|C_i^T K_i^T B_i^T P_i B_i L_{ij} C_j\|^2}{\delta} + \frac{\|\bar{A}_j^T P_i B_i L_{ij} C_j\|^2}{\delta} + \|C_j^T L_{ij}^T B_i^T P_i B_i L_{ij} C_j\|, \quad (7.169)$$

where δ is positive and real such that $\alpha_i > 0$ and $\lambda_{min}(Q_i)$ is the minimum eigenvalue of Q_i for $i = 1, 2, \ldots, N$.

Proof. The difference equation (7.153) can be described by

$$\begin{aligned} \Delta V_i(k) \leq\ & -\lambda_{min}(Q_i)\,\|x_i(k)\|^2 + \left\|2\bar{A}_i^T P_i B_i K_i C_i\right\| \|x_i(k)\|\, \|e_i(k)\| \\ & + \sum_{j\in N_i} \left\|2\bar{A}_i^T P_i \bar{A}_j\right\| \|x_i(k)\|\, \left\|x_j(k)\right\| \\ & + \sum_{j\in N_i} \left\|2\bar{A}_i^T P_i B_i L_{ij} C_j\right\| \|x_i(k)\|\, \|e_i(k)\| \\ & + \left\|C_i^T K_i^T B_i^T P_i B_i K_i C_i\right\| \|e_i(k)\|^2 \\ & + \sum_{j\in N_i} \left\|2C_i^T K_i^T B_i^T P_i \bar{A}_j\right\| \|e_i(k)\|\, \left\|x_j(k)\right\| \\ & + \sum_{j\in N_i} \left\|2C_i^T K_i^T B_i^T P_i B_i L_{ij} C_j\right\| \|e_i(k)\|\, \left\|e_j(k)\right\| \\ & + \sum_{j\in N_i} \left\|\bar{A}_j^T P_i \bar{A}_j\right\| \left\|x_j(k)\right\|^2 \end{aligned}$$

$$
\begin{aligned}
&+ \sum_{j \in N_i} \left\| 2\bar{A}_j^T P_i B_i L_{ij} C_j \right\| \left\| x_j(k) \right\| \left\| e_j(k) \right\| \\
&+ \sum_{j \in N_i} \left\| C_j^T L_{ij}^T B_i^T P_i B_i L_{ij} C_j \right\| \left\| e_j(k) \right\|^2 . \qquad (7.170)
\end{aligned}
$$

The Young's inequalities for any matrices E, F, and G with any positive real δ yield the following

$$
\|E\| \, \|F\| \, \|G\| \leq \delta \, \|F\|^2 + \frac{1}{\delta} \, \|E\|^2 \, \|G\|^2 . \qquad (7.171)
$$

Using (7.171), (7.170) can be rewritten as

$$
\begin{aligned}
\Delta V_i(x_i(k)) \quad \leq \quad & -\alpha_i \, \|x_i(k)\|^2 + \sum_{j \in N_i} \beta_{ij} \left\| x_j(k) \right\|^2 + \gamma_{ii} \, \|e_i(k)\|^2 \\
& + \sum_{j \in N_i} \gamma_{ij} \left\| e_j(k) \right\|^2 \qquad (7.172)
\end{aligned}
$$

with α_i, β_{ij}, γ_{ii}, and γ_{ij} as in (7.166)–(7.169). In addition, δ can be always found such that $\alpha_i > 0$ for $i = 1, 2, \ldots, N$.

By defining vectors

$$
\begin{aligned}
V_{vec}(x_i(k)) \quad &:= \quad [V_1(x_1(k)), V_2(x_2(k)), \ldots, V_N(x_N(k))]^T \\
\|x(k)\|_{vec} \quad &:= \quad \left[\|x_1(k)\|^2, \|x_2(k)\|^2, \ldots, \|x_N(k)\|^2 \right]^T \\
\|e(k)\|_{vec} \quad &:= \quad \left[\|e_1(k)\|^2, \|e_2(k)\|^2, \ldots, \|e_N(k)\|^2 \right]^T
\end{aligned}
$$

the inequality (3.32) can be compactly written as

$$
\Delta V_i(x_i(k)) \quad \leq \quad (-A + B) \, \|x(k)\|_{vec} + \Gamma \, \|e(k)\|_{vec} \qquad (7.173)
$$

with A, B, and Γ as in Lemma 7.5.

There exists a positive vector $\mu \in \mathbb{R}_+^n$ satisfying $\mu^T(-A + B) < 0$ if the spectral radius satisfies $r(A^{-1}B) < 1$ [258]. We chose the Lyapunov function to be $V(x(k)) := \mu^T V_{vec}(x_i(k))$. Thus, ΔV yields

$$
\begin{aligned}
\Delta V(x(k)) \quad &= \quad \mu^T \Delta V_{vec}(x_i(k)) \\
&\leq \quad \mu^T(-A + B) \, \|x(k)\|_{vec} + \mu^T \Gamma \, \|e(k)\|_{vec} . \qquad (7.174)
\end{aligned}
$$

Given that $\mu^T(-A + B) < 0$, we have

$$
\Delta V(x(k)) \leq -L \, \|x(k)\|_{vec} + J \, \|e(k)\|_{vec} , \qquad (7.175)
$$

where $L := \mu^T(A - B)$ and $J := \mu^T \Gamma$ are row vectors.

Let l_i and j_i be the entries of vectors L and J, respectively. Thus, we obtain

$$\begin{aligned}\Delta V(x(k)) &\leq \sum_{i\in N} l_i \|x_i(k)\|^2 + \sum_{i\in N} j_i \|e_i(k)\|^2 \\ &= -\sum_{i\in N}\left(l_i \|x_i(k)\|^2 - j_i \|e_i(k)\|^2\right),\end{aligned} \tag{7.176}$$

which implies asymptotic stability with $\sigma_i < \sqrt{\frac{l_i}{j_i}}$. □

Remark 7.9. The matrices P_i and Q_i, which are required to apply Lemma 7.5 and to calculate σ_i, can be calculated by solving the LMI (7.156) of Theorem 3.2, such that

$$P_i = X_i^{-1}, \quad Q_i = P_i - \bar{A}_i^T P_i \bar{A}_i. \tag{7.177}$$

7.5.6 Stability analysis under DoS attacks

We have discussed the selection of σ_i, which is related to the error bound to ensure the asymptotic stability of a distributed CPS using a round-robin protocol but under no attack, as discussed in the previous section. Unfortunately, the stability of the system is not guaranteed when the DoS attack affects the system even with the same round-robin protocol [101]. Our objective in this section is to discuss the stability of the distributed CCSs subject to DoS attack for a round-robin protocol.

Theorem 7.7. *We assume a distributed CCS (7.141) and a control input (7.142), where a round-robin protocol is used in the network communication with sampling interval Δ as in Assumption 7.7. The distributed system subjected to DoS attack with time and duration as in Assumptions 7.5 and 7.6 with arbitrary η and κ, and with τ_D and T, is asymptotically stable if*

$$\frac{1}{T} + \frac{\Delta_*}{\tau_D} < 1; \quad , \omega_1 < 1, \tag{7.178}$$

in which

$$\omega_1; = \min\{\frac{l_i - \sigma_i^2 j_i}{\lambda_{\max}(P_i)\mu_i}\}, \ \omega_2; = \frac{4\max\{j_i\}}{\min\{\mu_i \lambda_{\min}(P_i)\}} \tag{7.179}$$

and $\Delta_* = N\Delta$, l_i, j_i, μ_i, and σ_i are as in Lemma 7.5.

Proof. The proof of Theorem 7.7 is shown in three steps:

Step 1. Lyapunov function in periods without attacks. Referring to Assumption 7.7, (7.160) holds true for σ_i satisfying Lemma 7.5, and (7.176) is negative. Thus, the derivative of the Lyapunov function is described as

$$\Delta V(x(k)) \leq -\sum_{j\in N}\left(l_i - j_i\sigma_i^2\right)\|x_i(k)\|^2 \leq -\sum_{j\in N}\frac{l_i - j_i\sigma_i^2}{\lambda_{max}(P_i)\mu_i}\mu_i V_i$$

$$\triangleq \quad -\omega_1 V(x(k)), \tag{7.180}$$

where $\omega_1 ;= \min\{\frac{l_i - \sigma_i^2 j_i}{\lambda_{\max}(P_i)\mu_i}\}$. Thus, for $k \in [h_n + \tau_n, h_{n+1}[$ (DoS-free time), the Lyapunov function yields

$$V(x(k)) \leq (1-\omega_1)^{k-h_n-\tau_n} V(x(h_n+\tau_n)). \tag{7.181}$$

Step 2. Lyapunov function during attacks periods. Let us define z_m^i as the last successful sampling instant before the DoS attack occurs. By implementing $e_i(k)$ as defined above, we obtain

$$e_i(k) = x_i(z_m^i) - x_i(k) = x_i(h_n) - x_i(k) \tag{7.182}$$

and

$$\begin{aligned} \|e_i(k)\|^2 &\leq \|x_i(h_n)\|^2 + 2\,\|x_i(k)\|\,\|x_i(h_n)\| + \|x_i(k)\|^2 \\ &\leq 2\,\|x_i(h_n)\|^2 + 2\,\|x_i(k)\|^2 \end{aligned} \tag{7.183}$$

for $k \in H_n$. By summing up $\|e_i(k)\|^2$ for $i \in N$, we obtain

$$\sum_{i\in N} \|e_i(k)\|^2 \leq 2\sum_{i\in N} \|x_i(h_n)\|^2 + 2\sum_{i\in N} \|x_i(k)\|^2. \tag{7.184}$$

If $\sum_{i\in N} \|x_i(h_n)\|^2 \leq \sum_{i\in N} \|x_i(k)\|^2$, we have that $\sum_{i\in N} \|e_i(k)\|^2 \leq 4\sum_{i\in N} \|x_i(k)\|^2$, otherwise $\sum_{i\in N} \|e_i(k)\|^2 \leq 4\sum_{i\in N} \|x_i(h_n)\|^2$.

Recalling (7.176), it is simple to see that

$$\Delta V(x(k)) \leq \sum_{j\in N} j_i \,\|e_i(k)\|^2. \tag{7.185}$$

Thus, for all $k \in H_n$ (DoS-active time) in the case when $\sum_{i\in N} \|x_i(h_n)\|^2 \leq \sum_{i\in N} \|x_i(k)\|^2$, the difference of the Lyapunov function yields

$$\begin{aligned} \Delta V(x(k)) &\leq \max\{j_i\} \sum_{j\in N} \|e_i(k)\|^2 \leq 4\max\{j_i\} \sum_{j\in N} \|x_i(k)\|^2 \\ &\leq \frac{4\max\{j_i\}}{\min\{\mu_i \lambda_{min}(P_i)\}} \sum_{j\in N} \mu_i V(x_i(k)) \\ &\triangleq \omega_2 V(x(k)) \end{aligned} \tag{7.186}$$

with $\omega_2 := \frac{4\max\{j_i\}}{\min\{\mu_i \lambda_{min}(P_i)\}}$. On the other hand, for all $k \in H_n$ such that $\sum_{i\in N} \|x_i(h_n)\|^2 > \sum_{i\in N} \|x_i(k)\|^2$, we have

$$\Delta V(x(k)) \leq \omega_2 V(x(h_n)). \tag{7.187}$$

Thus, (7.186) and (7.187) imply that the Lyapunov function during H_n satisfies

$$V(x(k)) \leq (1+\omega_2)^{k-h_n} V(x(h_n)). \tag{7.188}$$

Step 3. Switching between stable and unstable modes. Let us consider a DoS attack with period τ_n, where the overall system has to wait an additional period with length $N\Delta$ at the end of this attack to have a full round of communications. The period where at least one subsystem transmission is not successful can be upper bounded by $\tau_n + N\Delta$. For all $\tau, k \in \mathbb{R}_{\leq 0}$ with $k \geq \tau$, the total length of time where communication is not possible over $[\tau, k[$, say $|\bar{\Xi}(\tau, k)|$, can be upper bounded by

$$|\bar{\Xi}(\tau,k)| \quad \leq \quad |\Xi(\tau,k)| + (1+n(\tau,k))\Delta_* \leq \kappa_* + \frac{k-\tau}{T_*}, \tag{7.189}$$

where $\Delta_* = N\Delta$, $\kappa_* := \kappa + (1+\eta)\Delta_*$, and $T_* = \frac{\tau_D T}{\tau_D + T\Delta_*}$.

In addition, we consider the additional waiting time caused by the protocol, and the Lyapunov function yields $V(x(k)) \leq (1-\omega_1)^{k-h_n-\tau_n-N\Delta} V(x(h_n + \tau_n + N\Delta))$ for $t \in [h_n + \tau_n + N\Delta, h_{n+1}[$ and $V(x(k)) \leq (1+\omega_2)^{k-h_n} V(x(h_n))$ for $t \in [h_n, h_n + \tau_n + N\Delta[$.

As a result, we can deal with the overall behavior of the closed-loop system as a switching system with two modes. Applying simple iterations to the Lyapunov functions with and without DoS attacks, we obtain

$$V(x(k)) \leq (1-\omega_1)^{\left[k-\kappa_*-(\frac{1}{T}+\frac{\Delta_*}{\tau_D})k\right]}(1+\omega_2)^{\left[\kappa_*+(\frac{1}{T}+\frac{\Delta_*}{\tau_D})k\right]} V(x(0)). \tag{7.190}$$

To ensure the stability of the last equation, (7.178) can be easily obtained. □

Remark 7.10. The resilience of the distributed systems depends on the largeness of ω_1 and the smallness of ω_2. To achieve this, we can try to find K_i and L_{ij} such that $\|B_i C_i K_i\|$ and $\|B_i C_j L_{ij}\|$ are small. On the other hand, the round-robin sampling interval also affects stability in the sense that it determines how quickly the overall system can restore communication. We can always apply smaller round-robin inter-sampling time to reduce the left-hand side of (7.178) at the expense of higher communication load.

7.5.7 Illustrative example II

The effectiveness of the proposed method presented in this chapter is shown by considering one of the most common CPSs, which is the quadruple-tank process which described in [315]. As shown in Fig. 7.16, the system consists of four tanks (two upper and two lower), and our objective is to control the level in the two lower tanks via two pumps. The process has two inputs (input voltage to the pumps) and two outputs (voltage from level measurement devices). The

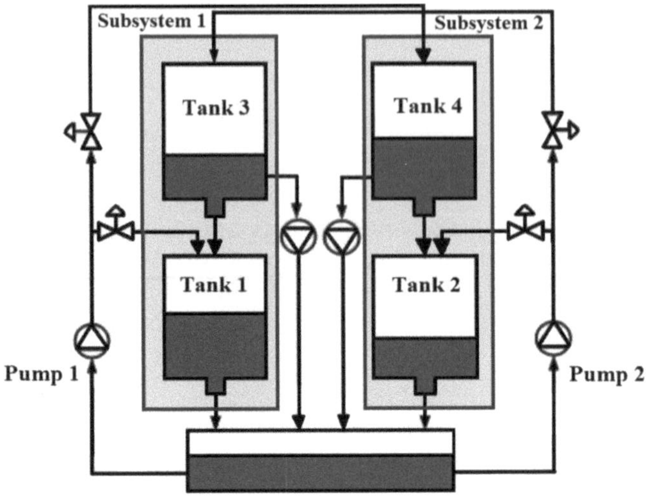

FIGURE 7.16 Schematic diagram of a quadruple-tank system.

system matrices for the linearized discrete-time state space model of the system are as follows:

$$
\begin{aligned}
A_1 &= \begin{bmatrix} 0.9998 & 0 \\ 0 & 0.9998 \end{bmatrix}, & B_1 &= \begin{bmatrix} 0.6359(10)^{-3} \\ 0.4559(10)^{-3} \end{bmatrix}, \\
H_{12} &= \begin{bmatrix} 0 & 0.0003 \\ 0 & 0 \end{bmatrix}, & C_1 &= \begin{bmatrix} 1 & 0 \end{bmatrix}, \\
A_2 &= \begin{bmatrix} 0.9999 & 0 \\ 0 & 0.9997 \end{bmatrix}, & B_2 &= \begin{bmatrix} 0.488(10)^{-3} \\ 0.6279(10)^{-3} \end{bmatrix}, \\
H_{21} &= \begin{bmatrix} 0 & 0.0002 \\ 0 & 0 \end{bmatrix}, & C_2 &= \begin{bmatrix} 1 & 0 \end{bmatrix}.
\end{aligned}
$$

Using YALMIP the controller gains were obtained:

$$
\begin{aligned}
K_1 &= -0.8476, \quad K_2 = -1.8838, \\
L_{12} &= -0.4756, \quad L_{21} = -1.3312.
\end{aligned} \tag{7.191}
$$

In addition, we found that

$$
\begin{aligned}
P_1 &= 10^{-5} \times \begin{bmatrix} 0.6421 & -0.5143 \\ -0.5143 & 0.9261 \end{bmatrix}, \\
P_2 &= 10^{-5} \times \begin{bmatrix} 0.8421 & -0.4895 \\ -0.4895 & 0.6357 \end{bmatrix},
\end{aligned}
$$

$$Q_1 = 10^{-8} \times \begin{bmatrix} 0.5514 & -0.1250 \\ -0.1250 & 0.3704 \end{bmatrix},$$

$$Q_2 = 10^{-8} \times \begin{bmatrix} 0.5583 & 0.1060 \\ 0.1060 & 0.3813 \end{bmatrix}. \tag{7.192}$$

Figs. 7.17 and 7.18 show that the system with the designed controller is stable in the nominal case.

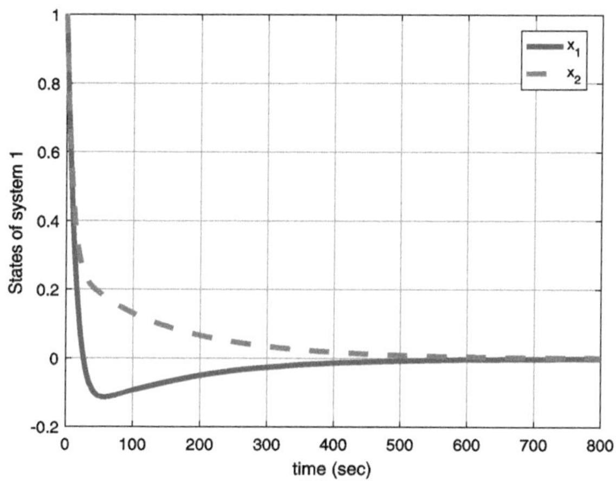

FIGURE 7.17 States of subsystem 1 in the nominal situation.

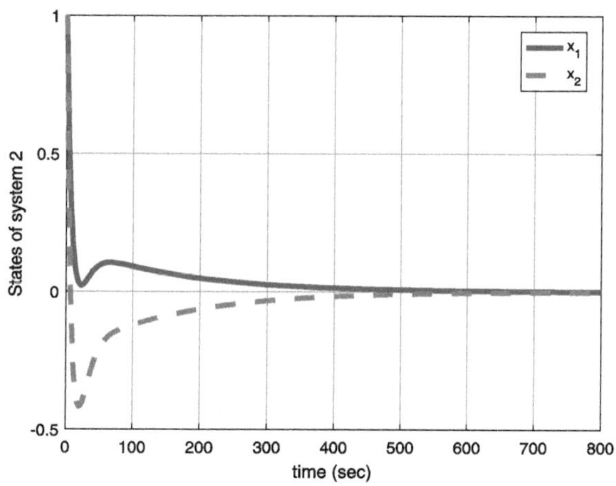

FIGURE 7.18 States of subsystem 2 in the nominal situation.

Using Lemma 7.1, we obtain:

$$A = 10^{-8} \times \begin{bmatrix} 0.2758 & 0 \\ 0 & 0.3010 \end{bmatrix}, \quad B = 10^{-9} \times \begin{bmatrix} 0 & 0.1059 \\ 0.1032 & 0 \end{bmatrix},$$

$$\Gamma = 10^{-6} \times \begin{bmatrix} 0.0278 & 0.0087 \\ 0.0630 & 0.1265 \end{bmatrix}$$

In addition, the parameters σ_1 and σ_2 are calculated to be (0.1570) and (0.1316), respectively. Thus, σ is selected to be (0.1). Based on Assumption 7.7, we select a round-robin sampling interval $\Delta = 0.01$ s. By applying these parameters, we found ω_1 and ω_2 to be $(1.2445(10)^4)$ and (0.2050), respectively.

The characteristics of the DoS attack were designed using Theorem 7.3. As shown in Figs. 7.19 and 7.20, the designed controller maintains the stability of the system in the presence of the DoS attack.

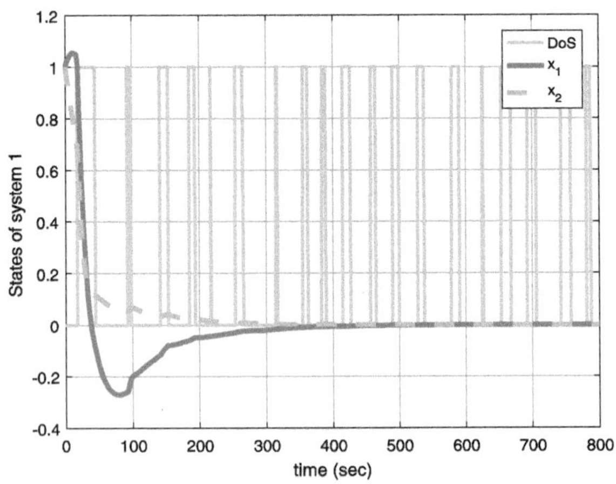

FIGURE 7.19 States of subsystem 1 under a DoS attack.

7.6 Notes

We investigated the stability of networked systems in the presence of DoS attacks. One contribution of this chapter is an explicit characterization of the frequency and duration of DoS attacks under which closed-loop stability can be preserved. The result is intuitive as it relates stability with the ratio of the *on*/*off* periods of jamming. An explicit characterization of sampling rules that achieve ISS was given. This characterization is flexible enough to allow the designer to choose from several implementation options that can be used to trade off between performance and communication resources.

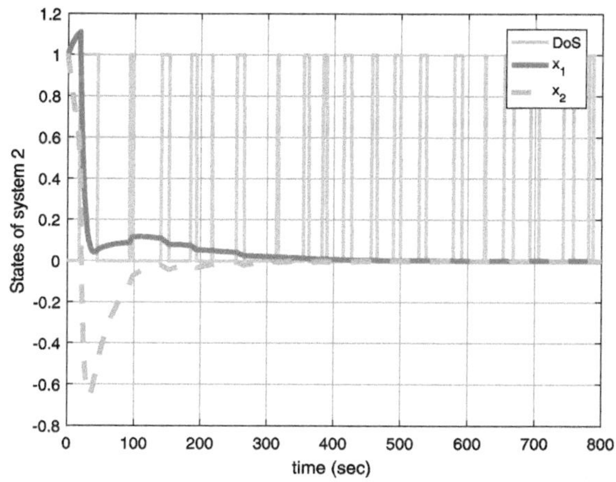

FIGURE 7.20 States of subsystem 2 under a DoS attack.

The results lend themselves to many possible extensions. For the framework considered here, identifying optimal attack and counter-attack strategies with respect to some prescribed performance objective is an interesting research venue. Moreover, we did not investigate the effect of possible limitations on the information, such as quantization and delays. As additional future research topics, we recommend the use of similar approaches to handle output feedback controllers as well as tackling problems of nonlinear systems. For the latter case, preliminary results have been reported in [104]. Finally, an interesting research line is to address the case where control and measurement channels can be interrupted *asynchronously*. In this respect the self-triggering logic described earlier, which relies on predictions of the process state, appears to be a convenient tool for updating the control action in case of DoS attacks in the measurement channel.

One of the main motivations for considering control over networks descends from problems of distributed coordination and control of large-scale systems [244,249,302–304]. Investigating our approach to control under DoS for self-triggered coordination problems, such as those in [304], also represents an interesting research venue.

A secure observer-based controller for discrete-time CPSs subject to both cyber (DoS and deception) and physical attacks was presented. The occurrence of cyber and physical attacks were considered as Bernoulli distributed white sequences with variable conditional probabilities. A sufficient condition was first derived where the observer-based control system is guaranteed to have the desired security level using the stochastic analysis techniques. Then the gains of the observer and controller were designed by solving a linear matrix inequality using YALMIP and MATLAB.

Finally, we discussed the stabilization of distributed CCSs affected by DoS attacks. First a static output feedback controller was designed to achieve the stability of a nominal distributed system. Then a simple and typical scenario where the communication sequence is purely round-robin was considered, and a bound of attack frequency and duration was calculated to ensure the stability of the distributed CCS. Finally, a numerical example was provided to demonstrate the feasibility of the proposed system.

Chapter 8

Secure group consensus

Contents

8.1 Couple-group consensus conditions under denial-of-service attacks

Multiagent systems (MASs) are networks of individual agents connected over a communication network for the purpose of information exchange to achieve a predefined control objective. The networks of MASs are often required to reach an agreement or consensus with respect to a common variable over a consensus protocol. Without doubt the simplest multiagent network is composed of two parts, a cyber layer and a physical layer, and thus MASs belong to the class of cyber-physical systems.

The prime motive in the study of MASs is the design of control algorithms to achieve coordination amongst the agents. Coordination control is an aspect of control theory primarily concerned with the design of algorithms and protocols that drive the agents within a multiagent network to demonstrate some form of intelligent coordination behavior with respect to some predefined objective.

Research in coordination control is mainly motivated by the desire to have artificial machines demonstrate intelligent coordination behavior exhibited by biological MASs such as bird swarms and fish schools. In this regard, coordi-

Cloud Control Systems. https://doi.org/10.1016/B978-0-12-818701-2.00016-0

nation control is divided into the following categories: consensus, swarming, formation, rendezvous, alignment, containment, and circumnavigation.

Consensus is achieved in a multiagent network when the states of all agents reach an agreement with respect to a common objective. All other forms of coordination control can be viewed as a special form of consensus as they all involve the states of the agents reaching an agreement. Many studies [362,319–324] have been carried out in consensus control for MASs including leader-follower, leaderless, pinning, finite-time, and bipartite, to mention a few.

8.1.1 Introduction

Recently, research efforts [325–328] are being focused on group consensus as a more practical extension of the general consensus. In group consensus the agents with the network are subdivided into different groups with each group having a decentralized control objective in harmony with the entire multiagent network.

On the other hand, the study of cyber-physical systems is involved with the design of resilient and robust control algorithms that can withstand different intruder attacks. Two main forms of attack in the literature are denial-of-service (DoS) and deception attacks. In DoS attacks the communication link between the interconnected systems is hijacked by an intruder for a short period of time, thereby preventing information exchange between the systems. In deception attacks false information is transmitted by the intruder over the hijacked communication network.

The authors in [329] studied the problem of distributed consensus control for MASs under DoS attacks. Their approach involved using both state-feedback and observer-based controllers to mitigate the effect of an intruder on the network. In addition, they derived sufficient conditions for consensus to be achieved during the attack period. In [330] the authors discussed secure consensus against deception attacks in synchronous networks. An algorithm was proposed to guarantee secure consensus, and it was shown to converge with an exponential rate through matrix analysis.

In [103] the authors characterize the relationship between closed-loop stability and the duration of DoS attacks. They concluded that the resilient nature of the control descends from adaptability of its sampling rate to occurrence of DoS. The authors in [331] solved the distributed average consensus problem for linear MASs under DoS attacks. They analyzed the frequency and duration of DoS attacks and proposed a distributed event-triggered control law for the MAS.

A novel distributed adaptive control architecture was proposed in [332] for networked MASs under the influence of exogenous disturbances and sabotaged sensors and actuators. The proposed controller also handles time-varying multiplicative actuator attacks on the followers with indirect links to the leaders in the network. The authors claim that the proposed controller guarantees uniform ultimate boundedness of the tracking error for each agent in the mean square sense.

In [333] the authors studied distributed tracking control problems for stochastic linear MASs under DoS attacks. The attacks considered were classified as connectivity-broken and connectivity-maintained attacks. Using average dwell time switching between stable and unstable modes based on graph analysis, some conditions were derived for robust mean square exponential consensus.

8.1.2 Algebraic graph theory

Graph theory is a standard framework for representing connections and interactions between networked, distributed, and multiagent systems. A graph $\mathcal{G}(\mathcal{V},\mathcal{E})$ is defined as a pair consisting of vertices $\mathcal{V}$ and edges $\mathcal{E}$. $\mathcal{V}(\mathcal{G})$ represents the set of vertices in $\mathcal{G}$, and $\mathcal{E}(\mathcal{G})$ is the edge set of $\mathcal{G}$. A graph is said to be undirected when the edge between any pair of vertices has no orientation. Conversely, in a directed graph or *digraph*, each edge $e \in \mathcal{E}(\mathcal{G})$ is directed between any pair of vertices, that is, the edge $e = v_i v_j$, originates at vertex v_i and terminates at vertex v_j.

$\Im(\chi)$ is used to denote imaginary parts of χ. $\Re(\chi)$ is used to denote real parts of χ. I_n represents an identity matrix of size n. $\det(A)$ is the determinant of matrix A. Some special matrices are used to describe the properties and information in a graph. These matrices include *degree*, *adjacency*, *incidence*, and *Laplacian* matrices. For a graph on n vertices and m edges the degree matrix $\Delta(\mathcal{G}) \in \mathcal{R}^{n\times n}$ is a diagonal matrix with elements on the diagonal representing the degree $d(v_i)$ of each vertex. $d(v_i)$ is the sum of edges incident to the vertex v_i. The adjacency matrix $\mathcal{A}(G)$ is an $n \times n$ matrix describing the adjacency relationship in $\mathcal{G}$. Each $a_{ij} \in \mathcal{A}(\mathcal{G})$ assumes 1 if $v_i v_j \in \mathcal{E}(\mathcal{G})$ and 0 otherwise. The Laplacian matrix in an undirected graph $\mathcal{L}(\mathcal{G}) = \Delta(\mathcal{G}) - \mathcal{A}(G)$. The incidence matrix $\mathcal{W}$ of a directed graph $\mathcal{D}$ is defined as $\mathcal{W} = [w_{ij}]$. $w_{ij} = -1$ if v_i is the tail of e_j, $w_{ij} = 1$ if v_i is the head of e_j, and $w_{ij} = 0$ if v_i is not adjacent to e_j. The Laplacian matrix of a directed graph $\mathcal{D}$ is $\mathcal{L}(\mathcal{D}) = \mathcal{W}(\mathcal{D})\mathcal{W}(\mathcal{D})^T$.

8.1.3 Consensus problem

Consider a network of MASs described by $\mathcal{G}$ consisting of n agents described by the first-order dynamics

$$\dot{x}_i(t) = u_i(t), \quad i = 1, 2, \ldots n, \tag{8.1}$$

where x_i and u_i represent the states and control inputs of each agent in the network.

Definition 8.1. The MAS described by $\mathcal{G}$ with dynamics (8.1) achieves consensus if for any $x_i(0)$

$$\lim_{t\to\infty} ||x_i(t) - x_j(t)|| = 0 \quad \forall i, j = 1, 2, \ldots n,$$

and the MAS asymptotically solves average-consensus problem when

$$\lim_{t\to\infty} ||x_i(t)|| = \frac{1}{n}\sum_{j=1}^{n} x_j(0) \quad \forall i, j = 1, 2, \ldots N.$$

Suppose now that the control protocol $u_i(t)$ is chosen as

$$u_i(t) = -\sum_{j=1}^{n} a_{ij}(x_i(t) - x_j(t)) \quad j = 1, 2, \ldots n, \tag{8.2}$$

the closed-loop system (8.1) under protocol (8.2) is given as

$$\dot{x} = -\mathcal{L}x. \tag{8.3}$$

Lemma 8.1. *[362] Suppose that $\mathcal{L} = [l_{ij}] \in \mathcal{R}^{n\times n}$ satisfies $l_{ij} < 0$, $i \neq j$ and $\sum_{j=1}^{n} l_{ij} = 0$, $i = 1, 2, ..., n$, then the following conditions are equivalent:*

- *$\mathcal{L}$ has a simple zero eigenvalue and all other eigenvalues have positive real parts;*
- *$\mathcal{L}x = 0$ implies that $x_1 = x_2 = \ldots x_n$;*
- *Consensus is reached asymptotically for the system $\dot{x} = -\mathcal{L}x$;*
- *The directed graph of $\mathcal{L}$ has a directed spanning tree;*
- *The rank of $\mathcal{L}$ is $n - 1$.*

8.1.4 Group consensus

Consider a network consisting of $n + m$ MASs belonging to subgroup χ_1 and χ_2 described by the first-order dynamics

$$\dot{x}_i(t) = ax_i + bu_i(t), \tag{8.4}$$

where n and m are the number of agents in subgroup χ_1 and χ_2, respectively.

Remark 8.1. When $a = 0$, $b = 1$, the dynamic in (8.4) is a single-integrator system defined by

$$\dot{x}_i(t) = u_i(t). \tag{8.5}$$

Definition 8.2. The network of MASs described by the dynamics in (8.4), achieves first-order couple-group consensus if the following conditions are satisfied:

$$\lim_{t\to\infty} \|x_i(t) - x_j(t)\| = 0, \ \forall i, j \in \chi_1,$$
$$\lim_{t\to\infty} \|x_i(t) - x_j(t)\| = 0, \ \forall i, j \in \chi_2.$$

Definition 8.3. A digraph $\mathcal{G}_i = \{\mathcal{V}_i, \mathcal{E}_i, \mathcal{A}_i\}$ is said to be a subgraph of $\mathcal{G} = \{\mathcal{V}, \mathcal{E}, \mathcal{A}\}$ if (a) $\mathcal{V}_i \subseteq \mathcal{V}$, (b) $\mathcal{E}_i \subseteq \mathcal{E}$.

Initially the following result is recalled:

Lemma 8.2. *[362] Let $\mathcal{G}_n$ be a graph on n vertices with m connected components. If $\mathcal{L}_n$ is the Laplacian of $\mathcal{G}$, then rank$(\mathcal{L}) = n - m$.*

Assumption 8.1. The MAS network described by the directed graph $\mathcal{G} = \{\mathcal{V}, \mathcal{E}, \mathcal{A}\}$ can be partitioned into two subgroups: $\mathcal{G}_1 = \{\mathcal{V}_1, \mathcal{E}_1, \mathcal{A}_1\}$ and $\mathcal{G}_2 = \{\mathcal{V}_2, \mathcal{E}_2, \mathcal{A}_2\}$. Therefore, the graph Laplacian $\mathcal{L}$ can be written as $\mathcal{L} = \begin{bmatrix} \mathcal{L}_{11} & \mathcal{L}_{12} \\ \mathcal{L}_{21} & \mathcal{L}_{22} \end{bmatrix}$. $\mathcal{L}_{11} \in \mathcal{R}^{n \times n}$ and $\mathcal{L}_{22} \in \mathcal{R}^{m \times m}$ are Laplacians of the subgraphs $\mathcal{G}_1$ and $\mathcal{G}_2$, respectively. $\mathcal{L}_{21} = \mathcal{L}_{12}^T$ is the Laplacian representing the interconnection between the two subgraphs. This means that the interconnection between the subgroups is undirected. In other words, the two subgroups can communicate in both directions with each other. In addition, $\mathcal{L}_{11}$ and $\mathcal{L}_{22}$ have eigenvalues μ_i^1 and μ_j^2, respectively, where $i = 1, 2, \ldots, n$ and $j = 1, 2, \ldots, m$.

Hence, the following preliminary results are established:

Lemma 8.3. *If the graph $\mathcal{G}$ has two connected components, the graph Laplacian $\mathcal{L}$ has exactly two simple zero eigenvalues and all other eigenvalues have positive real parts.*

Proof. Without loss of generality, we consider a graph $\mathcal{G}$ with two connected components $\mathcal{G}_1$ and $\mathcal{G}_2$; it follows from Lemma 8.2 that rank$(\mathcal{L}) = n - 2$. This implies that nullity $(\mathcal{L}) = 2$. Therefore, $\mathcal{L}$ has zero eigenvalues with geometric and algebraic multiplicity of 2. This suggests that each subgroup $\mathcal{G}_i$ has rank$(\mathcal{L}_i) = n_i - 1$, where n_i is the number of vertices in subgroup i. Therefore, the eigenvalues of $\mathcal{L}$ can be ordered as

$$\lambda_1 = \lambda_2 \leq \lambda_3 \leq \ldots \leq \lambda_{n+m},$$

where $\lambda_1 = \lambda_2 = 0$. □

Lemma 8.4. *Group consensus is achieved in a MAS network consisting of two subgroups if each of the subgraphs $\mathcal{G}_1$ and $\mathcal{G}_2$ representing each subgroup has a directed spanning tree.*

Proof. The proof of this lemma can be deduced directly from Lemma 8.1 and Lemma 8.3. Since the conditions in Lemma 8.1 are equivalent, it follows from Lemma 8.3 that if the MAS consists of m components and zero eigenvalues with algebraic multiplicity m, then each m component has a directed spanning tree. □

8.1.5 Attack model

In this paper we consider DoS attack models on the edges (or links) between connected agents.

Proposition 8.1. *If the MAS described by $\mathcal{G}$ with dynamics (8.1) is under DoS attack, then control protocol (8.2) can be redefined as*

$$u_i(t) = -\sum_{\substack{j=1\\ j\neq i\\ j\notin\Pi(t)}}^{n} a_{ij}(x_i(t) - x_j(t)) \;\; j = 1, 2, \ldots n, \tag{8.6}$$

where $\Pi(t)$ represents the attack matrix defining the set of nodes under DoS attack. Therefore, the closed-loop dynamics of the entire MAS network become [329]:

$$\dot{x}(t) = -(\mathcal{L} - \mathcal{L}^*(t))x(t) = \tilde{\mathcal{L}}(t). \tag{8.7}$$

Here $\mathcal{L}^*$ can be considered as Laplacian of attack modes. When no attack is present in the network, the entries of $\mathcal{L}^*$ are zero. The resulting graph Laplacian during the period of an attack is denoted $\tilde{\mathcal{L}}(t)$.

8.1.6 First-order group consensus under DoS attack

Consider a distributed consensus protocol for MASs under DoS attacks:

$$u_i(t) = \begin{cases} \alpha_1 \sum_{j=1, j\neq i, j\notin\Pi(t)}^{n} a_{ij}(x_j(t) - x_i(t)) \\ \quad + \beta_1 \sum_{j=n+1, j\neq i, j\notin\Pi(t)}^{n+m} a_{ij}x_j(t) \;\forall i \in \mathcal{G}_1 \\ \alpha_2 \sum_{j=n+1, j\neq i, j\notin\Pi(t)}^{n+m} a_{ij}(x_j(t) - x_i(t)) \\ \quad + \beta_2 \sum_{j=1, j\neq i, j\notin\Pi(t)}^{n} a_{ij}x_j(t) \;\forall i \in \mathcal{G}_2 \end{cases}. \tag{8.8}$$

Here $a_{ij} \geq 0$, $a_{ij} \in \mathcal{A}$ describes the interconnection between agents i, j; α_1, α_2 and β_1, β_2 can be considered as gains (or coupling) within and between subgroups, respectively; n, m are the numbers of agents in subgroup $\mathcal{G}_1$ and $\mathcal{G}_2$.

The MAS (8.4) under protocol (8.8) can be written as

$$\dot{x}(t) = \tilde{\mathcal{L}}(t)x(t) \tag{8.9}$$

$$\tilde{\mathcal{L}} = \begin{bmatrix} \tilde{\mathcal{L}}_{11} & \tilde{\mathcal{L}}_{12} \\ \tilde{\mathcal{L}}_{21} & \tilde{\mathcal{L}}_{22} \end{bmatrix}$$

$\tilde{\mathcal{L}}_{11} = aI_n - b\alpha_1(\mathcal{L}_{11} - \mathcal{L}_{11}^*(t))$, $\tilde{\mathcal{L}}_{12} = b\beta_1(\mathcal{L}_{12} - \mathcal{L}_{12}^*(t))$, $\tilde{\mathcal{L}}_{21} = b\beta_2(\mathcal{L}_{21} - \mathcal{L}_{21}^*(t))$, $\tilde{\mathcal{L}}_{22} = aI_m - b\alpha_2(\mathcal{L}_{22} - \mathcal{L}_{22}^*(t))$.

We define the error vector of the MAS network as

$$e_i(t) = \begin{cases} x_i - x_1 & \forall i = 2 \dots n \\ x_i - x_{n+1} & \forall i = n+2 \dots n+m \end{cases}. \tag{8.10}$$

Therefore, the error dynamics are defined as

$$\dot{e}(t) = \Psi(t) e(t), \tag{8.11}$$

where $\Psi(t)$ is defined as

$$\Psi(t) = \begin{bmatrix} \Psi_{11}(t) & \Psi_{12}(t) \\ \Psi_{21}(t) & \Psi_{22}(t) \end{bmatrix} \tag{8.12}$$

$\Psi_{11}(t) = aI_{n-1} - b\alpha_1(\hat{\mathcal{L}}_{11} - \hat{\mathcal{L}}^*_{11}(t))$, $\Psi_{12}(t) = b\beta_2(\hat{\mathcal{L}}_{12} - \hat{\mathcal{L}}^*_{12}(t))$, $\Psi_{21}(t) = b\beta_2(\hat{\mathcal{L}}_{21} - \hat{\mathcal{L}}^*_{21}(t))$, $\Psi_{22}(t) = aI_{m-1} - b\alpha_2(\hat{\mathcal{L}}_{22} - \hat{\mathcal{L}}^*_{22}(t))$ and $\hat{\mathcal{L}}_{11}, \hat{\mathcal{L}}^*_{11} \in \mathcal{R}^{(n-1)\times(n-1)}$ and $\hat{\mathcal{L}}_{22}, \hat{\mathcal{L}}^*_{22} \in \mathcal{R}^{(m-1)\times(m-1)}$ are defined as

$$\hat{\mathcal{L}}_{11} = \mathcal{L}_n + \mathbf{1}_{n-1}\alpha_n^T, \ \hat{\mathcal{L}}^*_{11} = \mathcal{L}^*_n + \mathbf{1}_{n-1}\alpha_n^T$$
$$\hat{\mathcal{L}}_{22} = \mathcal{L}_m + \mathbf{1}_{m-1}\alpha_m^T \ \hat{\mathcal{L}}^*_{22} = \mathcal{L}^*_m + \mathbf{1}_{m-1}\alpha_m^T$$

$$\mathcal{L}_n = \begin{bmatrix} d_2 & -a_{23} & \dots & -a_{2n} \\ -a_{32} & d_3 & \dots & -a_{3n} \\ \vdots & \vdots & \ddots & \vdots \\ -a_{n2} & -a_{n3} & \dots & d_n \end{bmatrix}$$

$$\mathcal{L}_m = \begin{bmatrix} d_{n+2} & -a_{n+2,n+3} & \dots & -a_{n+2,n+m} \\ -a_{n+3,n+2} & d_{n+3} & \dots & -a_{n+3,n+m} \\ \vdots & \vdots & \ddots & \vdots \\ -a_{n+m,n+2} & -a_{n+m,n+3} & \dots & d_{n+m} \end{bmatrix}$$

$$\alpha_n^T = \begin{bmatrix} a_{12} & a_{13} & \dots & a_{1n} \end{bmatrix}$$

$$\alpha_m^T = \begin{bmatrix} a_{n+1,n+2} & a_{n+1,n+3} & \dots & a_{n+1,n+m} \end{bmatrix}.$$

The system (8.4) or (8.9) achieves group consensus if (8.11) is asymptotically stable. Let $\beta_n = \begin{bmatrix} a_{21}, & a_{31}, & \dots, & a_{n1} \end{bmatrix}$ and $\beta_m = \begin{bmatrix} a_{n+2,n+1}, & a_{n+3,n+1}, & \dots, & a_{n+m,n+1} \end{bmatrix}$. It is then possible to write $\mathcal{L}_{11} = \begin{bmatrix} d_1 & -\alpha_n^T \\ -\beta_n & \mathcal{L}_n \end{bmatrix}$ and $\mathcal{L}_{22} = \begin{bmatrix} d_{n+1} & -\alpha_m^T \\ -\beta_m & \mathcal{L}_m \end{bmatrix}$. Using the transformation matrix $P_n = \begin{bmatrix} 1 & \mathbf{0}_{n-1}^T \\ \mathbf{1}_{n-1} & I_{n-1} \end{bmatrix}$ and

$P_m = \begin{bmatrix} 1 & \mathbf{0}_{m-1}^T \\ \mathbf{1}_{m-1} & I_{m-1} \end{bmatrix}$, we get

$$P_n^{-1}\mathcal{L}_{11}P_n = \begin{bmatrix} 0 & -\alpha_n^T \\ \mathbf{0}_{n-1} & \hat{\mathcal{L}}_{11} \end{bmatrix}, \tag{8.13}$$

$$P_m^{-1}\mathcal{L}_{11}P_m = \begin{bmatrix} 0 & -\alpha_m^T \\ \mathbf{0}_{m-1} & \hat{\mathcal{L}}_{22} \end{bmatrix}. \tag{8.14}$$

From (8.13) and (8.14) the eigenvalues of $\hat{\mathcal{L}}_{11}$ and $\hat{\mathcal{L}}_{22}$ are $\mu_{12}, \mu_{13}, \ldots \mu_{1n}$ and $\mu_{22}, \mu_{23}, \ldots \mu_{2m}$, respectively.

We recall the following:

Lemma 8.5. *Control protocol (8.8) makes the MAS described by $\mathcal{G}$ with (8.4) achieve group consensus asymptotically if and only if the error system (8.11) is stable, that is the eigenvalues of Ψ have negative real parts.*

And proceed to establish the following result:

Theorem 8.1. *The MAS described by $\mathcal{G}$ with (8.4) consisting of subgraphs $\mathcal{G}_1$ and $\mathcal{G}_2$ under protocol (8.8) without DoS attacks achieves group consensus if and only if each component contains a spanning tree and the following inequalities are satisfied simultaneously:*

$$b^2\gamma_b^2 + 4a^2 - 4ab\gamma_a + 4b^2(\alpha_1\alpha_2\gamma_c - \beta_1\beta_2\mu_l) > 0, \tag{8.15}$$

$$2a - b\gamma_a \neq 0, \tag{8.16}$$

where $\gamma_a = \alpha_1\Re(\mu_i^1) + \alpha_2\Re(\mu_j^2)$, $\gamma_b = \alpha_1\Im(\mu_i^1) + \alpha_2\Im(\mu_j^2)$, $\gamma_c = \Re(\mu_i^1)\Re(\mu_j^2) - \Im(\mu_i^1)\Im(\mu_j^2)$, $\gamma_d = \Re(\mu_i^1)\Im(\mu_j^2) + \Im(\mu_i^1)\Re(\mu_j^2)$; $\Re(.)$, $\Im(.)$ *represent real and imaginary parts, respectively;* μ_i, μ_j *are eigenvalues of* $\hat{\mathcal{L}}_{11}$ *and* $\hat{\mathcal{L}}_{22}$, *respectively; and* μ_l *is the maximum eigenvalue of* $\hat{\mathcal{L}}_{12}\hat{\mathcal{L}}_{21}$.

Proof. Let $\Pi = \det[sI_{n+m-2} - \Psi]$; then

$$\begin{aligned}
\Pi &= \det\begin{bmatrix} \Pi_1 & \Pi_2 \\ \Pi_3 & \Pi_4 \end{bmatrix}, \\
\Pi_1 &= (s-a)I_{n-1} + b\alpha_1\hat{\mathcal{L}}_{11}, \\
\Pi_2 &= -b\beta_1\hat{\mathcal{L}}_{12}, \\
\Pi_3 &= -b\beta_2\hat{\mathcal{L}}_{21}, \\
\Pi_4 &= (s-b)I_{m-1} + b\alpha_2\hat{\mathcal{L}}_{22}, \\
\Pi &= \prod_{i=1}^{n} f_i(s)\prod_{j=1}^{M} f_j(s) - \beta_1\beta_2\mu_l,
\end{aligned}$$

where $f_i(s) = s - a + b\alpha_1\mu_i$, $f_j(s) = s - a + b\alpha_2\mu_j$.

Now we define $g(s)$ as follows:

$$g(s) = (s - a + b\alpha_1\mu_i)(s - a + b\alpha_2\mu_j) - \beta_1\beta_2\mu_l.$$

Substituting $s = j\omega$, $\mu_i = Re(\mu_i) + jIm(\mu_i)$, and $\mu_j = Re(\mu_j) + jIm(\mu_j)$, it is possible to write $g(j\omega) = r(\omega) + jq(\omega)$, where $r(\omega)$ and $q(\omega)$ are defined as

$$\begin{aligned} r(\omega) &= -\omega^2 + a^2 - \omega b(\gamma_2) - ab\gamma_1 \\ &\quad + b^2(\alpha_1\alpha_2\gamma_3 - \beta_1\beta_2\mu_l), \qquad (8.17) \\ q(\omega) &= -\omega(2a - b\gamma_1) - ab\gamma_2 + b^2\alpha_1\alpha_2\gamma_4. \qquad (8.18) \end{aligned}$$

Let $\Delta = b^2\gamma_2^2 + 4a^2 - 4ab\gamma_1 + 4b^2(\alpha_1\alpha_2\gamma_3 - \beta_1\beta_2\mu_l)$. Therefore, the roots of $r(\omega)$ are

$$r_1 = \frac{-b\gamma_2 + \sqrt{\Delta}}{2}$$
$$r_2 = \frac{-b\gamma_2 - \sqrt{\Delta}}{2}.$$

And the root of $q(\omega)$ is

$$q_1 = \frac{b^2\alpha_1\alpha_2\gamma_4 - ab\gamma_2}{2a - b\gamma_1}. \qquad (8.19)$$

Clearly $\Delta > 0$ gives (8.15) and since the denominator of (8.19) cannot equal zero, we derive condition (8.16). □

Theorem 8.2. *The MAS described by $\mathcal{G}$ with (8.4) consisting of subgraphs $\mathcal{G}_1$ and $\mathcal{G}_2$ under protocol (8.8) in the presence of an attack in both subgroups χ_1 and χ_2 achieves group consensus if and only if each component contains a spanning tree and the following inequalities are satisfied simultaneously:*

$$\begin{aligned} & b^2(\gamma_1^*)^2 + 4a^2 - 4ab\gamma_1 + \\ 4b^2[\alpha_1\alpha_2[\eta_i\eta_j - \eta_i^*\eta_j^*] - \beta_1\beta_2\mu_l] &> 0, \qquad (8.20) \\ 2a - b\gamma_1 &\neq 0, \qquad (8.21) \end{aligned}$$

where $\gamma_1 = \alpha_1\eta_i + \alpha_2\eta_j$, $\gamma_1^* = \alpha_1\eta_i^* + \alpha_2\eta_j^*$, $\eta_i = \Re(\mu_i) - \Re(\mu_i^t)$, $\eta_i^* = \Im(\mu_i) - \Im(\mu_i^t)$, $\eta_j = \Re(\mu_j) - \Re(\mu_j^t)$, $\eta_j^* = \Im(\mu_j) - \Im(\mu_j^t)$; $\Re(.)$, $\Im(.)$ represent real and imaginary parts, respectively; and μ_l is the maximum eigenvalue of $\hat{\mathcal{L}}_{12}\hat{\mathcal{L}}_{21}$.

Proof. Let $\Pi = \det[sI_{n+m-2} - \Psi]$; then

$$\Pi = \det\begin{bmatrix} \Pi_1 & \Pi_2 \\ \Pi_3 & \Pi_4 \end{bmatrix}. \qquad (8.22)$$

Under the assumption that an attacker has influence only in subgroup χ_1, we have $\hat{\mathcal{L}}^t_{11} \neq 0$, $\hat{\mathcal{L}}^t_{12} = 0$, $\hat{\mathcal{L}}^t_{21} = 0$, $\hat{\mathcal{L}}^t_{22} \neq 0$; therefore,

$$\begin{aligned}
\Pi_1 &= (s-a)I_{n-1} + b\alpha_1(\hat{\mathcal{L}}_{11} - \hat{\mathcal{L}}^t_{11}), \\
\Pi_2 &= -b\beta_1\hat{\mathcal{L}}_{12}, \\
\Pi_3 &= -b\beta_2\hat{\mathcal{L}}_{21}, \\
\Pi_4 &= (s-b)I_{m-1} + b\alpha_2(\hat{\mathcal{L}}_{22} - \hat{\mathcal{L}}^t_{22}), \\
\Pi &= \prod_{i=1}^{n} f_i(s) \prod_{j=1}^{M} f_j(s) - b^2\beta_1\beta_2\mu_l,
\end{aligned}$$

where $f_i(s) = s - a + b\alpha_1(\mu_i - \mu^t_i)$ and $f_j(s) = s - a + b\alpha_2(\mu_j - \mu^t_j)$. Now we define $g(s)$ as follows:

$$\begin{aligned}
g(s) &= (s - a + b\alpha_1\mu_i - b\alpha_1\mu^t_i)(s - a + b\alpha_2\mu_j - b\alpha_1\mu^t_j) \\
&\quad -b^2\beta_1\beta_2\mu_l.
\end{aligned} \tag{8.23}$$

It is possible to write $g(\iota\omega) = r(\omega) + \iota q(\omega)$, where $r(\omega)$ and $q(\omega)$ are defined as follows:

$$\begin{aligned}
r(\omega) &= -\omega^2 + a^2 + \omega b\gamma_1^* - ab\gamma_1 \\
&\quad +b^2[\alpha_1\alpha_2[\eta_i\eta_j + \eta_i^*\eta_j^*] - \beta_1\beta_2\mu_l],
\end{aligned} \tag{8.24}$$

$$\begin{aligned}
q(\omega) &= -\omega(2a - b\gamma_1) + ab\gamma_1^* + \\
&\quad b^2\alpha_1\alpha_2(\eta_i\eta_j^* - \eta_i^*\eta_j).
\end{aligned} \tag{8.25}$$

Let $\Delta = b^2(\gamma_1^*)^2 + 4a^2 - 4ab\gamma_1 + 4b^2[\alpha_1\alpha_2[\eta_i\eta_j - \eta_i^*\eta_j^*] - \beta_1\beta_2\mu_l]$. Therefore, the roots of $r(\omega)$ are

$$r_1 = \frac{-b\gamma_1^* + \sqrt{\Delta}}{2}, \tag{8.26}$$

$$r_2 = \frac{-b\gamma_1^* - \sqrt{\Delta}}{2}, \tag{8.27}$$

and the root of $q(\omega)$ is

$$q_1 = \frac{b^2\alpha_1\alpha_2(\eta_i\eta_j^* + \eta_i^*\eta_j) + ab\gamma_1^*}{2a - b\gamma_1}. \tag{8.28}$$

□

Theorem 8.3. *The MAS described by $\mathcal{G}$ with (8.4) consisting of subgraphs $\mathcal{G}_1$ and $\mathcal{G}_2$ under protocol (8.8) in the presence of an attack in subgroup χ_1 achieves group consensus if and only if each component contains a spanning tree and the following inequalities are satisfied simultaneously:*

$$b^2(\gamma_2^*)^2 + 4a^2 - 4ab\gamma_2 +$$
$$4b^2[\alpha_1\alpha_2[\eta_i\Re(\mu_j) - \eta_i^*\Im(\mu_j)] - \beta_1\beta_2\mu_l] > 0, \tag{8.29}$$
$$2a - b\gamma_2 \neq 0, \tag{8.30}$$

where $\gamma_2 = \alpha_1\eta_i + \alpha_2\Re(\mu_j)$, $\gamma_2^* = \alpha_1\eta_i^* + \alpha_2\Im(\mu_j)$, $\eta_i = \Re(\mu_i) - \Re(\mu_i^t)$, $\eta_i^* = \Im(\mu_i) - \Im(\mu_i^t)$; $\Re(.)$, $\Im(.)$ *represent real and imaginary parts, respectively; and* μ_l *is the maximum eigenvalue of* $\hat{\mathcal{L}}_{12}\hat{\mathcal{L}}_{21}$.

Proof. Let $\Pi = \det[sI_{n+m-2} - \Psi]$; then

$$\Pi = \det\begin{bmatrix} \Pi_1 & \Pi_2 \\ \Pi_3 & \Pi_4 \end{bmatrix}. \tag{8.31}$$

Under the assumption that an attacker has influence only in subgroup χ_1, we have $\hat{\mathcal{L}}_{11}^t \neq 0$, $\hat{\mathcal{L}}_{12}^t = 0$, $\hat{\mathcal{L}}_{21}^t = 0$, $\hat{\mathcal{L}}_{22}^t = 0$; therefore,

$$\begin{aligned}
\Pi_1 &= (s-a)I_{n-1} + b\alpha_1(\hat{\mathcal{L}}_{11} - \hat{\mathcal{L}}_{11}^t),\\
\Pi_2 &= -b\beta_1\hat{\mathcal{L}}_{12},\\
\Pi_3 &= -b\beta_2\hat{\mathcal{L}}_{21},\\
\Pi_4 &= (s-b)I_{m-1} + b\alpha_2\hat{\mathcal{L}}_{22},\\
\Pi &= \prod_{i=1}^{n} f_i(s)\prod_{j=1}^{M} f_j(s) - b^2\beta_1\beta_2\mu_l,
\end{aligned}$$

where $f_i(s) = s - a + b\alpha_1(\mu_{i,1} - \mu_{i,1}^t)$ and $f_j(s) = s - a + b\alpha_2\mu_{j,2}$.

Now we define $g(s)$ as

$$\begin{aligned}
g(s) &= (s - a + b\alpha_1\mu_{i,1} - b\alpha_1\mu_i^t)(s - a + b\alpha_2\mu_j)\\
&\quad -b^2\beta_1\beta_2\mu_l.
\end{aligned} \tag{8.32}$$

It is possible to write $g(\iota\omega) = r(\omega) + \iota q(\omega)$, where $r(\omega)$ and $q(\omega)$ are defined as

$$\begin{aligned}
r(\omega) &= -\omega^2 + a^2 - \omega b\gamma_2^* - ab\gamma_2\\
&\quad +b^2(\alpha_1\alpha_2(\eta_i\Re(\mu_j) - \eta_i^*\Im(\mu_j)) - \beta_1\beta_2\mu_l), && (8.33)\\
q(\omega) &= -\omega(2a - b\gamma_2) - ab\gamma_2^*\\
&\quad +b^2\alpha_1\alpha_2(\eta_i\Im(\mu_j) + \eta_i^*\Re(\mu_j)). && (8.34)
\end{aligned}$$

Let $\Delta = b^2(\gamma_2^*)^2 + 4a^2 - 4ab\gamma_2 + 4b^2[\alpha_1\alpha_2[\eta_i\Re(\mu_j) - \eta_i^*\Im(\mu_j)] - \beta_1\beta_2\mu_l]$. Therefore, the roots of $r(\omega)$ are

$$r_1 = \frac{-b\gamma_2^* + \sqrt{\Delta}}{2}, \tag{8.35}$$

$$r_2 = \frac{-b\gamma_2^* - \sqrt{\Delta}}{2}, \tag{8.36}$$

and the root of $q(\omega)$ is

$$q_1 = \frac{b^2\alpha_1\alpha_2(\eta_i\Im(\mu_j) + \eta_i^*\Re(\mu_j)) + ab\gamma_2^*}{2a - b\gamma_2}. \tag{8.37}$$

□

Theorem 8.4. *The MAS described by $\mathcal{G}$ with (8.4) consisting of subgraphs $\mathcal{G}_1$ and $\mathcal{G}_2$ under protocol (8.8) in the presence of an attack in subgroup χ_2 achieves group consensus if and only if each component contains a spanning tree and the following inequalities are satisfied simultaneously:*

$$\begin{aligned} b^2(\gamma_3^*)^2 + 4a^2 - 4ab\gamma_3 + \\ 4b^2[\alpha_1\alpha_2[\Re(\mu_i)\eta_j - \Im(\mu_i)\eta_j^*] - \beta_1\beta_2\mu_l] > 0, \end{aligned} \tag{8.38}$$

$$2a - b\gamma_3 \neq 0, \tag{8.39}$$

where $\gamma_3 = \alpha_1\Re(\mu_i) + \alpha_2\eta_j$, $\gamma_3^* = \alpha_1\Im(\mu_i) + \alpha_2\eta_j^*$, $\eta_i = \Re(\mu_i) - \Re(\mu_i^t)$, $\eta_i^* = \Im(\mu_i) - \Im(\mu_i^t)$, $\eta_j = \Re(\mu_j) - \Re(\mu_j^t)$, $\eta_j^* = \Im(\mu_j) - \Im(\mu_j^t)$; $\Re(.)$, $\Im(.)$ *represent real and imaginary parts, respectively; and* μ_l *is the maximum eigenvalue of* $\hat{\mathcal{L}}_{12}\hat{\mathcal{L}}_{21}$.

Proof. Let $\Pi = \det[sI_{n+m-2} - \Psi]$; then

$$\Pi = \det\begin{bmatrix} \Pi_1 & \Pi_2 \\ \Pi_3 & \Pi_4 \end{bmatrix}. \tag{8.40}$$

Under the assumption that an attacker has influence in only subgroup χ_1, we have $\hat{\mathcal{L}}_{11}^t = 0$, $\hat{\mathcal{L}}_{12}^t = 0$ and $\hat{\mathcal{L}}_{21}^t = 0$, $\hat{\mathcal{L}}_{22}^t \neq 0$; therefore,

$$\begin{aligned} \Pi_1 &= (s-a)I_{n-1} + b\alpha_1\hat{\mathcal{L}}_{11}, \\ \Pi_2 &= -b\beta_1\hat{\mathcal{L}}_{12}, \\ \Pi_3 &= -b\beta_2\hat{\mathcal{L}}_{21}, \\ \Pi_4 &= (s-b)I_{m-1} + b\alpha_2(\hat{\mathcal{L}}_{22} - \hat{\mathcal{L}}_{22}^t), \\ \Pi &= \prod_{i=1}^{n} f_i(s)\prod_{j=1}^{M} f_j(s) - b^2\beta_1\beta_2\mu_l, \end{aligned}$$

where $f_i(s) = s - a + b\alpha_1\mu_{i,1}$ and $f_j(s) = s - a + b\alpha_2(\mu_{j,2} - \mu_{j,2}^{'t})$.

Now we define $g(s)$ as

$$\begin{aligned} g(s) = {} & (s - a + b\alpha_1\mu_{i,1})(s - a + b\alpha_2\mu_j - b\alpha_1\mu_{j,1}^{'t}) \\ & -b^2\beta_1\beta_2\mu_l. \end{aligned} \tag{8.41}$$

It is possible to write $g(\iota\omega) = r(\omega) + \iota q(\omega)$, where $r(\omega)$ and $q(\omega)$ are defined as follows:

$$\begin{aligned} r(\omega) &= -\omega^2 + a^2 - \omega b\gamma_3^* - ab\gamma_3 \\ &\quad + b^2(\alpha_1\alpha_2(\Re(\mu_i)\eta_j - \Im(\mu_i)\eta_j^*) - \beta_1\beta_2\mu_l), \end{aligned} \tag{8.42}$$

$$\begin{aligned} q(\omega) &= -\omega(2a - b\gamma_3) - ab\gamma_3^* \\ &\quad + b^2\alpha_1\alpha_2(\Re(\mu_i)\eta_j^* + \Im(\mu_i)\eta_j). \end{aligned} \tag{8.43}$$

Let $\Delta = b^2(\gamma_3^*)^2 + 4a^2 - 4ab\gamma_3 + 4b^2(\alpha_1\alpha_2(\Re(\mu_i)\eta_j^* - \Im(\mu_i)\eta_j) - \beta_1\beta_2\mu_l)$. Therefore, the roots of $r(\omega)$ are

$$r_1 = \frac{-b\gamma_3^* + \sqrt{\Delta}}{2}, \tag{8.44}$$

$$r_2 = \frac{-b\gamma_3^* - \sqrt{\Delta}}{2}, \tag{8.45}$$

and the root of $q(\omega)$ is

$$q_1 = \frac{b^2\alpha_1\alpha_2(\Re(\mu_i)\eta_j^* + \Im(\mu_i)\eta_j) + ab\gamma_3^*}{2a - b\gamma_3}. \tag{8.46}$$

□

8.1.7 Simulation studies

Consider the MAS described by the graph in Figure 8.1 from [334] with the following Laplacian (8.47) and adjacency (8.97) matrix:

$$\mathcal{L} = \begin{bmatrix} 4 & -3 & -1 & 1 & -1 \\ -3 & 3 & 0 & -1 & 1 \\ -1 & 0 & 1 & 0 & 0 \\ 1 & -1 & 0 & 3 & -3 \\ -1 & 1 & 0 & -3 & 3 \end{bmatrix} \tag{8.47}$$

$$\mathcal{D} = \begin{bmatrix} 0 & 3 & 1 & -1 & 1 \\ 3 & 0 & 0 & 1 & -1 \\ 1 & 0 & 0 & 0 & 0 \\ -1 & 1 & 0 & 0 & 3 \\ 1 & -1 & 0 & 3 & 0 \end{bmatrix}. \tag{8.48}$$

By simple computation, we obtain the eigenvalues of the Laplacian matrix as $\lambda = 0, 0, 1.3263, 4.3457, 8.3280$. The system comprises of two subgroups $\mathcal{G}_1$ and $\mathcal{G}_2$ with 3 and 2 agents, respectively. The following Laplacian matrices $\mathcal{L}_{11}$, $\mathcal{L}_{22}$, $\mathcal{L}_{12} = \mathcal{L}_{21}^T$ represent the links between each of the subgroups and the interconnection between the respective subgroups (see Fig. 8.1).

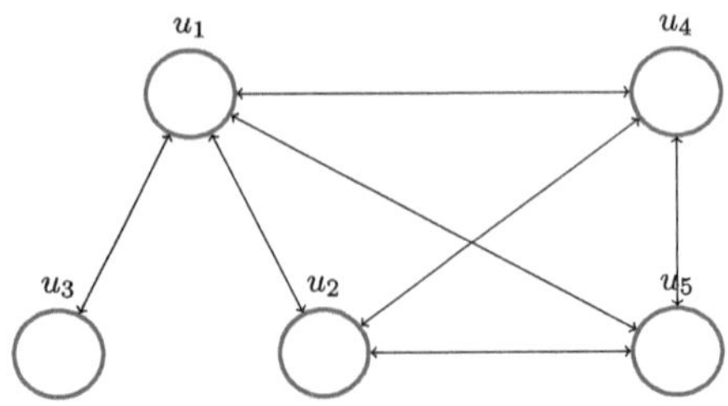

FIGURE 8.1 Graph showing interconnections between agents.

$$\mathcal{L}_{11} = \begin{bmatrix} 4 & -3 & -1 \\ -3 & 3 & 0 \\ -1 & 0 & 1 \end{bmatrix}, \quad \mathcal{L}_{22} = \begin{bmatrix} 3 & -3 \\ -3 & 3 \end{bmatrix} \tag{8.49}$$

$$\mathcal{L}_{12} = \begin{bmatrix} 1 & -1 \\ -1 & 1 \\ 0 & 0 \end{bmatrix} \tag{8.50}$$

$\mu_i = 0.0000, 1.3542, 6.6458$ and $\mu_j = 0, 6$ and $\mu_{max}(\mathcal{L}_{12}\mathcal{L}_{21}) = 4$.

We assume the simple case where we have a MAS with single-integrator dynamics that correspond to $a = 0$, $b = 1$. In Figs. 8.2 and 8.3 we present nominal cases without the influence of an attacker. Fig. 8.2 shows the case when the coupling strengths within and between subgroups are $\alpha_1 = \alpha_2 = 2$ and $\beta_1 = \beta_2 = 4$, while Fig. 8.3 represents the case with $\alpha_1 = \alpha_2 = 1$ and $\beta_1 = \beta_2 = 1$. Take note that in both instances the MAS achieves group consensus with respect to the conditions in (8.15) and as shown by simulation plots.

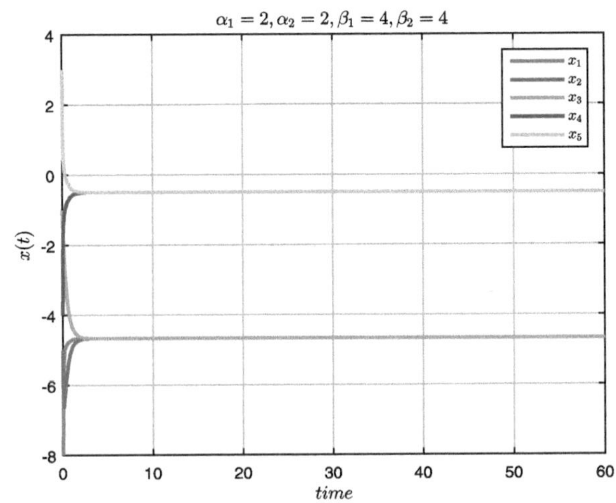

FIGURE 8.2 Nominal case with coupling gains $\alpha_1 = \alpha_2 = 2$, $\beta_1 = \beta_2 = 4$.

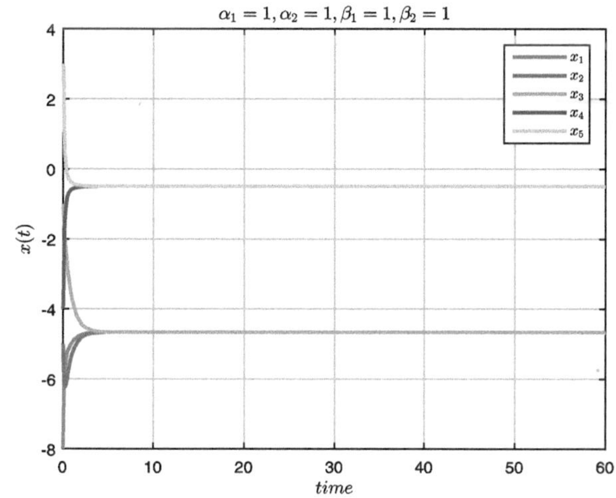

FIGURE 8.3 Nominal case with coupling gains $\alpha_1 = \alpha_2 = 1$, $\beta_1 = \beta_2 = 1$.

First we consider the case where the subgroup 1 is under attack. Specifically, we make the assumption that the intruder launches a DoS attack between agents 1 and 2. The graph Laplacian for the attack modes and the entire network are given by (8.51)–(8.52):

$$\mathcal{L} = \begin{bmatrix} 3 & 0 & -1 & 1 & -1 \\ 0 & 2 & 0 & -1 & 1 \\ -1 & 0 & 1 & 0 & 0 \\ 1 & -1 & 0 & 3 & -3 \\ -1 & 1 & 0 & -3 & 3 \end{bmatrix} \tag{8.51}$$

$$\mathcal{L}_{11}^{t} = \begin{bmatrix} 1 & 0 & 0 \\ 0 & 0 & 0 \\ 0 & 0 & 0 \end{bmatrix}. \tag{8.52}$$

The eigenvalues of the Laplacian matrices representing the entire network after the influence of the attacker and the attack mode are given as follows. In Fig. 8.4, we assume that the coupling strengths are $\alpha_1 = \alpha_2 = 2$ and $\beta_1 = \beta_2 = 4$. We can infer from the conditions in (8.40) and Fig. 8.4 that consensus is not achieved. Next we modify the coupling strengths to $\alpha_1 = \alpha_2 = 1$ and $\beta_1 = \beta_2 = 1$, and we can confirm by Fig. 8.5 and the established conditions in (8.29) that group consensus is achieved.

Now we consider the case where the attacker is operating within subgroup 2 interrupting the communication link between agents 4 and 5. For this case again we compute the Laplacian of the network and attack modes after the influence

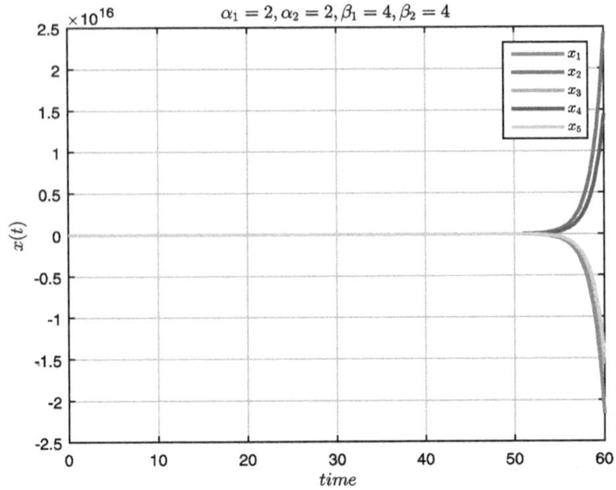

FIGURE 8.4 Attack in subgroup 1 with coupling gains $\alpha_1 = \alpha_2 = 2$, $\beta_1 = \beta_2 = 4$.

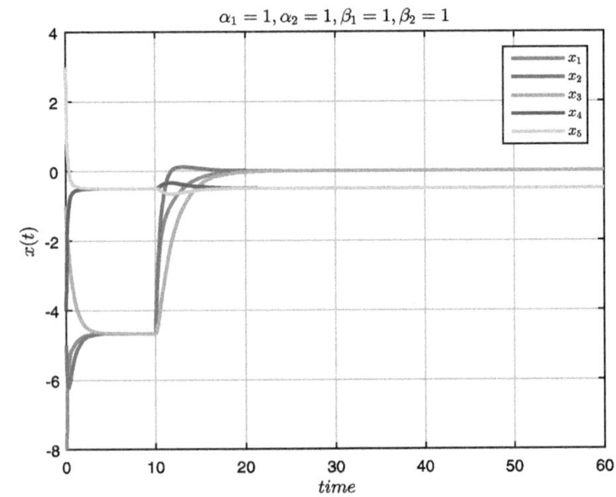

FIGURE 8.5 Attack in subgroup 1 with coupling gains $\alpha_1 = \alpha_2 = 1$, $\beta_1 = \beta_2 = 1$.

of the attacker as given by (8.53)–(8.54), and again we first assume the coupling strengths $\alpha_1 = \alpha_2 = 2$ and $\beta_1 = \beta_2 = 4$, and as we can observe that the conditions in (8.38) are violated and are as shown in Fig. 8.6, group consensus is not achieved and the system becomes unstable. Again, we readjust the coupling strengths to $\alpha_1 = \alpha_2 = 1$ and $\beta_1 = \beta_2 = 1$ and we can observe in Fig. 8.7 that although the link between agents 4 and 5 is interrupted, group consensus is achieved.

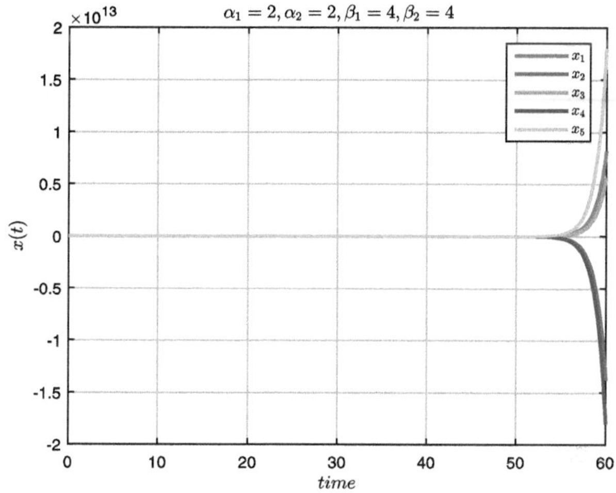

FIGURE 8.6 Attack in subgroup 2 with coupling gains $\alpha_1 = \alpha_2 = 2$, $\beta_1 = \beta_2 = 4$.

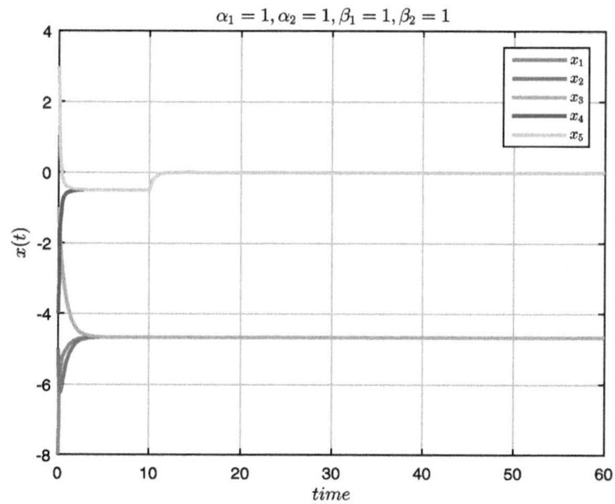

FIGURE 8.7 Attack in subgroup 2 with coupling gains $\alpha_1 = \alpha_2 = 1$, $\beta_1 = \beta_2 = 1$.

$$\mathcal{L} = \begin{bmatrix} 4 & -3 & -1 & 1 & -1 \\ -3 & 3 & 0 & -1 & 1 \\ -1 & 0 & 1 & 0 & 0 \\ 1 & -1 & 0 & 2 & 0 \\ -1 & 1 & -1 & 0 & 2 \end{bmatrix} \tag{8.53}$$

$$\mathcal{L}_{22}^{t} = \begin{bmatrix} 1 & -3 \\ -3 & 1 \end{bmatrix}. \tag{8.54}$$

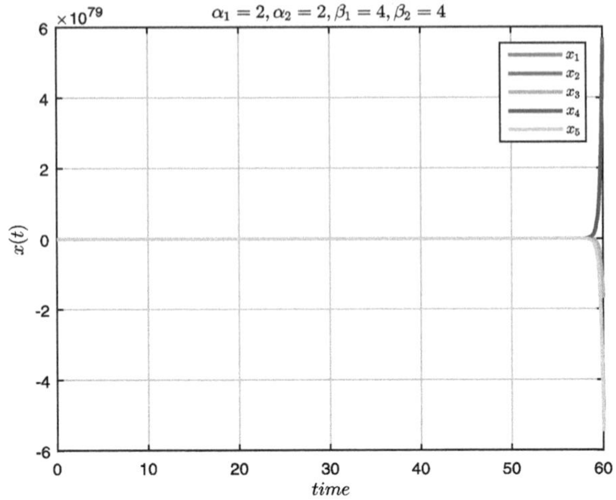

FIGURE 8.8 Attack in both subgroups with coupling gains $\alpha_1 = \alpha_2 = 2$, $\beta_1 = \beta_2 = 4$.

Finally, we examine the case where the attacker is able to influence both subgroups. In this scenario the communication link between agents 1 and 2 and between agents 4 and 5 are affected. The Laplacian matrix after the attack is given by (8.55). Yet again, we first assume the coupling strengths $\alpha_1 = \alpha_2 = 2$ and $\beta_1 = \beta_2 = 4$; we can observe that the conditions in (8.20) are violated and are as shown in Fig. 8.8; thus, group consensus is not achieved and the system becomes unstable. However, it is interesting to note that when we adjusted the coupling strengths to $\alpha_1 = \alpha_2 = 1$ and $\beta_1 = \beta_2 = 1$, the agents were able to reorganize themselves after the influence of the attack. As shown in Fig. 8.9, before the attack we can observe that agents 1, 2, and 3 form a single consensus as do agents 4 and 5. After the loss of communication link between agents 1 and 2, and between agents 4 and 5, we can observe that agents 2 and 4, and agents 1, 3, and 5 form a new consensus.

$$\mathcal{L} = \begin{bmatrix} 3 & 0 & -1 & 1 & -1 \\ 0 & 2 & 0 & -1 & 1 \\ -1 & 0 & 1 & 0 & 0 \\ 1 & -1 & 0 & 2 & 0 \\ -1 & 1 & 0 & 0 & 2 \end{bmatrix}. \tag{8.55}$$

8.2 Adaptive cluster consensus with unknown control coefficients

Multiagent systems are by definition a collection of multiple systems interconnected over a communication network, referred to as a *graph*, for the purpose of achieving some predefined objective. *Co-ordination control* is a branch of MAS

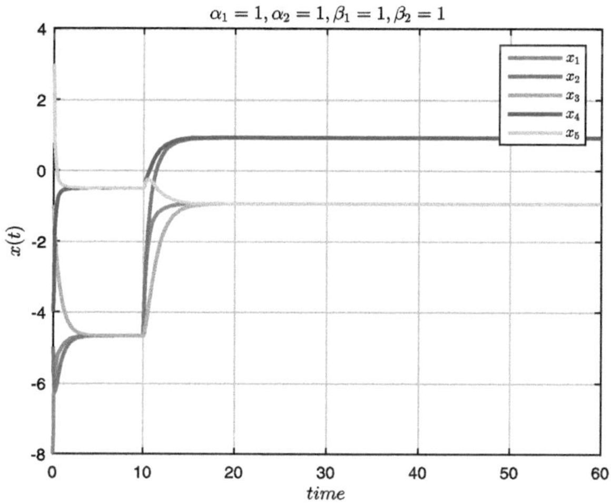

FIGURE 8.9 Attack in both subgroups with coupling gains $\alpha_1 = \alpha_2 = 1$, $\beta_1 = \beta_2 = 1$.

research primarily concerned with the design of complex control protocols to achieve synchronization amongst agents. Research in coordination control is subdivided into consensus, formation, rendezvous, alignment, swarming, containment, and circumnavigation.

In consensus control, the main objective is to drive the states of the agents to a *consensus* state. Broadly speaking, there are different types of consensus objectives that have been studied by some researchers including fixed (or finite) time, pinning, average, leader-follower, leaderless, adaptive, and group consensus.

Contrary to general consensus, the objective in cluster (or group) consensus is for agents to converge to two or more consensus states depending on the graph topology. Cluster consensus has been studied by numerous researchers [335, 336,326,337,338,327,340,341,325,342–348,328,349,350]. A class of problems in cluster consensus deal with deriving conditions under which cluster consensus is achieved. It is general knowledge that choosing proper inter- and intracluster gains in relation to the graph Laplacian could affect attaining cluster consensus.

8.2.1 Introduction

In [335] the authors studied cluster consensus problem for generic linear MASs over a directed communication graph. The paper investigates the question of how the interaction among clusters affects the cluster consensus without consideration for the magnitudes of the coupling strengths amongst agents. Conditions for cluster consensus were derived using mainly graph-theoretic approaches. It was proven that cluster consensus is achieved when the directed graph is acyclic.

Couple-group consensus problem for discrete-time MASs over directed graph topology was investigated in [336]. Some sets of algebraic conditions were established to guarantee couple-group consensus. In [326] the authors discuss couple-group consensus problem for a class of MASs with linear time invariant dynamics over directed communication graphs described by a continuous-time homogeneous Markov process. The authors derived algebraic conditions under which couple-group consensus can be established.

Group consensus problems for linearly coupled MASs with first- and second-order dynamics was investigated in [337]. The authors deduced that consensus is achievable for generally connected graphs without spanning trees. The L^1 group consensus problem was investigated in [338]. Using ergodicity theory and matrix analysis, the authors derived some L^1 group consensus conditions for MASs under switching topologies.

Some necessary and sufficient conditions for group consensus were derived in [339] using algebraic matrix theory and graph theory. In addition, the authors show that eigenvalues of both Laplacian and Euler rotation matrices play important roles in reaching group consensus.

In [341] some cluster leader-follower consensus conditions were derived for second-order MASs under impulsive effects and coupling delays. According the authors, the problem of switching topology was handled by impulsive stability and adaptive strategy.

Cluster consensus for MASs via intercluster nonidentical inputs was investigated in [325] for MASs over time-varying graph topologies. The authors subdivide the cluster consensus problem using the following terminologies: intracluster synchronization and intercluster separation. For the case of intracluster synchronization the concepts of spanning trees, scramblingness, and infinite stochastic matrix were extended. Nonidentical control inputs were used to maintain separation between clusters.

Cluster consensus problems were investigated in [342] for a class of generic heterogeneous linear MASs. The main objective of the study was to investigate the influence of intra- and intercluster coupling on reaching cluster consensus. It was deduced that in semiheterogeneous cases, cluster consensus is achieved when each agent suppresses the influence of intercluster coupling; instead, when all agents have completely different dynamics, cluster consensus is achieved when agents balance the effect of both intra- and intercluster couplings.

Output tracking distributed consensus control was studied in [351] for a class of unknown linear systems based on relative output information from neighboring agents. The dynamics of each agent were considered minimum-phase with a unity relative degree. The graph topology was assumed to be strongly connected. The proposed adaptive control protocols are independent of parameters of neighboring agents, and only require relative output of neighboring subsystems.

Cooperative output regulation problem for a linear heterogeneous MAS under directed communication graph was studied in [352]. The authors considered a scenario where the multiagent features an "exosystem" whose output is observable by a set of subsystems. In their investigation, they considered both nominal and uncertain instances. In the nominal instance a distributed adaptive observer based control protocol was designed to estimate the exogenous signal. The distributed adaptive observer and internal model control design principles were combined to derive the control law for the uncertain instance.

Leader-follower distributed adaptive control was proposed in [353] for a class of uncertain nonlinear systems based on output feedback design. The authors considered a directed graph topology and the adaptive control law was designed using relative output measurements. According to the authors the role of the leader in their study is similar to design approaches in model reference adaptive control.

Output feedback-based leader-follower consensus control was studied in [354] for a class of linear MASs on directed graphs. The leader node considered in this paper had a nonzero control input. A novel distributed adaptive output feedback protocol was designed without global information about the graph topology under the conditions that the agents are stabilizable and detectable.

Near-optimal adaptive distributed consensus for heterogeneous MASs with high nonlinearities was investigated in [355]. The control design approach employs sliding-mode auxiliary systems to reconstruct input-output relationships of the agents.

Adaptive consensus control was designed in [356] for first-order nonlinear MASs with dissimilar parameters and quantized state information. The control design was based on the edge Laplacian of the multiagent network. Under the assumptions that the control directions of the agents are unknown, a Nussbaum-type function was employed in the control protocol to help each agent determine the direction in an adaptive and co-operative manner.

Leader-follower consensus was also studied in [357] for a class of nonlinear MASs with unknown control directions. The authors designed distributed adaptive controllers for both first- and second-order nonlinear systems.

Authors in [358] studied adaptive consensus problem for a class of nonlinear MASs with unknown backlash-like hysteresis on undirected communication graphs. The control protocol was designed based on a backstepping procedure involving a distributed control scheme with a Nussbaum-type function. Uncertain nonlinear dynamics were neutralized using radial basis function neural networks.

An output feedback consensus problem was investigated in [359] for a class of high-order nonlinear MASs over directed communication topologies. Using a backstepping design, the Lyapunov theory, and neural networks, a distributed adaptive consensus protocol was designed for agents to track a desired trajectory.

Adaptive group consensus was investigated in [349] for a class of networked mechanical systems with Lagrangian dynamics on a directed acyclic network topology. A distributed adaptive consensus protocol based on neural networks was designed considering communication delays.

Distributed consensus control for a class of high-order nonlinear MASs with a Brunovsky-type model was investigated in [360] under the assumption of uncertain model parameters and unknown control directions. An adaptive control law was derived based on Nussbaum-type functions. According to the authors, the derived control law guarantees consensus when the direction of individual agents is unknown and unidentical.

A distributed consensus tracking problem was investigated in [361] for a class of MASs with unknown linear dynamics over a directed communication graph. Using output information among agents, a distributed adaptive control scheme involving a local observer, and an adaptive estimator was designed. The authors claim the control protocols are fully distributed requiring no global information about the graph Laplacian.

8.2.2 Algebraic graph theory

Graph theory is a mathematical framework for modeling interconnection between agents in a MAS. A graph $\mathcal{G}(\mathcal{V}, \mathcal{E})$ is defined as an ordered pair of a set of vertices $\mathcal{V}$ and edges $\mathcal{E}$. Based on the direction of information flow, a graph may be classified as undirected or directed. In an undirected graph, there is a bidirectional information flow between any pair of vertices; that is, any pair of agents can exchange information in either direction. Conversely, in a directed graph or *digraph*, each edge $e \in \mathcal{E}(\mathcal{G})$ is directed between any pair of vertices; that is, information flow between any pair of vertices can be bidirectional or unidirectional. Graphs can also be defined as simple, complete, or bipartite. In simple graphs, there are no self loops or multiple edges between vertices. In complete graphs, every pair of vertices is connected.

The information in a graph are defined using some matrices including *degree*, *adjacency*, *incidence*, and *Laplacian* matrix. For a graph on n vertices and m edges the degree matrix $\Delta(\mathcal{G}) \in \mathcal{R}^{n \times n}$ is a diagonal matrix, with elements on the diagonal representing the degree $d(v_i)$ of each vertex. $d(v_i)$ is the sum of edges incident to the vertex v_i. The adjacency matrix $\mathcal{A}(G)$ is a symmetric $n \times n$ matrix describing the adjacency relationship in $\mathcal{G}$. Each $a_{ij} \in \mathcal{A}(\mathcal{G})$ assumes 1 if $v_i v_j \in \mathcal{E}(\mathcal{G})$ and 0 otherwise. The Laplacian matrix in an undirected graph $\mathcal{L}(\mathcal{G}) = \Delta(\mathcal{G}) - \mathcal{A}(G)$. The incidence matrix $\mathcal{W}$ of a directed graph $\mathcal{D}$ is defined as $\mathcal{W} = [w_{ij}]$. $w_{ij} = -1$ if v_i is the tail of e_j, $w_{ij} = 1$ if v_i is the head of e_j and $w_{ij} = 0$ if v_i is not adjacent to e_j. The Laplacian matrix of a directed graph $\mathcal{D}$ is $\mathcal{L}(\mathcal{D}) = \mathcal{W}(\mathcal{D})\mathcal{W}(\mathcal{D})^T$.

8.2.3 Consensus

Consider a network of MASs described by $\mathcal{G}$ consisting of n agents described by the first-order dynamics

$$\dot{x}_i(t) = f_i(x_i, u_i), \quad i = 1, 2, \ldots n, \tag{8.56}$$

where x_i and u_i represent the states and control inputs of each agent in the network.

Definition 8.4. The MAS described by $\mathcal{G}$ with dynamics (8.56) achieves consensus if for any $x_i(0)$

$$\lim_{t \to \infty} ||x_i(t) - x_j(t)|| = 0 \quad \forall i, j = 1, 2, \ldots n.$$

Lemma 8.6. *[362] If $\mathcal{L} = [l_{ij}] \in \mathcal{R}^{n \times n}$ satisfies $l_{ij} < 0$, $i \neq j$ and $\sum_{j=1}^{n} l_{ij} = 0$, $i = 1, 2, ..., n$, then the following conditions are equivalent:*

- *$\mathcal{L}$ has a simple zero eigenvalue and all other eigenvalues have positive real parts;*
- *$\mathcal{L}x = 0$ implies that $x_1 = x_2 = \ldots x_n$;*
- *Consensus is reached asymptotically for the system $\dot{x} = -\mathcal{L}x$;*
- *The directed graph of $\mathcal{L}$ has a directed spanning tree;*
- *The rank of $\mathcal{L}$ is $n - 1$.*

8.2.4 Group consensus

Definition 8.5. The network of MASs described by the dynamics in (8.56), achieve cluster consensus if the following conditions are satisfied:

$$\lim_{t \to \infty} \|x_i(t) - x_j(t)\| = 0, \ \forall i, j \in \mathcal{G}_p.$$

Definition 8.6. A digraph $\mathcal{G}_i = \{\mathcal{V}_i, \mathcal{E}_i, \mathcal{A}_i\}$ is said to be a subgraph of $\mathcal{G} = \{\mathcal{V}, \mathcal{E}, \mathcal{A}\}$ if (a) $\mathcal{V}_i \subseteq \mathcal{V}$, (b) $\mathcal{E}_i \subseteq \mathcal{E}$.

The following result is recalled:

Lemma 8.7. *[362] Let $\mathcal{G}_n$ be a graph on n vertices with m connected components. If $\mathcal{L}_n$ is the Laplacian of $\mathcal{G}$, then* $\text{rank}(\mathcal{L}) = n - m$.

8.2.5 Single-integrator linear dynamics

Consider the MAS described by the single-integrator dynamics

$$\dot{x}_i \quad = \quad b_i u_i, \tag{8.57}$$

where b_i is an unknown constant and u_i is the control input. We defined the local intracluster error ζ_i^p and intercluster exchange η_i^p for agent i within each subgraph as

$$\zeta_i^p = \sum_{j \in \mathcal{G}_p} a_{ij}(x_i - x_j), \tag{8.58}$$

$$\eta_i^p = \sum_{j \notin \mathcal{G}_p} a_{ij}x_j. \tag{8.59}$$

Theorem 8.5. *The MAS defined by (8.57) on graph $\mathcal{G}$ with subgraphs $\mathcal{G}_p$ satisfying conditions in Lemmas 8.6 and 8.7 achieves P-group consensus under control protocols (8.60) and adaptive laws (8.61) and (8.62)*

$$u_i = \theta_i \zeta_i + \beta_i \eta_i, \tag{8.60}$$

$$\dot{\theta}_i = -sign(b_i)(\zeta_i(\zeta_i + \eta_i) + \nu_i \zeta_i^2), \tag{8.61}$$

$$\dot{\beta}_i = -sign(b_i)(\eta_i(\zeta_i + \eta_i) + \mu_i \eta_i^2), \tag{8.62}$$

where $\theta_i = \frac{k_i^1}{b_i}$ and $\beta_i = \frac{k_i^2}{b_i}$ are unknown control gains (or coupling strengths).

Proof. Consider the Lyapunov function

$$V = \sum_{p=1}^{P} V_p, \tag{8.63}$$

where V_p is defined as

$$V_p = \frac{1}{2}x_p^T \mathcal{L} x_p + \frac{1}{2}\sum_{i \in \mathcal{G}_p} |b_i|\theta_i^2 + \frac{1}{2}\sum_{i \in \mathcal{G}_p} |b_i|\beta_i^2. \tag{8.64}$$

The time derivative of (8.64) gives

$$\dot{V}_p = x_p^T \mathcal{L} \dot{x}_p + \sum_{i \in \mathcal{G}_p} |b_i|\theta_i\dot{\theta}_i + \sum_{i \in \mathcal{G}_p} |b_i|\beta_i\dot{\beta}_i \tag{8.65}$$

$$\dot{V}_p = \sum_{i \in \mathcal{G}_p} (\zeta_i + \eta_i) b_i u_i + \sum_{i \in \mathcal{G}_p} |b_i|\theta_i\dot{\theta}_i + \sum_{i \in \mathcal{G}_p} |b_i|\beta_i\dot{\beta}_i$$

$$\dot{V}_p = \sum_{i \in \mathcal{G}_p} (\zeta_i + \eta_i) b_i (\theta_i\zeta_i + \beta_i\eta_i) + \sum_{i \in \mathcal{G}_p} |b_i|\theta_i\dot{\theta}_i + \sum_{i \in \mathcal{G}_p} |b_i|\beta_i\dot{\beta}_i$$

$$\begin{aligned}\dot{V}_p = {} & \sum_{i \in \mathcal{G}_p} (\zeta_i + \eta_i) b_i \theta_i\zeta_i + \sum_{i \in \mathcal{G}_p} (\zeta_i + \eta_i) b_i \beta_i\eta_i \\ & + \sum_{i \in \mathcal{G}_p} |b_i|\theta_i\dot{\theta}_i + \sum_{i \in \mathcal{G}_p} |b_i|\beta_i\dot{\beta}_i.\end{aligned}$$

Adding and subtracting $\sum_{i\in\mathcal{G}_p}\mu_i\eta_i^2, \sum_{i\in\mathcal{G}_p}\nu_i\zeta_i^2$ gives

$$\begin{aligned}
\dot{V}_p &= \sum_{i\in\mathcal{G}_p}(\zeta_i+\eta_i)b_i\theta_i\zeta_i + \sum_{i\in\mathcal{G}_p}(\zeta_i+\eta_i)b_i\beta_i\eta_i \\
&\quad + \sum_{i\in\mathcal{G}_p}|b_i|\theta_i\dot{\theta}_i + \sum_{i\in\mathcal{G}_p}|b_i|\beta_i\dot{\beta}_i + \sum_{i\in\mathcal{G}_p}\mu_i\eta_i^2 \\
&\quad + \sum_{i\in\mathcal{G}_p}\nu_i\zeta_i^2 - \sum_{i\in\mathcal{G}_p}\mu_i\eta_i^2 - \sum_{i\in\mathcal{G}_p}\nu_i\zeta_i^2
\end{aligned}$$

$$\begin{aligned}
\dot{V}_p &= -\sum_{i\in\mathcal{G}_p}\mu_i\eta_i^2 - \sum_{i\in\mathcal{G}_p}\nu_i\zeta_i^2 + \sum_{i\in\mathcal{G}_p}(\zeta_i+\eta_i)b_i\theta_i\zeta_i \\
&\quad + \sum_{i\in\mathcal{G}_p}(\zeta_i+\eta_i)b_i\beta_i\eta_i + \sum_{i\in\mathcal{G}_p}|b_i|\theta_i\dot{\theta}_i + \\
&\quad \sum_{i\in\mathcal{G}_p}|b_i|\beta_i\dot{\beta}_i + \sum_{i\in\mathcal{G}_p}\mu_i b_i\beta_i\eta_i^2 + \sum_{i\in\mathcal{G}_p}\nu_i b_i\theta_i\zeta_i^2
\end{aligned}$$

$$\begin{aligned}
\dot{V}_p &= -\sum_{i\in\mathcal{G}_p}\mu_i\eta_i^2 - \sum_{i\in\mathcal{G}_p}\nu_i\zeta_i^2 \\
&\quad + \sum_{i\in\mathcal{G}_p}b_i(sign(b_i)\dot{\theta}_i + \zeta_i(\zeta_i+\eta_i) + \nu_i\zeta_i^2)\theta_i \\
&\quad + \sum_{i\in\mathcal{G}_p}b_i(sign(b_i)\dot{\beta}_i + \eta_i(\zeta_i+\eta_i) + \mu_i\eta_i^2)\beta_i .
\end{aligned}$$

Applying update laws (8.61) and (8.62), the Lyapunov function becomes

$$\dot{V}_p = -\sum_{i\in\mathcal{G}_p}\mu_i\eta_i^2 - \sum_{i\in\mathcal{G}_p}\nu_i\zeta_i^2 < 0. \tag{8.66}$$

□

8.2.6 Single integrator with nonlinear dynamics

Consider the MAS described by the following single-integrator dynamics

$$\dot{x}_i = a_i\psi_i(x_i) + b_i u_i, \tag{8.67}$$

where a_i and b_i are unknown constants and $\psi_i(x_i)$ is a known nonlinear function.

Theorem 8.6. *The MAS defined by (8.67) on graph $\mathcal{G}$ with subgraphs $\mathcal{G}_p$ satisfying conditions in Lemmas 8.6 and 8.7 achieves P-group consensus under control protocols (8.68) and adaptive laws (8.69) and (8.70)*

$$\begin{aligned}
u_i &= \theta_i \zeta_i + \beta_i \eta_i + \gamma_i \theta_i \psi_i(x_i), &(8.68)\\
\dot{\theta}_i &= -sign(b_i)(\zeta_i + \gamma_i \psi_i(x_i)(\zeta_i + \eta_i) + \nu_i \zeta_i^2), &(8.69)\\
\dot{\beta}_i &= -sign(b_i)(\eta_i(\zeta_i + \eta_i) + \mu_i \eta_i^2), &(8.70)\\
\dot{\gamma}_i &= -sign(b_i)(\psi_i(x_i)(\zeta_i + \eta_i)), &(8.71)
\end{aligned}$$

where $\theta_i = \frac{k_i^1}{b_i}$, $\beta_i = \frac{k_i^2}{b_i}$, *and* $\gamma_i = \frac{a_i}{b_i}$ *are unknown control gains (or coupling strengths).*

Proof. Consider the Lyapunov function

$$V = \sum_{p=1}^{P} V_p, \tag{8.72}$$

where V_p is defined as

$$\begin{aligned}
V_p &= \frac{1}{2} x_p^T \mathcal{L} x_p + \frac{1}{2} \sum_{i \in \mathcal{G}_p} |b_i| \theta_i^2 + \frac{1}{2} \sum_{i \in \mathcal{G}_p} |b_i| \beta_i^2 &(8.73)\\
&\quad + \frac{1}{2} \sum_{i \in \mathcal{G}_p} |b_i| \gamma_i^2.
\end{aligned}$$

The time derivative of (8.64) gives

$$\begin{aligned}
\dot{V}_p &= x_p^T \mathcal{L} \dot{x}_p + \sum_{i \in \mathcal{G}_p} |b_i| \theta_i \dot{\theta}_i + \sum_{i \in \mathcal{G}_p} |b_i| \beta_i \dot{\beta}_i \\
&\quad + \sum_{i \in \mathcal{G}_p} |b_i| \gamma_i \dot{\gamma}_i &(8.74)\\
\dot{V}_p &= \sum_{i \in \mathcal{G}_p} (\zeta_i + \eta_i)(a_i \psi_i(x_i) + b_i u_i) + \sum_{i \in \mathcal{G}_p} |b_i| \theta_i \dot{\theta}_i \\
&\quad + \sum_{i \in \mathcal{G}_p} |b_i| \beta_i \dot{\beta}_i + \sum_{i \in \mathcal{G}_p} |b_i| \gamma_i \dot{\gamma}_i &(8.75)\\
\dot{V}_p &= \sum_{i \in \mathcal{G}_p} (\zeta_i + \eta_i) a_i \psi_i(x_i) + \sum_{i \in \mathcal{G}_p} (\zeta_i + \eta_i) b_i u_i + \sum_{i \in \mathcal{G}_p} |b_i| \theta_i \dot{\theta}_i \\
&\quad + \sum_{i \in \mathcal{G}_p} |b_i| \beta_i \dot{\beta}_i + \sum_{i \in \mathcal{G}_p} |b_i| \gamma_i \dot{\gamma}_i.
\end{aligned}$$

Substituting $u_i = \theta_i \zeta_i + \beta_i \eta_i + \gamma_i \theta_i \psi_i(x_i)$ gives

$$\dot{V}_p = \sum_{i \in \mathcal{G}_p} (\zeta_i + \eta_i) a_i \psi_i(x_i) + \sum_{i \in \mathcal{G}_p} (\zeta_i + \eta_i) b_i \theta_i \zeta_i$$

$$+\sum_{i\in\mathcal{G}_p}(\zeta_i+\eta_i)b_i\beta_i\eta_i+\sum_{i\in\mathcal{G}_p}(\zeta_i+\eta_i)b_i\gamma_i\theta_i\psi_i(x_i)$$
$$+\sum_{i\in\mathcal{G}_p}|b_i|\theta_i\dot{\theta}_i+\sum_{i\in\mathcal{G}_p}|b_i|\beta_i\dot{\beta}_i+\sum_{i\in\mathcal{G}_p}|b_i|\gamma_i\dot{\gamma}_i.$$

Substituting $a_i=\gamma_i b_i$ gives

$$\begin{aligned}\dot{V}_p &= \sum_{i\in\mathcal{G}_p}(\zeta_i+\eta_i)\gamma_i b_i\psi_i(x_i)+\sum_{i\in\mathcal{G}_p}(\zeta_i+\eta_i)b_i\theta_i\zeta_i\\ &\quad+\sum_{i\in\mathcal{G}_p}(\zeta_i+\eta_i)b_i\beta_i\eta_i+\sum_{i\in\mathcal{G}_p}(\zeta_i+\eta_i)b_i\gamma_i\theta_i\psi_i(x_i)\\ &\quad+\sum_{i\in\mathcal{G}_p}|b_i|\theta_i\dot{\theta}_i+\sum_{i\in\mathcal{G}_p}|b_i|\beta_i\dot{\beta}_i+\sum_{i\in\mathcal{G}_p}|b_i|\gamma_i\dot{\gamma}_i.\end{aligned}$$

Adding and subtracting $\sum_{i\in\mathcal{G}_p}\mu_i\eta_i^2$, $\sum_{i\in\mathcal{G}_p}\nu_i\zeta_i^2$ gives

$$\begin{aligned}\dot{V}_p &= \sum_{i\in\mathcal{G}_p}(\zeta_i+\eta_i)\gamma_i b_i\psi_i(x_i)+\sum_{i\in\mathcal{G}_p}(\zeta_i+\eta_i)b_i\theta_i\zeta_i\\ &\quad+\sum_{i\in\mathcal{G}_p}(\zeta_i+\eta_i)b_i\beta_i\eta_i+\sum_{i\in\mathcal{G}_p}(\zeta_i+\eta_i)b_i\gamma_i\theta_i\psi_i(x_i)\\ &\quad+\sum_{i\in\mathcal{G}_p}|b_i|\theta_i\dot{\theta}_i+\sum_{i\in\mathcal{G}_p}|b_i|\beta_i\dot{\beta}_i+\sum_{i\in\mathcal{G}_p}|b_i|\gamma_i\dot{\gamma}_i\\ &\quad-\sum_{i\in\mathcal{G}_p}\mu_i\eta_i^2-\sum_{i\in\mathcal{G}_p}\nu_i\zeta_i^2+\sum_{i\in\mathcal{G}_p}\mu_i b_i\beta_i\eta_i^2+\sum_{i\in\mathcal{G}_p}\nu_i b_i\theta_i\zeta_i^2,\\ \dot{V}_p &= -\sum_{i\in\mathcal{G}_p}\mu_i\eta_i^2-\sum_{i\in\mathcal{G}_p}\nu_i\zeta_i^2\\ &\quad+\sum_{i\in\mathcal{G}_p}b_i[(\zeta_i+\eta_i)\psi_i(x_i)+sign(b_i)\dot{\gamma}_i]\gamma_i\\ &\quad+\sum_{i\in\mathcal{G}_p}b_i[\eta_i(\zeta_i+\eta_i)+\mu_i\eta_i^2]\beta_i\\ &\quad+\sum_{i\in\mathcal{G}_p}b_i[\zeta_i(\zeta_i+\eta_i)+\gamma_i(\zeta_i+\eta_i)\phi_i(x_i)+\nu_i\zeta_i^2]\beta_i.\end{aligned}$$

Applying update laws (8.69), (8.70), and (8.71), the Lyapunov function becomes

$$\dot{V}_p = -\sum_{i\in\mathcal{G}_p}\mu_i\eta_i^2-\sum_{i\in\mathcal{G}_p}\nu_i\zeta_i^2<0. \tag{8.76}$$

□

8.2.7 Linear double-integrator dynamics

Consider the MAS described by the linear dynamics

$$\dot{x}_i = v_i, \tag{8.77}$$
$$\dot{v}_i = b_i u_i, \tag{8.78}$$

where b_i is an unknown constant.

Theorem 8.7. *The MAS defined by (8.77) on graph $\mathcal{G}$ with subgraphs $\mathcal{G}_p$ satisfying conditions in Lemmas 8.6 and 8.7 achieves P-group consensus under control protocols (8.79) and adaptive laws (8.80) and (8.81)*

$$u_i = \theta_i(v_i + \zeta_i) + \beta_i \eta_i, \tag{8.79}$$
$$\dot{\theta}_i = -sign(b_i)((2v_i + \zeta_i)(\zeta_i + \eta_i) + v_i \zeta_i^2), \tag{8.80}$$
$$\dot{\beta}_i = -sign(b_i)(\eta_i(\zeta_i + \eta_i) + \mu_i \eta_i^2), \tag{8.81}$$

where $\theta_i = \frac{k_i^1}{b_i}$ and $\beta_i = \frac{k_i^2}{b_i}$ are unknown control gains (or coupling strengths).

Proof. Consider the Lyapunov function

$$V = \sum_{p=1}^{P} V_p, \tag{8.82}$$

where V_p is defined as

$$V_p = \frac{1}{2}x_p^T \mathcal{L} x_p + \frac{1}{2}v_p^T \mathcal{L} v_p + \frac{1}{2}\sum_{i\in\mathcal{G}_p} |b_i|\theta_i^2 + \frac{1}{2}\sum_{i\in\mathcal{G}_p} |b_i|\beta_i^2. \tag{8.83}$$

The time derivative of (8.64) gives

$$\dot{V}_p = x_p^T \mathcal{L}\dot{x}_p + v_p^T \mathcal{L}\dot{v}_p + \sum_{i\in\mathcal{G}_p} |b_i|\theta_i\dot{\theta}_i + \sum_{i\in\mathcal{G}_p} |b_i|\beta_i\dot{\beta}_i$$

$$\dot{V}_p = \sum_{i\in\mathcal{G}_p} (\zeta_i + \eta_i)v_i + \sum_{i\in\mathcal{G}_p} (\zeta_i + \eta_i)b_i u_i + \sum_{i\in\mathcal{G}_p} |b_i|\theta_i\dot{\theta}_i + \sum_{i\in\mathcal{G}_p} |b_i|\beta_i\dot{\beta}_i.$$

Substituting $u_i = \theta_i v_i + \theta_i \zeta_i + \beta_i \eta_i$ gives

$$\dot{V}_p = \sum_{i\in\mathcal{G}_p} (\zeta_i + \eta_i)v_i + \sum_{i\in\mathcal{G}_p} (\zeta_i + \eta_i)b_i\theta_i(v_i + \zeta_i) + \sum_{i\in\mathcal{G}_p} (\zeta_i + \eta_i)b_i\beta_i\eta_i + \sum_{i\in\mathcal{G}_p} |b_i|\theta_i\dot{\theta}_i + \sum_{i\in\mathcal{G}_p} |b_i|\beta_i\dot{\beta}_i.$$

Adding and subtracting $\sum_{i\in\mathcal{G}_p}\mu_i\eta_i^2$, $\sum_{i\in\mathcal{G}_p}\nu_i\zeta_i^2$ gives

$$\begin{aligned}
\dot{V}_p &= \sum_{i\in\mathcal{G}_p}(\zeta_i+\eta_i)v_i + \sum_{i\in\mathcal{G}_p}(\zeta_i+\eta_i)b_i\theta_i(v_i+\zeta_i) \\
&\quad + \sum_{i\in\mathcal{G}_p}(\zeta_i+\eta_i)b_i\beta_i\eta_i + \sum_{i\in\mathcal{G}_p}|b_i|\theta_i\dot{\theta}_i + \sum_{i\in\mathcal{G}_p}|b_i|\beta_i\dot{\beta}_i \\
&\quad + \sum_{i\in\mathcal{G}_p}\mu_i\eta_i^2 + \sum_{i\in\mathcal{G}_p}\nu_i\zeta_i^2 - \sum_{i\in\mathcal{G}_p}\mu_i\eta_i^2 - \sum_{i\in\mathcal{G}_p}\nu_i\zeta_i^2 \\
\dot{V}_p &= -\sum_{i\in\mathcal{G}_p}\mu_i\eta_i^2 - \sum_{i\in\mathcal{G}_p}\nu_i\zeta_i^2 + \sum_{i\in\mathcal{G}_p}(\zeta_i+\eta_i)b_i\theta_i v_i \\
&\quad + \sum_{i\in\mathcal{G}_p}(\zeta_i+\eta_i)b_i\theta_i(v_i+\zeta_i) + \sum_{i\in\mathcal{G}_p}(\zeta_i+\eta_i)b_i\gamma_i\eta_i \\
&\quad + \sum_{i\in\mathcal{G}_p}|b_i|\theta_i\dot{\theta}_i + \sum_{i\in\mathcal{G}_p}|b_i|\beta_i\dot{\beta}_i + \sum_{i\in\mathcal{G}_p}|b_i|\beta_i\dot{\beta}_i \\
&\quad + \sum_{i\in\mathcal{G}_p}\mu_i b_i\beta_i\eta_i^2 + \sum_{i\in\mathcal{G}_p}\nu_i b_i\theta_i\zeta_i^2 \\
\dot{V}_p &= -\sum_{i\in\mathcal{G}_p}\mu_i\eta_i^2 - \sum_{i\in\mathcal{G}_p}\nu_i\zeta_i^2 \qquad (8.84) \\
&\quad + \sum_{i\in\mathcal{G}_p}b_i[sgn(b_i)\dot{\theta}_i + (2v_i+\zeta_i)(\zeta_i+\eta_i) + \nu_i\zeta_i^2]\theta_i \\
&\quad + \sum_{i\in\mathcal{G}_p}b_i[sgn(b_i)\dot{\beta}_i + \eta_i(\zeta_i+\eta_i) + \mu_i\eta_i^2]\beta_i.
\end{aligned}$$

Applying update laws (8.80) and (8.81), the Lyapunov function becomes

$$\dot{V}_p = -\sum_{i\in\mathcal{G}_p}\mu_i\eta_i^2 - \sum_{i\in\mathcal{G}_p}\nu_i\zeta_i^2 < 0. \qquad (8.85)$$

□

8.2.8 Nonlinear dynamics

Consider the MAS described by the single-integrator dynamics

$$\begin{aligned}
\dot{x}_i &= v_i, \\
\dot{v}_i &= a_i\psi_i(x_i) + b_i u_i, \qquad (8.86)
\end{aligned}$$

where a_i and b_i are unknown constants and $\psi_i(x_i)$ is a known nonlinear function.

Theorem 8.8. *The MAS defined by (8.86) on graph $\mathcal{G}$ with subgraphs $\mathcal{G}_p$ satisfying conditions in Lemmas 8.6 and 8.7 achieves P-group consensus under control protocols (8.87) and adaptive laws (8.88), (8.89), and (8.90)*

$$u_i = \theta_i(\zeta_i + v_i) + \beta_i \eta_i + \gamma_i \theta_i \psi_i, \tag{8.87}$$

$$\dot{\theta}_i = -sign(b_i)(2v_i + \zeta_i + \gamma_i \phi_i(x_i))(\zeta_i + \eta_i) + v_i \zeta_i^2), \tag{8.88}$$

$$\dot{\beta}_i = -sign(b_i)(\eta_i(\zeta_i + \eta_i) + \mu_i \eta_i^2), \tag{8.89}$$

$$\dot{\gamma}_i = -sign(b_i)(\psi_i(x_i)(\zeta_i + \eta_i)), \tag{8.90}$$

where $\theta_i = \dfrac{k_i^1}{b_i}$ *and* $\beta_i = \dfrac{k_i^2}{b_i}$ *are unknown control gains (or coupling strengths).*

Proof. Consider the Lyapunov function

$$V = \sum_{p=1}^{P} V_p, \tag{8.91}$$

where V_p is defined as

$$\begin{aligned} V_p &= \frac{1}{2} x_p^T \mathcal{L} x_p + \frac{1}{2} v_p^T \mathcal{L} v_p + \frac{1}{2} \sum_{i \in \mathcal{G}_p} |b_i| \theta_i^2 \\ &\quad + \frac{1}{2} \sum_{i \in \mathcal{G}_p} |b_i| \beta_i^2 + \frac{1}{2} \sum_{i \in \mathcal{G}_p} |b_i| \gamma_i^2. \end{aligned} \tag{8.92}$$

The time derivative of (8.92) gives

$$\begin{aligned} \dot{V}_p &= x_p^T \mathcal{L} \dot{x}_p + v_p^T \mathcal{L} \dot{v}_p + \sum_{i \in \mathcal{G}_p} |b_i| \theta_i \dot{\theta}_i \\ &\quad + \sum_{i \in \mathcal{G}_p} |b_i| \beta_i \dot{\beta}_i + \sum_{i \in \mathcal{G}_p} |b_i| \gamma_i \dot{\gamma}_i \end{aligned} \tag{8.93}$$

$$\begin{aligned} \dot{V}_p &= \sum_{i \in \mathcal{G}_p} (\zeta_i + \eta_i) v_i + \sum_{i \in \mathcal{G}_p} (\zeta_i + \eta_i)(a_i \psi_i(x_i) + b_i u_i) \\ &\quad + \sum_{i \in \mathcal{G}_p} |b_i| \theta_i \dot{\theta}_i + \sum_{i \in \mathcal{G}_p} |b_i| \beta_i \dot{\beta}_i + \sum_{i \in \mathcal{G}_p} |b_i| \gamma_i \dot{\gamma}_i. \end{aligned}$$

Substituting $u_i = \theta_i v_i + \theta_i \zeta_i + \beta_i \eta_i$ gives

$$\begin{aligned} \dot{V}_p &= \sum_{i \in \mathcal{G}_p} (\zeta_i + \eta_i) v_i + \sum_{i \in \mathcal{G}_p} (\zeta_i + \eta_i) a_i \psi_i(x_i) \\ &\quad + \sum_{i \in \mathcal{G}_p} (\zeta_i + \eta_i) b_i \theta_i (v_i + \gamma_i) + \sum_{i \in \mathcal{G}_p} b_i (\zeta_i + \eta_i) \beta_i \eta_i \end{aligned}$$

$$+\sum_{i\in\mathcal{G}_p}(\zeta_i+\eta_i)b_i\theta_i\gamma_i\psi_i(x_i)+\sum_{i\in\mathcal{G}_p}|b_i|\theta_i\dot{\theta}_i$$
$$+\sum_{i\in\mathcal{G}_p}|b_i|\beta_i\dot{\beta}_i+\sum_{i\in\mathcal{G}_p}|b_i|\gamma_i\dot{\gamma}_i.$$

Adding and subtracting $\sum_{i\in\mathcal{G}_p}\mu_i\eta_i^2$, $\sum_{i\in\mathcal{G}_p}\nu_i\zeta_i^2$ gives

$$\begin{aligned}\dot{V}_p &= \sum_{i\in\mathcal{G}_p}(\zeta_i+\eta_i)v_i+\sum_{i\in\mathcal{G}_p}(\zeta_i+\eta_i)a_i\psi_i(x_i)\\ &+\sum_{i\in\mathcal{G}_p}(\zeta_i+\eta_i)b_i\theta_i(v_i+\gamma_i)+\sum_{i\in\mathcal{G}_p}b_i(\zeta_i+\eta_i)\beta_i\eta_i\\ &+\sum_{i\in\mathcal{G}_p}(\zeta_i+\eta_i)b_i\theta_i\gamma_i\psi_i(x_i)+\sum_{i\in\mathcal{G}_p}|b_i|\theta_i\dot{\theta}_i\\ &+\sum_{i\in\mathcal{G}_p}|b_i|\beta_i\dot{\beta}_i+\sum_{i\in\mathcal{G}_p}|b_i|\gamma_i\dot{\gamma}_i\\ &+\sum_{i\in\mathcal{G}_p}\mu_i\eta_i^2+\sum_{i\in\mathcal{G}_p}\nu_i\zeta_i^2-\sum_{i\in\mathcal{G}_p}\mu_i\eta_i^2-\sum_{i\in\mathcal{G}_p}\nu_i\zeta_i^2.\end{aligned}$$

Substituting $a_i=\gamma_i b_i$ gives

$$\begin{aligned}\dot{V}_p &= \sum_{i\in\mathcal{G}_p}(\zeta_i+\eta_i)\theta_i b_i v_i+\sum_{i\in\mathcal{G}_p}(\zeta_i+\eta_i)\gamma_i b_i\psi_i(x_i)\\ &+\sum_{i\in\mathcal{G}_p}(\zeta_i+\eta_i)b_i\theta_i(v_i+\gamma_i)+\sum_{i\in\mathcal{G}_p}b_i(\zeta_i+\eta_i)\beta_i\eta_i\\ &+\sum_{i\in\mathcal{G}_p}(\zeta_i+\eta_i)b_i\theta_i\gamma_i\psi_i(x_i)+\sum_{i\in\mathcal{G}_p}|b_i|\theta_i\dot{\theta}_i\\ &+\sum_{i\in\mathcal{G}_p}|b_i|\beta_i\dot{\beta}_i+\sum_{i\in\mathcal{G}_p}|b_i|\gamma_i\dot{\gamma}_i\\ &+\sum_{i\in\mathcal{G}_p}\mu_i\eta_i^2+\sum_{i\in\mathcal{G}_p}\nu_i\zeta_i^2-\sum_{i\in\mathcal{G}_p}\mu_i\eta_i^2-\sum_{i\in\mathcal{G}_p}\nu_i\zeta_i^2\end{aligned}$$

$$\begin{aligned}\dot{V}_p &= -\sum_{i\in\mathcal{G}_p}\mu_i\eta_i^2-\sum_{i\in\mathcal{G}_p}\nu_i\zeta_i^2+\sum_{i\in\mathcal{G}_p}(\zeta_i+\eta_i)\theta_i b_i v_i\\ &+\sum_{i\in\mathcal{G}_p}(\zeta_i+\eta_i)\gamma_i b_i\psi_i(x_i)+\sum_{i\in\mathcal{G}_p}(\zeta_i+\eta_i)b_i\theta_i(v_i+\gamma_i)\\ &+\sum_{i\in\mathcal{G}_p}b_i(\zeta_i+\eta_i)\beta_i\eta_i+\sum_{i\in\mathcal{G}_p}(\zeta_i+\eta_i)b_i\theta_i\gamma_i\psi_i(x_i)\end{aligned}$$

$$
\begin{aligned}
&+\sum_{i\in\mathcal{G}_p}|b_i|\theta_i\dot{\theta}_i+\sum_{i\in\mathcal{G}_p}|b_i|\beta_i\dot{\beta}_i+\sum_{i\in\mathcal{G}_p}|b_i|\gamma_i\dot{\gamma}_i\\
&+\sum_{i\in\mathcal{G}_p}\mu_i b_i\theta_i\eta_i^2+\sum_{i\in\mathcal{G}_p}\nu_i b_i\beta_i\zeta_i^2
\end{aligned}
$$

$$
\begin{aligned}
\dot{V}_p \;=\;& -\sum_{i\in\mathcal{G}_p}\mu_i\eta_i^2-\sum_{i\in\mathcal{G}_p}\nu_i\zeta_i^2 \qquad (8.94)\\
&+\sum_{i\in\mathcal{G}_p}b_i[sgn(b_i)\dot{\theta}_i+(2v_i+\zeta_i+\gamma_i\phi_i(x_i))(\zeta_i+\eta_i)+\nu_i\zeta_i^2]\theta_i\\
&+\sum_{i\in\mathcal{G}_p}b_i[sgn(b_i)\dot{\beta}_i+\eta_i(\zeta_i+\eta_i)+\mu_i\eta_i^2]\beta_i\\
&+\sum_{i\in\mathcal{G}_p}b_i[sgn(b_i)\dot{\gamma}_i+\phi_i(x_i)(\zeta_i+\eta_i)]\gamma_i.
\end{aligned}
$$

Applying update laws (8.88), (8.89), and (8.90), the Lyapunov function becomes

$$
\dot{V}_p = -\sum_{i\in\mathcal{G}_p}\mu_i\eta_i^2-\sum_{i\in\mathcal{G}_p}\nu_i\zeta_i^2<0. \qquad (8.95)
$$

□

8.2.9 Simulation studies

For simulation purposes we consider the MAS consisting of five agents described by the graph in Fig. 8.10 with the following Laplacian (8.96) and Adjacency (8.97) matrices:

$$
\mathcal{L}=\begin{bmatrix}4&-3&-1&1&-1\\-3&3&0&-1&1\\-1&0&1&0&0\\1&-1&0&3&-3\\-1&1&0&-3&3\end{bmatrix}, \qquad (8.96)
$$

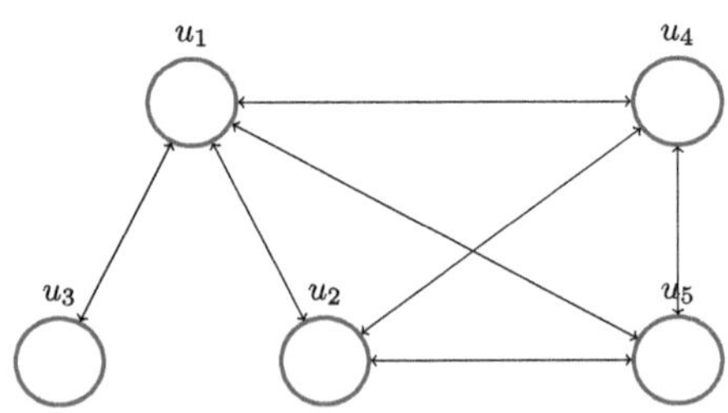

FIGURE 8.10 Graph showing interconnections between agents.

$$\mathcal{D} = \begin{bmatrix} 0 & 3 & 1 & -1 & 1 \\ 3 & 0 & 0 & 1 & -1 \\ 1 & 0 & 0 & 0 & 0 \\ -1 & 1 & 0 & 0 & 3 \\ 1 & -1 & 0 & 3 & 0 \end{bmatrix}. \tag{8.97}$$

The eigenvalues of the Laplacian matrix are $\lambda = 0, 0, 1.3263, 4.3457, 8.3280$. Based on Lemma 8.6 the presence of two zero eigenvalues indicates the presence of two subgraphs (or subgroups) $\mathcal{G}_1$ and $\mathcal{G}_2$. We conducted a simulation in four categories based on the adaptive laws proposed in the previous section.

8.2.10 Single integrator with linear dynamics

Figs. 8.11–8.13 present simulation plots for the MAS described by (8.57) under control protocol (8.60) and adaptive laws (8.61)–(8.62). The unknown control coefficients b_i were chosen as $b_1 = 1$, $b_2 = -1$, $b_3 = 2$, $b_4 = -1$, and $b_5 = 3$. The initial conditions used in the simulation were chosen as $x(0) = [-1\ 1\ 2\ -3\ -4]$, $v(0) = [1\ 1\ 1\ 1\ 1]$, $\theta(0) = [1\ 1\ 1\ 1\ 1]$, and $\beta(0) = [1\ 1\ 1\ 1\ 1]$. Fig. 8.11 shows the state responses $x_i(t)$ of each agent. Figs. 8.12 and 8.13 show the responses of the intracluster and intercluster gains, θ_i and β_i, respectively.

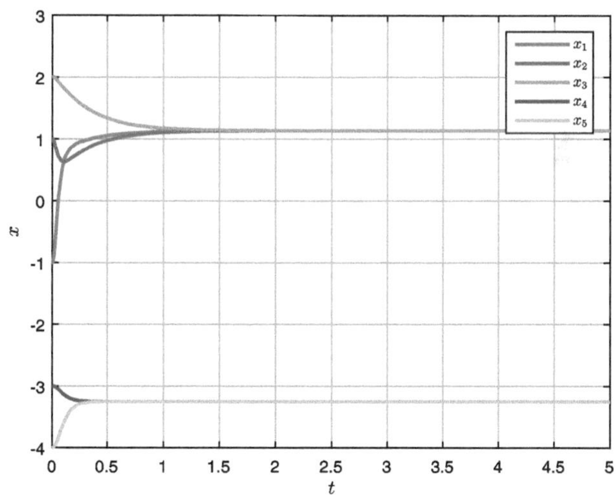

FIGURE 8.11 Single integrator with linear dynamics: state trajectory of $x(t)$.

8.2.11 Single integrator with nonlinear dynamics

In Figs. 8.14–8.17 we present simulation plots for MAS described by (8.67) under control protocol (8.68) and adaptive laws (8.69)–(8.71). The unknown control coefficients b_i were chosen as $b_1 = 1$, $b_2 = -1$, $b_3 = 2$, $b_4 = -1$,

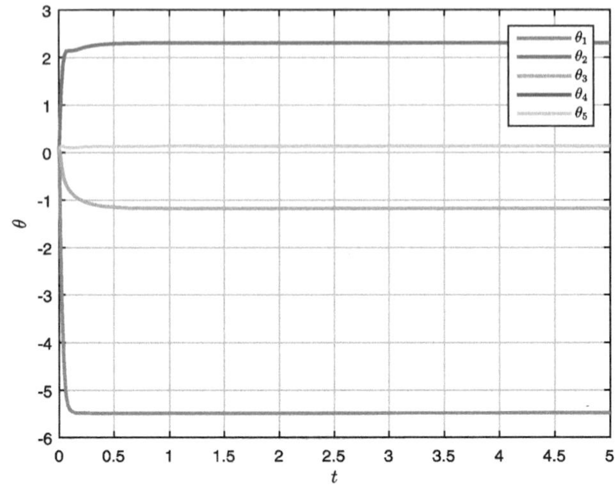

FIGURE 8.12 Single integrator with linear dynamics: state trajectory of $\theta(t)$.

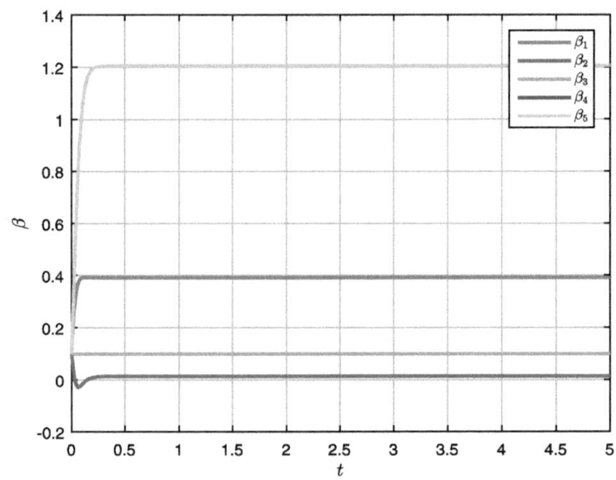

FIGURE 8.13 Single integrator with linear dynamics: state trajectory of $\beta(t)$.

and $b_5 = 3$. The unknown constants a_i were selected as $a_1 = -0.8$, $a_2 = 1$, $a_3 = -0.3$, $a_4 = 1$, and $a_5 = 0.2$. The known nonlinear functions $\phi_i(x_i)$ were chosen as $\phi_1(x_1) = sin(x_1)$, $\phi_2(x_2) = x_2^2$, $\phi_3(x_3) = cos(x_3)$, $\phi_4(x_4) = sin(x_4)$, and $\phi_5(x_5) = x_5^2$. The initial conditions used in the simulation were chosen as $x(0) = [10 \ -1 \ 5 \ -10 \ -4]$, $v(0) = [1\ 1\ 1\ 1\ 1]$, $\theta(0) = [1\ 1\ 1\ 1\ 1]$, $\beta(0) = [1\ 1\ 1\ 1\ 1]$, and $\gamma(0) = [1\ 1\ 1\ 1\ 1]$. Fig. 8.14 shows the state responses $x_i(t)$ of each agent. Figs. 8.15 and 8.16 show the responses of the intracluster and intercluster gains, θ_i and β_i, respectively. Fig. 8.17 shows the response of γ_i.

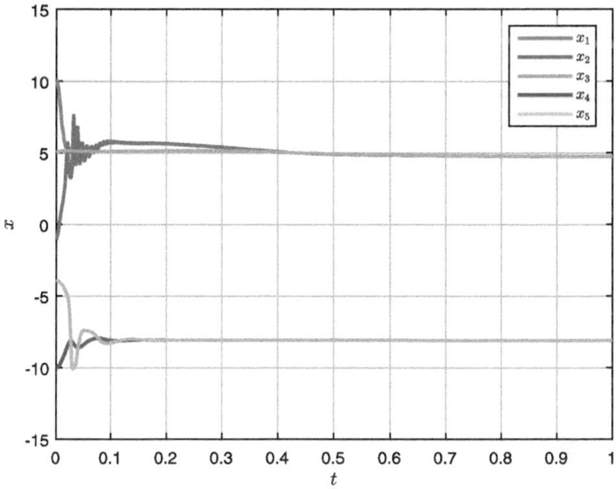

FIGURE 8.14 Single integrator with nonlinear dynamics: state trajectory of $x(t)$.

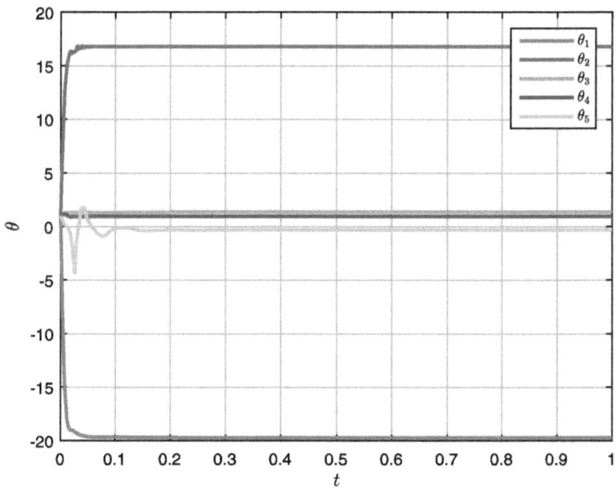

FIGURE 8.15 Single integrator with nonlinear dynamics: state trajectory of $\theta(t)$.

8.2.12 Double integrator with linear dynamics

In Figs. 8.18–8.21 we present simulation plots for MAS described by (8.77) under control protocol (8.79) and adaptive laws (8.81)–(8.81). The unknown control coefficients b_i were chosen as $b_1 = 1$, $b_2 = -1$, $b_3 = 2$, $b_4 = -1$, and $b_5 = 3$. The initial conditions used in the simulation were chosen as $x(0) = [10\ -1\ 5\ -10\ -4]$, $v(0) = [1\ 1\ 1\ 1\ 1]$, $\theta(0) = [1\ 1\ 1\ 1\ 1]$, $\beta(0) = [1\ 1\ 1\ 1\ 1]$, and $\gamma(0) = [1\ 1\ 1\ 1\ 1]$. Figs. 8.18 and 8.19 show the state responses $x_i(t)$ and

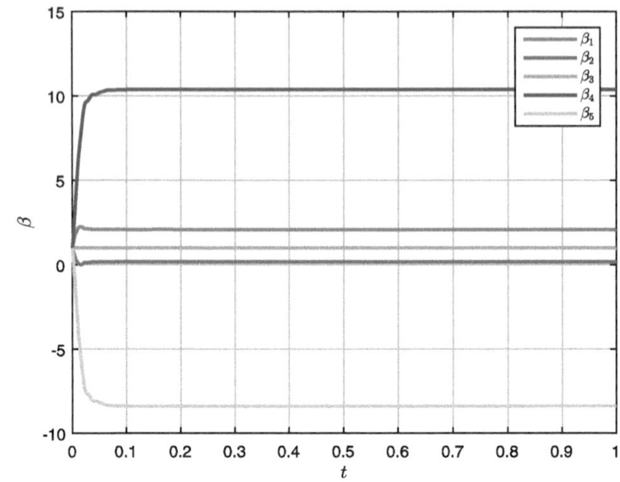

FIGURE 8.16 Single integrator with nonlinear dynamics: state trajectory of $\beta(t)$.

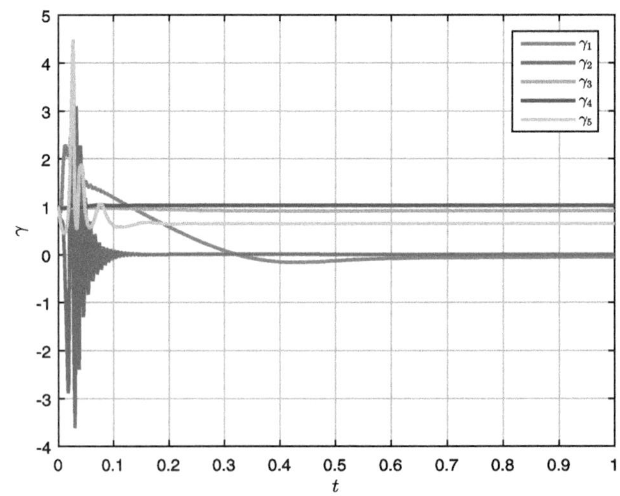

FIGURE 8.17 Single integrator with nonlinear dynamics: state trajectory of $\gamma(t)$.

$v_i(t)$ of each agent. Figs. 8.20 and 8.21 show the responses of the intracluster and intercluster gains, θ_i and β_i, respectively.

8.2.13 Double integrator with nonlinear dynamics

In Figs. 8.22–8.26 we present simulation plots for MAS described by (8.86) under control protocol (8.87) and adaptive laws (8.88)–(8.90). The unknown control coefficients b_i were chosen as $b_1 = 1$, $b_2 = -1$, $b_3 = 2$, $b_4 = -1$,

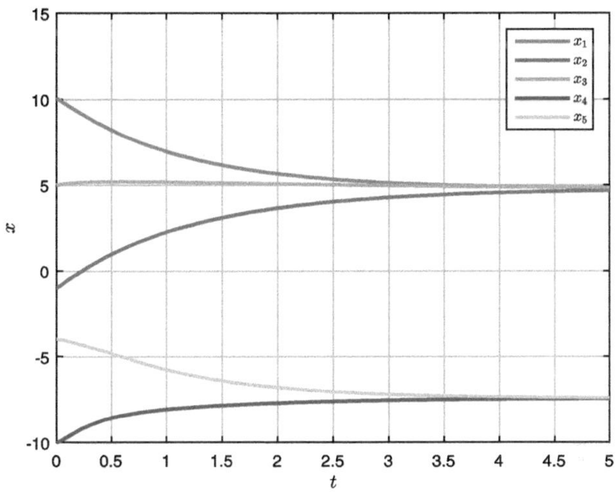

FIGURE 8.18 Double integrator with linear dynamics: state trajectory of $x(t)$.

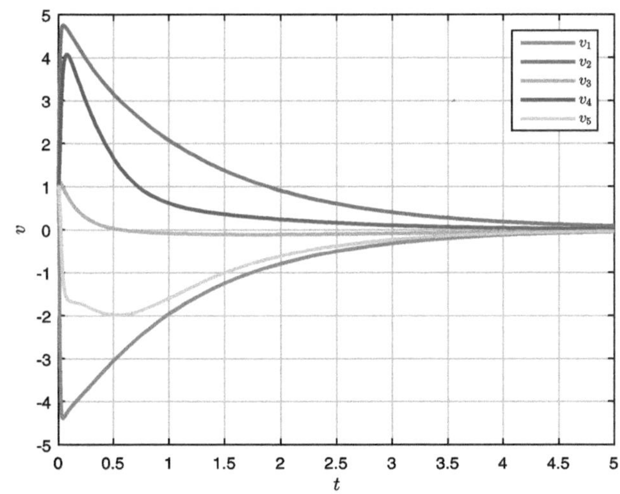

FIGURE 8.19 Double integrator with linear dynamics: state trajectory of $v(t)$.

and $b_5 = 3$. The unknown constants a_i were selected as $a_1 = -0.8$, $a_2 = 1$, $a_3 = -0.3$, $a_4 = 1$, and $a_5 = 0.2$. The known nonlinear functions $\phi_i(x_i)$ are chosen as $\phi_1(x_1) = sin(x_1)$, $\phi_2(x_2) = x_2^2$, $\phi_3(x_3) = cos(x_3)$, $\phi_4(x_4) = sin(x_4)$, and $\phi_5(x_5) = x_5^2$. The initial conditions used in the simulation were chosen as $x(0) = [10\ -1\ 5\ -10\ -4]$, $v(0) = [1\ 1\ 1\ 1\ 1]$, $\theta(0) = [1\ 1\ 1\ 1\ 1]$, $\beta(0) = [1\ 1\ 1\ 1\ 1]$, and $\gamma(0) = [1\ 1\ 1\ 1\ 1]$. Figs. 8.22 and 8.23 show the state responses $x_i(t)$ and $v_i(t)$ of each agent. Figs. 8.24 and 8.25 show the responses of the intraclus-

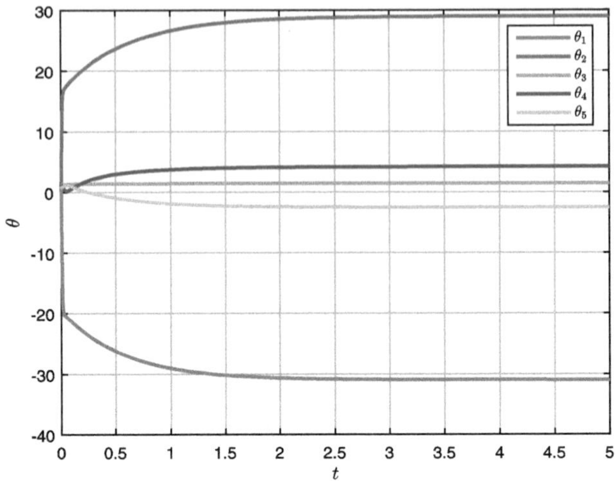

FIGURE 8.20 Double integrator with linear dynamics: state trajectory of $\theta(t)$.

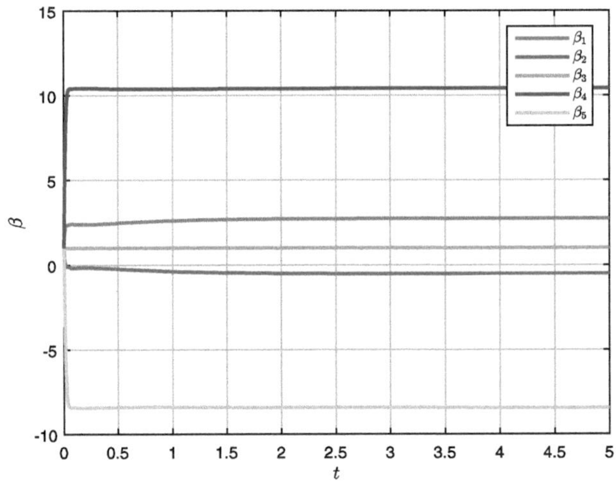

FIGURE 8.21 Double integrator with linear dynamics: state trajectory of $\beta(t)$.

ter and intercluster gains, θ_i and β_i, respectively. Fig. 8.17 shows the response of γ_i.

8.3 Notes

This chapter presented a state filtering for linear stochastic discrete-time systems subject to deception attacks and data losses on the control signals transmitted by the controller to the plant. A bias state-dependent intermittent unknown input disabled at the occurrence time of data losses was used to derive a fixed dimen-

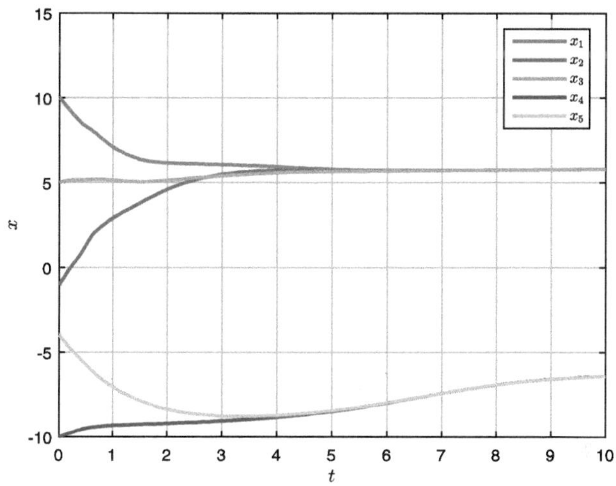

FIGURE 8.22 Double integrator with nonlinear dynamics: state trajectory of $x(t)$.

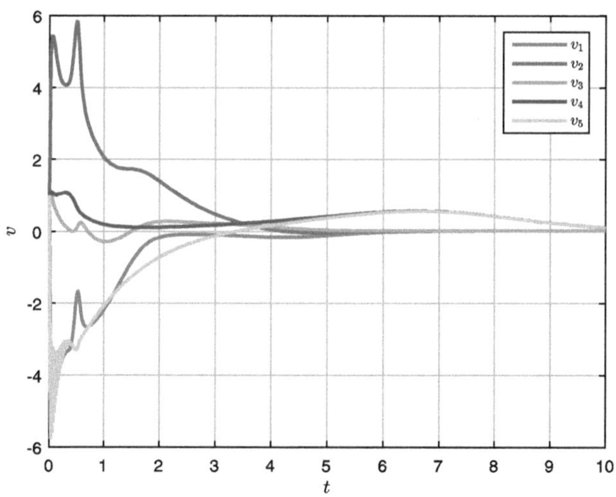

FIGURE 8.23 Double integrator with nonlinear dynamics: state trajectory of $v(t)$.

sional augmented state model of the plant allowing a direct application of the intermittent unknown input Kalman filter.

Next we analyzed the influence of DoS attacks on group consensus in MASs. We derived some necessary consensus conditions using the information about the graph Laplacian, and coupling strengths within and between subgroups for cases when the MAS is under DoS attacks. Overall, three distinct scenarios were examined: when the DoS attack is within subgroup 1, subgroup 2, and both

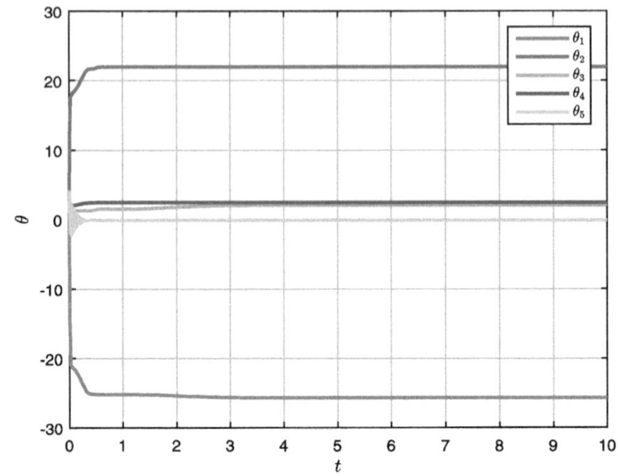

FIGURE 8.24 Double integrator with nonlinear dynamics: state trajectory of $\theta(t)$.

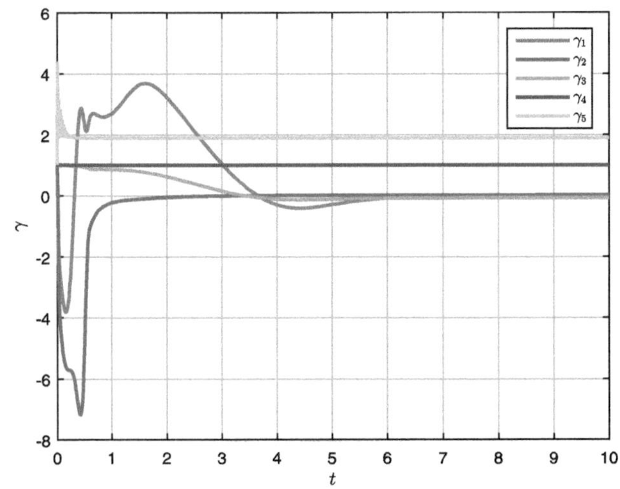

FIGURE 8.25 Double integrator with nonlinear dynamics: state trajectory of $\beta(t)$.

subgroups. Based on the derived conditions and simulation examples, we arrive at the following conclusions:

- Based on the established conditions, the coupling strengths of the network can be designed to withstand DoS attacks provided the graph remains connected after the influence of an attack.
- It is possible to design a control technique to modify the coupling strengths of the network after an attack has been detected to improve the resilience of the network and maintain consensus.

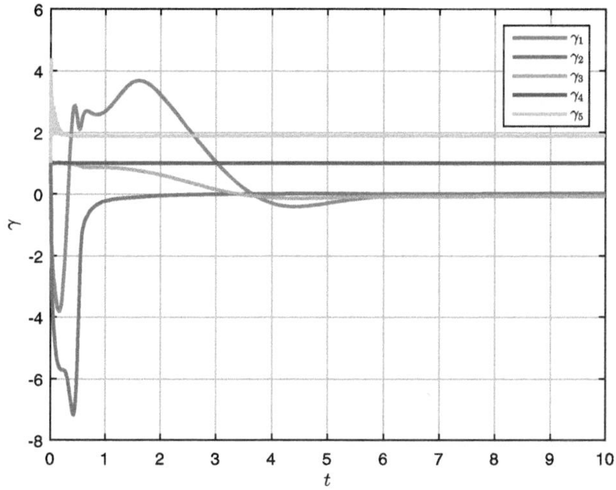

FIGURE 8.26 Double integrator with nonlinear dynamics: state trajectory of $\gamma(t)$.

- The agents in the MAS can reorganize themselves after the influence of an attack and "renegotiate" a new consensus with other agents that do not belong to the same subgroup.

Then we presented some distributed adaptive protocols for cluster consensus for single- and double-integrator MAS with linear and nonlinear dynamics. The proposed protocols simplify the design of control protocols for MAS by eliminating the need to compute optimal gains under which consensus can be achieved. The proposed approach in this chapter uses relative errors between agents to update the gains of both inter- and intracluster couplings.

Chapter 9

Cybersecurity for the electric power system

Contents

9.1 Problem description

An increasing demand for reliable energy and numerous technological advancements have motivated the development of a smart electric grid. The smart grid will expand the current capabilities of the grid's generation, transmission, and distribution systems to provide an infrastructure capable of handling future requirements for distributed generation, renewable energy sources, electric vehicles, and the demand-side management of electricity. The US Department of Energy (DOE) has identified seven properties required for the smart grid to meet future demands [363]. These requirements include attack resistance, self-healing, consumer motivation, power quality, generation and storage accommodation, market availability, and asset optimization.

While technologies such as phasor measurement units (PMU), wide area measurement systems, substation automation, and advanced metering infras-

Cloud Control Systems. https://doi.org/10.1016/B978-0-12-818701-2.00017-2

tructures (AMI) will be deployed to help achieve these objectives, they also present an increased dependency on cyber resources which may be vulnerable to attack [364]. Recent US Government Accountability Office (GAO) investigations into the grid's cyber infrastructure have questioned the adequacy of the current security posture [365]. The North American Electric Reliability Corporation (NERC) has recognized these concerns and introduced compliance requirements to enforce baseline cybersecurity efforts throughout the bulk power system [366]. Additionally, current events have shown attackers using increasingly sophisticated attacks against industrial control systems, while numerous countries have acknowledged that cyber attacks have targeted their critical infrastructures [367], [368].

A comprehensive approach to understanding security concerns within the grid must utilize cyber-physical system (CPS) interactions to appropriately quantify attack impacts [369] and evaluate effectiveness of countermeasures. This paper highlights CPS security for the power grid as the functional composition of the following:

1) the physical components and control applications;
2) the cyber infrastructures required to support necessary planning, operational, and market functions;
3) the correlation between cyber attacks and the resulting physical system impacts;
4) the countermeasures to mitigate risks from cyber threats.

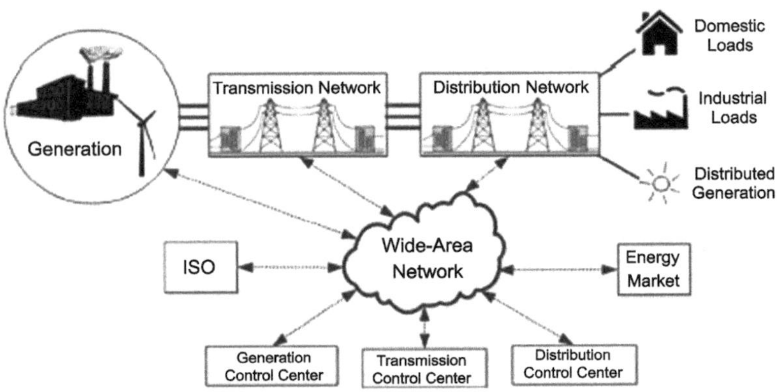

FIGURE 9.1 Power grid cyber-physical infrastructure.

Fig. 9.1 shows a CPS security view of the power grid. The cyber systems, consisting of electronic field devices, communication networks, substation automation systems, and control centers, are embedded throughout the physical grid for efficient and reliable generation, transmission, and distribution of power. The control center is responsible for real-time monitoring, control, and operational decision making. Independent system operators (ISOs) perform coordination between power utilities, and dispatch commands to their control centers. Utilities that participate in power markets also interact with the ISOs to support market functions based on real-time power generation, transmission, and demand.

9.2 Risk assessment methodology

The complexity of the cyber-physical relationship can present unintuitive system dependencies. Performing accurate risk assessments requires the development of models that provide a basis for dependency analysis and quantifies the resulting impacts. This association between the salient features within the cyber and physical infrastructures will assist in the risk review and mitigation processes. This paper presents a coarse assessment methodology to illustrate the dependency between the power applications and supporting infrastructure. An overview of the methodology is presented in Fig. 9.2.

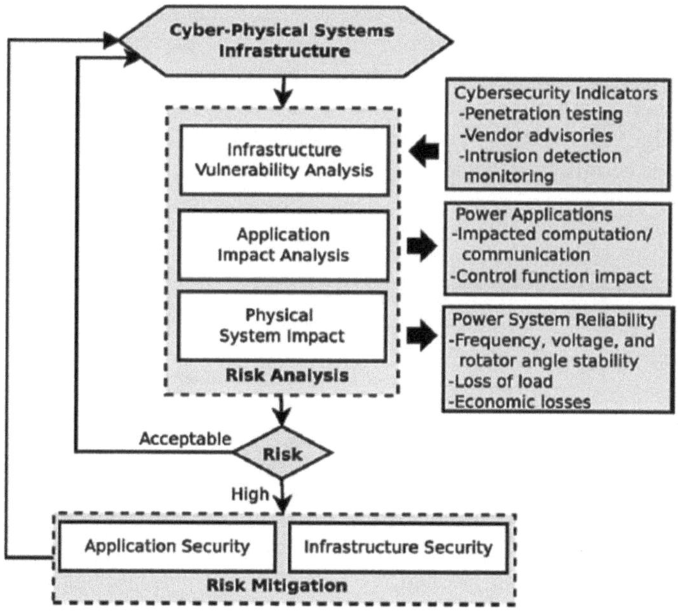

FIGURE 9.2 Risk assessment methodology.

9.2.1 Risk analysis

The initial step in the risk analysis process is the infrastructure vulnerability analysis. Numerous difficulties are encountered when determining cyber vulnerabilities within control system environments due to the high availability requirements and dependencies on legacy systems and protocols [370].

9.2.2 Risk mitigation

Mitigation activities should attempt to minimize unacceptable risk levels. This can be performed through the deployment of a more robust supporting infrastructure or power applications as discussed in Sections III and IV. Understanding opportunities to focus on specific or combine approaches may present novel mitigation strategies.

9.3 Power system control security

A power system is functionally divided into generation, transmission, and distribution. In this section we present a classification of control loops in the power system that identifies communication signals and protocols, machines and devices, computations, and control actions associated with select control loops in each functional classification. The section also sheds light on the potential impact of cyber attacks directed at these control loops on system-wide power grid stability.

Control centers receive measurements from sensors that interact with field devices (transmission lines, transformers, etc.). The algorithms running in the control center process these measurements to make operational decisions. The decisions are then transmitted to actuators to implement these changes on field devices. Fig. 9.3 shows a generic control loop that represents this interaction between the control center and the physical system.

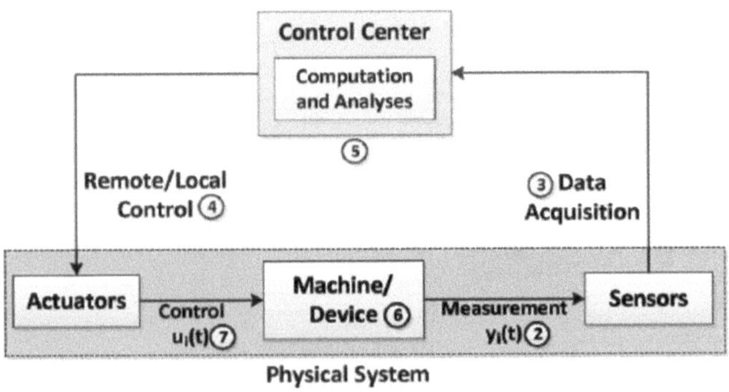

FIGURE 9.3 A typical power system control loop.

The smart grid can provide an efficient way of supplying and consuming energy by providing two-way energy flow and communication [73]. It can integrate multiple renewable distributed energy resources (DERs) which are environmently friendly, have low greenhouse gas emission, and effectively alleviate transmission power losses. The associated connectivity and advanced information/communication infrastructure make the smart grid susceptible to cyber attacks [73], [372]. Statistics in the energy sector show that more than 150 cyber attacks happened in 2013 and 79 in 2014 [73]. As a result, the power outage cost is about $80 billion per year in the United States. Usually, the utility operators amortize it by increasing the energy tariff, which is unfortunately transferred to consumer expenses [373]. The renewable microgrid incorporating DERs can be a potential solution, but it needs to be properly monitored as its generation pattern depends on the weather and surrounding conditions. One of the smart grid features is that it can integrate multiple microgrids and monitor them using reliable communication networks.

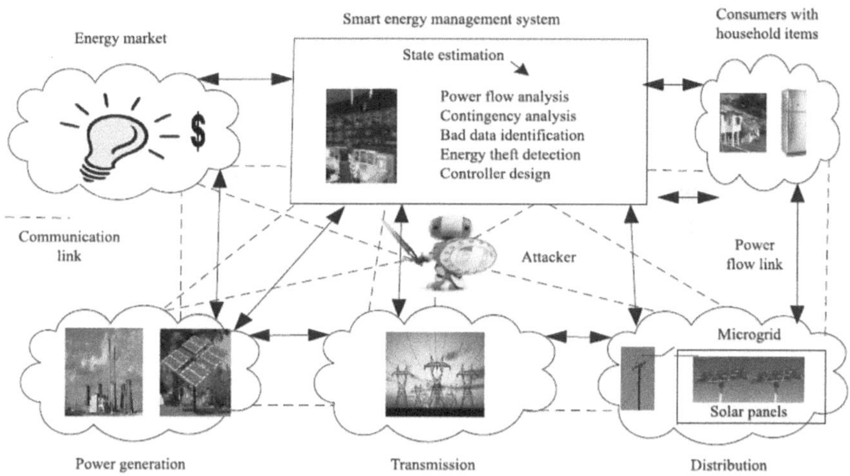

FIGURE 9.4 Flow of electricity and information through different sections of a smart grid [3].

The generation pattern of a microgrid varies on a time-place basis, which means that its operating condition should be closely monitored. Therefore, the microgrid state estimation is an important function in the smart grid energy management system (EMS). As shown in Fig. 9.4, the system state estimation is an essential task for the monitoring and control of the power network. In order to monitor the grid information, the utility company deploys a set of sensors around the smart grid. The communication infrastructure is used to send grid information from sensors to the EMS. The accurately estimated states can also be used in other functions of EMS such as contingency analysis, bad data detection, energy theft detection, stability analysis, and optimal power dispatch [374]. However, it is not economical or even feasible to measure all states, so the state estimation is also a key task in this regard [375]. More importantly, cyber attacks can cause major social, economical, and technical problems such as blackouts in power systems, tampering with smart meter reading, and changing the forecasted load profiles [373]. These types of catastrophic phenomena are much easier to commit in microgrids, so they create problems in a smart grid that are much more serious than in a traditional grid [384]. Therefore, the system state estimation under cyber attacks for smart grids has drawn significant interest in the energy industry, and in signal processing-based information and communication companies. Many studies have been carried out to investigate cyber attacks in smart grid state estimations. To begin with, most of the state estimation methods use the weighted least-squares (WLS) technique under cyber attacks [66,385,386]. A chi-square detector is also used to expose these attacks. Even though it is easy to implement this approach for nonlinear systems, it is computationally intensive and it cannot eliminate the attacks properly [66]. To this end, the WLS-based L_1 optimization method is explored in [377]. Furthermore, a new detection scheme to detect the false data injection attack is proposed in [21]. It employs a Kullback–Leibler method to calculate the distance between the probability distributions derived from the observation variations. A sequential detection of false data injection in smart grids is investigated in [21]. It adopts a centralized detector based on the generalized likelihood ratio and cumulative sum algorithm. Note that this detector usually depends on the parametric inferences and so is inapplicable to the nonparametric inferences [241]. A semidefinite programming-based AC power system state estimation is proposed in [380]. Thereafter, a Kalman filter (KF) based microgrid energy theft detection algorithm is presented in [90].

A great deal of effort has been devoted towards the power system state estimation under the condition of unreliable communication channels. Generally, the attackers limit their energies to jamming channels in order to achieve the desired goals [252]. The sensor data scheduling for state estimation with energy constraints is studied in [378]. In this research the sensor has to decide whether to send its data to a remote estimator or not based on its energy and estimation error covariance matrix. This idea is further extended in [381], where both the sensor and attacker have energy constraints for sending information. The considered attack is on the communication channel between a sensor and a remote estimator. Basically the sensor aims to minimize the average estimation error covariance matrix, while attackers try to maximize it. An iterative game theory is used to solve the optimization problem. Due to the motivation of unknown attacking patterns, the authors in [252] and [301] investigated how the attacker can design the attacking policy so the estimation performance can be deteriorated. Then the optimal scheduling strategy based on average estimation error covariance is proposed to avoid this kind of attack.

Many feedback control algorithms have been proposed to regulate the system. The linear quadratic Gaussian (LQG) detecting techniques for cyber integrity attacks on the sensors of a control system is proposed in [252] and [473]. It shows that the residual error based chi-squared detection technique is not suitable when the attacker does not know the system dynamics. Based on this analysis, they consider the cyber attack model as an independent and identically distributed (i.i.d.) Gaussian distribution, and then the LQG objective function is modified. In the end, they developed sufficient conditions to detect the false alarm probability, and proposed an optimization algorithm to minimize it. In [382] a new strategy is recommended for designing a communication and control infrastructure in a distribution system based on the virtual microgrid concept. It is known that designing a state feedback control framework for a general case of polynomial discrete-time system is quite challenging because the solution is nonconvex. Thus, the convex optimization-based controller design has gained growing interest in the research community.

A comprehensive approach to understanding security concerns within the grid must utilize CPS interactions to appropriately quantify attack impacts [369] and evaluate the effectiveness of countermeasures. This section highlights CPS security for the power grid as the functional composition of the following:

1) the physical components and control applications;

2) the cyber infrastructures required to support necessary planning, operations, and market functions;

3) the correlation between cyber attacks and the resulting physical system impacts;

4) the countermeasures to mitigate risks from cyber threats.

The cyber systems of a smart grids consisting of electronic field devices, communication networks, substation automation systems, and control centers are embedded throughout the physical grid for efficient and reliable generation, transmission, and distribution of power. The control center is responsible for real-time monitoring, control, and operational decision making. Independent system operators (ISOs) perform coordination between power utilities, and dispatch commands to their control centers. Utilities that participate in power markets also interact with the ISOs to support market functions based on real-time power generation, transmission, and demand.

9.3.1 Model of microgrid system

Consider N microsources connected to the main grid. For simplicity we assume that N = 4 solar panels are connected through the IEEE-4 bus test feeder, as shown in Fig. 9.5 [301]. Here, the input voltages are denoted by

$$v_p = [v_{p1}\ v_{p2}\ v_{p3}\ v_{p4}]^T,$$

where v_{pj} is the j-th DER input voltage. The four microsources are connected to the power network at the corresponding points of common coupling (PCCs)

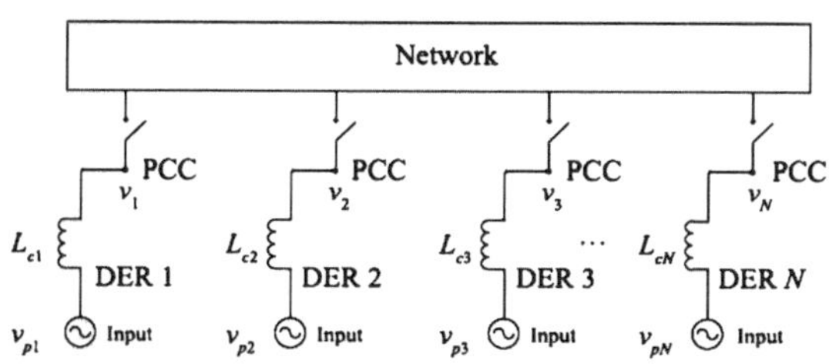

FIGURE 9.5 Microsources connected to the power network.

whose voltages are denoted by

$$v_s = [v_{s1} \; v_{s2} \; v_{s3} \; v_{s4}]^T,$$

where v_j is the j-th point of common coupling (**PCC**) voltage.

By applying Laplace transformation the nodal voltage equation can be obtained,

$$\mathbf{Y}(s)v_s(s) = \frac{1}{s}\mathbf{L_c}^{-1} v_p(s), \tag{9.1}$$

where $\mathbf{L_c} = diag[L_{c1} \; L_{c2} \; L_{c3} \; L_{c4}]$ and $\mathbf{Y}(s)$ is the admittance matrix of the entire power network incorporating four microsources [20]. Now we can convert the transfer function form into the linear state-space model [20]. Given v_{ref} and v_{pref} as the PCC reference voltage and the reference control effort, the discrete-time linear dynamic system can be derived as

$$x(k+1) = A_d x(k) + B_d u(k) + w_d(k), \tag{9.2}$$

where $x(k) = v_s - v_{ref}$ is the PCC state voltage deviation, $u(k) = v_p - v_{pref}$ is the DER control input deviation, and $w_d(k)$ is the zero-mean process noise whose covariance matrix is Q_w. Following [20], typical values of the state matrix A_d and input matrix B_d for a prescribed discretization step Δt are

$$A_d = \begin{bmatrix} 175.9 & 176.8 & 511 & 103.6 \\ -350 & 0 & 0 & 0 \\ -544.2 & -474.8 & -408.8 & -828.8 \\ -119.7 & -554.6 & -968.8 & -1077.5 \end{bmatrix}, \tag{9.3}$$

$$B_d = \begin{bmatrix} 175.9 & 176.8 & 511 & 103.6 \\ -350 & 0 & 0 & 0 \\ -544.2 & -474.8 & -408.8 & -828.8 \\ -119.7 & -554.6 & -968.8 & -1077.5 \end{bmatrix}. \tag{9.4}$$

In the following the observation model and attack process are explored.

9.3.2 Observation model and cyber attack

The measurements of the microgrid states are obtained by a set of sensors and can be modeled as

$$y(k) = Cx(k) + v(k), \tag{9.5}$$

where $y(k)$ is the measurement, C is the measurement matrix, and $v(k)$ is a zero-mean sensor measurement noise whose covariance matrix is R_v. Generally, the objective of attackers is to insert false data into the observations as

$$y(k) \quad = \quad Cx(k) + v(k) + a(k), \tag{9.6}$$

where $a(k)$ is the false data inserted by the attacker [1–3]. We consider that the attackers have complete access to the system infrastructure so that they can hijack, record, and manipulate data according to their best interest. In this work the cyber attack pattern is similar to those illustrated in [1], [2], and [22]. Fig. 9.6 shows the observation model and cyber attack process in the context of smart grid state estimations.

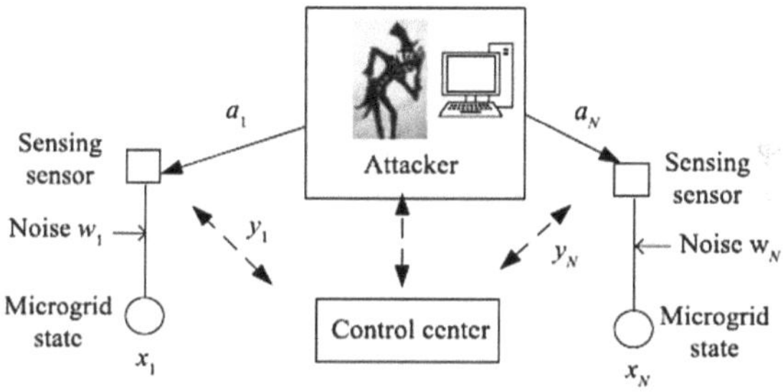

FIGURE 9.6 Observation model with a cyber attack in the microgrid.

The channel code is used to secure the system states in the signal processing research community. Motivated by the convolutional coding concept [243], [383] the microgrid state-space and observation models are regarded as the outer code. Then the standard uniform quantizer performs quantization to get the sequence of bits b(k), which is encoded by a recursive systematic convolutional (RSC) channel code which is regarded as the inner code. The main reason for using an RSC code is to mitigate impairments and introduce redundancy in the system to protect the grid information. Generally speaking, an RSC code is characterized by three parameters: the codeword length n, the message length l, and the constraint length m (i.e., (n; l;m)). The quantity l=n refers to the code rate, which indicates the number of parity bits added to the data stream. The constraint length specifies m1 memory elements, which represent the number of bits in the encoder memory that affect the RSC generation output bits. If the constraint length m increases, the encoding process intrinsically needs a longer time to execute the logical operations. Other advantages of the RSC code compared with the convolutional and turbo encoder include its reduced computation complexity, systematic output features, and no error floor [24]. From this point of view this paper considers a (2, 1, 3) RSC code and (1 0 1, 1 1 1) code generator polynomial in the feedback process. According to the RSC features, the code rate is 1=2 and there are two memories in the RSC process. As shown in Fig. 9.7, this RSC code produces two outputs and can convert an entire data stream into one single codeword [25]. The codeword is then passed through the binary phase shift keying (BPSK) to obtain $s(k)$, which is passed through the additive white Gaussian noise (AWGN) channel. Fig. 9.7 shows the proposed cyber attack protection procedure in the context of smart grids. In the end the received signal is

$$r(k) = s(k) + e(k), \tag{9.7}$$

where $e(k)$ is the AWGN. The received signal is followed by the log-maximum a posteriori (Log-MAP) decoding for this dynamic system. The Log-MAP works recursively from the forward path to the backward path to recover the state information [243]. The Log-MAP output information is sent for demodulation and dequantization processes, followed by the state estimation scheme (see Fig. 9.7).

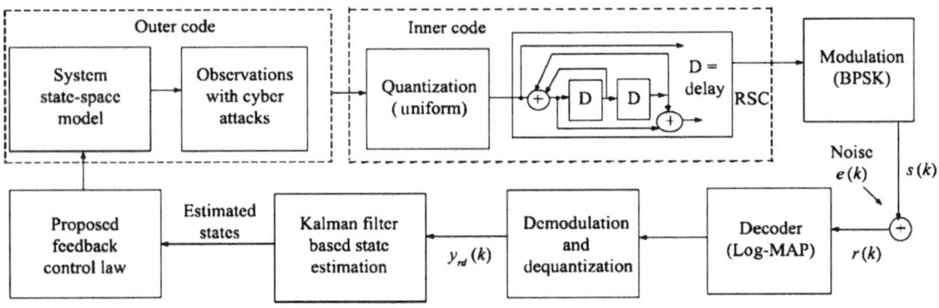

FIGURE 9.7 Cyber attack protection in smart grids.

9.3.3 Cyber attack minimization in smart grids

A recursive Kalman filter estimator (RKFE) is constructed to operate on observation information to produce the optimal state estimation. The forecasted system state estimate is expressed as [26]

$$\hat{x}^r(k) = A_d\hat{x}(k-1) + B_d u(k-1), \tag{9.8}$$

where $\hat{x}(k-1)$ is the state last-step estimate. Then the forecasted error covariance matrix is given by

$$P^r(k) = A_d P(k-1) A_d^T + Q_w(k-1), \tag{9.9}$$

where $P(k-1)$ is the last-step estimated error covariance matrix. The observation innovation residual $d(k)$ is given by

$$d(k) = y_{rd}(k) - C\hat{x}^r(k), \tag{9.10}$$

where $y_{rd}(k)$ is the dequantized and demodulated output bit sequence. The Kalman gain matrix can be written as

$$K(k) = P^r(k)C^T[CP^r(k)C^T + R_v(k)]^{-1}. \tag{9.11}$$

This yields the updated state estimation as

$$\hat{x}(k) = \hat{x}^r(k) - K(k)d(k), \tag{9.12}$$

along with the updated estimated error covariance matrix:

$$P(k) = P^r(k) - K(k)CP^r(k). \tag{9.13}$$

After estimating the system state, the proposed control strategy is applied to regulate the microgrid states as shown in the next section.

9.3.4 Stabilizing feedback controller

Given the availability of the microgrid state estimate information, we move to regulate the microgrid dynamics using the state feedback controller

$$u(k) = Fx(k) \tag{9.14}$$

by minimizing the quadratic cost function

$$J = \mathbb{E}[\lim_{N \longrightarrow \infty} \frac{1}{N} \sum_{j=0}^{N-1} \{x^T(j)Q_x x(j) + u^T(j)R_x u(j)\}], \tag{9.15}$$

where $\mathbb{E}(.)$ denotes the expectation operator and F is the state feedback gain matrix, and $Q_x > 0$ and $R_x > 0$ are state weighting and control weighting matrices. From (9.2) and (9.14) the closed loop system is

$$x(k+1) = (A_d + B_d F)x(k) + w_d(k). \tag{9.16}$$

Using the properties of the trace operator $Tr(.)$ and (9.14), it is easy to see that

$$\begin{aligned} J &= \mathbb{E}[\lim_{N \longrightarrow \infty} \frac{1}{N} \sum_{j=0}^{N-1} Tr(Q_x x(j)x^T(j) + F^T R_x F x(j)x^T(j))] \\ &= Tr(Q_x + F^T R_x F)P \\ P &= \mathbb{E}[\lim_{N \longrightarrow \infty} \frac{1}{N} \sum_{j=0}^{N-1} x(j)x^T(j)]. \end{aligned} \tag{9.17}$$

Algebraic manipulation yields

$$\begin{aligned} P &= \mathbb{E}[\lim_{N \longrightarrow \infty} \frac{1}{N} \sum_{j=0}^{N-1} x(j)x^T(j)] \\ &- (A_d + B_d F)P(A_d + B_d F)^T + Q_w. \end{aligned} \tag{9.18}$$

It follows that there exists a stabilizing matrix P_o satisfying $P_o < P$ such that

$$(A_d + B_d F)P_o(A_d + B_d F)^T - P_o + Q_w < 0, \tag{9.19}$$

which is a nonlinear matrix inequality in F and P_o. Applying Schur complements, (9.19) can be cast into the following minimization problem over linear matrix inequality (LMI),

$$\min_{P_o, W_o} \quad Tr(Q_x + P_o^{-1} W_o^T R_x W_o$$
$$\begin{bmatrix} Q_w - P_o & A_d P_o + B_d W_o \\ \bullet & -P_o \end{bmatrix} < 0, \tag{9.20}$$

and the controller gain is expressed as $F = W_o P_o^{-1}$ (see Fig. 9.8).

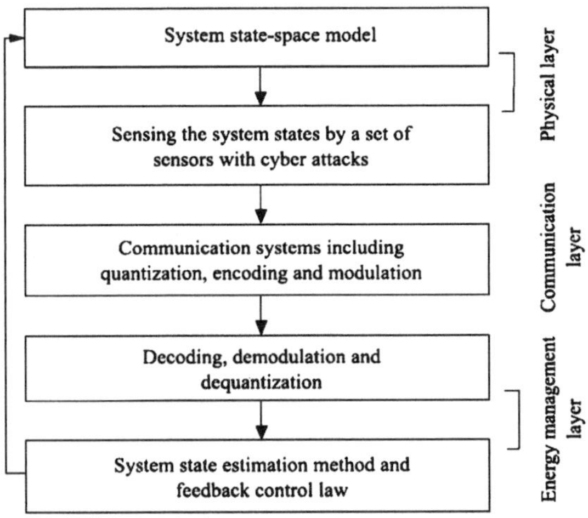

FIGURE 9.8 System level diagram for system state estimation and control.

9.4 Security of a smart grid infrastructure

The electric grid is arguably the world's largest engineered system. It is vital to human life, and its reliability is a major and often understated accomplishment of humankind. It is the motor of the economy and the major driver of progress. In its current state the grid consists of four major components:

1) generation, which produces electric energy in different ways (e.g., burning fossil fuels, inducing nuclear reactions, and harnessing water, wind, solar, and tidal forces);

2) transmission, which moves electricity via a very high voltage infrastructure;

3) distribution, which steps down current and spreads it out for consumption;

4) consumption (industrial, commercial, and residential), which uses the electric energy in a multitude of ways.

Given the wide variety of systems, their numerous owners, and a diverse range of regulators, a number of weaknesses have emerged. Outages are often recognized only after consumers report. Matching generation to demand is challenging because utilities do not have clear-cut methods to predict demand and to request demand reduction (load shedding). As a consequence, they need to overgenerate power for peak demand, which is expensive and contributes to greenhouse gas (GhG) emissions. For similar reasons it is difficult to incorporate variable generation, such as wind and solar power, into the grid. Last, there is a dearth of information available for consumers to determine how and when to use energy.

9.4.1 Introduction

To address these challenges the smart grid concept has evolved. The smart grid uses communications and information technologies to provide better "situational awareness" to utilities regarding the state of the grid. The smart grid provides numerous benefits [387], [388], [389], [390]. Using intelligent communications, load shedding can be implemented so that peak demand can be flattened, which reduces the need to bring additional (and expensive) generation plants online. Using information systems to perform predictive analysis, including when wind and solar resources will produce less power, the utilities can keep power appropriately balanced. As new storage technologies emerge on the utility scale, incorporation of these devices will likewise benefit from intelligent demand prediction. Last, the ability for consumers to receive and respond to price signals will help them manage their energy costs, while helping utilities avoid building additional generation plants.

With all these approaches, the smart grid enables a drastic cost reduction for both power generation and consumption.

Dynamic pricing and distributed generation with local generators can significantly reduce the electricity bill. Fig. 9.9A shows how to use electricity during off-peak periods when the price is low. Conversely, Fig. 9.9B shows load shedding during peak times and utilization of energy storage to meet customer demand. The effect of peak demand reduction by "demand management" is shown in Fig. 9.10. Pilot projects in the US states of California and Washington [387] indicate that scheduling appliances based on price information can reduce electricity costs by 10% for consumers. More advanced smart grid technologies promise to provide even larger savings.

To establish the smart grid vision, widespread sensing and communications between all the grid components (generation, transmission, distribution, storage) and consumers must be created and managed by information technology systems. Furthermore, sophisticated estimation, control, and pricing algorithms need to be implemented to support the increasing functionality of the grid while

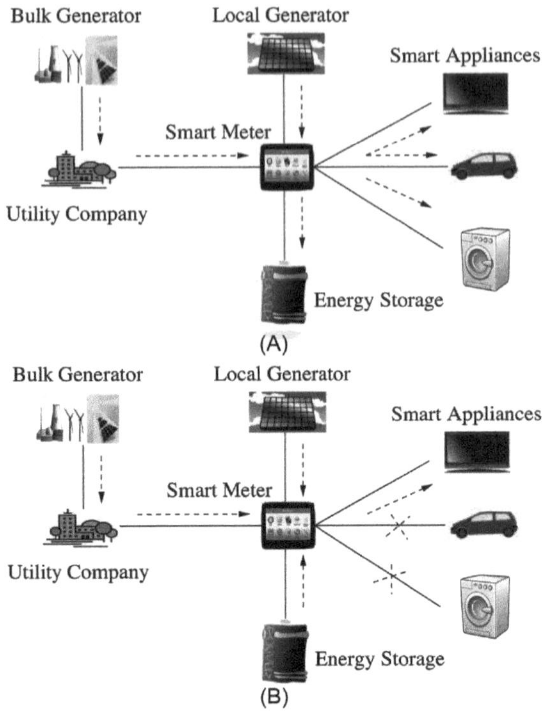

FIGURE 9.9 During off-peak time periods, inexpensive electric power can be used without restrictions (e.g., diverted to energy storage). During peak time periods, some appliances are temporarily turned off, and stored energy is used. (A) Power usage during off-peak time period. (B) Power usage during peak time period.

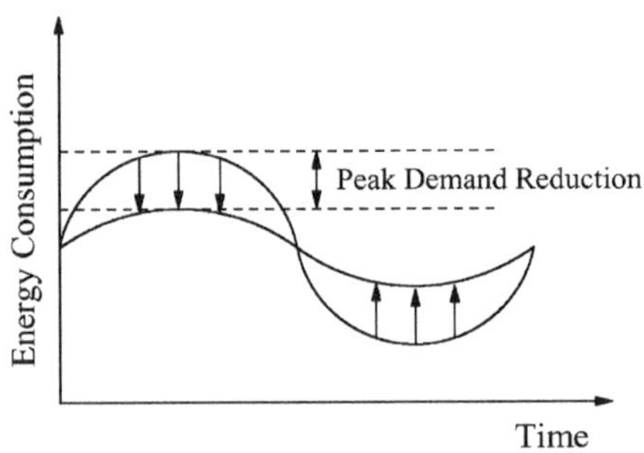

FIGURE 9.10 The peak demand for electricity will be reduced by the use of smart appliances, local generators, and/or local energy storage.

maintaining reliable operations. It is the greatly increased incorporation of IT systems that supports the vision, but unfortunately also creates exploitable vulnerabilities for the grid and its users.

9.4.2 A cyber-physical approach to smart grid security

A wide variety of motivations exist for launching an attack on the power grid, ranging from economic reasons (e.g., reducing electricity bills) and pranks all the way to terrorism (e.g., threatening people by controlling electricity and other life-critical resources). The emerging smart grid, while benefiting the benign participants (consumers, utility companies), also provides powerful tools for adversaries.

The smart grid will reach every house and building, giving potential attackers easy access to some of the grid components. While incorporating information technology (IT) systems and networks, the smart grid will be exposed to a wide range of security threats [391]. Its large scale also makes it nearly impossible to guarantee security for every single subsystem. Furthermore, the smart grid will not only be large, but also very complex. It needs to connect different systems and networks, from generation facilities and distribution equipment to intelligent end points and communication networks that are possibly deregulated and owned by several entities. It can be expected that the heterogeneity, diversity, and complexity of smart grid components may introduce new vulnerabilities, in addition to the common ones in interconnected networks and stand-alone microgrids [389]. To make the situation even worse, the sophisticated control, estimation, and pricing algorithms incorporated in the grid may also create additional vulnerabilities.

The first-ever control system malware called Stuxnet was found in July 2010. This malware, targeting vulnerable supervisory control and data acquisition (SCADA) systems, raises new questions about power grid security [392]. SCADA systems are currently isolated, preventing external access. Malware, however, can spread using Universal Serial Bus (USB) drives and can be specifically crafted to sabotage SCADA systems that control electric grids. Furthermore, increasingly interconnected smart grids will unfortunately provide external access, which in turn can lead to compromise and infection of components.

Many warnings concerning the security of smart grids have appeared [393–395,281,396,397], and some guidelines have been published, for example by the US National Institute of Standards and Technology (NISTIR 7628 [389] and NIST SP 1108 [398]). The NIST guidelines argue that a new approach to security, bringing together cybersecurity and system theory under the umbrella of cyber-physical system security, is needed to address the requirements of complex and large-scale infrastructures like the smart grid. In these systems cyber attacks can cause disruptions that transcend the cyber realm and affect the physical world. Stuxnet is a clear example of a cyber attack used to induce physical consequences. Conversely, physical attacks can affect the cyber system. For example the integrity of an electricity meter can be compromised by using a shunt to bypass it. Secrecy can be broken by placing a compromised sensor beside a legitimate one. As physical protection of all assets of large-scale physical systems, such as the smart grid, is economically infeasible, there is a need to develop methods and algorithms that can detect and counter hybrid

TABLE 9.1 Taxonomy of attacks and consequences in cyber and physical systems.

Attack	Consequence on cyber	Consequence on physical
Cyber	Eavesdropping of private information	Stuxnet
Physical	Meter bypassing	Instability due to physical destruction

attacks. Based on the discussions at the Army Research Office workshop on CPS security in 2009, we classify current attacks on CPSs into four categories and provide examples to illustrate our classification in Table 9.1. Although cybersecurity and system theory have achieved remarkable success in defending against pure cyber or pure physical attacks, neither of them alone is sufficient to ensure smart grid security against hybrid attacks. Cybersecurity is not equipped to provide an analysis of the possible consequences of attacks on physical systems. System theory is usually concerned with properties such as performance, stability, and safety of physical systems. Its theoretical framework, while well consolidated, does not provide a complete modeling of the IT infrastructure.

In what follows, we propose combining system theory and cybersecurity to ultimately build a science of CPS security. To move towards this goal, it is important to develop cyber-physical security models capable of integrating dynamic systems and threat models within a unified framework. We believe that cyber-physical security will be able to address problems that cannot be currently solved, but will also provide new improved solutions for detection, response, reconfiguration, and restoration of system functionalities while keeping the system operating. We also believe that some existing modeling formalisms can be used as a starting point towards a systematic treatment of cyber-physical security. Game theory [21] can capture the adversarial nature of the interaction between an attacker and a defender. Networked control systems [148] aim to integrate computing and communication technologies with system theory, providing a common modeling framework for CPSs. Finally, hybrid dynamic systems [399] can capture the discrete nature of events such as attacks on control systems.

9.4.3 Cybersecurity approaches

This section outlines cybersecurity approaches to smart grid security. It starts by presenting a dynamic model of smart power grid, then outlines the cybersecurity requirements before addressing some counter measures and providing plausible attack model. Then we examine the system communication architecture, system and device security and present several technical approaches to upgrade the performance. Detailed points and remarks are provided to illuminate the crucial role of cybersecurity.

9.4.4 System model

As Fig. 13.4 shows, smart grids consist of four components: generation, transmission, distribution, and consumption. In the consumption component, customers use electric devices (e.g., smart appliances, electric vehicles), and their electricity consumption will be measured by an enhanced metering device, called a smart meter. The smart meter is one of the core components of the advanced metering infrastructure (AMI) [400]. The meter can be connected to and interact with a gateway of a home-area network (HAN) or a business-area network (BAN). As a simple illustration, we denote a smart meter in the figure as a gateway of a HAN. A neighbor-area network (NAN) is formed under one substation where multiple HANs are hosted. Finally, a utility company may leverage a wide-area network (WAN) to connect distributed NANs (see Fig. 9.11).

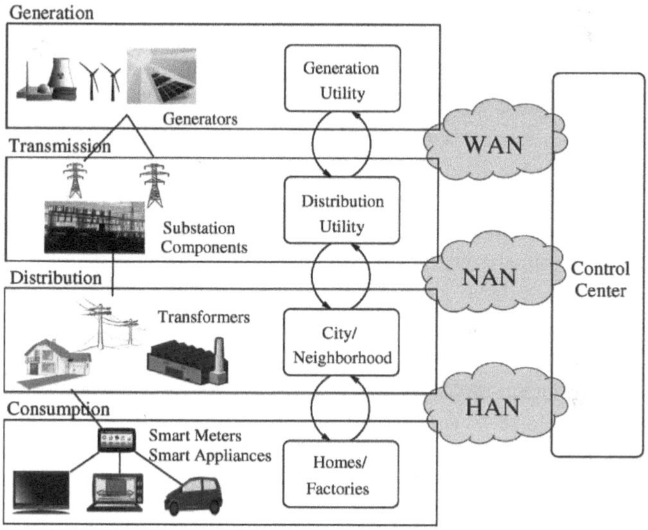

FIGURE 9.11 A cybersecurity view of the smart grid.

9.4.5 Cybersecurity requirements

In this section, we analyze the information security requirements for smart grids. In general, information security requirements for a system include three main security properties: confidentiality, integrity, and availability. Confidentiality prevents an unauthorized user from obtaining secret or private information. Integrity prevents an unauthorized user from modifying the information. Availability ensures that the resource can be used when requested.

As shown in Fig. 9.12, price information, meter data, and control commands are the core information exchanged in the smart grids that we consider in this chapter. While more types of information are exchanged in reality, these core information types provide a comprehensive sample of security issues.

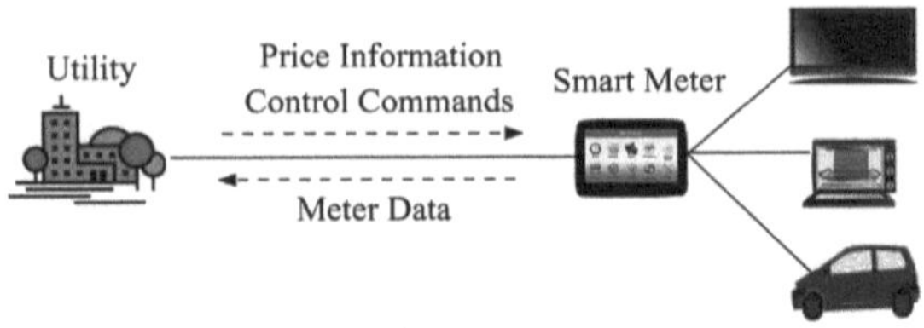

FIGURE 9.12 Information flows to and from a smart meter, including price information, control commands, and meter data.

We now examine the importance of protecting the core information types with respect to the main security properties. The degree of importance for price information, control commands, and meter data is equivalent to the use cases of NISTIR 7628 [389], to which we added the degree of importance for software. The most important requirement for protecting smart grids are outlined below.

- Confidentiality of power usage: Confidentiality of meter data is important because power usage data provides information about the usage patterns for individual appliances, which can reveal personal activity through nonintrusive appliance monitoring [188]. Confidentiality of price information and control commands are not important in cases where it is public knowledge. Confidentiality of software should not be critical because the security of the system should not rely on the secrecy of the software, but only on the secrecy of the keys, according to Kirchhoff principle [401].
- Integrity of data, commands, and software: Integrity of price information is critical. For instance, negative prices injected by an attacker can cause an electricity utilization spike as numerous devices would simultaneously turn on to take advantage of the low price. Although integrity of meter data and commands is important, their impact is mostly limited to revenue loss. On the other hand, integrity of software is critical since compromised software or malware can control any device and grid component.
- Availability against DoS/DDoS attacks: Denial-of-service (DoS) attacks are resource consumption attacks that send fake requests to a server or a network. Distributed DoS (DDoS) attacks are accomplished by utilizing distributed attacking sources such as compromised smart meters and appliances. In smart grids the availability of information and power is a key aspect [371]. More specifically the availability of price information is critical, due to serious financial and possibly legal implications. Moreover, outdated price information can adversely affect demand. The availability of commands is also important, especially when turning a meter back on after completing the payment of an electric bill. On the other hand, availability of meter data (e.g., power usage) may not be as critical because the data can usually be read at a later point.

From the above discussion, we can summarize the importance of data, commands, and software, which are shown in Table 9.2. "High" risk implies that a property of certain information is very important and critical, and "medium" and

TABLE 9.2 The importance of security properties for data, commands, and software.

	Price information	Control command	Meter data	Software
Confidentiality	Low	Low	Medium	Low
Integrity	High	High	High	High
Availability	High	High	Low	N/A

"low" risks classify properties that are important and noncritical, respectively. This classification enables the prioritization of risks in order to focus on the most critical aspects first. For example, integrity of price information is more important than its confidentiality; consequently, we need to focus on efficient cryptographic authentication mechanisms before encryption.

9.4.6 Attack model

To launch an attack an adversary must first exploit entry points, and upon successful entry an adversary can deliver specific cyber attacks on the smart grid infrastructure. In the following sections we describe this attacker model in detail.

9.4.6.1 Attack entry points

In general, strong perimeter defense is used to prevent external adversaries from accessing information or devices within the trusted grid zone. Unfortunately, the size and complexity of grid networks provide numerous potential entry points as follows:

- Inadvertent infiltration through infected devices: Malicious media or devices may be inadvertently infiltrated inside the trusted perimeter by personnel. For example, USB memory sticks have become a popular tool used to circumvent perimeter defenses: a few stray USB sticks left in public spaces are picked up by employees and plugged into previously secure devices inside the trusted perimeter, enabling malware on the USB sticks to immediately infect the devices. Similarly, devices used both inside and outside the trusted perimeter can get infected with malware when outside the system, and infiltrate that malware when used inside. Common examples are corporate laptops that are used at private homes over the weekend.
- Network-based intrusion: Perhaps the most common mechanism to penetrate a trusted perimeter is through a network-based attack vector. Exploiting poorly configured firewalls for both misconfigured inbound and faulty outbound rules is a common entry point, enabling an adversary to insert a malicious payload onto the control system.

Backdoors and holes in the network perimeter may be caused by components of the IT infrastructure with vulnerabilities or misconfigurations. Networking devices at the perimeter (e.g., fax machines, forgotten but still connected modems) can be manipulated to bypass proper access control mechanisms. In particular, dial-up access to remote terminal units (RTUs) is used for remote management, and an adversary can directly dial into modems attached to field equipment where many units do not require a password for authentication or have unchanged default passwords. Furthermore, adversaries can exploit vulnerabilities of the devices and install backdoors for future access to the prohibited area. Exploiting trusted peer utility links is another potential network-based entry point.

An attacker could wait for a legitimate user to connect to the trusted control system network via a virtual private network (VPN) and then hijack that VPN connection. The network-based intrusions described above are particularly dangerous because they enable a remote adversary to enter the trusted control-system network.

- Compromised supply chain: An attacker can preinstall malicious codes or backdoors into a device prior to shipment to a target location, called supply chain attacks. Consequently, the need for security assurance in the development and manufacturing process for sourced software, firmware, and equipment is critical to safeguarding the cyber supply chain involving technology vendors and developers.
- Malicious insider: An employee or legitimate user who is authorized to access system resources can perform actions that are difficult to detect and prevent. Privileged insiders also have intimate knowledge of the deployed defense mechanisms, which they can often easily circumvent. Trivial accessibility to smart grid components will increase the possibility of escalating an authorized access to a powerful attack.

9.4.6.2 *Adversary actions*

Once the adversaries gain access to the power control network, they can perform a wide range of attacks. Table 9.3 lists actions that an adversary can perform to violate the main security properties (confidentiality, integrity, availability) for the core types of information. We classify more specific cyber attacks that lead to either cyber or physical consequences.

TABLE 9.3 **Threat type classification as caused by attacking security properties.**

	Price information	Control command	Meter data	Software
Confidentiality	Leakage of price info.	Exposure of control structure	Unauthorized access to meter data	Theft of proprietary software
Integrity	Incorrect price info.	Change of control commands	Incorrect meter data	Malicious software
Availability	Unavailability of price info.	Inability to control grid	Unavailability of billing info.	N/A

Cyber consequences:

- Malware spreading and controlling devices: An adversary can develop malware and spread it to infect smart meters [402] or company servers. Malware can be used to replace or add any function to a device or a system such as sending sensitive information or controlling devices.
- Vulnerabilities in common protocols: Smart grid components will use existing protocols, inheriting the vulnerabilities on the protocols. Common protocols may include transmission control protocol (TCP), internet protocol (IP), and remote procedure call (RPC).
- Access through database links: Control systems record their activities onto a database on the control system network then mirror logs into the business network. A skilled attacker can gain access to the database on the business network, and the business network gives a path to the control system network. Modern database architectures allow this type of attack if they are improperly configured.
- Compromising communication equipment: An attacker can potentially reconfigure or compromise some of the communication equipment, such as multiplexers.
- Injecting false information on price and meter data: An adversary can send packets to inject false information on current or future prices, or send incorrect meter data to a utility company. Results of injecting false prices, such as negative pricing, will be power shortage or other significant damage on the target region. Results of sending incorrect data include reduced electric bills for economic damage due to the loss of revenue of a utility company. Fake information can also give huge financial impacts on electricity markets [397].
- Eavesdropping attacks: An adversary can obtain sensitive information by monitoring network traffic, which results in privacy breaches by stealing power usage, disclosure of the controlling structure of smart grids, and future price information. This eavesdropping can be used to gather information to perpetrate further crimes. For example, an attacker can gather and examine network traffic to deduce information from communication patterns, and even encrypted communication can be susceptible to traffic analysis attacks.
- Modbus security issues: A SCADA protocol of noteworthy concern is the Modbus protocol [403], which is widely used in industrial control applications such as in water, oil, and gas infrastructures. The Modbus protocol defines the message structure and communication rules used by process control systems to exchange SCADA information for operating and controlling industrial processes. Modbus is a simple client–server protocol that was originally designed for low-speed serial communication in process control networks. Given that the Modbus protocol was not designed for highly security-critical environments, several attacks are possible:
  1. Broadcast message spoofing involves sending fake broadcast messages to slave devices;

2. Baseline response replay involves recording genuine traffic between a master and a field device, and replaying some of the recorded messages back to the master;
3. Direct slave control involves locking out a master and controlling one or more field devices;
4. Modbus network scanning involves sending benign messages to all possible addresses on a Modbus network to obtain information about field devices;
5. Passive reconnaissance involves passively reading Modbus messages or network traffic;
6. Response delay involves delaying response messages so that the master receives out-of-date information from slave devices;
7. Rogue interlopers attack a computer with the appropriate (serial or Ethernet) adapters to an unprotected communication link.

Physical consequences:

- Interception of SCADA frames: An attacker can use a protocol analysis tool for sniffing network traffic to intercept SCADA Distributed Network Protocol 3.0 (DNP3) frames and collect unencrypted plaintext frames that would provide valuable information, such as source and destination addresses. This intercepted data, which include control and setting information, could then be used at a later date on another SCADA system or intelligent equipment device (IED), at worst shutting services down or at best causing service disruptions.
- Malware targeting industrial control systems: An attacker can successfully inject worms into vulnerable control systems and reprogram industrial control systems. A well-known example is Stuxnet, as discussed earlier.
- DoS/DDoS attacks on networks and servers: An adversary can launch a DoS/DDoS attack against various grid components including smart meters, networking devices, communication links, and utility business servers. If the attack is successful, then electricity cannot be controlled in the target region. Furthermore, power supply can be stopped from the result of the attack.
- Sending fake commands to smart meters in a region: An adversary can send fake commands to a device or a group of devices in a target region. For example, sending disconnect messages to smart meters in a region will stop power delivery to that region. In addition, invalid switching of electric devices can result in unsafe connections which may lead to setting the target place on fire. Thus, insecure communication in smart grids may be able to threaten human life.

The attacks mentioned above are not exhaustive, but they serve to illustrate risks to help develop secure grid systems. Additional examples of SCADA threats are available on the website of US-CERT.[1]

1. http://www.us-cert.gov/control_systems/csvuls.html.

9.4.7 Countermeasures

9.4.7.1 Key management

Key management is a fundamental approach for information security. Shared secret keys or authentic public keys can be used to achieve secrecy and authenticity for communication. Authenticity is especially important to verify the origin which in turn is important for access control.

The key setup in a system defines the root of trust. For example, a system based on public and private keys may define the public key of a trust center as the root of trust. The trust center's private key is used to sign certificates and to delegate trust to other public keys. In a symmetric-key system, each entity and the trust center would set up shared secret keys and establish additional trust relationships among other nodes by leveraging the trust center, as in Kerberos.

The challenge in this space is key management across a very broad and diverse infrastructure. As a recent NIST report documents [389], several dozen secure communication scenarios are required, ranging from communication between the power distributor and the smart meter for communication between equipment and field crews. For all these communication scenarios, keys need to be set up to ensure secrecy and authenticity. In addition to the tremendous diversity of equipment, there is also a wide variety of stakeholders: government, corporations, and consumers. Even secure email communication among different corporations is a challenge today; even so, the secure communication between equipment from one corporation and a field crew from another poses numerous additional challenges. By adding a variety of key management operations to the mix (e.g., key refresh, key revocation, key backup, key recovery), the complexity of key management becomes truly formidable. Moreover, business, policy, and legal aspects also need to be considered as a message signed by a private key can hold the key owner liable for the contents. A recent publication from NIST provides a good guideline for designing cryptographic key management systems to support an organization [404], but the diverse requirements of smart grid infrastructures are not considered.

9.4.8 Secure communication architecture

Designing a highly resilient communication architecture for a smart grid is critical to mitigating attacks while achieving high-level availability. Here are the required components:

- Network topology design: A network topology represents the connectivity structure among nodes, which can have an impact on the robustness against attacks [405]. Thus, connecting networking nodes to be highly resilient under attack can be the basis for building a secure communication architecture.
- Secure routing protocol: A routing protocol on a network is used to build logical connectivity among nodes, and the simplest way to prevent communication is by attacking the routing protocol. By compromising a single router

and by injecting bogus routes, all communications in the entire network can come to a standstill. Thus, we need to consider the security of a routing protocol running on top of a network topology.

- Secure forwarding: An adversary who controls a router can alter, drop, and delay existing data packets or inject new packets. Thus, securing individual routers and detecting malicious behavior will be required to achieve secure forwarding.
- End-to-end communication: From the end-to-end perspective, secrecy and authenticity of data are the most crucial properties. Secrecy prevents an eavesdropper from learning the data content, while authenticity (sometimes referred to as integrity) enables the receiver to verify that the data indeed originated from the sender, thus preventing an attacker from altering the data.

While numerous protocols exist (e.g., SSL/TLS, IPsec, SSH), some low-power devices may need lightweight protocols to perform the associated cryptography.

- Secure broadcasting: Many smart grid environments rely on broadcast communication. Especially for price dissemination, authenticity of the information is important because an adversary could inject a negative cost and cause electricity use to spike when numerous devices simultaneously turn on to take advantage of the low price.
- DoS defense: Given all the above mechanisms, an adversary can still prevent communication by mounting a DoS attack. For example, if an adversary controls many end points after compromising them, they can be used to send data to flood the network. Hence, enabling communication under these circumstances is crucial, for example to perform network management operations to defend against the attack. Moreover, electricity itself, rather than communication networks, can be a target of DoS attacks [406].
- Jamming defense: To prevent an external adversary from jamming the wireless network, jamming detection mechanisms can be used to detect attacks and raise alarms. A multitude of methods to counter jamming attacks have been developed [407], enabling operation during jamming.

9.4.9 System and device security

An important area is to address vulnerabilities that enable exploitation through software-based attacks, where an adversary either exploits a software vulnerability to inject malicious code into a system or where a malicious insider uses administrative privileges to install and execute malicious code. The challenge in such an environment is to obtain the "ground truth" when communicating with a potentially compromised system: Is the response sent by legitimate code or by malware? An illustration of this problem is when we attempt to run a virus scanner on a potentially compromised system. If the virus scanner returns the result that no virus is present, is that really because no virus could be identified or is it because the virus has disabled the virus scanner? A related problem is that current virus scanners contain an incomplete list of virus signatures, and the absence of a virus detection could occur because the virus scanner does not yet recognize the new virus.

In the context of smart grids, researchers have proposed several techniques to provide prevention and detection mechanisms against malware. McLaughlin et al. have proposed diversity for embedded firmware [408] to avoid an apocalyptic scenario where malware pervasively compromises equipment; each device executes different software, thus avoiding common vulnerabilities.

A promising new approach to provide remote code verification is a technology called attestation. Code attestation enables an external entity to inquire the software that is executing on a system in a way that prevents malware from hiding. Since attestation reveals a signature of executing code, even unknown malware will alter that signature and can thus be detected. In this direction, LeMay et al. have studied hardware-based approaches for attestation [409], [410]. Software-based attestation is an approach that does not rely on specialized hardware, but makes some assumptions that the verifier can uniquely communicate with the device under verification [411]. Shah et al. have demonstrated the feasibility of this concept on SCADA devices [412].

9.4.10 System-theoretic approaches

In this section we focus on system-theoretic approaches to the real-time security of smart grids, which encompasses two main parts: contingency analysis (CA) and system monitoring [413]. Fig. 9.13 shows a typical system-theoretic view of an IEEE 14-bus system. The focus of such a view is the physical interactions between the components in the grid, while the cyber view focuses on the modeling of IT infrastructures.

Suppose the grid consists of N buses. Let us define the active power flow, reactive power flow, the voltage magnitude, and phase angle for each bus as P_i, Q_i, V_i, and θ_i, respectively.[2] Let us define vectors P, Q, V, and θ as the collections of P_i, Q_i, V_i, and θ_i, respectively.

The relationship between node current I_k and voltage $V_k e^{j\theta_k}$ is given by the linear equation [202]

$$I_k = \sum_{i=1}^{N} Y_{ki} V_i e^{j\theta_i},$$

where Y_{ki} is the admittance between bus k and i. As a result the active and reactive power at node k are given by

$$P_k + jQ_k = V_k e^{j\theta_k} \times \overline{I_k} = V_k e^{j\theta_k} \sum_{i=1}^{N} \overline{Y_{ki} V_i e^{j\theta_i}}, \tag{9.21}$$

2. We assume that bus N is the reference bus and its phase angle is 0.

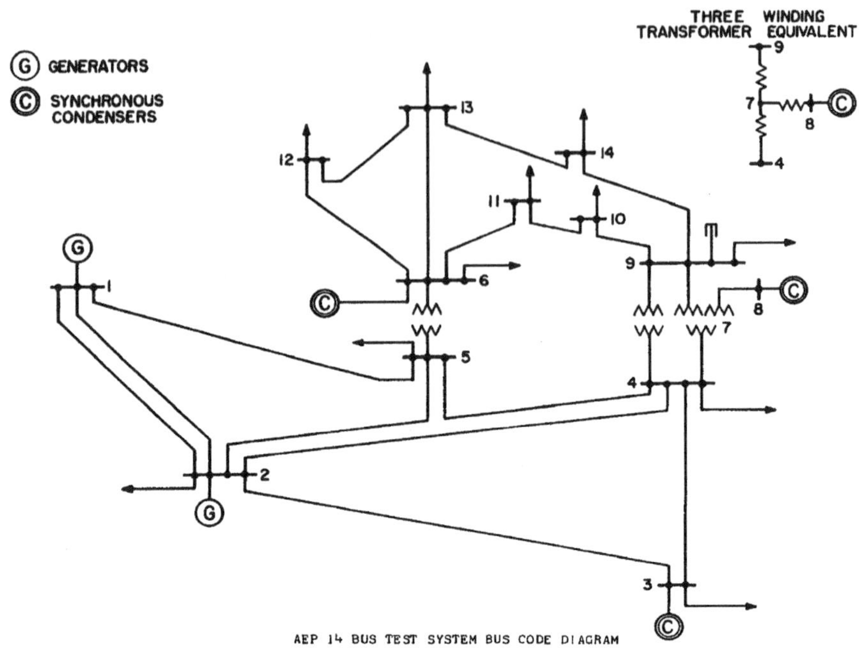

FIGURE 9.13 A typical system-theoretic view of an IEEE standard 14-bus system.

where $\overline{I_k}$ means complex conjugate. It can be seen that V and θ are the states of the system since they completely determine power flow P and Q. Let us define the state[3] x as $x = [V', \theta_1, \cdots, \theta_{N-1}]' \in \mathbb{R}^{2N-1}$. The remote terminal units (RTUs) provide the system's measurements. Let us denote as $z \in \mathbb{R}^m$ the collection of all measurements assumed to satisfy the following equation,

$$z = h(x) + v, \tag{9.22}$$

where $h : \mathbb{R}^{2N-1} \to \mathbb{R}^m$ represents the sensor model and $v \in \mathbb{R}^m$ denotes the measurement noise, which is further assumed to be Gaussian with mean 0 and covariance R.

Here we briefly introduce the WLS estimator [414], as it is widely used in practice. Define the estimated state as $\hat{x}$, and the residue vector as $r = z - h(\hat{x})$, which measures the inconsistency between state estimation $\hat{x}$ and measurements z. A WLS estimator tries to find the best estimation $\hat{x}$ with minimum inconsistency. In particular, the WLS estimator computes $\hat{x}$ based on the following minimization problem:

$$\hat{x} = \text{argmin}_{\hat{x}}\, r^T R^{-1} r. \tag{9.23}$$

3. The state does not include θ_N as its phase angle is assumed to be 0.

9.4.11 Security requirements

The US Department of Energy (DoE) Smart Grid System Report [204] summarizes six characteristics of the smart grid, which were developed from the seven characteristics of "Characteristics of the Modern Grid" [205] published by the National Energy Technology Laboratory (NETL). With respect to security, the most important characteristic identified by the DoE is to *operate resiliently even during disturbances, attacks, and natural disasters*.

In real-time security settings, the following properties are essential for the resilience of smart grids:

1. The power system should withstand a prespecified list of contingencies;
2. The accuracy of state estimation should degrade gracefully with respect to sensor failures or attacks.

The first property is passive and prevention based. The second property enables the detection of attacks or abnormalities and helps the system operator actively mitigate the damage.

9.4.12 Attack model

A contingency can usually be modeled as a change in vectors P, Q, V, θ (such as a loss of a generator) or as a change in the admittance Y_{ki} (such as an opening transmission line). For system monitoring, corrupted measurements can be modeled as an additional term in (9.24), i.e.,

$$z^a = z + u = h(x) + v + u, \tag{9.24}$$

where $u = [u_1, \cdots, u_m]' \in \mathbb{R}^m$ and $u_i \neq 0$ only if the sensor i is corrupted.

9.4.13 Countermeasures

Contingency analysis checks if the steady-state system is outside the operating region for each contingency [413]. However, the number of potential contingencies is high for large power grids. Due to real-time constraints, it is impossible to evaluate each contingency. As a result, in practice, usually only "$N-1$" contingencies are evaluated by considering single failure cases instead of multiple cases. Moreover, the list of possible contingencies is usually screened and ranked. After that a selected number of contingencies is evaluated. If a violation occurs, the system needs to determine the control actions that can mitigate or completely eliminate the violation.

9.4.14 Bad data detection

A bad data detector such as $\mathcal{X}^2$ or the largest normalized residue detector [414] identifies the corruption in measurement z by checking the residue vector r. For uncorrupted measurements it is expected that the residue vector r will be small

TABLE 9.4 Comparison between cybersecurity and system-theoretic security.

	Cybersecurity	System-theoretic security
System model	WAN/NAN/HAN model	Power flow model Sensor model
Requirements	Confidentiality Integrity Availability	Robust to prespecified contingency Accurate state estimation
Attack model	DoS attack Networked-based intrusion ...	Contingencies Sensor failures, false data injection
Countermeasures	Key management Secure communication System and device security	Contingency analysis Bad data detection

since z should be consistent with (9.22). However, this detection scheme has an inherent vulnerability as different z vectors can generate the same residue r. By exploiting this vulnerability, Liu et al. [281] show that an adversary can inject a stealthy input u into the measurements to change the state estimate $\hat{x}$ and fool the bad data detector at the same time. Sandberg et al. [415] consider how to find a sparse stealthy u, which enables the adversary to launch an attack with the minimum number of compromised sensors. To counter such a vulnerability, Kosut et al. [416] suggest using the prior knowledge of the state x to help detect malicious sensors.

9.4.15 The need for cyber-physical security

Table 9.4 summarizes the information from previous sections. The cybersecurity approaches focus on the IT infrastructures of the smart grid, while system-theoretic approaches focus more on the physical aspects. We argue that pure cyber or system-theoretic approaches are insufficient to guarantee the security of the smart grid for the following reasons:

1. *The system and attack models of both approaches are incomplete:* Cybersecurity does not model the physical system. Therefore, cybersecurity can hardly defend against physical attacks. For example, cybersecurity protects the integrity of measurement data by using secure devices and communication protocols.
   However, the integrity of sensors can be broken by modifying the physical state of the system locally, for example shunt connectors can be placed in parallel with a meter to bypass it and cause energy theft. In that case, no purely cybersecurity method can be employed to effectively detect and counter such attacks since the cyber portion of the system is not compromised. Thus, even the goals of cybersecurity cannot be achieved by pure

cyber approaches in CPSs. Moreover, cybersecurity is not well equipped to predict the effect of cyber attacks and countermeasures on the physical system.

For example, the DoS attacks can cause drops of measurement data and control command, which can lead to instability of the grid. A countermeasure to DoS attacks is to isolate some of the compromised nodes from the network, which may result in even more severe stability issues. Thus, an understanding of the physical system is crucial even for defending against cyber attacks. On the other hand, the system-theoretic model does not model the whole IT infrastructure, but usually just a high level abstraction. As a result of this oversimplification of the cyber world, it is difficult to analyze the effect of cyber attacks on physical systems. For example, in DoS attacks, some control commands may be dropped due to limited bandwidth. However, the effect of the lossy communication cannot be evaluated in a pure power flow model.

2. *The security requirements of the two approaches are incomplete and the security of the smart grid requires both of them:* System level concerns, such as stability, safety, and performance, have to be guaranteed in the event of cyber attacks. Cybersecurity metrics do not currently include the above-mentioned metrics. On the other hand, system theory is not concerned with secrecy of information. Furthermore, it usually treats integrity and availability of information as intermediate steps to achieve stability, safety, or better performance. In the design of a secure smart grid it is important to identify a set of metrics that combines and addresses the concerns of the two communities.
3. *The countermeasures of both approaches have drawbacks:* System-theoretic methods will not be able to detect any attack until it acts on the physical system. Furthermore, since system theory is based on approximate models and is subject to unknown disturbances, there will always be a discrepancy between the observed and the expected behavior. Most of the attacks can bypass system-theoretic intrusion detection algorithms with a small probability, which could be detrimental. Last, contingency analysis generally focuses on $N-1$ contingencies, which is usually enough for independent equipment failures. However, as we integrate the IT infrastructures into the smart grid, it is possible that several contingencies will happen simultaneously during an attack.

   On the other hand, cyber countermeasures alone are not sufficient to guarantee the security of the smart grid. History has so far taught us that cybersecurity is not always impenetrable. As operational continuity is essential, the system must be built to withstand and operate even in the event of zero-day vulnerabilities or insider threats, resorting to rapid reconfiguration to provide graceful degradation of performance in the face of an attack. As a large blackout can happen in a few minutes [208], it is questionable that pure cybersecurity approaches can react fast enough to withstand zero-day vulnerability exploits or insider attacks.

As demonstrated earlier, both cyber and system-theoretic approaches are essential for the security of smart grids. In this section we use two examples to show how the combination of cyber and system-theoretic approaches together can provide a better security level than traditional methods. In the first example, we show how system-theoretic countermeasures can be used to defend against a replay attack, which is a cyber attack on the integrity of the measurement data. In the second example, we show how system theory can guide cybersecurity investment strategies.

9.4.16 Defense against replay attacks

In this example we consider the defense against a replay attack, where an adversary records a sequence of sensor measurements and replays the sequence afterwords. Replay attacks are cyber attacks that break the integrity, or more precisely the freshness, of measurement data. It is worth mentioning that *Stuxnet* [417] employed a replay attack of this type to cover its goal of damaging the centrifuges in a nuclear facility by inducing excessive vibrations or distortions. While acting on the physical system, the malware was reporting old measurements indicating normal operations. This integrity attack, clearly conceived and operated in the cyber realm, exploited four zero-day vulnerabilities to break the cyber infrastructures and it remained undiscovered for several months after its release. Therefore, a pure cyber approach to preventing replay attacks may not be able to react fast enough before the system is damaged.

Next we develop the concept of physical authentication, a methodology that can detect such attacks independently of the type of attack used to gain access to the control system. This algorithm [142] was developed long before Stuxnet appeared. We give a summary below.

To achieve greater generality, the method is presented for a generic control system. We assume the sensors are monitoring a system with the following state dynamics,

$$x_{k+1} = Fx_k + Bu_k + w_k, \tag{9.25}$$

where $x_k \in R^n$ is the vector of state variables at time k, $w_k \in R^n$ is the process noise at time k, and x_0 is the initial state. We assume w_k, x_0 are independent Gaussian random variables, $x0 \sim \mathcal{N}(\bar{x}_0, \Sigma)$, $w_k \sim \mathcal{N}(0, Q)$.

For each sampling period k, the true measurement equation of the sensors can be written as

$$z_k = Hx_k + v_k, \tag{9.26}$$

where $z_k \in R^m$ is a collection of all the measurements from the sensors at time k and $v_k \sim \mathcal{N}(0, R)$ is the measurement noise independent of x_0 and w_k.

We assume that an attacker records a sequence of measurements from time T_0 to time $T_0 + T - 1$ and replays it from time $T_0 + T$ to time $T_0 + 2T - 1$, where $T_0 \geq 0$, $T \geq 1$. As a result, the corrupted measurements z_k^a received by the system operator are

$$z_k^a = \begin{cases} z_k, & 0 \leq k \leq T_0 + T - 1 \\ z_{k-T}, & T_0 + T \leq k \leq T_0 + 2T - 1. \end{cases} \tag{9.27}$$

Our goal is to design an estimator, a controller, and a detector such that

1. the system is stable when there is no replay attack;
2. the detector can detect the replay attack with a high probability.

We propose the following design of a fixed gain estimator, a fixed gain controller with random disturbance and a $\mathcal{X}^2$ detector. In particular, our estimator takes the form

$$\hat{x}_{k+1} = F\hat{x}_k + Bu_k + Kr_{k+1}, \qquad \hat{x}_0 = \bar{x}_0, \tag{9.28}$$

where K is the observation gain matrix and the residue r_k is computed as

$$r_{k+1} = z_{k+1}^a - C(F\hat{x}_k + Bu_k). \tag{9.29}$$

Our controller takes the form

$$u_k = L\hat{x}_k + \Delta u_k, \tag{9.30}$$

where L is the control gain matrix and Δu_ks are independent and identically distributed (i.i.d.) Gaussian noise generated by the controller, with zero mean and covariance $\mathcal{Q}$.

It can be easily shown that the residue r_k is a Gaussian random variable with zero mean when there is no replay. As a result, with high probability it cannot be far away from 0. Therefore, we design our filter to trigger an alarm at time k based on the following event,

$$\left\{g_k = r_k' \mathcal{P} r_k \geq \text{ threshold}\right\}, \tag{9.31}$$

where $\mathcal{P}$ is a predefined weight matrix. Fig. 9.14 shows the diagram of the proposed system.

We first consider the stability of the proposed system. It is well known that without Δu_k the closed-loop system without replay is stable if and only if both $F - KCF$ and $F + BL$ are stable. Moreover, we can easily prove that adding Δu_k does not affect the stability of the system since Δu_k is i.i.d. Gaussian distributed. Hence, to ensure that the system is closed-loop stable without replay, we only need to make $F - KCF$ and $F + BL$ stable, which can be easily done as long as the system is both detectable and stabilizable.

Now we want to show our system design can successfully detect replay attacks. Consider the residue r_k, where $T_0 + T \leq k \leq T_0 + 2T - 1$; we can then

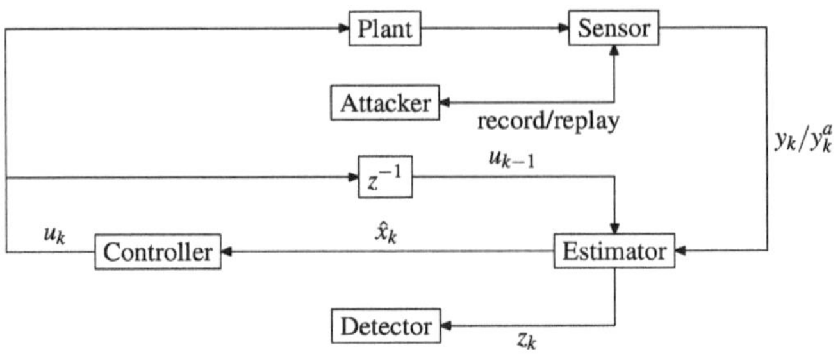

FIGURE 9.14 System diagram.

prove that

$$\begin{aligned} r_k &= r_{k-T} + C\mathcal{A}^{k-T_0-T}(I - KC)\left(\hat{x}_{T_0} - \hat{x}_{T_0+T}\right) \\ &+ \sum_{i=0}^{k-T-T_0-1} C\mathcal{A}^i B(\Delta u_{k-T-1-i} - \Delta u_{k-1-i}), \end{aligned}$$

where $\mathcal{A} = (F + BL)(I - KC)$. The second term above converges to 0 exponentially quickly if $\mathcal{A}$ is stable. As a result, if we do not introduce any random control disturbance (i.e., $\Delta u_k = 0$), then the third term vanishes and the residue r_k under replay attack converges to the residue r_{k-T} when no replay attack is presented. Therefore, the detection rate of the replay attack will be the same as the false alarm rate. In other words, the detector cannot distinguish between healthy and corrupted measurements. However, if $\Delta u_k \neq 0$, then the third term will always be presented and therefore the detector can detect replay attacks with a probability higher than the false alarm rate.

It is worth mentioning that the role of $\Delta u_k = 0$ is similar to an authentication signal on the measurements. When the system is under normal operation, it is expected that the measurement z_k will reflect the random disturbances $\Delta u_k = 0$. On the other hand, when the replay begins, z_k and $\Delta u_k = 0$ become independent of each other. Therefore, the integrity and freshness of the measurements can be protected by checking the correlation between z_k and $\Delta u_k = 0$. This technique is cyber-physical as it uses the physics of the system to authenticate data coming from the cyber portion.

We now provide a numerical example to illustrate the performance of our detection algorithm. We impose the following parameters: $F = B = \mathcal{Q} = R = \mathcal{P} = 1$, $K = 0.9161$, $L = -0.618$. It is possible to verify that $\mathcal{A} = 0.0321 < 1$. The threshold of the filter is chosen such that the false alarm rate is 1%. We assume that the recording starts at time 1 and replay starts at time 11. Fig. 9.15 shows different detection rates over time as $\mathcal{Q}$ increases. It can be seen that the detection fails when there is no disturbance. Moreover, a larger disturbance can increase the performance of the detector.

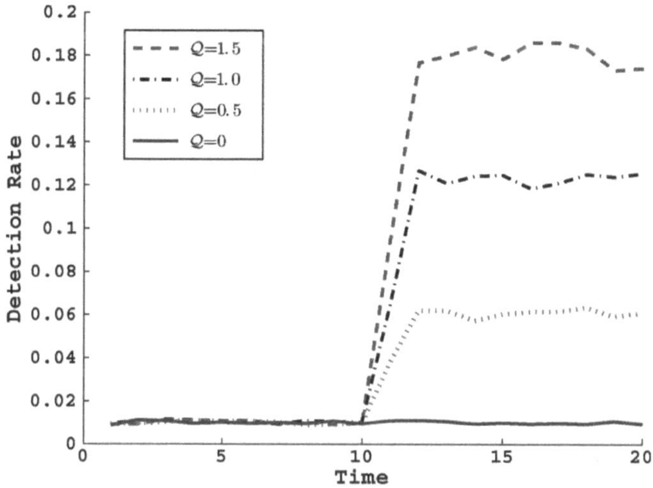

FIGURE 9.15 Detection rate over time.

9.4.17 Cybersecurity investment

In this example, we show how system theory can be used to expose the critical assets to protect and thus provide important insights towards the allocation of security investments. In particular, we consider how to deploy secure sensors to help detect corrupted measurements. We assume the true measurements of the sensors follow a linearized model of (9.22)

$$z = Hx + v, \tag{9.32}$$

where $z \in \mathbb{R}^m$ and $x \in \mathbb{R}^{2N-1}$, and $H \in \mathbb{R}^{m\times(2N-1)}$ is assumed to be of full column rank. For linearized models, (9.23) can be solved analytically as

$$\hat{x}(z) = (H'R^{-1}H)^{-1}H'R^{-1}z = Kz. \tag{9.33}$$

Therefore, the residue can be calculated explicitly as

$$r(z) = z - H\hat{x}(z) = (I - HK)z = Sz, \tag{9.34}$$

where $S = I - HK$.

Suppose that an attacker is able to modify the readings of a subset of sensors. As a result, the corrupted measurements take the form

$$z^a = z + u = Hx + v + u, \tag{9.35}$$

where $u = [u_1, \cdots, u_m]' \in \mathbb{R}^m$ indicates the error introduced by the attacker, and $u_i \neq 0$ only if sensor i is compromised.

An attack is called stealthy if the residue r does not change during the attack. In mathematical terms, a stealthy attack u satisfies $r(z) = r(z + u)$. Since $r(z)$ is linear with respect to z, we can simplify the above equation to

$$r(u) = Su = 0 \tag{9.36}$$

without loss of generality.

As shown by Liu et al. [281], the $\mathcal{X}^2$ detectors fail to detect a stealthy input u. Any detector based on r is ineffective against stealthy attacks as they do not change the residue r. On the other hand, a stealthy attack can introduce estimation error to $\hat{x}$.

To defend against these attacks we deploy secure devices, such as tamper resistant devices, to protect the sensors. To this end, we define a sensor i to be secure if it cannot be compromised; in other words, the corresponding u_i is guaranteed to be 0. Let us also define the set of secure sensors to be $S_e \subseteq \{1, \cdots, m\}$. An attack u is feasible if and only if $u_i = 0$ for all $i \in S_e$.

Our security goal is to deploy the minimum number of secure sensors such that the system can detect the compromised nodes. In other words, we want to find the smallest set S_e such that there is no nonzero feasible and stealthy u.

This problem is of practical importance in the smart grid as the current insecure sensors can only be replaced gradually by secure sensors due to the scale of the grids. As a result, it is crucial to know which set of sensors to replace first to achieve better security.

Let us define $\Gamma(S_e) = \text{diag}(\gamma_1, \cdots, \gamma_m)$, where $\gamma_i = 1$ if and only if $i \in S_e$. A set S_e is called observable if and only if $\Gamma(S_e)H$ is of full column rank. In other words, if a vector $p \in \mathbb{R}^{2N-1} \neq 0$, then $\Gamma(S_e)Hp \neq 0$. The following theorem relates the observability of secure sensor set S_e with the existence of a feasible and stealthy attack u.

Theorem 9.1. *The only feasible and stealthy attack is $u = 0$ if and only if S_e is observable.*

Proof. First suppose that S_e is observable and u is stealthy and feasible. As a result, $\Gamma(S_e)u = 0$. On the other hand, since u is stealthy, $S_e = 0$, which implies that

$$HKu = (I - S)u = u.$$

Therefore,

$$\Gamma(S_e)HKu = \Gamma(S_e)u = 0.$$

Since $\Gamma(S_e)H$ is full column rank, we know that $Ku = 0$, which implies that $HKu = 0$. Thus,

$$u = (I - HK + HK)u = Su + HKu = 0.$$

On the other hand, suppose that S_e is not observable. Find $x \neq 0$ such that $\Gamma(S_e)Hx = 0$. Choose $u = Hx$. Since H is full column rank, $u \neq 0$. Moreover, $\Gamma(S_e)u = \Gamma(S_e)Hx = 0$. Hence, u is feasible. Finally

$$\begin{aligned} Su &= (I - HK)u = u - H(H'R^{-1}H)^{-1}H'R^{-1}u \\ &= Hx - H(H'R^{-1}H)^{-1}H'R^{-1}Hx = 0, \end{aligned}$$

which implies that u is stealthy. □

Therefore, finding the smallest S_e such that there is no nonzero feasible, and stealthy u is equivalent to finding the smallest observable S_e, which can be achieved using the following theorem:

Theorem 9.2. *If S_e is observable and rank$(\Gamma(S_e)) > 2N - 1$, then there exists an observable S_e', which is a proper subset of S_e.*

Proof. Let $H' = [H_1, \cdots, H_m]$, where $H_i \in \mathbb{R}^{2N-1}$. Since S_e is observable, rank$(\gamma_1 H_1, \cdots, \gamma_m H_m) = 2N - 1$. Without loss of generality, let us assume that $S_e = \{1, \cdots, l\}$. Thus, $\gamma_1 = \cdots = \gamma_l = 1$ and $\gamma_{l+1} = \cdots = \gamma_m = 0$, where $l > 2N - 1$. Since $H_i \in \mathbb{R}^{2N-1}$, $H_1, \cdots, H_l$ are not linearly independent. Hence, there exist $\alpha_1, \cdots, \alpha_l \in \mathbb{R}$ that are not all zero such that $\alpha_1 H_1 + \cdots + \alpha_l H_l = 0$. Without loss of generality, let us assume that $\alpha_l \neq 0$. Therefore,

$$\text{span}(H_1, \ldots, H_{l-1}) = \text{span}(H_1, \ldots, H_l) = \mathbb{R}^{2N-1},$$

which implies that $S_e' = \{1, \ldots, l-1\}$ is observable. □

It is easy to see that rank$(\Gamma(S_e))$ must be no less than $2N - 1$ to make S_e observable. As a result, we can use the procedure described in the proof of Theorem 9.2 to find the smallest observable set. Analyses of this kind are essential to prioritize security investments.

Remark 9.1. It is worth noticing that the attacks we discussed in this section are cyber attacks that have physical consequences. The replay attack itself can render the system unstable if the original system is open-loop unstable or it can enable future attacks on the physical system, as in the case of Stuxnet. The stealthy integrity attack can cause large estimation errors and potentially damage the system.

Furthermore, our approaches to security are hybrid in nature. In the first example, we used system-theoretic models and countermeasures to detect replay attacks, which is a cyber attack. Our detection algorithm complements the pure cybersecurity approaches and provides an additional layer of protection. In the second example, we used a system-theoretic model of the grid to develop an optimal cybersecurity countermeasure to integrity attacks. The results illustrate that combining cybersecurity and system theory can provide a better level of security for the smart grid.

9.5 Notes

With the proliferation of remote management and control of CPSs, security plays a critically important role because the convenience of remote management can be exploited by adversaries for nefarious purposes from the comfort of their own homes.

Compared to current cyber infrastructures, the physical component of cyber-physical infrastructures adds significant complexity that greatly complicates security. On the one hand, the increased complexity will require more effort from the adversary to understand the system, but on the other hand, this increased complexity also introduces numerous opportunities for exploitation. From the perspective of the defender, more complex systems require dramatically more effort to analyze and defend because of the state-space explosion when considering combinations of events.

Current approaches to secure cyber infrastructures are certainly applicable to securing CPSs: techniques for key management, secure communication (offering secrecy, authenticity, and availability), secure code execution, intrusion detection systems, etc. Unfortunately, these approaches are largely unaware of the physical aspects of CPSs.

System-theoretic approaches already consider physical aspects in more detail than the traditional security and cryptographic approaches. These approaches model malicious behaviors as either component failures, external inputs, or noise; analyze their effects on the system; and design detection algorithms or counter measures to the attacks. The strength of model-based approaches lies in a unified framework to model, analyze, detect, and counter various kinds of cyber and physical attacks. However, the physical world is modeled with approximations and is subject to noise, which can result in a deviation of any model to the reality. Therefore, system-theoretic approaches are nondeterministic compared to information security.

As discussed in this chapter, CPS security demands additional security requirements, such as continuity of power delivery and accuracy of dynamic pricing introduced by the physical system. These requirements are usually closely related to the models and states of the system, which are difficult to address by information security alone. Therefore, information-based security and system-theory-based security are both essential to securing CPSs, offering exciting research challenges for many years to come.

Chapter 10

Resilient design under cyber attacks

Contents

10.1 Introduction

In cyber-physical systems (CPSs) the strong dependence on the cyber infrastructure in the overall system increases the risk of cyber attacks such as injection attacks. The security of CPSs needs to be considered and designed carefully [379]. Cyber-physical security has now developed into a comprehensive discipline that extends classic fault detection, and complements cybersecurity [47]. The aim in this area is to build a resilient mechanism to make the system aware of any attacks in progress and to adjust its dynamics if needed to guarantee the desired performance even with attacks or failures.

Much interesting research has been done in this field and we can provide only a brief summary. Numerous papers have emphasized modeling and detecting adversarial attacks [105,418–420]. In particular, Pasqualetti et al. [421] characterize the detectability of an attack using the system output. Fawzi et al. [422] give the threshold value of the number of channels that when attacked can be successfully corrected. Zhang et al. [147] provide a stability condition for a system under a Bernoulli Denial-of-Service (DoS) attack. Li et al. [115] formulate a game-theory framework and provide optimal strategies for the attacker and the plant using Markov chain theory for the case when the channel between sensor and estimator is jammed. Mo and Sinopoli [72] analyze the effect of replay attacks and the trade-off between linear quadratic Gaussian (LQG) performance and the accuracy of a failure detector.

How to guarantee system performance even in the presence of an attack still remains an open challenge. Specifically, most of the existing work focuses on

Cloud Control Systems. https://doi.org/10.1016/B978-0-12-818701-2.00018-4

detecting an ongoing attack under the assumption that once detected the attacker will be removed from the system. However, often the plant must be run even in the presence of an attack by altering the controller suitably if needed. This would require a joint design of an attack detector and a controller that switches to maintain suitable performance both in the absence of and in the presence of an attack. Such an architecture has not yet received sufficient attention in the literature. While some recent work [423–425] has considered this architecture in the context of fault detection, these works typically do not consider security threats and impose restrictions on the form and evolution of the disturbance signals. Thus, these architectures are more useful for situations when assumptions attached on attack signals about linearity or stochastic dynamics stand (see [421], [426], [105], [427]). A design that simultaneously guarantees system security and control performance has not been explored much. Dissipativity theory in general, and passivity in particular, provide a fundamental perspective for the design and analysis of dynamical systems based on generalized energy concept. It should be noted that passivity implies stability under weak assumptions.

Consider a continuous system with dynamics given by

$$\begin{aligned} \dot{x} &= f(x, u), \\ y &= h(x, u), \end{aligned} \tag{10.1}$$

where $x \in \mathcal{X} \subseteq \mathbb{R}^n$, $u \in \mathcal{U} \subseteq \mathbb{R}^m$, and $y \in \mathcal{Y} \subseteq \mathbb{R}^m$ are the system state, and input and output spaces, respectively; here f and h are smooth mappings of appropriate dimensions.

Definition 10.1. [428] A state-space system (10.1) is said to be dissipative with respect to supply rate $\omega(u(t), y(t))$ if there exists a nonnegative storage function $V(x) : \mathcal{X} \to \mathbb{R}_{\geq 0}$ satisfying $V(0) = 0$, such that for all $x_0 \in \mathcal{X}$, all $t_1 \geq t_0$, and $u \in \mathcal{U}$,

$$V(x(t_1)) \leq V(x(t_0)) + \int_{t_0}^{t_1} \omega(u(t), y(t))dt, \tag{10.2}$$

where $x(t_0) = x_0$ and $x(t_1) = \phi(t_1, t_0, x_0, u)$.

Definition 10.2. [428], [429]. Suppose system (10.1) is dissipative. It is called:

1. *passive,* if (10.2) holds for

$$\omega(u, y) = u^T y; \tag{10.3}$$

2. *QSR-dissipative*, if (10.2) holds and there exist matrices $Q = Q^T$, S, $R = R^T$, such that

$$\omega(u, y) = u^T Ru + 2y^T Su + y^T Qy; \tag{10.4}$$

3. $\mathcal{L}_2$ stable with finite gain $\gamma > 0$, if the system is dissipative with supply rate given by (10.4), where $R = \gamma^2 I$, $S = 0$, $Q = -I$, such that with $\beta \leq 0$,

$$y^T y \geq -\beta + \gamma^2 u^T u. \tag{10.5}$$

10.2 Problem statement

In this chapter we consider the detection of a data injection attack and operation of the plant even after an attack has been detected by switching the controller. Through an appropriate passivation approach, local passivity and exponential stability are guaranteed even under attack. Because passivity is compositional, this provides a preliminary setup for possible passivity-based control of large-scale interconnected networked systems. The overall system framework is shown in Fig. 10.1.

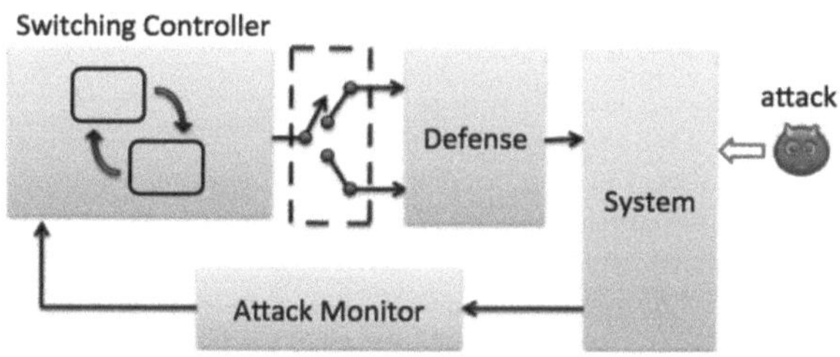

FIGURE 10.1 System framework.

We consider a linear model (10.6) for the plant and we assume that the attacker knows the parameters $\{A, B, C, D\}$ for the system under attack (see (10.6)). The intelligent attacker intends to corrupt the system state and the measurements based on this knowledge. To this end, the attacker injects data through the external control inputs. Conversely, the controller seeks to monitor the measured output to identify if an attacker is present. If an attack is detected, the controller switches to another configuration to maintain performance in spite of the attack.

One possible strategy for designing a controller that guarantees a desired level of performance in spite of the presence of an attack is to design using a nonswitching H_∞ controller. However, this procedure may lead to a design that is too conservative when an attack is only rarely present. Instead, we design a controller using a passivity framework that ensures that the passivity levels of the closed-loop system are guaranteed even when an attack is present. For this, we use the input–output transformation M-matrix as introduced in [430] that does not require knowledge of the passivity levels of either the plant or the controller.

10.2.1 System model

Consider a system with the dynamics given as follows,

$$\begin{aligned}\dot{x}(t) &= Ax(t) + Bu(t) + w(t),\\ y(t) &= Cx(t) + Du(t),\end{aligned} \tag{10.6}$$

where $x(t) \in \mathbb{R}^n$, $u(t) \in \mathbb{R}^m$, $y(t) \in \mathbb{R}^p$ are the state, and system input and output, respectively; and $w(t)$ is the unknown external control input that the attacker may possibly inject into the system. This term is set to 0 when the system is in normal operation. We refer to the signal $w(t)$ as the attack signal. We assume that the system input signal is smooth. With a switching controller the evolution of the system is described in Fig. 10.2.

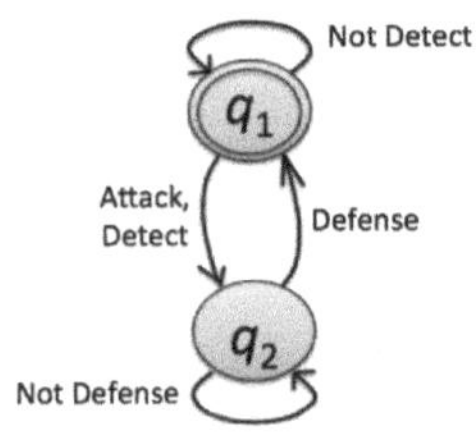

FIGURE 10.2 A hybrid automaton framework. Event *attack* is triggered if the attacker injects an attack signal into the system. Event *detect* is triggered if the attack monitor successfully detects the existence of attack signal. Event *defense* is triggered if the system becomes passive by switching the controller.

10.2.2 Attack monitor

Lemma 10.1. *(Following [421].) An attack is undetectable by any attack monitor if there exist initial conditions x_1, x_2 and an attack signal w such that, for all $t \geq 0$, the input injected by this attack generates zero-dynamics on the plant as*

$$y(x_1, w, t) = y(x_2, 0, t). \tag{10.7}$$

Remark 10.1. In view of the fundamental limitation of an attack monitor as illustrated in [421], we limit our consideration to detectable attacks. We use a modified Luenberger observer for attack monitoring.

Assume that the unknown external control input that the attacker injects into the system is expressed as

$$w(t) = \hat{w}(t) + \Delta w, \tag{10.8}$$

where $\hat{w}(t)$ is the estimate of $w(t)$ and Δw is the corresponding estimation error. Let L denote the observer gain matrix. The classic Luenberger type disturbance observer can be written as

$$\dot{\hat{w}}(t) = -L\hat{w}(t) + L(\dot{x} - Ax - Bu), \tag{10.9}$$

from which we can get the equalities

$$\dot{\hat{w}}(t) = L\Delta w \tag{10.10}$$

and

$$\Delta\dot{w} = \dot{w}(t) - L\Delta w. \tag{10.11}$$

Since a nonlinear dependency on x is not realizable in practice (see [431]), we need to modify the design for the monitor as follows:

Theorem 10.1. *Consider system (10.6) and assume that the attacks are detectable. The attack detection filter can be obtained as*

$$\begin{aligned} \dot{\epsilon} &= -l(x)\epsilon + l(x)(-Ax - Bu - \rho(x)), \\ \hat{w} &= \epsilon + \rho(x), \\ \frac{d}{dt}\rho(x) &= l(x)\dot{x}, \end{aligned} \tag{10.12}$$

where ϵ is the internal state variable of the monitor, $\rho(x)$ is a nonlinear function to be designed, and $l(x)$ is the gain of the modified detection filter that can possibly depend on x. The output of the detection filter is the residual signal

$$v(t) = \hat{w}(t). \tag{10.13}$$

Proof. We assume the internal state variable

$$\epsilon = \hat{w} - \rho(x), \tag{10.14}$$

where $\rho(x)$ can be determined by the modified detection filter gain $l(x)$ as

$$\frac{d}{dt}\rho(x) = l(x)\dot{x}.$$

Then according to (10.6), (10.8), (10.10), and (10.14), we have

$$\begin{aligned} \dot{\epsilon} &= \dot{\hat{w}} - \dot{\rho}(x) \\ &= l(x)\Delta w - l(x)\dot{x} \\ &= -l(x)[\epsilon + \rho(x)] + l(x)[Ax + Bu + \rho(x)] - l(x)\dot{x} \\ &= -l(x)\epsilon + l(x)(-Ax - Bu - \rho(x)). \end{aligned}$$

Thus, the modified attack detection filter (10.12) can be achieved. □

The error dynamics is given by

$$\begin{aligned}\Delta\dot{w} &= \dot{w}(t) - \dot{\hat{w}}(t) = \dot{w} - \dot{\epsilon} - \frac{d\rho(x)}{dt} \\ &= l(x)[\hat{w} - \rho(x)] + l(x)[Ax + Bu + \rho(x)] + \dot{w} - l(x)\dot{x} \\ &= \dot{w}(t) - l(x)[w(t) - \hat{w}(t)].\end{aligned}$$

Thus, we obtain that

$$\Delta\dot{w} = \dot{w}(t) - l(x)\Delta w. \tag{10.15}$$

Remark 10.2. Note that the modified attack monitor does not need the derivative term $\dot{x}$, yet it has similar error dynamics as the basic disturbance observer error dynamics. This attack monitor not only detects the existence of an attack, but can also track its trajectory. We can also use the detected signal as an estimated value of the unexpected input.

The following theorem gives us the criteria to design the filter gain $l(x)$.

Theorem 10.2. *If there exists an invertible matrix X and a positive definite matrix Γ such that with the gain $l(x)$,*

$$l(x)^T X^T X + X^T X l(x) \geq \Gamma, \tag{10.16}$$

and further the derivative of $w(t)$ is negligible compared with Δw as in estimation error dynamics (10.8) (i.e., $\dot{w}(t) \approx 0$), then the designed disturbance observer is exponential stable.

Proof. Consider the candidate Lyapunov function as

$$W(\Delta w, x) = (X\Delta w)^T(X\Delta w) = \Delta w^T X^T X \Delta w. \tag{10.17}$$

We can see that the scalar function W is positive definite. With the Eqs. (10.15), (10.16), and (10.17), when $\dot{w}(t) \approx 0$ we get

$$\begin{aligned}\dot{W}(\Delta w, x) &= \Delta\dot{w}^T X^T X \Delta w + \Delta w^T X^T X \Delta\dot{w} \\ &= (\dot{w} - l(x)\Delta w)^T X^T X \Delta w \\ &\quad + \Delta w^T X^T X(\dot{w} - l(x)\Delta w) \\ &= -\Delta w^T (l(x)^T X^T X - X^T X l(x))\Delta w.\end{aligned}$$

Since the condition in Theorem 10.2 is satisfied, $\dot{W}(\Delta w, x)$ is negative definite for all Δw and $lim_{t\to\infty}\Delta w = 0$ for all $\forall\Delta w \in \mathbb{R}^n$. Moreover, from (10.15) the disturbance tracking error will converge exponentially to zero for all $\forall\Delta w \in \mathbb{R}^n$. It implies that $\hat{w}$ will exponentially approach w if the detection gain $l(x)$ is chosen with (10.16) regardless of x. □

10.2.3 Switching the controller

Once an attack is detected, the controller switches to a new structure that ensures that the system continues to be stable and performs well. For this purpose, we use the framework of transformation M-matrix that is shown in Fig. 10.3. The parameters m_{11}, m_{12}, m_{21}, and m_{22} are chosen such that closed-loop system guarantees the desired passivity level even when no a priori knowledge of passivity indices of the system and controller is available.

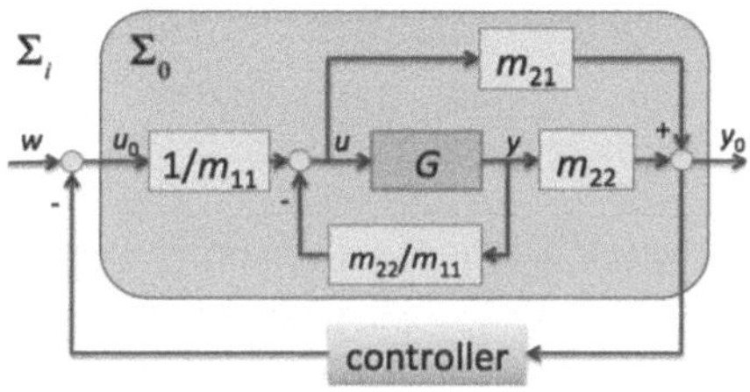

FIGURE 10.3 System framework with transformation M-matrix.

We proceed as follows. Consider the unforced system [432]

$$\begin{aligned} \dot{x}(t) &= \Sigma_{i=1}^{2}\sigma_i A_i x(t) + \Sigma_{i=1}^{2}\sigma_x w(t), \\ y(t) &= \Sigma_{i=1}^{2}\sigma_i C_i x(t), \end{aligned} \tag{10.18}$$

where $\sigma_i(t) = \begin{cases} 1, & \text{when } \Sigma_i \text{ is active} \\ 0, & \text{otherwise} \end{cases}$. We define the indicator function $\sigma(t) = [\sigma_1(t) \quad \sigma_2(t)]^T$ with $\sigma_1(t) + \sigma_2(t) = 1$.

The optimal control goal is to guarantee the passivity of the closed-loop switching system. If the passivation transformation M is chosen appropriately, such that the hybrid automaton is passive, we can say that event *defense* is triggered.

A quadratic cost function is defined as

$$L_r(u, w) = \int |y(t)|^2 dt - \gamma^2 \int |w(t)|^2 dt. \tag{10.19}$$

An optimal control policy [433] guarantees the desired system performance

$$\frac{\|y\|}{\|w\|} \leq \gamma, \tag{10.20}$$

where $\|w(t)\|^2 = \int |w(t)|^2 dt$, $\|y(t)\|^2 = \int |y(t)|^2 dt$. The following theorem illustrates the condition that automaton mode Σ_i is stable when that mode is active:

Theorem 10.3. *For fixed $\gamma \geq 0$, the system has stable performance (10.20) under attack, if there exists $\beta_{ij} \geq 0$ and symmetric a positive definite matrix P_i, $i, j \in \overline{M} = \{1, 2\}$ such that*

$$\begin{bmatrix} A^T P_i + P_i A + \gamma^{-1} C^T C + \Sigma_{i=1}^2 \beta_{ij}(P_j - P_i) & P_i \\ P_i & -\gamma I \end{bmatrix} < 0. \tag{10.21}$$

Proof. Choose a transition law between Σ_1 and Σ_2 mode as

$$\sigma(t) = \arg\min_{i \in \overline{\mathcal{M}}} \theta(t) P_i \theta(t), \tag{10.22}$$

where $\theta(t)^T = [x(t)^T \quad w(t)^T]$. Choose a candidate Lyapunov function for the hybrid system as

$$V(t, x(t)) = x(t)^T (\Sigma_{i=1}^2 \sigma_i(t) P_i) x(t). \tag{10.23}$$

Depending on whether the transition between modes happens or not, we consider two cases:

1. When $\sigma(t + \Delta t) = \sigma(t) = i$, (i.e., there is no transition between two modes). Based on (10.18) and (10.23),

$$\begin{aligned} \dot{V} &= \dot{x}^T (\Sigma_{i=1}^2 \sigma_i P_i) x(t) + x(t)^T (\Sigma_{i=1}^2 \sigma_i P_i) \dot{x} \\ &= [\Sigma A x + \Sigma w]^T (\Sigma \sigma P_i) x + x^T (\Sigma \sigma P_i) [\Sigma A x + \Sigma w] \\ &= \theta(t)^T \begin{bmatrix} A^T P_i + P_i A & P_i \\ P_i & 0 \end{bmatrix} \theta(t). \end{aligned}$$

2. When $\sigma(t + \Delta t) = \sigma(\tilde{t}) = j$, $\sigma(t) = i \neq j$ (i.e., the transition between modes occurs), using (10.22),

$$\begin{aligned} \dot{V}(t, x) &= \lim_{\Delta t \to 0} \frac{x(\tilde{t})^T P_j x(\tilde{t}) - x(t)^T P_i x(t)}{\Delta t} \\ &\leq \lim_{\Delta t \to 0} \frac{x(\tilde{t})^T P_i x(\tilde{t}) - x(t)^T P_i x(t)}{\Delta t} \\ &= \begin{bmatrix} A^T P_i + P_i A & P_i \\ P_i & 0 \end{bmatrix} \theta. \end{aligned}$$

Define $L(y(t), w(t)) = \gamma w(t)^T w(t) - \gamma^{-1} y(t)^T y(t)$, based on (10.18),

$$\begin{aligned} & \dot{V}(t, x) - L(y, w) \\ \leq\ & \theta^T \begin{bmatrix} A^T P_i + P_i A & P_i \\ P_i & 0 \end{bmatrix} \theta - \gamma w^T w + \gamma^{-1} y^T y \end{aligned}$$

$$= \theta^T \begin{bmatrix} A^T P_i + P_i A + \gamma^{-1} C^T C + \Sigma\beta(P_j - P_i) & P_i \\ P_i & -\gamma I \end{bmatrix} \theta.$$

If (10.21) is satisfied, $\dot{V}(t,x) - L(y,w) < \theta^T(\Sigma\sigma_i(P_i - P_j))\theta$. Based on (10.22), we get

$$\dot{V}(t,x) < L(y,w), \text{ for } \forall t. \tag{10.24}$$

1. When $w = 0$, $\dot{V}(t,x) < L(y,w) = -\gamma^{-1,} y(t)^T y(t)$, the system is stable.
2. Under zero initial condition, we have $\int_0^\infty \dot{V}(t,x)dt < \int_0^\infty w(t)dt$. Since $\int_0^\infty \dot{V} dt = V(\infty)$, this implies that

$$\int_0^\infty |y(t)|^2 dt < \gamma^2 \int_0^\infty |w(t)|^2 dt, \tag{10.25}$$

and optimal performance γ is satisfied. □

Under condition (10.21), Σ_i is finite-gain stable. The objective is to obtain the criteria for how to choose the passivation parameters in Fig. 13.4 such that the closed-loop hybrid system is passive, without any prior knowledge of the passivity levels for plant G and the preset controller H. The following theorem discusses how to select the M-matrix in order to render active automaton mode Σ_i be passive.

Theorem 10.4. *Consider system Σ_i with (10.21) satisfied. Assume the predesigned controller H is passive. If we select a passivation transformation M-matrix with*

$$\begin{aligned} \frac{m_{11}m_{12}}{\gamma_1^2} + m_{12}m_{22} &\geq 0, \\ m_{12}m_{21} + m_{11}m_{22} &= 0, \end{aligned} \tag{10.26}$$

then the interconnected feedback system Σ_i under attack is passive.

Proof. Now $u_0 = w - y_2$, $u_1 = \frac{1}{m_{11}}u_0 - \frac{m_{12}}{m_{11}}y_1$, $y_0 = u_2 = m_{21}u_1 + m_{22}y_1$. We have

$$w = u_0 + y_2 = m_{11}u_1 + m_{12}y_1 + y_2. \tag{10.27}$$

Since system G is finite-gain stable under (10.21), let its finite gain be $\gamma_1 > 0$. We have supply rate

$$\omega(u_1, y_1) = \gamma_1^2 u_1^T u_1 - y_1^T y_1 \geq 0. \tag{10.28}$$

When the interconnected system is passive, we have

$$\omega(w, y_0) = w^T y_0$$

$$
\begin{aligned}
&= m_{11}m_{21}u_1^T u_1 + m_{12}m_{22}y_1^T y_1 \\
&\quad +(m_{12}m_{21} + m_{11}m_{22})u_1^T y_1 \\
&\quad +m_{21}y_2^T u_1 + m_{22}y_2^T y_1 \\
&= \begin{bmatrix} u_1^T & y_1^T \end{bmatrix} \begin{bmatrix} m_{11}m_{21} & m_{11}m_{22} \\ m_{12}m_{21} & m_{12}m_{22} \end{bmatrix} \begin{bmatrix} u_1 \\ y_1 \end{bmatrix} \\
&+ \; y_2^T u_2 \geq 0.
\end{aligned}
$$

The interconnected system is passive if (10.26) is satisfied with the assumption that the predesigned controller H is passive. □

Theorem 10.5. *Consider the automaton framework in Fig. 10.2. If we select a passivation transformation M-matrix with criteria (10.26), passivity of the overall hybrid automaton is guaranteed under condition (10.21).*

Proof. If we select a transformation M-matrix with (10.26), each active mode Σ_i is passive under condition (10.21). Similarly, consider inactive mode Σ_j, $j \neq i$, we have $u_0 = m_{11}u_1 + m_{12}y_1$, $y_0 = m_{21}u_1 + m_{22}y_1$, and $w = u_0 + y_2$.

Arbitrarily choose an M-matrix with $m_{11} \neq 0$. There exists a supply rate

$$
\omega(w, y_0) = y_0^T y_0 = u_2^T u_2 \geq 0. \tag{10.29}
$$

Each mode Σ_i is dissipative when it is inactive. Moreover, since the accumulated energy flows from active subsystem i to inactive subsystem j at each switching instant t_{i_k}, $\forall i, k$ is finite without external supply, which means that

$$
V_0(x(t_0)) + \sum_{i_k=1}^{\infty}[V_{i_k}(x(t_{i_k})) - V_{i_{k-1}}(x(t_{i_k}))] \leq \infty, \tag{10.30}
$$

and according to Definition 3.3 in [434] the hybrid automaton is passive. □

10.2.4 Simulation results I

Example 10.1. Consider a cyber-physical vehicle as in Fig. 10.4.

Let the dynamics of the vehicle dynamic system be [435].

$$
\begin{bmatrix} \ddot{\phi} \\ \ddot{\delta} \\ \dot{\phi} \\ \dot{\delta} \end{bmatrix} = \begin{bmatrix} -2.11 & -6.61 & 9.48 & -357.05 \\ 73.54 & -61.70 & 11.71 & -757.81 \\ 1 & 0 & 0 & 0 \\ 0 & 1 & 0 & 0 \end{bmatrix} \begin{bmatrix} \dot{\phi} \\ \dot{\delta} \\ \phi \\ \delta \end{bmatrix} + \begin{bmatrix} T_\phi & T_\delta & 0 & 0 \end{bmatrix}^T u + \begin{bmatrix} 0 & 8 & 0 & 0 \end{bmatrix}^T w.
$$

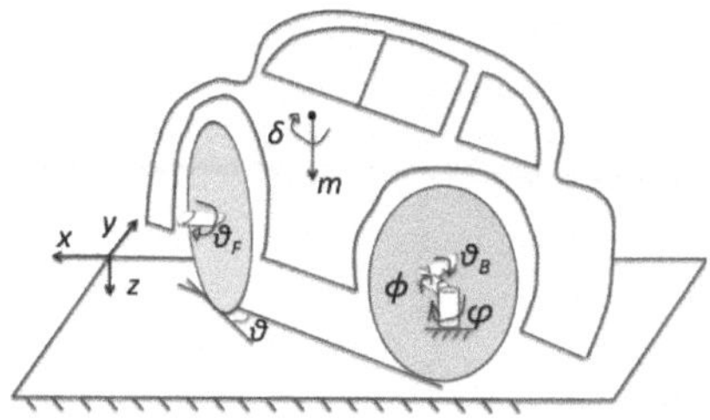

FIGURE 10.4 A vehicle dynamic system.

$$y(t) = \begin{bmatrix} 20 & 8 & 0 & 1 \end{bmatrix} \begin{bmatrix} \dot{\phi} & \dot{\delta} & \phi & \delta \end{bmatrix}^T. \tag{10.31}$$

The lean rotation ϕ is the angular rotation about the x-axis; the steering angle δ is the rotation of the front tires with respect to rear tires about the steering axis; T_ϕ represents the right lean torque; and T_δ is an action-reaction steering torque. Here we set $T_\phi = 1.2$ and $T_\delta = 10$. w is the unknown external input that is injected by an attacker. Assume the forward speed $V = 20\ \mathrm{m\,s^{-1}}$.

In this example, first we choose the detector gain independently of x, based on Theorem 2.1, as $l(x) = X^{-1} = 2$. The attack detector can be designed as

$$\begin{aligned}
\dot{\epsilon} &= -2\epsilon + \begin{bmatrix} 0.22 & 13.22 & -18.96 & 714.1 \\ -147.07 & 119.39 & -23.42 & 1515.62 \\ -2 & 0 & -4 & 0 \\ 0 & -2 & 0 & -4 \end{bmatrix} \begin{bmatrix} \dot{\phi} \\ \dot{\delta} \\ \phi \\ \delta \end{bmatrix} \\
&\quad - \begin{bmatrix} -2.4 \\ -20 \\ 0 \\ 0 \end{bmatrix} u, \\
\hat{w} &= \epsilon + \rho(x), \\
\frac{d}{dt}\rho(x) &= l(x)\dot{x} = 2\dot{x}.
\end{aligned} \tag{10.32}$$

We implement the designed attack monitor. The attacker generates an aperiodic rectangular signal with sample time of 5 seconds. The simulation result for the designed attack monitor is shown in Fig. 10.5.

The first part of Fig. 10.5 represents the dynamic response of the system under irregular pulse injected by the attacker. The blue dashed line represents the dynamic of angular lean velocity $\dot{\phi}$ and the red solid line represents the response of angular steering velocity $\dot{\phi}$ for the initial condition $(\dot{\phi}, \dot{\delta}) = (8\ \text{rad/s}, 10\ \text{rad/s})$.

In the second part of Fig. 10.5, the blue dashed line represents the real signal the attacker injected to the system and the red solid line represents the output of

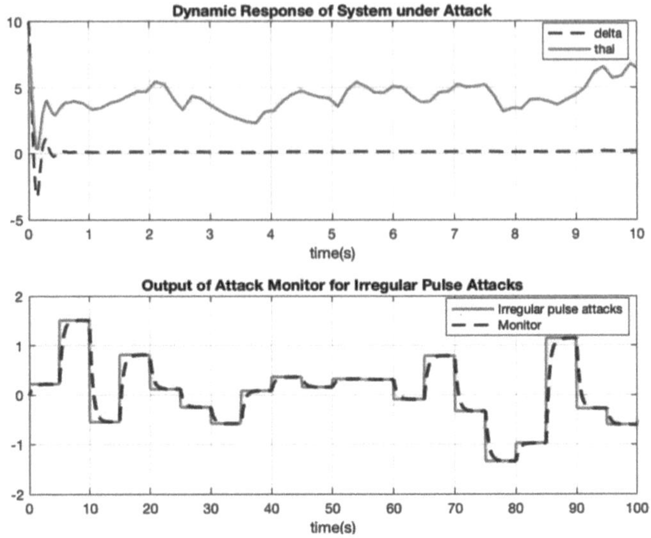

FIGURE 10.5 Dynamic response of a system under attack.

the attack monitor. We can see that the monitor design tracks the unknown input well.

Furthermore, by choosing $\beta_{21} = 2$, condition (10.21) is satisfied. We get $\gamma = 1.008$; we select the transformation M-matrix as $m_{11} = 1$, $m_{12} = 1.2$, $m_{21} = -1.2$, and $m_{22} = 6$ such that condition (10.26) is satisfied. We set the stable state feedback controller with feedback gain as

$$K = \begin{bmatrix} 0.1728 & 0.0074 & -0.0008 & 2.1023 \end{bmatrix}.$$

The defense is realized by M-matrix passivation and the dynamic response of the system after correction is shown in the first part of Fig. 10.6. The second part of Fig. 10.6 represents the supply rate of the system after the defense mechanism is triggered.

The first part of Fig. 10.6 depicts the dynamic response of $\dot{\phi}$ after correction as the blue dashed line, and the response of $\dot{\delta}$ after the defense mechanism is triggered as the red solid line. As we can see, the velocities become smoother after correction. At the mean time, the angular lean velocity becomes more gentle such that the driving process is robust in spite of the irregular force applied to the vehicle. The discomfort of driver and passengers in a vehicle that is under a cyber attack has been reduced. The second part of Fig. 10.6 depicts the supply rate of the system after correction. We define the supply rate as the inner product of the injected input and output of the system. The output supply rate of the system shown here is always positive after the settling time. This demonstrates that the system under attack is passive with the designed passivation M-matrix.

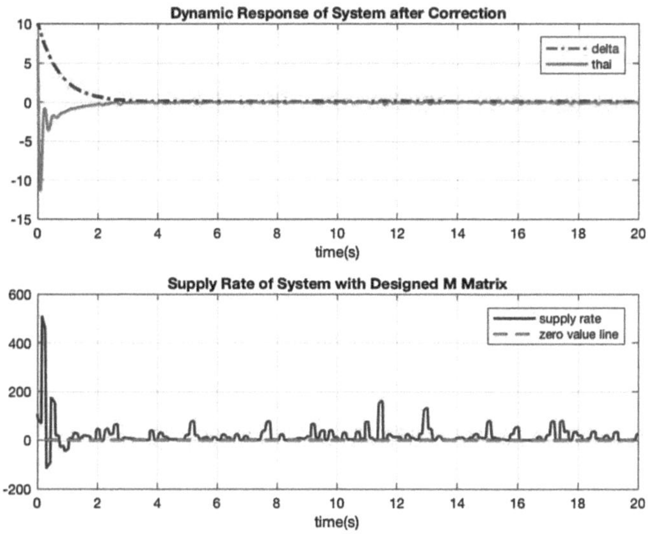

FIGURE 10.6 Dynamic response and supply rate after correction.

10.3 Secure control subject to stochastic attacks

Cyber-physical systems are the integration of communication, computation, and control for achieving the desired performance of physical systems. With its wide range of applications, such as sustainable and blackout-free electricity generation and distribution, CPSs have attracted the interest of researchers [316]. Other applications for CPSs include clean and energy-aware buildings and cities, smart medical and healthcare systems, transportation networks, chemical process control, smart grids, water and gas distribution networks, emergency management [38].

On the other hand, security issues increase the challenges in control of CPSs because CPSs have a high possibility of being affected by several cyber attacks without providing notification of failure. These attacks can lead to a disruption of the physical system; for example, the disarrangement of coordination packets in medium-access control layers could be a result of malware injected by an adversary. Moreover, in order to destroy the normal operation an attacker can illegally obtain access to the supervision centers while obtaining the encryption key. That means, system dynamics can be disturbed arbitrarily by an attacker, and when there is no security protection in hardware or software strategies the attacker has the ability to induce perturbations [58].

The communication among the components of control systems (i.e., sensors, actuators, and controllers) occurs through a common network medium. This network needs to be secured to prohibit vulnerability of attack by adversaries during data transmission. These attacks could lead the system to instability or drive the plant to undesired operations, as mentioned before. Thus, considering security issues is very important when designing controllers for such a system.

From a control security viewpoint, cyber attacks can be classified as two main types: 1) DoS attacks, which are strategies that are often used to occupy the communication resources and to prohibit transmitting measurement or control signals, and 2) deception attacks, also called false data injection (FDI) attacks, are the modification of data integrity for the transmitted packets among some cyber parts of the CPS.

Control of CPSs under cyber attack is one of the main issues in control engineering and it has attracted a great deal of research. Most of the literature only considers one kind of attack, such as [99], [100], [102], [103], and [104] for DoS attacks and [122], [136], and [121] for deception attacks. Some of the literature considers two kinds of attacks, randomly occurring DoS and deception attacks were both considered when designing an event-based security control system [98]. In [91] the optimal control problem was investigated for a class of networked control systems (NCSs) subject to DoS, deception, and physical attacks using a delta operator approach and by applying ϵ-Nash equilibrium. A resilient linear quadratic Gaussian control strategy for NCSs affected by zero dynamic attacks was designed [137]. Dynamic programming was applied for the control strategy and a power transmission strategy was designed using value iteration methods for a class of CPSs subject to DoS attacks [317]. It should be noted that the literature considered the attack to be random variable with a constant value, which does not fully represent the practical situations where the attack could be designed and implemented with variable conditional probabilities.

10.3.1 Problem formulation and preliminaries

A CPS composed of plant, observer-based controller, and communication network is considered here, and is shown in Fig. 7.5. The system can be affected by cyber attacks in the forward (plant to observer) or backward (observer to plant) communication represented by (A_1) and (A_2), respectively, in Fig. 10.7. The

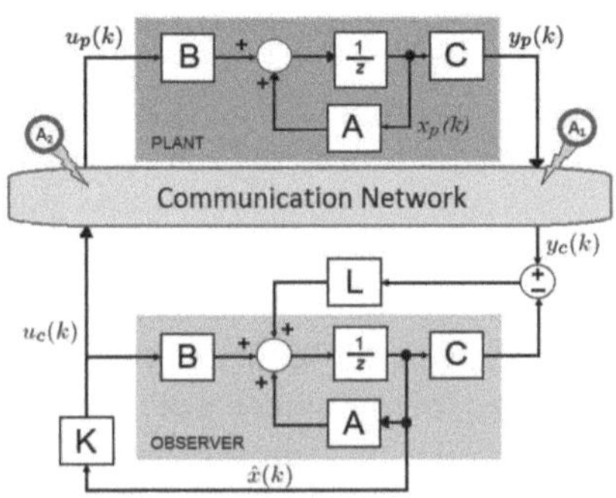

FIGURE 10.7 Attacks on a cyber-physical system (CPS).

discrete-time linear time-invariant model of the plant is

$$x(k+1) = Ax(k) + Bu_p(k), \quad y_p(k) = Cx(k), \tag{10.33}$$

where $x(k) \in \Re^n$ is the plant's state vector, $u_p(k) \in \Re^m$ is the control input, and $y_p(k) \in \Re^p$ is the output vector. A, B, and C are the known matrices of the plant of appropriate dimensions. The measurement signal after passing the network is described by

$$\begin{aligned} y_c(k) &= \alpha(k)[y_p(k) + \beta(k)(-y_p(k) + \zeta_y(k))] \\ &\quad +(1-\alpha(k))y_p(k-\tau_k^f), \end{aligned} \tag{10.34}$$

where τ_k^f stands for *forward delay* with a Bernoulli distribution caused by DoS attack in the forward path; $\alpha(k)$ and $\beta(k)$ are Bernoulli distributed white sequences exhibiting the occurrence of forward DoS and deception attacks, respectively; and $\zeta_y(k)$ is the signal that affects the system in the forward deception attack.

The observer-based controller below is implemented while considering an attack that occurs on the forward path:

$$\begin{aligned} &Observer: \\ &\hat{x}(k+1) = A\hat{x}(k) + Bu_c(k) + L(y_c(k) - \hat{y}_c(k)) \\ &\hat{y}_c(k) = C\hat{x}(k) \end{aligned} \tag{10.35}$$

$$\begin{aligned} &Controller: \\ &u_c(k) = K\hat{x}(k) \\ &u_p(k) = \gamma(k)[u_c(k) + \delta(k)(-u_c(k) + \zeta_u(k))] \\ &\quad +(1-\gamma(k))u_c(k-\tau_k^b). \end{aligned} \tag{10.36}$$

Here $\hat{x}(k) \in \Re^n$ is the estimate of the states (10.35), $\hat{y}_c(k) \in \Re^p$ is the observer output, $L \in \Re^{n\times p}$ is the observer gain, $K \in \Re^{m\times n}$ is the controller gain, and τ_k^b is the *backward delay* caused by the backward DoS attack.

The stochastic variables $\gamma(k)$ and $\delta(k)$, mutually independent of $\alpha(k)$ and $\beta(k)$, are also Bernoulli distributed white sequences exhibiting the occurrence of DoS and deception backward attacks, respectively; $\zeta_u(k)$ is the signal that affects the system in the backward deception attack.

Here we assume τ_k^b and τ_k^f to be bounded time-varying variables as follows:

$$\tau_f^- \le \tau_k^f \le \tau_f^+, \quad \tau_b^- \le \tau_k^b \le \tau_b^+. \tag{10.37}$$

Fig. 10.8 shows the different types of attacks affecting the system on forward and backward paths, and the probability associated with each case. Here it is assumed that only one type of attack will happen at a time (i.e., either DoS attack ($j = 1, \cdots, 3$) or deception attack ($j = 4, \cdots, 7$) will take place).

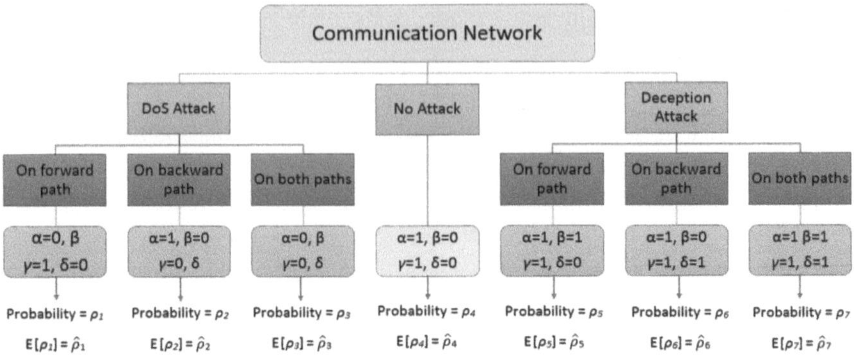

FIGURE 10.8 Types of attack.

The estimation error is defined as $e(k) = x(k) - \hat{x}(k)$. Thus, we can obtain

$$\begin{aligned} x(k+1) &= Ax(k) + \gamma(k)(1-\delta(k))BKx(k) \\ &\quad + \gamma(k)\delta(k)B\zeta_u(k) + (1-\gamma(k))BKx(k-\tau_k^b) \\ &\quad - \gamma(k)(1-\delta(k))BKe(k) \\ &\quad - (1-\gamma(k))BKe(k-\tau_k^b) \end{aligned} \tag{10.38}$$

$$\begin{aligned} e(k+1) &= (-BK+LC)x(k) + (A+BK-LC)e(k) \\ &\quad + \gamma(k)(1-\delta(k))BKx(k) \\ &\quad + \gamma(k)\delta(k)B\zeta_u(k) + BK(1-\gamma(k))x(k-\tau_k^b) \\ &\quad - \gamma(k)(1-\delta(k))BKe(k) \\ &\quad - BK(1-\gamma(k))e(k-\tau_k^b) \\ &\quad - \alpha(k)(1-\beta(k))LCx(k) - \alpha(k)\beta(k)L\zeta_y(k) \\ &\quad - (1-\alpha(k))LCx(k-\tau_k^f). \end{aligned} \tag{10.39}$$

By defining $\xi(k) = [x^T(k) \quad e^T(k)]^T$, system (10.38) and (10.39) can be formulated as

$$\xi(k+1) = \bar{A}_j\xi(k) + \bar{B}_j\xi(k-\tau_k^f) + \bar{C}_j\xi(k-\tau_k^b) + \bar{D}_j\zeta(k), \tag{10.40}$$

where $\zeta(k) = [\zeta_u^T(k) \quad \zeta_y^T(k)]^T$, $\{\bar{A}_j,\ \bar{B}_j,\ \bar{C}_j,\ \bar{D}_j,\ j = 1, \cdots, 7\}$, and j is an index representing each situation in Fig. (10.8) and their values as follows:

$$\bar{A}_1 = \begin{bmatrix} A+BK & -BK \\ LC & A-LC \end{bmatrix},$$

$$\bar{A}_2 = \begin{bmatrix} A & 0 \\ -BK & A+BK-LC \end{bmatrix},$$

$$
\begin{aligned}
\bar{A}_3 &= \begin{bmatrix} A & 0 \\ [4pt] -BK+LC & A+BK-LC \end{bmatrix}, \\
\bar{A}_4 &= \begin{bmatrix} A+BK & -BK \\ 0 & A-LC \end{bmatrix}, \\
\bar{A}_5 &= \begin{bmatrix} A+BK & -BK \\ LC & A-LC \end{bmatrix}, \\
\bar{A}_6 &= \begin{bmatrix} A & 0 \\ -BK & A+BK-LC \end{bmatrix}, \\
\bar{A}_7 &= \begin{bmatrix} A & 0 \\ -BK+LC & A+BK-LC \end{bmatrix}, \\
\bar{B}_1 &= \bar{B}_3 = \begin{bmatrix} 0 & 0 \\ -LC & 0 \end{bmatrix}, \\
\bar{B}_j &= \mathbf{0}, \qquad j=2,4,\cdots,7, \\
\bar{C}_2 &= \bar{C}_3 = \begin{bmatrix} BK & -BK \\ BK & -BK \end{bmatrix}, \\
\bar{C}_j &= \mathbf{0}, \qquad j=1,4,\cdots,7, \\
\bar{D}_j &= \mathbf{0}, \qquad j=1,\cdots,4, \\
\bar{D}_5 &= \begin{bmatrix} 0 & 0 \\ 0 & -L \end{bmatrix}, \ \bar{D}_6 = \begin{bmatrix} B & 0 \\ B & 0 \end{bmatrix}, \\
\bar{D}_7 &= \begin{bmatrix} B & 0 \\ B & -L \end{bmatrix}.
\end{aligned} \tag{10.41}
$$

Remark 10.3. The following should be noted from (10.41):

1. When there is no attack or when there is a DoS attack, then

$$
\bar{A}_j + \bar{B}_j + \bar{C}_j = \begin{bmatrix} A+BK & -BK \\ 0 & A-LC \end{bmatrix}, \quad j=1,\cdots,4. \tag{10.42}
$$

   The result of $\bar{A}_j + \bar{B}_j + \bar{C}_j$, $j=1,\cdots,4$ represents the fundamental matrix of system (10.40).
2. Part of the fundamental matrix in (10.42) is changed due to a signal injected by the attacker in the case of the deception attack.

Remark 10.4. A deception attack is considered an arbitrary bounded energy signal with the following characteristic:

$$\zeta^T \zeta < \eta^2. \tag{10.43}$$

The objective in this chapter is to build an observer-based controller, as formulated in (10.35) and (10.36), to guarantee the exponential stability in the mean square of the closed-loop system (10.40). Our method is inspired by switched time-delay systems [376]. To simplify the expressions, we define each probability as ρ_j and its expected value as $\mathbb{E}[\rho_j]$ for $j = 1, \cdots, 7$ as shown in Fig. 10.8.

10.3.2 Design results

The stability analysis and controller synthesis problems for the closed-loop system (10.40) is investigated in this section. We discuss the stability analysis problem to obtain a sufficient condition to guarantee the exponential stability in the mean square of system (10.40) with the given observer-based controller (10.36) and (10.37). By expansion of the work in [376], the main theorem will be established using the candidate Lyapunov function

$$V(\xi(k)) = \sum_{i=1}^{5} V_i(\xi(k)), \tag{10.44}$$

where

$$\begin{aligned}
V_1(\xi(k)) &= \sum_{j=1}^{7} \xi^T(k) P \xi(k), \; P > 0 \\
V_2(\xi(k)) &= \sum_{j=1}^{7} \sum_{i=k-\tau_k^f}^{k-1} \xi^T(i) Q_j \xi(i), \; Q_j = Q_j^T > 0 \\
V_3(\xi(k)) &= \sum_{j=1}^{7} \sum_{i=k-\tau_k^b}^{k-1} \xi^T(i) Q_j \xi(i) \\
V_4(\xi(k)) &= \sum_{j=1}^{7} \sum_{\ell=-\tau_f^+ +2}^{-\tau_f^- +1} \sum_{i=k+\ell-1}^{k-1} \xi^T(i) Q_j \xi(i) \\
V_5(\xi(k)) &= \sum_{j=1}^{7} \sum_{\ell=-\tau_b^+ +2}^{-\tau_b^- +1} \sum_{i=k+\ell-1}^{k-1} \xi^T(i) Q_j \xi(i).
\end{aligned} \tag{10.45}$$

Let us define real scalars $\upsilon > 0$ and $\varrho > 0$ with the following characteristics:

$$\upsilon \|\xi\|^2 \le V(\xi(k)) \le \varrho \|\xi(k)\|^2. \tag{10.46}$$

Theorem 10.6. *For a given controller and observer gains K and L, system (10.40) is exponentially stable if there exist matrices $P > 0$, $Q_j^T = Q_j > 0$, $j = 1, \cdots, 7$ and matrices F_i, U_i, and Z_i, $i = 1, 2$, satisfying the following LMIs:*

$$\Upsilon_j = \begin{bmatrix} \Upsilon_{1j} & \Upsilon_{2j} & \Upsilon_{3j} \\ \bullet & \Upsilon_{4j} & \mathbf{0} \\ \bullet & \bullet & \Upsilon_{5j} \end{bmatrix} < 0, \tag{10.47}$$

$$\begin{aligned}
\Upsilon_{1j} &= \begin{bmatrix} \Psi_j + \Phi_{j1} & -F_1 + U_1^T & -F_2 + U_2^T \\ \bullet & -U_1 - U_1^T - \hat{\rho}_j Q_j & 0 \\ \bullet & \bullet & -U_2 - U_2^T - \hat{\rho}_j Q_j \end{bmatrix}, \\
\Upsilon_{2j} &= \begin{bmatrix} -F_1 + Z_1^T - \Phi_{j2} & -F_2 + Z_2^T - \Phi_{j3} \\ -U_1 - Z_1^T & 0 \\ 0 & -U_2 - Z_2^T \end{bmatrix}, \\
\Upsilon_{3j} &= \begin{bmatrix} \Phi_{j7} \\ 0 \\ 0 \end{bmatrix}, \\
\Upsilon_{4j} &= \begin{bmatrix} -Z_1 - Z_1^T + \Phi_{j4} & \Phi_{j5} \\ \bullet & -Z_2 - Z_2^T + \Phi_{j6} \end{bmatrix}, \\
\Upsilon_{5j} &= \Phi_{j8} - \eta I,
\end{aligned} \tag{10.48}$$

where

$$\begin{aligned}
\Psi_j &= -P + \hat{\rho}_j(\tau_f^+ - \tau_f^- + \tau_b^+ - \tau_b^- + 2)Q_j \\
&\quad + F_1 + F_1^T + F_2 + F_2^T, \\
\Phi_{j1} &= (\bar{A}_j + \bar{B}_j + \bar{C}_j)^T \hat{\rho}_j P (\bar{A}_j + \bar{B}_j + \bar{C}_j), \\
\Phi_{j2} &= (\bar{A}_j + \bar{B}_j + \bar{C}_j)^T \hat{\rho}_j P \bar{B}_j, \\
\Phi_{j3} &= (\bar{A}_j + \bar{B}_j + \bar{C}_j)^T \hat{\rho}_j P \bar{C}_j, \\
\Phi_{j4} &= \bar{B}_j^T \hat{\rho}_j P \bar{B}_j, \\
\Phi_{j5} &= \bar{B}_j^T P \bar{C}_j, \\
\Phi_{j6} &= \bar{C}_j^T \hat{\rho}_j P \bar{C}_j, \\
\Phi_{j7} &= (\bar{A}_j + \bar{B}_j + \bar{C}_j)^T \hat{\rho}_j P \bar{D}_j, \\
\Phi_{j8} &= \bar{D}^T \hat{\rho}_j P \bar{D}_j.
\end{aligned}$$

Proof. Let $y(k) = x(k+1) - x(k)$. Thus,

$$\xi(k-\tau_k^f) = \xi(k) - \sum_{i=k-\tau_k^f}^{k-1} y(i), \tag{10.49}$$

$$\xi(k-\tau_k^b) = \xi(k) - \sum_{i=k-\tau_k^b}^{k-1} y(i), \tag{10.50}$$

and system (10.40) can be represented as

$$\xi(k+1) = (\bar{A}_j + \bar{B}_j + \bar{C}_j)\xi(k) - \bar{B}_j\lambda(k) - \bar{C}_j\epsilon(k) + \bar{D}_j\zeta(k), \tag{10.51}$$

where

$$\lambda(k) = \sum_{i=k-\tau_k^f}^{k-1} y(i), \quad \epsilon(k) = \sum_{i=k-\tau_k^b}^{k-1} y(i).$$

By evaluation of the difference of $V_1(\xi(k))$ along the solution of system (10.51), this leads to

$$\begin{aligned}
\mathbb{E}[\Delta V_1(\xi(k))] &= \mathbb{E}[V_1(\xi(k+1))] - V_1(\xi(k)) \\
&= \sum_{j=1}^{7}\Big[\xi^T(k)[\Phi_{j1} - P]\xi(k) \\
&\quad -2\xi^T(k)\Phi_{j2}\lambda(k) - 2\xi^T(k)\Phi_{j3}\epsilon(k) \\
&\quad +\lambda^T(k)\Phi_{j4}\lambda(k) + 2\lambda^T(k)\Phi_{j5}\epsilon(k) \\
&\quad +\epsilon^T(k)\Phi_{j6}\epsilon(k) + 2\xi^T(k)\Phi_{j7}\zeta(k) \\
&\quad +\zeta^T(k)(\Phi_{j8} - \eta I)\zeta(k)\Big].
\end{aligned} \tag{10.52}$$

A straightforward computation gives

$$\begin{aligned}
\mathbb{E}[\Delta V_2(\xi(k)] &= \sum_{j=1}^{7}\hat{\rho}_j\Big[\sum_{i=k+1-\tau_{k+1}^f}^{k} \xi^T(i)Q_j\xi(i) - \sum_{i=k-\tau_k^f}^{k-1} \xi^T(i)Q_j\xi(i)\Big] \\
&= \xi^T(k)Q\xi(k) - \xi(k-\tau_k^f)Q_j\xi(k-\tau_k^f) \\
&\quad + \sum_{i=k+1-\tau_{k+1}^f}^{k-1} \xi^T(i)Q_j\xi(i) - \sum_{i=k+1-\tau_k^f}^{k-1} \xi(i)Q_j\xi(i).
\end{aligned} \tag{10.53}$$

In view of

$$\begin{aligned}
&\sum_{i=k+1-\tau_{k+1}^{f}}^{k-1} \xi^T(i)Q_j\xi(i) \\
=& \sum_{i=k+1-\tau_{k+1}^{f}}^{k-\tau_k^f} \xi^T(i)Q_j\xi(i) + \sum_{i=k+1-\tau_k^f}^{k-1} \xi^T(i)Q_j\xi(i) \\
\leq& \sum_{i=k+1-\tau_k^f}^{k-1} \xi^T(i)Q_j\xi(i) + \sum_{i=k+1-\tau_f^+}^{k-\tau_f^-} \xi^T(i)Q_j\xi(i), \qquad (10.54)
\end{aligned}$$

we readily obtain

$$\begin{aligned}
&\mathbb{E}[\Delta V_2(\xi(k))] \leq \\
&\sum_{j=1}^{7} \hat{\rho}_j \Bigg[\xi^T(k)Q_j\xi(k) - \xi^T(k-\tau_k^f)Q_j\xi(k-\tau_k^f) \\
&\quad + \sum_{i=k+1-\tau_f^+}^{k-\tau_f^-} \xi^T(i)Q_j\xi(i) \Bigg]. \qquad (10.55)
\end{aligned}$$

By applying the same procedure, we have

$$\begin{aligned}
&\mathbb{E}[\Delta V_3(\xi(k))] \leq \\
&\sum_{j=1}^{7} \hat{\rho}_j \Bigg[\xi^T(k)Q_j\xi(k) - \xi^T(k-\tau_k^b)Q_j\xi(k-\tau_k^b) \\
&\quad + \sum_{i=k+1-\tau_b^+}^{k-\tau_b^-} \xi^T(i)Q_j\xi(i) \Bigg]. \qquad (10.56)
\end{aligned}$$

Finally,

$$\begin{aligned}
\mathbb{E}[\Delta V_4(\xi(k))] =& \sum_{j=1}^{7} \hat{\rho}_j \Bigg[\sum_{\ell=-\tau_f^+ +2}^{-\tau_f^- +1} [\xi^T(k)Q_j\xi(k) \\
&-\xi^T(k+\ell-1)Q_j\xi(k+\ell-1)] \Bigg] \\
=& \sum_{j=1}^{7} \hat{\rho}_j \Bigg[(\tau_f^+ - \tau_f^-)\xi^T(k)Q_j\xi(k)
\end{aligned}$$

$$- \sum_{i=k+1-\tau_f^+}^{k-\tau_f^-} \xi^T(i)Q_j\xi(i)\Bigg] \tag{10.57}$$

$$\mathbb{E}[\Delta V_5(\xi(k))] = \sum_{j=1}^{7} \hat{\rho}_j \Bigg[(\tau_b^+ - \tau_b^-)\xi^T(k)Q_j\xi(k)$$

$$- \sum_{i=k+1-\tau_b^+}^{k-\tau_b^-} \xi^T(i)Q_j\xi(i)\Bigg]. \tag{10.58}$$

It follows from (10.49) and (10.50) that

$$\xi(k) - \xi(k-\tau_k^f) - \lambda(k) = 0, \tag{10.59}$$

$$\xi(k) - \xi(k-\tau_k^b) - \epsilon(k) = 0, \tag{10.60}$$

so for any matrices F_i, U_i and Z_i, $i = 1, 2$, with appropriate dimensions, we can use the following formulas:

$$2[\xi^T(k)F_1 + \xi^T(k-\tau_k^f)U_1 + \lambda^T(k)Z_1] \times [\xi(k) - \xi(k-\tau_k^f) - \lambda(k)] = 0 \tag{10.61}$$

$$2[\xi^T(k)F_2 + \xi^T(k-\tau_k^b)U_2 + \epsilon^T(k)Z_2] \times [\xi(k) - \xi(k-\tau_k^b) - \epsilon(k)] = 0. \tag{10.62}$$

The combination of (10.52)–(10.62) leads to

$$\begin{aligned}
\mathbb{E}[\Delta V(\xi(k))] \le \sum_{j=1}^{7} \Bigg[& \xi^T(k)\Psi_j\xi(k) \\
& + \sum_{j=1}^{7} \xi^T(k)(-2F_1 + 2U_1^T)\xi(k-\tau_k^f) \\
& + \xi^T(k)(-2F_2 + 2U_2^T)\xi(k-\tau_k^b) \\
& + \xi^T(k)(-2F_1 + 2Z_1^T - 2\Phi_{j2})\lambda(k) \\
& + \xi^T(k)(-2F_2 + 2Z_2^T - 2\Phi_{j3})\epsilon(k) \\
& + \xi^T(k-\tau_k^f)(-U_1 - U_1^T - \hat{\rho}_j Q_j)\xi(k-\tau_k^m) \\
& + \xi^T(k-\tau_k^f)(-2U_1 - 2Z_1^T)\lambda(k) \\
& + \xi^T(k-\tau_k^b)(-U_2 - U_2^T - \hat{\rho}_j Q_j)\xi(k-\tau_k^a) \\
& + \xi^T(k-\tau_k^b)(-2U_2 - 2Z_2^T)\epsilon(k) \\
& + \lambda^T(k)(-Z_1 - Z_1^T + \Phi_{j4})\lambda(k)
\end{aligned}$$

$$+\epsilon^T(k)(-Z_2 - Z_2^T + \Phi_{j5})\epsilon(k)$$
$$+\lambda^T(k)\Phi_{j6}\epsilon(k)$$
$$\left. +2\xi^T\Phi_{j7}\zeta + \zeta^T(\Phi_{j8} - \eta I)\zeta \right]$$
$$= \sum_{j=1}^{7}\left[\Omega^T(k)\widetilde{\Upsilon}_j\Omega(k)\right], \tag{10.63}$$

where

$$\Omega(k) = \left[\xi^T(k) \;\; \xi^T(k - \tau_k^f) \;\; \xi^T(k - \tau_k^b) \right.$$
$$\left. \lambda^T(k) \;\; \epsilon^T(k) \;\; \zeta^T(k)\right]^T \tag{10.64}$$

and $\widetilde{\Upsilon}_j$ corresponds to Υ_j in (10.47) by Schur complements. If $\Upsilon_j < 0$, $j = 1, \cdots, 7$ holds, then

$$\mathbb{E}[V(\xi(k+1)) - V(\xi(k))] = \sum_{j=1}^{7}\left[\Omega^T(k)\widetilde{\Upsilon}_j\Omega(k)\right]$$
$$\leq \sum_{j=1}^{7}\left[-\lambda_{min}(\widetilde{\Upsilon}_j)\Omega^T(k)\Omega(k)\right]$$
$$< -\sum_{j=1}^{7}\left[\beta_j\Omega^T(k)\Omega(k)\right], \tag{10.65}$$

where

$$0 < \beta_j < min\left[\lambda_{min}(\Upsilon_j), max\{\lambda_{max}(P),\ \lambda_{max}(Q_j)\}\right].$$

Inequality (10.63) implies that $\mathbb{E}[V(\xi(k+1)) - V(\xi(k))] < -\phi V(\xi(k))$, $0 < \phi < 1$. Proceeding further we have

$$||\xi(k)||^2 \leq \frac{\upsilon}{\kappa}||\xi(0)||^2(1-\phi)^k + \frac{\lambda}{\upsilon\phi}.$$

Thus, the exponential stability of system (10.40) can be verified. □

Corollary 10.1. *For the case of DoS attack alone, for a given controller gain K and observer gain L, system (10.40) is exponentially stable if there exist matrices $P > 0$, $Q_j^T = Q_j > 0$, $j = 1, \cdots, 3$ and matrices F_i, U_i and Z_i, $i = 1, 2$, satisfying the following LMIs:*

$$\Upsilon_j \;\; = \;\; \begin{bmatrix} \Upsilon_{1j} & \Upsilon_{2j} \\ \bullet & \Upsilon_{3j} \end{bmatrix} < 0, \tag{10.66}$$

$$
\begin{aligned}
\Upsilon_{1j} &= \begin{bmatrix} \Psi_j + \Phi_{j1} & -F_1 + U_1^T & -F_2 + U_2^T \\ \bullet & -U_1 - U_1^T - \hat{\rho}_j Q_j & 0 \\ \bullet & \bullet & -U_2 - U_2^T - \hat{\rho}_j Q_j \end{bmatrix}, \\
\Upsilon_{2j} &= \begin{bmatrix} -F_1 + Z_1^T - \Phi_{j2} & -F_2 + Z_2^T - \Phi_{j3} \\ -U_1 - Z_1^T & 0 \\ 0 & -U_2 - Z_2^T \end{bmatrix}, \\
\Upsilon_{3j} &= \begin{bmatrix} -Z_1 - Z_1^T + \Phi_{j4} & \Phi_{j5} \\ \bullet & -Z_2 - Z_2^T + \Phi_{j6} \end{bmatrix},
\end{aligned} \tag{10.67}
$$

where

$$
\begin{aligned}
\Psi_j &= -P + \hat{\rho}_j(\tau_f^+ - \tau_f^- + \tau_b^+ - \tau_b^- + 2)Q_j \\
&\quad + F_1 + F_1^T + F_2 + F_2^T, \\
\Phi_{j1} &= (\bar{A}_j + \bar{B}_j + \bar{C}_j)^T \hat{\rho}_j P (\bar{A}_j + \bar{B}_j + \bar{C}_j), \\
\Phi_{j2} &= (\bar{A}_j + \bar{B}_j + \bar{C}_j)^T \hat{\rho}_j P \bar{B}_j, \\
\Phi_{j3} &= (\bar{A}_j + \bar{B}_j + \bar{C}_j)^T \hat{\rho}_j P \bar{C}_j, \\
\Phi_{j4} &= \bar{B}_j^T \hat{\rho}_j P \bar{B}_j, \\
\Phi_{j5} &= \bar{B}_j^T P \bar{C}_j, \\
\Phi_{j6} &= \bar{C}_j^T \hat{\rho}_j P \bar{C}_j.
\end{aligned}
$$

Proof. The proof of Corollary 10.1 can be obtained by applying the same procedure as for proof of Theorem 10.6. □

Theorem 10.7. *For a given delay bounds τ_f^+, τ_f^-, τ_b^+, τ_b^- and $\hat{\rho}_j$, $j = 1, \cdots, 7$, system (10.40) is exponentially stable if there exist matrices X, Y_1, Y_2, $\Xi_j > 0$, $j = 1, \cdots, 7$ and matrices H_i, M_i and R_i, $i = 1, 2$, satisfying the following LMI:*

$$
\left[\begin{array}{cccc} \widehat{\Upsilon}_{1j} & \widehat{\Upsilon}_{2j} & \mathbf{0} & \\ \bullet & \widehat{\Upsilon}_{3j} & \mathbf{0} & \widehat{\Theta}_j \\ \bullet & \bullet & -\eta I & \\ \hdashline & \bullet & & -\hat{\rho}_j \widehat{X} \end{array}\right] < 0, \tag{10.68}
$$

where

$$
\widehat{X} = \begin{bmatrix} X & 0 \\ 0 & X \end{bmatrix}, \tag{10.69}
$$

$$
\widehat{\Psi}_j = -\widehat{X} + \hat{\rho}_j(\tau_f^+ - \tau_f^- + \tau_b^+ - \tau_b^- + 2)\Xi_j + H_1 + H_1^T + H_2 + H_2^T,
$$

$$\widehat{\Upsilon}_{1j} = \begin{bmatrix} \widehat{\Psi}_j & -H_1 + M_1^T & -H_2 + M_2^T \\ \bullet & -M_1 - M_1^T - \hat{\rho}_j \Xi_j & 0 \\ \bullet & \bullet & -M_2 - M_2^T - \hat{\rho}_j \Xi_j \end{bmatrix},$$

$$\widehat{\Upsilon}_{2j} = \begin{bmatrix} -H_1 + R_1^T & -H_2 + R_2^T \\ -M_1 - R_1^T & 0 \\ 0 & -M_2 - R_2^T \end{bmatrix},$$

$$\widehat{\Upsilon}_{3j} = \begin{bmatrix} -R_1 - R_1^T & 0 \\ \bullet & -R_2 - R_2^T \end{bmatrix},$$

$$\widehat{\Theta}_j = \begin{bmatrix} \widehat{\Theta}_{1j} & 0 & 0 & \widehat{\Theta}_{4j} & \widehat{\Theta}_{5j} & \widehat{\Theta}_{6j} \end{bmatrix}^T, \tag{10.70}$$

$$\widehat{\Theta}_{1j} = \begin{bmatrix} XA^T + Y_1^T B^T & 0 \\ XA^T & XA^T - Y_2^T \end{bmatrix}, \ j = 1, \cdots, 4,$$

$$\widehat{\Theta}_{15} = \begin{bmatrix} XA^T + Y_1^T B^T & Y_2^T \\ XA^T & XA^T \end{bmatrix},$$

$$\widehat{\Theta}_{16} = \begin{bmatrix} XA^T & -Y_1^T B^T \\ XA^T & XA^T - Y_2^T \end{bmatrix},$$

$$\widehat{\Theta}_{17} = \begin{bmatrix} XA^T & -Y_1^T B^T + Y_2^T \\ XA^T & XA^T \end{bmatrix},$$

$$\widehat{\Theta}_{4j} = \begin{bmatrix} 0 & Y_2^T \\ 0 & Y_2^T \end{bmatrix}, \ j = 1, 3,$$

$$\widehat{\Theta}_{5j} = \begin{bmatrix} -Y_1^T B^T & -Y_1^T B^T \\ 0 & 0 \end{bmatrix}, \ j = 2, 3,$$

$$\widehat{\Theta}_{65} = \begin{bmatrix} 0 & 0 \\ 0 & -XL^T \end{bmatrix},$$

$$\widehat{\Theta}_{66} = \begin{bmatrix} XB^T & XB^T \\ XB^T & XB^T \end{bmatrix},$$

$$\widehat{\Theta}_{67} = \begin{bmatrix} XB^T & XB^T \\ XB^T & XB^T - XL^T \end{bmatrix},$$

$$\widehat{\Theta}_{4j} = 0, \ j = 2, 4, \cdots, 7, \qquad \widehat{\Theta}_{5j} = 0, \ j = 1, 4, \cdots, 7,$$

$$\widehat{\Theta}_{6j} = 0, \ j = 1, \cdots, 4,$$

with $K = Y_1 X^{-1}$ *and* $L = Y_2 X^{-1} C^{\dagger}$.

Proof. By defining

$$\Theta_j = \left[(\bar{A}_j + \bar{B}_j + \bar{C}_j) \quad 0 \quad 0 \quad -\bar{B}_j \quad -\bar{C}_j \quad \bar{D}_j \right]^T$$

we can describe matrix inequality (10.47) as:

$$\Upsilon_j = \widetilde{\Upsilon}_j + \Theta_j P \Theta_j^T < 0, \tag{10.71}$$

$$\widetilde{\Upsilon}_j = \begin{bmatrix} \widetilde{\Upsilon}_{1j} & \widetilde{\Upsilon}_{2j} & \mathbf{0} \\ \bullet & \widetilde{\Upsilon}_{3j} & \mathbf{0} \\ \bullet & \bullet & -\eta I \end{bmatrix} < 0,$$

$$\widetilde{\Upsilon}_{1j} = \begin{bmatrix} \Psi_j & -F_1 + U_1^T & -F_2 + U_2^T \\ \bullet & -U_1 - U_1^T - \hat{\rho}_j Q_j & 0 \\ \bullet & \bullet & -U_2 - U_2^T - \hat{\rho}_j Q_j \end{bmatrix},$$

$$\widetilde{\Upsilon}_{2j} = \begin{bmatrix} -F_1 + Z_1^T & -F_2 + Z_2^T \\ -U_1 - Z_1^T & 0 \\ 0 & -U_2 - Z_2^T \end{bmatrix},$$

$$\widetilde{\Upsilon}_{3j} = \begin{bmatrix} -Z_1 - Z_1^T & 0 \\ \bullet & -Z_2 - Z_2^T \end{bmatrix}. \tag{10.72}$$

Select $\widehat{X} = P^{-1}$, and by applying Schur complements, we formulate matrix Υ_j in (10.71) as follows:

$$\left[\begin{array}{ccc|c} \widetilde{\Upsilon}_{1j} & \widetilde{\Upsilon}_{2j} & \mathbf{0} & \\ \bullet & \widetilde{\Upsilon}_{3j} & \mathbf{0} & \Theta_j \\ \bullet & \bullet & -\eta I & \\ \hline & \bullet & & -\hat{\rho}_j \widehat{X} \end{array}\right] < 0. \tag{10.73}$$

Multiplying the matrix inequality in (10.72) on the right and left by $diag[\widehat{X}, \widehat{X}, \widehat{X}, \widehat{X}, \widehat{X}, I, I]$ and applying (10.68) and

$$\begin{aligned} &\Xi_j = \widehat{X} Q_j \widehat{X}, \quad H_i = \widehat{X} F_i \widehat{X}, \\ &M_i = \widehat{X} U_i \widehat{X}, \quad R_i = \widehat{X} Z_i \widehat{X}, \quad j = 1, \cdots, 7, \quad i = 1, 2, \end{aligned}$$

matrix inequality (10.68) subject (10.70) can be obtained. □

Corollary 10.2. *For the case of DoS attack alone, for a given delay bounds τ_f^+, τ_f^-, τ_b^+, τ_b^- and $\hat{\rho}_j$, $j = 1, \cdots, 4$, system (10.40) is exponentially stable if there exist matrices $0 < X$, Y_1, Y_2, $0 < \Xi_j$, $j = 1, \cdots, 4$ and matrices H_i, M_i and*

R_i, $i = 1, 2$, satisfying the following LMIs:

$$\left[\begin{array}{cc:c} \widehat{\Upsilon}_{1j} & \widehat{\Upsilon}_{2j} & \widehat{\Theta}_j \\ \bullet & \widehat{\Upsilon}_{3j} & \\ \hdashline & \bullet & -\hat{\rho}_j \widehat{X} \end{array}\right] < 0, \tag{10.74}$$

$$\widehat{X} = \left[\begin{array}{cc} X & 0 \\ 0 & X \end{array}\right], \tag{10.75}$$

$$\begin{aligned}
\widehat{\Psi}_j &= -X + \hat{\rho}_j(\tau_f^+ - \tau_f^- + \tau_b^+ - \tau_b^- + 2)\Xi_j \\
&\quad + H_1 + H_1^T + H_2 + H_2^T,
\end{aligned}$$

$$\widehat{\Upsilon}_{1j} = \left[\begin{array}{ccc} \widehat{\Psi}_j & -H_1 + M_1^T & -H_2 + M_2^T \\ \bullet & -M_1 - M_1^T - \hat{\rho}_j \Xi_j & 0 \\ \bullet & \bullet & -M_2 - M_2^T - \hat{\rho}_j \Xi_j \end{array}\right],$$

$$\widehat{\Upsilon}_{2j} = \left[\begin{array}{cc} -H_1 + R_1^T & -H_2 + R_2^T \\ -M_1 - R_1^T & 0 \\ 0 & -M_2 - R_2^T \end{array}\right],$$

$$\widehat{\Upsilon}_{3j} = \left[\begin{array}{cc} -R_1 - R_1^T & 0 \\ \bullet & -R_2 - R_2^T \end{array}\right], \tag{10.76}$$

$$\widehat{\Theta}_j = \left[\begin{array}{ccccc} \widehat{\Theta}_{1j} & 0 & 0 & \widehat{\Theta}_{4j} & \widehat{\Theta}_{5j} \end{array}\right]^T,$$

$$\widehat{\Theta}_{1j} = \left[\begin{array}{cc} XA^T + Y_1^T B^T & 0 \\ XA^T & XA^T - Y_2^T \end{array}\right], \quad j = 1, \cdots, 4,$$

$$\widehat{\Theta}_{15} = \left[\begin{array}{cc} XA^T + Y_1^T B^T & Y_2^T \\ XA^T & XA^T \end{array}\right],$$

$$\widehat{\Theta}_{16} = \left[\begin{array}{cc} XA^T & -Y_1^T B^T \\ XA^T & XA^T - Y_2^T \end{array}\right],$$

$$\widehat{\Theta}_{4j} = \left[\begin{array}{cc} 0 & Y_2^T \\ 0 & Y_2^T \end{array}\right], \quad j = 1, 3,$$

$$\widehat{\Theta}_{5j} = \left[\begin{array}{cc} -Y_1^T B^T & -Y_1^T B^T \\ 0 & 0 \end{array}\right], \quad j = 2, 3,$$

$$\begin{aligned}
&\widehat{\Theta}_{4j} = 0, \quad j = 2, 4, \cdots, 6, \\
&\widehat{\Theta}_{5j} = 0, \quad j = 1, 4, \cdots, 6,
\end{aligned} \tag{10.77}$$

with $K = Y_1 X^{-1}$ and $L = Y_2 X^{-1} C^{\dagger}$.

Proof. The proof of Corollary 10.2 can be obtained by applying the same procedure as for proof of Theorem 10.7 with

$$\Theta_j = \begin{bmatrix} (\bar{A}_j + \bar{B}_j + \bar{C}_j) & 0 & 0 & -\bar{B}_j & -\bar{C}_j \end{bmatrix}^T.$$

□

10.3.3 Simulation results II

The effectiveness of the proposed method presented in this section is shown by solving the control problem of the autonomous underwater vehicle (AUV) mentioned in [435]. The objective of the controller is to guarantee a stable prescribed motion of the AUV. The discrete-time model of the AUV is

$$A = \begin{bmatrix} -0.14 & -0.69 & 0 \\ -0.19 & -0.048 & 0 \\ 0 & 1 & 0 \end{bmatrix}; \quad B = \begin{bmatrix} 0.056 \\ -0.23 \\ 0 \end{bmatrix};$$

$$C = \begin{bmatrix} 1 & 0 & 0 \end{bmatrix}.$$

The initial states value is assumed to be $x_1(0) = x_2(0) = x_3(0) = 1$. Using YALMIP the gains of the controller and estimator (10.35) and (10.40) were obtained:

$$K = \begin{bmatrix} 0.00388 & 0.16755 & 0.00001 \end{bmatrix},$$

$$L = \begin{bmatrix} -0.038 & -0.076 & -162.02 \end{bmatrix}^T.$$

Three attack scenarios were considered and their states and estimation errors were obtained using MATLAB®/Simulink as follows:

1. Systems without attack, Figs. 10.9–10.10;
2. Systems under DoS and physical attacks, Figs. 10.11–10.12;
3. Systems under deception and physical attacks, Figs. 10.13–10.14.

As shown in Figs. 10.9–10.14, the designed observer-based controller shows stability in the states and a small error in estimating the states under all possibilities of attacks. It should also be noted in Figs. 10.10, 10.12, and 10.14 that there is a high peak of error on the estimation at a certain time, while it is caused by the initial error on estimation in the first scenario and the occurrences of high value of attacks in the second and third scenarios, the overall observer performance and the stability of the states are not affected.

10.4 Notes

In this chapter we adopted a hybrid automaton approach to describe the dynamic transitions between the nominal CPS and the system under attack. The unified

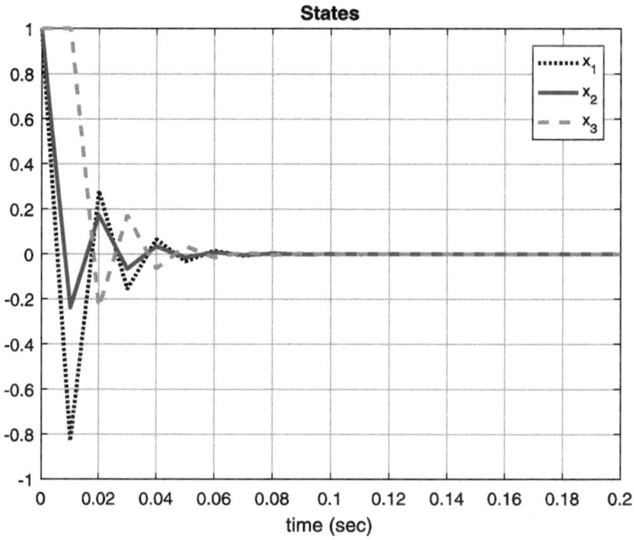

FIGURE 10.9 States with no attack.

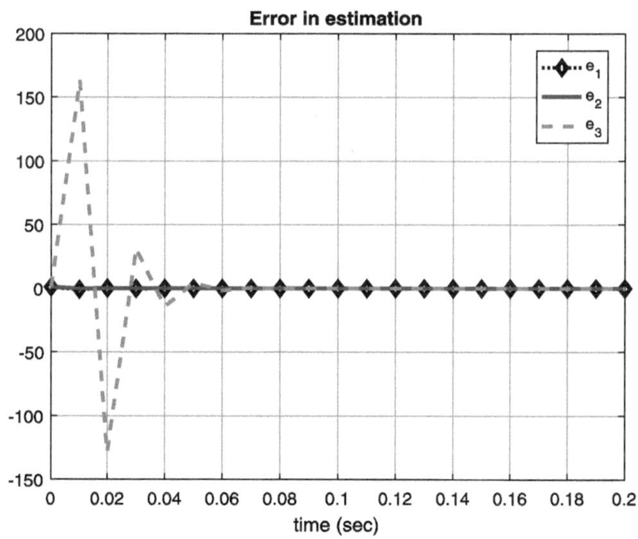

FIGURE 10.10 Estimation error of states with no attack.

modeling framework of CPS under attack is described as a time-invariant system subject to unknown input. The monitor designed in this chapter is capable of detecting exogenous attacks and triggering the discrete event, or attack, in the hybrid automaton. The defense mechanism is achieved via a passivation trans-

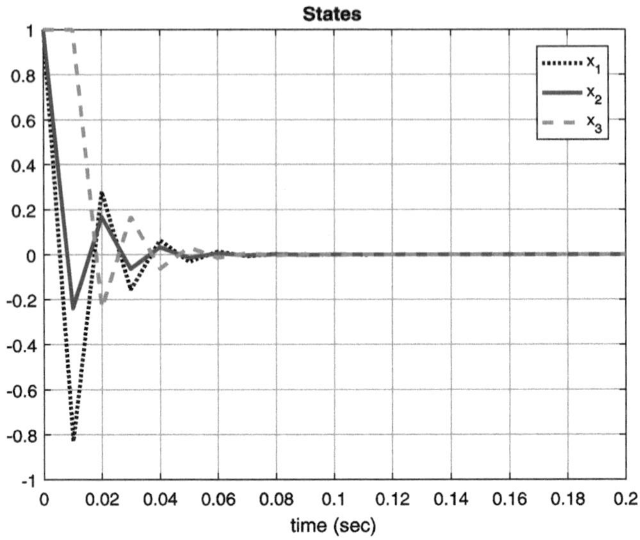

FIGURE 10.11 States with DoS attack.

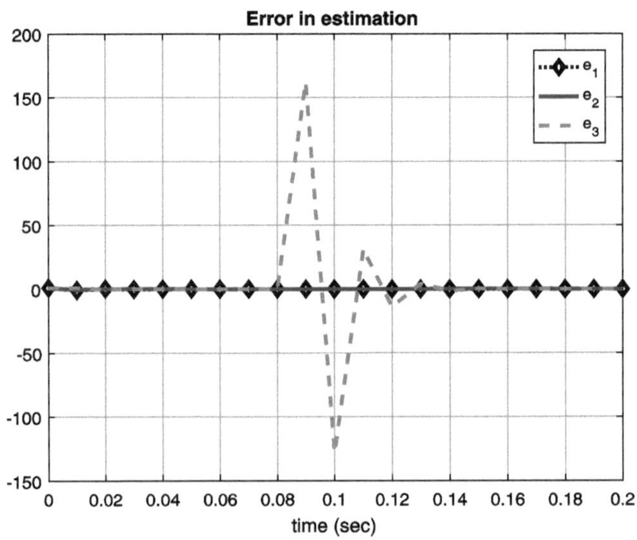

FIGURE 10.12 Estimation error of states with DoS attack.

formation M-matrix design. Passivity is guaranteed for the hybrid automaton under attack.

We then proposed and studied an improved observer-based stabilizing controller for CPSs under DoS and deception attacks. The occurrences of DoS and deception attacks are modeled as Bernoulli distributed white sequences

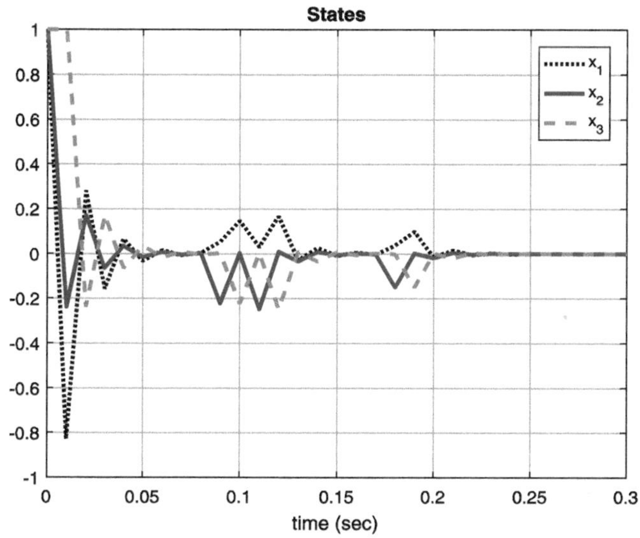

FIGURE 10.13 States with deception attack.

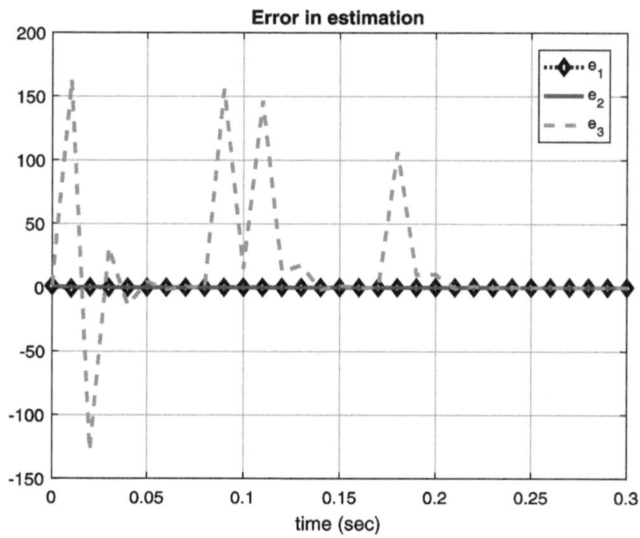

FIGURE 10.14 Estimation error of states with deception attack.

with variable conditional probabilities. The criterion was formulated in terms of linear matrix inequalities. Detailed simulation experiments of representative systems have shown the applicability of the proposed methodology and its ability to keep the system within the desired stability conditions. As a future work, we are planning to test the proposed design using a real prototype.

Chapter 11

Safety assurance under stealthy cyber attacks

Contents

11.1 Introduction

We look at Cyber-Physical Control Systems (CPCS), which consist of logical elements such as embedded computers and physical elements connected by communication channels such as the Internet. Cybersecurity is one of the major concerns for many CPSs [436]. The methods proposed to solve the cybersecurity problem for CPSs can be categorized into two classes: information security, which is mainly focused on encryption and data security, and secure control theory, which studies how cyber attacks affect the control system's physical dynamics [437]. The safety tools that only use information security are not sufficient for the secure control of CPSs because they cannot describe the system's macro-behavior. Therefore, they need to be complemented with secure control theory.

Secure control theory is based on an attack model, which is a challenging task due to the uncertain and erratic nature of cyber attacks. To the best of our knowledge the various cyber attack models for CPSs can be reduced to two kinds. The first model, Denial-of-Service (DoS) attacks, refers to obstructing the communication between networked agents [438]. These attacks can jam the communication channels, attack the routing protocols, etc. As a countermeasure, [90] provides a secure control scheme in the presence of DoS attacks. Another research group applied the game theory approach to achieve a robust and resilient control against DoS attacks [23]. The second kind of attack model is deception attack, which represent the injection of false information from sensors or controllers. In this attack the attacker can obtain the secret key or compromise some cyber elements in order to falsify the data. This kind of attack has mainly

Cloud Control Systems. https://doi.org/10.1016/B978-0-12-818701-2.00019-6

been studied in the application of electric power distribution [423] and some other applications can be found in [276], [127], and [439].

This section focuses on deception attacks in the area of secure control theory. Instead of identifying a specific deception attack model, we study the system's response to the possible deception attacks. While a CPS is being attacked the monitoring system can compare a sequence of the compromised data to the expected output of the healthy system in order to detect the cyber attacks. However, the monitoring system cannot detect all feasible deception attacks since the attacker with a good knowledge of the system can design a sophisticated deception attack that is nearly or absolutely undetectable. If the system does not know that there is a cyber attack, it cannot protect itself against the attack effectively.

In what follows, we study the stealthy deception attack problem. The CPS is modeled as a stochastic linear system with Gaussian noise. The noise measurements are fed into a steady-state Kalman filter (KF) which is used for cyber attack detection and diagnosis. Malicious stealthy deception attacks then try to drive the state estimate of the KF away from the actual state without being detected. Compared with previous studies which mainly focus on the design of stealthy deception attacks [423], [127], we derive the conditions of the system dynamics to evaluate the impact of feasible stealthy deception attacks. Considering the three kinds of stealthy deception attacks according to the attacker's ability to compromise the system, these conditions identify under which systems the attacker has a capability to perform the stealthy deception attacks causing unbounded estimation errors. Such an analysis can provide design criteria to improve the security of the CPS against malicious cyber attacks. In addition, our research considers attacks on both sensor components and control units of the CPSs. Thus, the cybersecurity problem considered in this chapter is more general than those in most of the existing research that considers cyber attacks either on sensors or actuators alone.

11.2 Cloud system model subject to cyber attacks

The dynamics of the plant in a cloud control system (CCS) under a cyber attack are modeled as a linear time-invariant system. In the CCS actuators and sensors are usually connected by communication channels, which are susceptible to malicious data injection by cyber attacks.

The CCS model subject to cyber attacks is given by

$$\begin{aligned} x_a(k+1) &= Ax_a(k) + Bu(k) + B_c a_c(k) + w(k) \\ y_a(k) &= Cx_a(k) + B_o a_o(k) + v(k), \end{aligned} \tag{11.1}$$

where $x_a(k) \in \Re^n$ (the subscript a means the system is under attack), $u(k) \in \Re^p$, and $y_a(k) \in \Re^m$ are the system's state, input, and output respectively, and $w(k) \in \Re^q$ and $v(k) \in \Re^r$ are the process and measurement noise. It is assumed that the input $u(k)$ is known and $w,\ v$ are Gaussian white noise with constant covariance matrices Q and R respectively. A, B, C, and E_1 are the system matrices of appropriate dimensions, and $k \in \{0, 1, ..., N\}$ denotes the discrete-time index, taking values from the time horizon $[0, N]$. The attack sequences $a_c(k) \in \Re^s$ and $a_o(k) \in \Re^q$ are injected into the actuators and sensors with the attack matrices B_c and B_o of compatible dimensions. Note that the system matrix pairs (A, B) and (C, A) satisfy the controllability and observability conditions.

Here our framework includes attacks on both sensor components and control units of the CPS. Thus, the cybersecurity problem considered in this chapter is more general than those in most existing research that considers cyber attacks either on sensors or actuators alone. In a real application the attack matrices are dependent on the physical structure of sensors and actuators. Thus, the attack matrices B_c and B_o should satisfy

$$span\,\{B_c\} \subseteq span\,\{B\}, \quad span\,\{B_o\} \subseteq span\,\{C\}, \tag{11.2}$$

where "span" denotes the column space of the matrix. If the attacker is allowed to access all the input and output components of the given system, it is possible to assign the matrices as $B_c = B$ and $B_o = C$. To monitor the CPS state, we consider the steady-state KF represented by

$$\begin{aligned} \hat{x}_a(k+1) &= A\hat{x}_a(k) + Bu(k) \\ &+ L[y_a(k+1) - CA\hat{x}_a - CBu(k)], \end{aligned} \tag{11.3}$$

where L is the steady-state Kalman gain given by

$$L = \Sigma_p C^T (C\Sigma_p C^T + R)^{-1} \tag{11.4}$$

and Σ_p is the predicted error covariance matrix, which is the solution to the following discrete-time algebraic Riccati equation:

$$\begin{aligned} -\Sigma_p &+ A\Sigma_p A^T - A\Sigma_p C^T (C\Sigma_p C^T + R)^{-1} C\Sigma_p A^T \\ &+ Q = 0. \end{aligned} \tag{11.5}$$

Since an injected cyber attack can cause faulty behaviors of the CPS, existing fault diagnosis algorithms can be used for cyber attack detection. Specifically, we implement an attack detection scheme that evaluates the residuals generated by the steady-state KF. The residual is defined as

$$r(k+1) := y_a(k+1) - CA\hat{x}_a - CBu(k). \tag{11.6}$$

Note that without cyber attacks the residual has a zero-mean Gaussian distribution with a constant covariance matrix $\Sigma_r = (C\Sigma_p C^T + R)$. Therefore, cyber attacks can be diagnosed by testing the following two incompatible statistical hypotheses:

$$\mathcal{H}_0 : r(k) \in \mathcal{N}(0, \Sigma_r), \quad \mathcal{H}_0 : r(k) \notin \mathcal{N}(0, \Sigma_r).$$

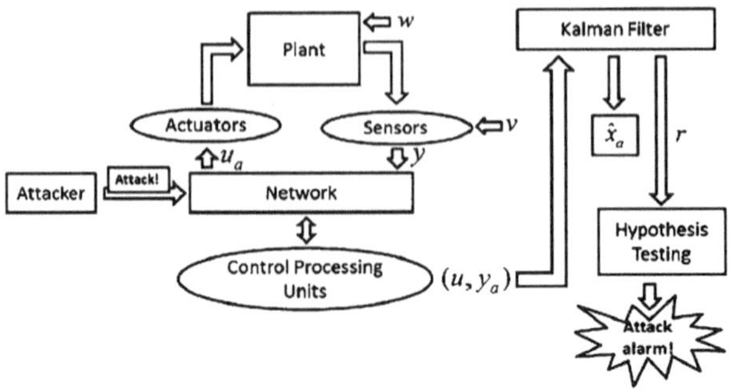

FIGURE 11.1 Deception attack detection mechanism.

Here $\mathcal{N}(\rho; \Sigma)$ is the probability density function (pdf) of the Gaussian random variable with mean ρ and covariance Σ. Fig. 11.1 illustrates a typical deception attack detection mechanism. Among the various hypothesis testing algorithms [127] we use the Sequential Probability Ratio Test (SPRT) since it is widely used, can generally accommodate other common tests like the chi-square test [439], and is suitable for carrying out the vulnerability analysis against stealthy cyber attacks. The SPRT computes the cumulative $W \in \mathbb{R}$ of the log-likelihood ratio $\Psi \in \mathbb{R}$ recursively as follows:

$$W(k) := W(k-1) + \Psi(k) \tag{11.7}$$

$$\Psi(k) := log \frac{p_a[r(k)|r(k-1)...r(0)]}{p[r(k)|r(k-1)...r(0)]}. \tag{11.8}$$

Alternatively, the statistical hypothesis testing can be implemented using the SPRT, cumulative sum (CUSUM), etc.

In the following we consider the above hypothesis test by checking the power of residuals, $r^T(k)\Sigma_r^{-1}r(k)$, which is known as Compound Scalar Testing [440],

$$\begin{cases} \text{Accept} & \mathcal{H}_0 \quad \text{If,} \quad r^T(k)\Sigma_r^{-1}r(k) \leq h \\ \text{Accept} & \mathcal{H}_1 \quad \text{If,} \quad r^T(k)\Sigma_r^{-1}r(k) > h \end{cases}, \tag{11.9}$$

where the threshold value $h > m$, with m the dimension of the measurements used to avoid the high false alarm rate. If there are no faults or attacks, $r^T(k)\Sigma_r^{-1}r(k)$ follows a χ^2 distribution with m degrees of freedom, and yields $\mathbb{E}[r^T(k)\Sigma_r^{-1}r(k)] = m$. In this case $\mathcal{H}_0$ is accepted. If $r^T(k)\Sigma_r^{-1}r(k) > h$, $\mathcal{H}_1$ is accepted and the algorithm declares a fault in the system which may be induced by cyber attacks. Therefore, an adversary who wants a deception attack to be undetectable should avoid causing a large increase in the power of residuals, thereby triggering an alarm.

11.3 Stealthy deception attack design

In this section, we investigate how an attacker can design a deception attack sequence for the CPCS described in Section II, which causes an unbounded estimation error without being detected by the system's detection scheme. The following analysis characterizes the security of the system. In the following, we consider two distinct problems:

- First security problem (FSP) over infinite time horizon;
- Second security problem (SSP) over finite time horizon T.

First security problem (FSP) over infinite time horizon: This can be phrased as follows:

For a given CPS (11.1) we look for an attack pair sequence $(a_c(k), a_o(k))$, $\forall k \in \{1, 2, \cdots\}$ *from all the possible attack pairs that cause an increase in estimation error without a large increase in the power,* $r^T(k)\Sigma^{-1}r(k)$, *which can be detected by the monitoring scheme.*

Remark 11.1. Note that since the estimation error and residual are random variables, we consider the increase in the mean value, that is, $\mathbb{E}[e^T(k)e(k)] > 0$ and $\mathbb{E}[r^T(k)\Sigma^{-1}r(k)]\ h$.

Assumption. To consider the worst-case security problem, we assume the attacker has a perfect knowledge of the system model.

Before providing an answer to the above problem, we classify the deception attacks into the following three cases according to the attacker's capability:

- **Case 1:** The attackers are able to compromise the data only in the actuators, that is, $B_o = 0$.
- **Case 2:** The attackers are able to compromise the data only from sensors, that is, $B_c = 0$.
- **Case 3:** The attackers are able to compromise the data both from sensors and controllers, that is, $B_o \neq 0$ and $B_c \neq 0$.

It should be noted that compared with Case 1 and Case 2, where only a single attack sequence $(a_c(k), a_o(k))$ can be injected into the system, both attack sequences are injected in Case 3 and can interact with each other. The deception attack of each case requires a specific condition in order to cause the unbounded estimation error for an infinite time horizon without triggering the alarm of the monitoring system. We discuss these conditions for the three cases below.

11.3.1 Actuators are compromised

First we examine Case 1 stealthy deception attacks through the following lemma:

Lemma 11.1. *Without compromising the sensors, that is,* $B_o = 0$, *attackers can never induce an infinite estimation error in any time horizon.*

Proof. Suppose there exists an attack sequence causing the unbounded estimation error, which is still undetectable by the monitoring system under the $B_o = 0$ condition. Then the mean of the residual

$$\mathbb{E}[r(k+1)] = -CA\mathbb{E}[e_a(k)] + C\,B_c a_c(k), \tag{11.10}$$

should be bounded, while the mean of the estimation error,

$$\mathbb{E}[e(k+1)] = -A\mathbb{E}[e_a(k)] - C\,B_c a_c(k) + K\mathbb{E}[r(k+1)], \tag{11.11}$$

grows indefinitely as $k \longrightarrow \infty$. However, multiplying by C in (11.11) yields

$$C\mathbb{E}[e(k+1)] = -\mathbb{E}[r(k+1)] + CK\mathbb{E}[r(k+1)]. \tag{11.12}$$

Since the pair $(C,\ A)$ is observable, (11.12) implies that the infinite error is induced by the residual. This remains valid only by violating the boundedness condition of the residual. Lemma 11.1 is proved by contradiction. □

Lemma 11.1 demonstrates that the stealthy deception attack on actuators can cause only finite estimation errors while not being detected.

11.3.2 Sensors are compromised

We now derive a condition for Case 2 stealthy deception attacks, which is given by the following theorem:

Theorem 11.1. *Define the estimation error under the deception attack $e_a(k) := \hat{x}_a(k) - x_a(k)$. Then for the prescribed CPS there exist attack pair sequences $(a_c(k), a_o(k)), \forall k \in \{1, 2, \cdots\}$ such that*

$$\lim ||e(k)|| \longrightarrow \infty, \quad -\mathbb{E}[r^T(k)\Sigma^{-1}r(k)] \;\leq\; h, \forall\, k \tag{11.13}$$

if and only if system matrix A is unstable and at least one eigenvector ξ corresponding to the unstable mode satisfies

$$\xi \in span\{Q_{oa}\}, \quad C\xi \in span\{B_o\}, \tag{11.14}$$

where Q_{oa} is the controllability matrix associated with the pair $(A - KCA,\ KB_o)$.

Proof. To begin with, we prove the necessary condition. From (11.1) and (11.3) with $B_c = 0$, the mean error dynamics are given by

$$\begin{aligned}\mathbb{E}[e(k+1)] &= A\mathbb{E}[e(k)] + K(B_o a_o(k+1) - CA\mathbb{E}[e_a(k)]) \\ &= (A - KCA)\mathbb{E}[e(k)] + KB_o a_o(k+1).\end{aligned} \tag{11.15}$$

On taking the norm of (11.15), we reach the following inequality:

$$||\mathbb{E}[e(k+1)]|| \leq ||A||||\mathbb{E}[e(k)]|| + ||K||||(B_o a_o(k+1) - CA\mathbb{E}[e_a(k)])||. \tag{11.16}$$

The last term of the above inequality should be bounded. Thus, (11.16) can be transformed into a cumulative form such that

$$||\mathbb{E}[e(k+1)]|| \leq \sum_{j=1}^{k} ||A||^{j-1} \, ||K|| \, \gamma, \tag{11.17}$$

where $\gamma \geq ||(B_o a_o(k+1) - CA\mathbb{E}[e_a(k)])||$ is finite. From (11.17) the mean of the error goes to infinity when $||A|| \geq 1$. Recall that for every matrix $P > 0$ the following properties hold:

$$||A^T P A|| = ||A||||P||||A||, \quad ||A||||P||||A|| - ||P|| \geq 0. \tag{11.18}$$

From (11.18) there is no positive definite matrix P satisfying $A^T P A - P < 0$, and hence the system matrix A is unstable by the Lyapunov theorem.

Using this fact, the mean of the error can be written in a decomposed form such that

$$||\mathbb{E}[e(k)]|| = \sum_{j}^{k} c_j(k)\xi_j + \eta(k). \ \forall j \in \{j \mid \lim_{k \longrightarrow \infty} c_j(k) = \infty\}, \tag{11.19}$$

where ξ_j is an unstable eigenvector with coefficients $c_j(k)$ and $\eta(k)$ containing the remaining bounded vector components of the error. Then, we only need to track each $c_j(k)\xi_j$ because it is the only component that eventually goes to infinity. Hence, (11.15) can be rewritten as

$$\begin{aligned} c_j(k+1)\xi &= \lambda_j \, c_j(k)\xi_j + K(B_o a_o(k+1) - C\lambda_j \, c_j(k)\xi_j) \\ &= (A - KCA) \, c_j(k)\xi_j + K B_o a_o(k+1), \end{aligned} \tag{11.20}$$

where λ_j is the corresponding unstable eigenvalue. In order to bound $||(B_o a_o(k+1) - C\lambda_j \, c_j(k)\xi_j)||$, it is quite clear that $C\xi_j \in B_o$. In addition, since $(A - KCA)$ is stable, ξ_j should lie on the controllable subspace of the pair $(A - KCA, KB_o)$ so that the error goes to infinity, thus proving the necessary condition. On the other hand, the sufficient condition is easily proved in the following way. Suppose ξ^* is an unstable eigenvector satisfying (11.14). Then we can design the attack sequence $a_o(k)$ recursively such that

$$B_o a_o^*(k+1) = CA\mathbb{E}[e_a(k)], \quad B_o a_o^*(1) = \epsilon C\xi^*, \tag{11.21}$$

where ϵ is an arbitrarily chosen value from the interval

$$\epsilon^2 < \sqrt{\frac{h-m}{\xi^{*T} C^T \Sigma^{-1} C \xi^*}}. \tag{11.22}$$

Applying (11.21) to (11.15) gives an induced mean error at each time. By induction, the attack sequence (11.21) yields the error and residual that comply with (11.13) and the proof is completed. □

An important observation is in order. Theorem 11.1 indicates that the deception attack under the system condition (11.14) eventually causes an infinite error without being detected, though the error has a limited reachable region in a finite time window. Once the given CPS is evaluated as vulnerable to Case 2 stealthy deception attacks, we can evaluate this reachable error region through the stochastic optimal control problem.

Second security problem (SSP) over finite time horizon: This can be phrased as follows:

For a finite time horizon T, we look for an attack sequence $(a_o(k)), \forall k \in \{1, 2, \cdots, T\}$ from all the possible attack pairs which are the solution to the following constrained optimization problem:

$$\begin{aligned} &\max_{a_o} \; \|e_a(T)\| \\ &\textit{Subject to} \quad \mathbb{E}[r^T(k)\Sigma^{-1} r(k)] \le h \\ &\textit{and} \quad \textit{the Mean error dynamics } (11.15). \end{aligned} \tag{11.23}$$

The optimization problem (11.23) can be transformed into a deterministic optimization problem by taking a mean at each term, and therefore the **SSP over finite time horizon** can be solved either analytically or numerically.

11.3.3 Both actuators and sensors are compromised

The general situation in which both sensors and actuators are compromised is now addressed. We examine Case 3 stealthy deception attacks, which are more complex than the previous two cases. The result is summarized by the following theorem:

Theorem 11.2. *For the given CPS (11.1), there exists a nonzero attack pair sequence $(a_c(k), a_o(k)), \forall k \in \{1, 2, \cdots\}$ that can cause the unbounded error while maintaining $r(k) \in \mathcal{N}(0, \Sigma_r)$ if and only if*

$$span\{C Q_{oa}\} \subseteq span\{B_o\}, \tag{11.24}$$

where Q_{oa} is the controllability matrix associated with the pair (A, B_c).

Proof. Rewriting the residual in (11.6) as

$$\begin{aligned} r(k+1) &= \Psi(k) + CE_1 w(k) + E_2 v(k+1) \\ \Psi(k) &= -CAe_a(k) + CB_c a_c + B_o a_o(k+1), \end{aligned} \tag{11.25}$$

under the steady state of the KF we have

$$Cov[e_a(k),\ e_a(k)] \equiv Cov[e(k), e(k)]$$

for any k regardless of $a_c(k)$ and $a_o(k)$, which is denoted by Σ:

$$\Sigma = \Sigma_p - \Sigma_p C^T (C\Sigma_p C^T + E_2 R E_2^T)^{-1} C\Sigma_p \tag{11.26}$$

and thus,

$$\mathbb{E}[\Psi(k)\Psi^T(k)] = \mathbb{E}[\Psi(k)]\mathbb{E}[\Psi^T(k)] + CA\Sigma A^T C^T. \tag{11.27}$$

It follows from (11.27) and squaring the residual that

$$\begin{aligned} \mathbb{E}[r(k+1)r^T(k+1)] &= \mathbb{E}[\Psi(k)\Psi^T(k)] + CA\Sigma A^T C^T + E_2 R E_2 \\ &= \mathbb{E}[\Psi(k)]\mathbb{E}[\Psi^T(k)] + \Sigma_r. \end{aligned} \tag{11.28}$$

Enforcing $\mathbb{E}[\Psi(k)] \equiv 0$ will ensure that $\mathbb{E}[r(k)] \equiv 0$ and $\mathbb{E}[r(k+1)r^T(k+1)] = \Sigma_r$. This implies that the attack pair sequence satisfies

$$B_o a_o(k+1) = CA\mathbb{E}[e_a(k)] - CB_c a_c. \tag{11.29}$$

It is therefore necessary that the column space of B_o contains C multiplied by any vectors on the controllable subspace of the pair $(A,\ B_c)$. In turn, this leads to (11.24) by application of the Cayley Hamilton Theorem.

With regard to sufficiency, since $span\{CQ_{oa}\} \subseteq span\{B_o\}$, there exists a nonzero attack pair sequence satisfying (11.29). The condition $\mathbb{E}[\Psi(k)]$ always holds, and thus the residual keeps its statistical property, which establishes the desired result. □

Corollary 11.1. *If the attack matrix B_o is of full column rank, a Case 3 stealthy deception attack sequence to the CPS satisfying the condition (11.24) is designed as*

$$a_o(k) = -B_o^+ C \sum_{s=1}^{k-1} A^{k-1-s} B_c a_c \tag{11.30}$$

for arbitrary sequence $a_c(k)$, where $B_o^+ = [B_o^T B_o]^{-1} B_o^T$ denotes the Moore–Penrose pseudo inverse of the matrix B_o.

Proof. The mean error dynamic when $B_o \neq 0$ and $B_c \neq 0$ is

$$\mathbb{E}[e_a(k+1)] = (A - KCA)\mathbb{E}[e_a(k)] - (B_c - KCB_c)a_c(k) + KB_o a_o(k+1). \tag{11.31}$$

The substitution of (11.29) for $a_o(k)$ into (11.31) yields a simplified form:

$$\mathbb{E}[e_a(k+1)] = A\mathbb{E}[e_a(k)] - B_c a_c(k). \tag{11.32}$$

Since $\mathbb{E}[e_a(0)] = \mathbb{E}[e(k)] = 0$, it follows that (11.32) results in

$$\mathbb{E}[e_a(k)] = -\sum_{s=1}^{k-1} A^{k-1-s} B_c a_c. \tag{11.33}$$

Then (11.30) can be derived by using (11.29) and (11.33). □

Corollary 11.2. *If the attacker has the capability of full access to control units in the given CPS and (11.24) is satisfied, that is, $(A,\ B_c)$ is controllable, the estimation error can be driven to any state in a stealthy way such that*

$$a_c(k) = -B_c^T A^{T(N-1)-k} W_c^{-1}(N)\, e_f, \tag{11.34}$$

with $a_o(k)$ from (11.30), $\forall k \in \{1, 2, \cdots, N\}$ and

$$W_c(N) = \sum_{s=1}^{N-1} (A^{(N-1)-s}) B_c B_c^T (A^{T(N-1)-s})$$

is the discrete-time controllability Gramian associated with the pair $(A,\ Bc)$, and e_f is the attacker's objective error state over a finite time interval $[0,\ N]$.

Proof. By Corollary 11.1, $a_o(k)$ in (11.30) guarantees undetectability. Applying the above $a_c(k)$ into (11.33) results in $\mathbb{E}[e_a(N)] = e_f$, which completes the proof. □

Remark 11.2. It is significant to summarize the characteristics of the stealthy deception attack in the individual cases:

Case 1: Injected attack sequences can cause only the limited estimation error to avoid being detected.

Case 2: Attack sequences gradually falsify the data so that the residual power does not become large enough to trigger an attack alarm. Attackers can eventually induce the infinite estimation error as time goes to infinity, while not being detected.

Case 3: Attackers can manipulate the estimation error insofar as u_a is physically admissible. Moreover, since the resulting residuals are not changed in their statistical property, any residual-based monitoring systems are of no use against this type of attack.

From these results the attacker can design a stealthy deception attack in the vulnerable control subspaces. The defender, on the other hand, can find such vulnerable subspaces and reduce or eliminate them by increasing the integrity of major communication channels, installing redundant network agents, etc.

11.3.4 Application to UAV navigation systems

In this section we illustrate the effects of stealthy deception attacks in the Unmanned Aerial Vehicle (UAV) navigation system. Today, combining data from different sensors to improve the performance of the overall sensing system becomes necessary in various applications. For UAV navigation, two major sensing systems are used: Inertial Navigation Systems (INS) and Global Positioning Systems (GPS) [441]. Although the INS is the more commonly used navigation sensor in UAVs, it has cumulative errors that continuously increase with time. Hence, reliable and effective additional information to reduce these errors, such as GPS measurement, is required. Fig. 11.2 shows the general architecture of the INS-GPS integrated UAV navigation system.

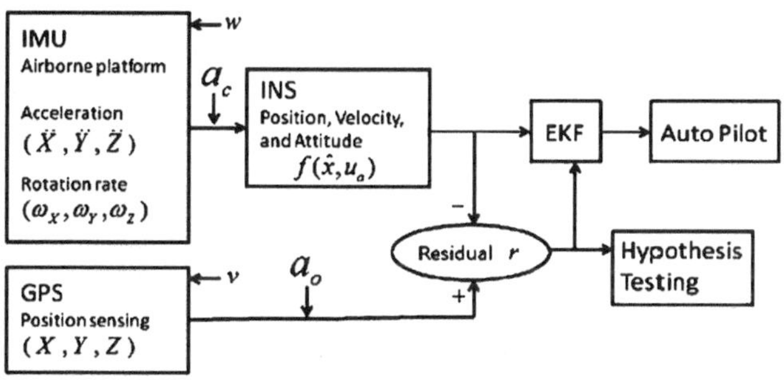

FIGURE 11.2 UAV navigation system architecture.

A commonly adopted technique for combining these two types of measurements is an extended Kalman filter (EKF) where the nonlinear dynamics are linearized around each estimated state. In computer simulation, we consider a simplified UAV dynamics and navigation system by eliminating the attitude and rotational motions of the UAV. Then the Inertial Measurement Unit (IMU) is assumed to provide only the accelerations $(\ddot{X},\ \ddot{Y},\ \ddot{Z})$.

Thus, the observation equations of the IMU and GPS are as follows:

$$
\begin{aligned}
u_a(k) &= \begin{bmatrix} 0 & C_I \end{bmatrix}\begin{bmatrix} x(k) \\ \dot{x}(k) \end{bmatrix} + B_c a_c(k) + E_1 w(k), \\
y_a(k) &= \begin{bmatrix} C_G & 0 \end{bmatrix}\begin{bmatrix} x(k) \\ \dot{x}(k) \end{bmatrix} + B_o a_o(k) + E_2 v(k),
\end{aligned}
$$

$$C_I(k) = \begin{bmatrix} 0 & 1 & 0 & 0 & 0 & 0 \\ 0 & 0 & 0 & 1 & 0 & 0 \\ 0 & 0 & 0 & 0 & 0 & 1 \end{bmatrix}, \quad C_G = \begin{bmatrix} 1 & 0 & 0 & 0 & 0 & 0 \\ 0 & 0 & 1 & 0 & 0 & 0 \\ 0 & 0 & 0 & 0 & 1 & 0 \end{bmatrix}, \tag{11.35}$$

where

$$x = \begin{bmatrix} X & \dot{X} & Y & \dot{Y} & Z & \dot{Z} \end{bmatrix}$$

is the UAV position and velocity, and

$$\begin{aligned} u_a &= \begin{bmatrix} \ddot{X} & \ddot{Y} & \ddot{Z} \end{bmatrix}^T, \\ y_a &= \begin{bmatrix} X & Y & Z \end{bmatrix}^T \end{aligned}$$

are the compromised measurements from the IMU and GPS, respectively. Then the filter dynamics are given by

$$\begin{aligned} \hat{x}_a(k+1) &= A_I\hat{x}_a(k) + B_I u_a(k) - K\,(y_a(k+1) - C_G A_I \hat{x}_a(k) \\ &- C_G B_I u_a(k)), \\ A_I &= \begin{bmatrix} 1 & T_s & 0 & 0 & 0 & 0 \\ 0 & 1 & 0 & 0 & 0 & 0 \\ 0 & 0 & 1 & T_s & 0 & 0 \\ 0 & 0 & 0 & 1 & 0 & 0 \\ 0 & 0 & 0 & 0 & 1 & T_s \\ 0 & 0 & 0 & 0 & 0 & 1 \end{bmatrix}, \\ B_I &= \begin{bmatrix} \frac{T_s^2}{2} & 0 & 0 \\ T_s & 0 & 0 \\ 0 & \frac{T_s^2}{2} & 0 \\ 0 & T_s & 0 \\ 0 & 0 & \frac{T_s^2}{2} \\ 0 & 0 & T_s \end{bmatrix}, \end{aligned}$$

where A_I and B_I are the dynamics of the INS, and T_s is a sampling time. In computer simulation, we consider three different stealthy deception attack scenarios associated with Case 3, which is the most dangerous case for the UAV navigation system. For this purpose, we let both IMU data and GPS signals be attackable simultaneously, that is, $B_c = C_I$ and $B_o = C_G$. In view of Corollaries 11.1 and 11.2, the attacker can spoof the UAV's position and velocity to arbitrary states.

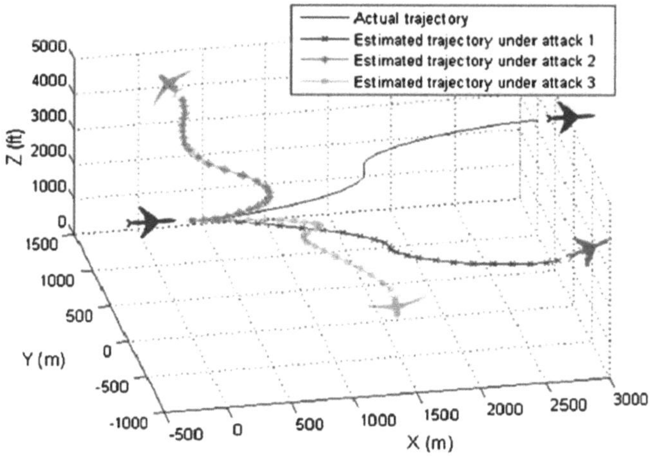

FIGURE 11.3 Trajectory estimates under three different attacks.

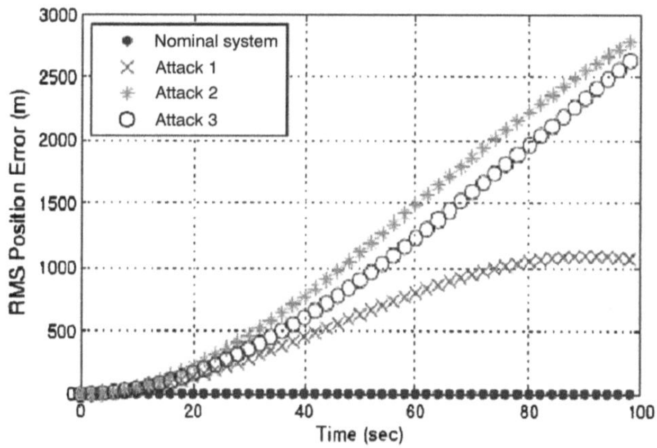

FIGURE 11.4 Estimation error under three different attacks.

Fig. 11.3 demonstrates the effects of three different Case 3 stealthy deception attacks on the 3D tracking of the UAV. Here the estimated UAV trajectories under the three different attacks significantly deviate from the actual UAV trajectory without being detected (i.e., the UAV can be hijacked to any place that the attacker wants).

Figs. 11.4 and 11.5 compare the values of the position estimation error and the residual power $r^T(k)\Sigma^{-1}r(k)$ of the nominal system and the system under the three different attacks shown in Fig. 11.3, which are computed by the average of the Monte Carlo simulations with 1000 runs. The results show each attack can easily cause the large estimation error, and the empirical test statistics are around $\mathbb{E}[r^T(k)\Sigma^{-1}r(k)] = m$ regardless of attacks. This demonstrates that the

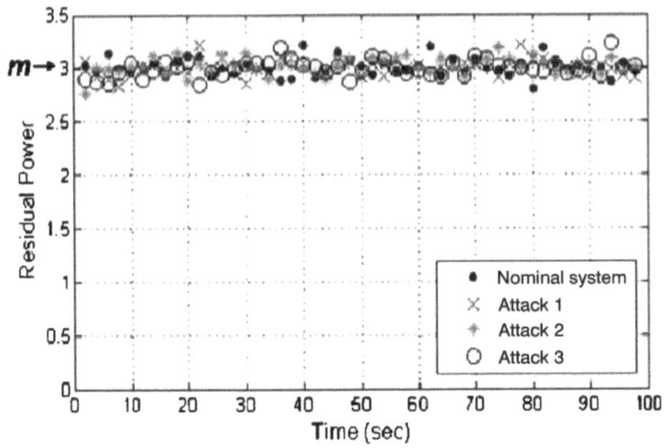

FIGURE 11.5 Test statistics under three different attacks.

residuals under attack have no significant change in their statistical properties, which implies that the attacks are undetectable via the residual tests.

11.4 Notes

In this chapter, we studied the cybersecurity of a class of linear state estimators which are widely used in Air Traffic Control (ATC) applications. The cyber attacks can inject false data into the noisy observations such that the estimators give incorrect state estimates, which is a dangerous scenario in safety-critical ATC operations. We derived the conditions under which the deception attacks fail the state estimators while successfully bypassing the monitoring system. Using the proposed conditions, we not only evaluate the vulnerability level of a given CPS, but also develop secure system design methodologies against stealthy deception attacks. The developed conditions are tested with numerical simulation for a UAV navigation system.

Chapter 12

A unified game approach under DoS attacks

Contents

12.1 Introduction

Recent years have witnessed the wide application of network technology to the control system. Integrated with communication channels, networked control systems (NCSs) have increased mobility and interoperability, and low maintenance and installation costs [442]. The exposure to public networks means that the control system is the target of potential cyber attacks. Being the connection between information and the real world, control systems targeted by cyber attacks can have serious incidents, which have been verified during the past decade [443], [444].

By targeting different components of the control system the attacker can launch various types of attacks. Most of these control system-oriented attacks can be categorized as deception attacks [445], [446], which compromise data integrity and denial-of-service (DoS) attacks, which in turn compromise data availability. Compared with deception attacks, DoS attacks that lead to congestion in the communication network require little prior knowledge of the control system [242], but they can seriously threaten the control systems operating in real time. For example, the control system using deadline corrective control may be driven to instability under DoS attacks [113]. Hence, it has been listed as the most financially expensive of security incidents [88] and raises major concerns. In the literature, some arguably representative theoretic tools have

Cloud Control Systems. https://doi.org/10.1016/B978-0-12-818701-2.00020-2

been employed in securing the NCS under DoS attacks, which are algebraic graph methods [447], game theory [105], and control methods such as the linear quadratic Gaussian (LQG) optimal control method [448], model predictive control method [449], and event-trigger control method [450].

There have been two major lines of research on securing the NCS under a DoS attack, which can be categorized as the attack-tolerant control methods and attack-compensation control methods. For the first category control strategies have been exploited that can tolerate unwanted network-induced phenomena caused by DoS attacks, such as extra packet dropout or time delays [449], [451]. Note that overlong time delays or excessive loss of packets may exceed the ability of the attack-tolerant control method [448], [449], and hence the control system may still be driven to instability. Thus, some attack-compensation control methods have been employed to compensate for the control performance degradation caused by DoS attacks via contradicting undesirable attack-induced phenomena. In [113], [105], and [452] the intrusion detection system (IDS) was deployed in the cyber layer, which can defend against DoS attacks and improve the performance of the underlying control system. [453] developed data-sending strategies to contradict the packet dropout induced by DoS attacks. [91] introduced the concept of security investment into the filed of resilient control, where the degradation of control performance can be minimized by finding the appropriate investment strategies. Unfortunately, to the best of the authors' knowledge, all these resilient control schemes were developed in the continuous or discrete domain, and therefore they ignore the possible numerical illness for the sampled system with a high sampling rate [312]. The purpose of this chapter is therefore to reduce this gap.

In this chapter a novel resilient control method against DoS attacks is proposed for NCSs by recurring to the delta operator, which is a well-recognized tool for overcoming numerical illness in the discrete-time system with a fast sampling rate [454]. Two control structures are considered for NCSs, which are the multiple-tasking optimal control (MTOC) and central-tasking optimal control (CTOC) structures. The algorithms used to obtain the optimal compensation strategy (defense strategy) and optimal attack strategy are provided. Specifically, this chapter provides three main contributions:

1. The existence conditions and analytical expressions for the MTOC and CTOC strategies are derived in terms of backward recursions in the delta domain.
2. The attacking scenario is considered where the defender and attacker possess asymmetric information. From the defender's perspective the inverse game approach is proposed, which can compensate for the attack-induced performance loss.
3. An energy-constraint attacker is considered, and the attack strategies are provided, which aim to drive the control system out of the safety zone using the lowest attacking intensity.

Notation. Some standard notations are used throughout. For a matrix M, M^T denotes its transpose. $M > 0$ ($M < 0$, respectively) means that M is positive definite (negative definite, respectively). $M \geq 0$ ($M \leq 0$, respectively) means that M is a nonnegative (nonpositive definite, respectively) matrix. $\{M^i\}_{i=1}^S$ denotes the set of matrices from M^1 to M^S. $\mathbb{E}\{.\}$ indicates taking the expected value of ".". $\mathbb{R}^n$ denotes the n-dimensional Euclidean space. $\mathrm{diag}(x)$ is a diagonal matrix with the diagonal entries given by the elements of the vector x.

12.2 Problem description

12.2.1 Model of NCS subject to DoS attack

Consider the dynamics of the delta-domain NCS under DoS attack as follows:

$$\delta x_k = A_\delta x_k + \sum_{i=1}^{S} \alpha_k^i B_\delta^i u_k^i. \tag{12.1}$$

Here $x_k = x_{KT_s}$, T_s is the sampling interval, and A_δ and B_δ^i are matrices in the delta domain with appropriate dimensions. The variable $\alpha_k^i = \alpha_k^{FN} \alpha_k^{BN}$, $i \in \mathbf{S} := \{1, 2, \cdots, S\}$ indicates the effect of DoS attacks on the control system. When DoS attacks are launched, there is a chance that the kth sensor packet is dropped and the "zero-control" input strategy is applied. We assume that α_k^i is the random variable that is distributed according to the Bernoulli distribution. Suppose that α_k^i is independent and identically distributed (i.i.d.) and, for $i \neq j, i, j \in \mathbf{S}$, α_k^i is independent of α_k^j. Let us denote

$$\mathbb{P}\{\alpha_k^i = 0\} = \alpha^i, \quad \mathbb{P}\{\alpha_k^i = 1\} = 1 - \alpha^i = \bar{\alpha}^i$$
$$\forall i \in \mathbf{S}, k \in \mathbf{K} := \{1, 2, \ldots, K\}.$$

Note that α^i here is viewed as intensity of attack (IoA). A more intense DoS attack will lead to higher IoA α^i. Let us define the strategy set of the DoS attacker to be $\mathcal{G} := \{\alpha^i\}_{i=1}^S$.

Remark 12.1. **1.** System (12.1) is actually a delta-domain extension of the widely used discrete-domain model describing the control system under DoS attacks; see, e.g., [113], [448], and [91].

2. Resilient control under DoS attacks should be able to address the problem of packet dropout which is also common in a traditional NCS [113], [448], [451], [91]. However, it is worth mentioning that the packet dropout rate caused by inherent communication failure is much lower compared with that induced by a malicious DoS attack [442]. This puts forward a higher requirement for resilient control, which should be able to tolerate serious congestion in the communication channel, not just occasionally occurring information loss.

12.2.2 MTOC and CTOC design

In this section, we consider a NCS with the structures of MTOC and CTOC. The CTOC structure requires that all the controllers should be coordinated to reach a common objective, while those of the MTOC do not cooperate with each other and only minimize the individual cost function. Thus, MTOC can be used to model some complex and distributed systems where each decision maker is selfish and only pursues its own interest. The CTOC structure, on the other hand, aims to address problems where the central coordination and cooperation among decision makers are possible. Note that the controller design of MTOC falls within the framework of noncooperative dynamic game in the delta domain with each controller (player) minimizing the cost-to-go function,

$$J^i = \mathbb{E}\left\{x_K^T Q_\delta^K x_K + T_s \sum_{k=0}^{K-1} \Psi_0\right\}, \quad i \in \mathbf{S}, \tag{12.2}$$

where $\Psi_0 = x_k^T Q_\delta^i x_k + \alpha_k^i u_k^{i\,T} R_\delta^i u_k^i$. We assume that $Q_\delta^K \geq 0$, $Q_\delta^i \geq 0$, and $R_\delta^i > 0$ for all $i \in \mathbb{S}$. In contrast, for CTOC there is a central coordination such that all players cooperate to minimize a common cost-to-go function defined as $\tilde{J} = \sum_{i=1}^S \eta^i J^i$, where $\eta^i > 0$ is the weighting factor on player i's cost-to-go function and satisfies the normalization condition $\sum_{i=1}^S \eta^i = 1$. For both MTOC and CTOC a transmission control protocol (TCP) is applied where each packet is acknowledged, and the information set is defined as $\mathcal{I}_0 = \{x_0\}$, $\mathcal{I}_k = \{x_1, x_2, \cdots, x_k, \alpha_0, \alpha_1, \cdots, \alpha_{k-1}\}$. The admissible strategies μ_k^i for MTOC and $\tilde{\mu}_k^i$ for CTOC are seen as functional to map information set $\mathcal{I}_k$ to u_k^i (i.e., $u_k^i = \mu_k^i(\mathcal{I}_k)$ or $u_k^i = \tilde{\mu}_k^i(\mathcal{I}_k)$). Let us further define

$$\mu^i := \{\mu_0^i, \mu_1^i, \ldots, \mu_{K-1}^i\}, \quad \tilde{\mu}^i := \{\tilde{\mu}_0^i, \tilde{\mu}_1^i, \ldots, \tilde{\mu}_{K-1}^i\}.$$

To this end, two optimization problems should be addressed. For Problem 12.1, all the players are selfish and noncooperative. Let μ^{-1} denote the collection of strategies of all players except player i, (i.e., $\mu^{-1} = \{\mu^1, \cdots, \mu^{i-1}, \mu^{i+1}, \mu^S\}$). Player i is faced with minimizing its own associated cost function by solving the following dynamic optimization problem.

Problem 12.1. Find control strategies $\mu^{i*}, i \in \mathbf{S}$ such that the following optimization problem is solved for all $i \in \mathbf{S}$:

$$\begin{aligned}
(\text{OC(i)}) \quad & \min_{\mu^i} \; J^i(\mu^i, \mu^{-i*}) \\
& := \mathbb{E}\left\{x_K^T Q_\delta^K x_K + T_s \sum_{k=0}^{K-1} \Psi_0\right\} \\
\text{s.t.} \quad & \delta x_k = A_\delta x_k + \sum_{i=1}^S \alpha_k^i B_\delta^i u_k^i.
\end{aligned} \tag{12.3}$$

If the optimization is carried out for each i, Nash equilibrium (NE) will be obtained which is $J^{i*}(\mu^{i*}, \mu^{-i*}) \leq J^i(\mu^i, \mu^{-i})$. The corresponding total cost achieved is given by $J^* = \sum_{i=1}^{S} \eta^i J^{i*}$. In the same vein, the associate optimization problem of CTOC is shown as follows:

Problem 12.2. Find control strategies $\tilde{\mu}^* := \{\tilde{\mu}^{i*}, \tilde{\mu}^{-i*}\}$ such that the following optimization problem is solved:

$$\text{(COC)} \quad \min_{\tilde{\mu}} \quad \tilde{J}(\tilde{\mu}^i, \tilde{\mu}^{-i}) \tag{12.4}$$

$$:= \sum_{i=1}^{S} \eta^i \mathbb{E}\left\{x_K^T Q_\delta^K x_K + T_s \sum_{k=0}^{K-1} \Psi_0\right\}$$

$$\text{s.t.} \quad \delta x_k = A_\delta x_k + \sum_{i=1}^{S} \alpha_k^i B_\delta^i u_k^i.$$

This minimization is essentially an optimal control problem. The optimal value $\tilde{J}^*(\tilde{\mu}^{i*}, \tilde{\mu}^{-i*}) \leq \tilde{J}(\tilde{\mu}^i, \tilde{\mu}^{-i})$ will be obtained if the optimization problem is solved.

12.2.3 Defense and attack atrategy design

In this chapter, we assume that the defender has the freedom to participate in choosing the weighting matrices $Q_\delta^i, i \in \mathbf{S}$ to compensate for the performance degradation of NCS with the MTOC (CTOC, respectively) structure [456]. We define the set of defense strategies as $\mathcal{F} := \{Q_\delta^i\}_{i=1}^S$ and the optimal defense strategies are obtained by solving the following problem:

Problem 12.3. For the MTOC structure, find the optimal defense strategy $\mathcal{F}^*$ that is the solution of

$$\min_{\mathcal{F}} \quad \left| J_{\mathcal{G}_1}^* / \tilde{J}_{\mathcal{G}_0}^* - 1 \right|,$$

while for the CTOC structure, find the optimal defense strategy $\mathcal{F}^*$ that is the solution of

$$\min_{\mathcal{F}} \quad \left| \tilde{J}_{\mathcal{G}_1}^* / \tilde{J}_{\mathcal{G}_0}^* - 1 \right|,$$

where $\mathcal{G}_0 = \{\alpha^i = 0\}_{i=1}^S$ and $\mathcal{G}_1 = \{0 < \alpha^i < 1\}_{i=1}^S$.

It is worth mentioning that $J_{\mathcal{G}_1}^* > \tilde{J}_{\mathcal{G}_1}^* > \tilde{J}_{\mathcal{G}_0}^* > 0$, since both the noncooperative behavior and packet dropout phenomenon can lead to performance degradation of the control system. On the other hand, for DoS attackers, a case is considered where the attacker is fully aware of the defender's compensation strategy $\mathcal{F}^*$ (i.e., the attacker and defender possess asymmetric information). To save the attacking cost, the DoS attacker aims to drive the underlying NCS out of the safety zone using an attacking intensity $\mathcal{G} = \{\alpha^i\}_{i=1}^S$ that is as low as possible. The optimal attack strategies $\mathcal{G}^*$ are obtained by solving the following problem:

Problem 12.4. For the given weighting factors $\{\rho^i\}_{i=1}^S$ and the MTOC structure, find the optimal attacking strategy $\mathcal{G}^*$ that is the solution of the following optimization problem:

$$\begin{aligned} &\min \alpha^1 \\ &\text{s.t.} \quad \left| J^*_{\mathcal{G}_1} / \tilde{J}^*_{\mathcal{G}_0} - 1 \right| > \mathcal{S}_o, \quad \alpha^1 = \rho^2 \alpha^2 = \cdots = \rho^S \alpha^S, \end{aligned} \tag{12.5}$$

while for the CTOC structure, find the optimal attacking strategy $\mathcal{G}^*$ that is the solution of the following optimization problem:

$$\begin{aligned} &\min \ \alpha^1 \\ &\text{s.t.} \quad \left| \tilde{J}^*_{\mathcal{G}_1} / \tilde{J}^*_{\mathcal{G}_0} - 1 \right| > \mathcal{S}_o, \quad \alpha^1 = \rho^2 \alpha^2 = \cdots = \rho^S \alpha^S, \end{aligned} \tag{12.6}$$

where $\mathcal{S}_o$ is a scalar representing the safety zone. Without loss of generality, we assume that $\rho^i = 1, i = 2, \cdots, S$ subsequently.

12.3 MTOC and CTOC control strategies

In this section the conditions and analytical form of optimal control strategies for the MTOC and CTOC structures are provided. Some preliminary notations are provided as follows:

$$\begin{aligned}
\Theta_0 &= (T_s A_\delta + I) - T_s \sum_{j \neq i}^{S} \bar{\alpha}^j B_\delta^j L_k^j \\
\Theta_1 &= A_\delta - \sum_{i=1}^{S} \bar{\alpha}^i B_\delta{}^i L_k^i \\
\Theta_2 &= (T_s A_\delta + I) x_k + T_s \sum_{i=1}^{S} \alpha_k^i B_\delta^i u_k^i \\
\Theta_3 &= (T_s A_\delta + I) x_k + T_s \sum_{j \neq i}^{S} \bar{\alpha}^j B_\delta^j u_k^j \\
\Omega_0 &= \sum_{i=1}^{S} \eta^i Q_\delta^i \\
\Omega_1 &= (T_s A_\delta + I) \\
\Omega_2 &= \operatorname{diag}\Big(\Big[\bar{\alpha}^1(1-\bar{\alpha}^1) B_\delta^{1^T} \tilde{P}_{k+1} B_\delta^1, \bar{\alpha}^2(1-\bar{\alpha}^2) \\
&\quad B_\delta^{2^T} \tilde{P}_{k+1} B_\delta^2, \ldots, \bar{\alpha}^S(1-\bar{\alpha}^S) B_\delta^{1^T} \tilde{P}_{k+1} B_\delta^S\Big]\Big).
\end{aligned}$$

The following theorem is shown first to provide solution to Problem 12.1.

Theorem 12.1. *For an NCS with an MTOC structure and a given attack strategy $\mathcal{G}$, the following conclusions are presented:*

1. *There exists a unique NE if*

$$R_\delta^i + T_s B_\delta^{i\,T} P_{k+1}^i B_\delta^i > 0 \tag{12.7}$$

and the matrix Φ_k is invertible, where $\Phi_k(i,i) = R_\delta^i + T_s B_\delta^{i\,T} P_{k+1}^i B_\delta^i$, $\Phi_k(i,j) = T_s \bar{\alpha}^j B_\delta^{i\,T} P_{k+1}^i B_\delta^j$.

2. *Under condition 1 the optimal control strategies of MTOC are given by $u_k^i = \mu_k^{i*}(\mathcal{I}_k) = -L_k^i x_k$ for all $i \in \mathbf{S}$, where*

$$L_k^i = (R_\delta^i + T_s B_\delta^{i\,T} P_{k+1}^i B_\delta^i)^{-1} B_\delta^{iT} P_{k+1}^i \Theta_0. \tag{12.8}$$

3. *The backward iterations are carried out with $P_K^i = Q_\delta^K$ and*

$$\begin{aligned} -\delta P_k^i &= Q_\delta^i + \bar{\alpha}^i L_k^{i\,T} R_\delta^i L_k^i \\ &\quad + T_s \Theta_1^T P_{k+1}^i \Theta_1 + \Theta_1^T P_{k+1}^i + P_{k+1}^i \Theta_1 \\ &\quad + T_s \sum_{j=1}^{S} \left(\left(\bar{\alpha}^j - \bar{\alpha}^{j2} \right) L_k^{j\,T} B_\delta^{j\,T} P_{k+1}^i B_\delta^j L_k^j \right) \end{aligned} \tag{12.9}$$

$$P_k^i = P_{k+1}^i - T_s \delta P_k^i. \tag{12.10}$$

4. *Under condition 1 the NE values under the MTOC are $J^{i^*} = x_0^T P_0^i x_0$, $i \in \mathbf{S}$, where x_0 is the initial value.*

Proof. An induction method is employed here. The claim is clearly true for $k = K$ with parameters $P_K^i = Q_\delta^K$. Let us suppose that the claim is now true for $k+1$, and the cost function at $k+1$ is constructed as

$$V^i(x_{k+1}) = \mathbb{E}\{x_{k+1}^T P_{k+1}^i x_{k+1}\} \tag{12.11}$$

with $P_{k+1}^i > 0$. According to ([454], Lemma 1) the above equation can be rewritten in the delta domain as

$$\begin{aligned} V^i(x_{k+1}) &= T_s \delta (x_k^T P_k^i x_k) + x_k^T P_k^i x_k \\ &= T_s \delta x_k^T P_{k+1}^i x_k + T_s x_k^T P_{k+1}^i \delta x_k \\ &\quad + T_s^2 \delta x_k P_{k+1}^i \delta x_k + x_k^T P_{k+1}^i x_k. \end{aligned}$$

Using the dynamic programming method, the cost at time k is obtained by

$$V^i(x_k) = \min_{u_k^i} \mathbb{E} \left\{ T_s x_k^T Q_\delta^i x_k + T_s \alpha_k^i u_k^{i\,T} R_\delta^i u_k^i + V^i(x_{k+1}) \right\}. \tag{12.12}$$

The cost-to-go function $V^i(x_k)$ is strictly convex of u_k^i since the second derivative of (12.12) yields $R_\delta^i + T_s B_\delta^{i^T} P_{k+1}^i B_\delta^i > 0$. The optimal control strategies can be obtained by solving $\partial V^i(x_k)/\partial u_k^i = 0$, which is set up for all players. Thus, there exists a unique NE if the invertible matrix Φ_k satisfies

$$\Phi_k \bar{L}_k = \Pi_k \tag{12.13}$$

with $\Phi_k(i,i) = R_\delta^i + T_s B_\delta^{i^T} P_{k+1}^i B_\delta^i$, $\Phi_k(i,j) = T_s \bar{\alpha}^j B_\delta^{i^T} P_{k+1}^i B_\delta^j$, $\Pi_k(i,i) = B_\delta^{i^T} P_{k+1}^i (T_s A_\delta + I)$, and $\Pi_k(i,j) = 0$, $\bar{L}_k = \left[L_k^{1T}, L_k^{2T}, \cdots, L_k^{ST}\right]^T$. Substituting $u_k^{i^*} = -L_k^i x_k$ into (12.12), we have (12.9). This completes the proof. □

Next the following theorem is given to provide solutions to Problem 12.2:

Theorem 12.2. *First, let us define*

$$\begin{aligned}
\tilde{R}_\delta &:= diag\left(\left[\bar{\alpha}^1 \eta^1 R_\delta^1, \bar{\alpha}^2 \eta^2 R_\delta^2, \ldots, \bar{\alpha}^S \eta^S R_\delta^S\right]\right)\\
\tilde{B}_\delta &:= \left[\bar{\alpha}^1 B_\delta^1, \bar{\alpha}^2 B_\delta^2, \ldots, \bar{\alpha}^S B_\delta^S\right]\\
\Xi &:= \tilde{R}_\delta + diag\left(\left[T_s \bar{\alpha}^1 (1-\bar{\alpha}^1) B_\delta^{1^T} \tilde{P}_{k+1} B_\delta^1,\right.\right.\\
&\quad T_s \bar{\alpha}^2 (1-\bar{\alpha}^2) B_\delta^{2^T} \tilde{P}_{k+1} B_\delta^2, \ldots, T_s \bar{\alpha}^S (1-\bar{\alpha}^S)\\
&\quad \left.\left. B_\delta^{S^T} \tilde{P}_{k+1} B_\delta^S\right]\right) + T_s \tilde{B}_\delta^T \tilde{P}_{k+1} \tilde{B}_\delta.
\end{aligned}$$

For an NCS with a CTOC structure and a given attack strategy $\mathcal{G}$, the following conclusions are presented.

1. *If we have an invertible matrix $\Xi > 0$, there exists a unique optimal solution.*
2. *Let us denote $\tilde{u}_k = [u_k^{1T}, u_k^{2T}, \ldots, u_k^{ST}]^T$. Under condition 1 the optimal control strategy of CTOC $\tilde{\mu}^*$ is given by $\tilde{u}_k = \tilde{\mu}^*(\mathcal{I}) = -\tilde{L}_k x_k$, where*

$$\tilde{L}_k = \Xi^{-1} \tilde{B}_\delta^T \tilde{P}_{k+1} \Omega_1. \tag{12.14}$$

3. *The backward recursions are carried out with $\tilde{P}_K = \sum_{i=1}^S \eta^i Q_\delta^K$ and*

$$\begin{aligned}
-\delta \tilde{P}_k &= \Omega_0 + T_s A_\delta^T \tilde{P}_{k+1} A_\delta + \tilde{P}_{k+1} A_\delta \qquad (12.15)\\
&\quad + A_\delta^T \tilde{P}_{k+1} - \Omega_1^T \tilde{P}_{k+1} \tilde{B}_\delta \Xi^{-1} \tilde{B}_\delta^T \tilde{P}_{k+1} \Omega_1\\
\tilde{P}_k &= \tilde{P}_{k+1} - T_s \delta \tilde{P}_k. \qquad (12.16)
\end{aligned}$$

4. *Under condition 1 the optimal value with the CTOC structure is $\tilde{J}^* = x_0^T \tilde{P}_0 x_0$, where x_0 is the initial value.*

Proof. The proof is similar to that of Theorem 12.1 and is omitted here. □

Remark 12.2. Note that Theorems 12.1 and 12.2 differ from most of the existing literature on the control of delta operator system due mainly to the following:

1. By recurring to the dynamic programming method, the optimal solution can be obtained instead of suboptimal results obtained using the LMI method [312].
2. The results in Theorems 12.1 and 12.2 are readily extended to develop time-variant optimal control strategies for time-variant system, while the LMI method normally obtains time-invariant control strategies [312], [454].

12.4 Defense and attack strategies

In this section, an intelligent but resource-limited attacker is considered. The attacker is intelligent in the sense that it can capture the defender's strategies and adopt countermeasures (i.e., the attacker and defender possess asymmetric information). These interactions can be described by a Stackelberg game [113], where the attacker acts as the leader and the defender acts as the follower. The defender aims to minimize the performance degradation caused by the attacker. On the other hand, being fully aware of the defender's strategies, the attacker aims to drive the control system out of the safety zone while minimizing the attacking intensity. The details of the Stackelberg game and the derivation of the corresponding Stackelberg solutions are shown below.

12.4.1 Development of defense strategies

According to the previous section, an NCS with a CTOC structure under no DoS attack is viewed as the nominal/ideal case. Thus, the desirable cost-to-go function yields $\tilde{J}^*_{\mathcal{G}_0}$. On the other hand, we denote the desired feedback gains to be $\{\hat{L}^i_k\}_{i=1}^S$, which are "desirable" in the sense that they can adapt to the network communication environment better. The desirable feedback gains can be determined by using Theorems 12.1 and 12.2, or via a system identification method from the experimental data [455]. In what follows we provide algorithms to achieve both desirable cost-to-go function and feedback gains.

Theorem 12.3. *For the given $\mathcal{G}_1 = \{\alpha^i\}_{i=1}^S$ and desired strategies $\{\hat{L}^i_k\}_{i=1}^S$:*

1. *If positive matrices $\{Q^i_\delta\}_{i=1}^S$ and $\{P^i_0, P^i_1, \ldots, P^i_{K-1}\}_{i=1}^S$ exist such that (12.8) and (12.9) hold with $L^i_k = \hat{L}^i_k$, then $\{u^i_k = \hat{L}^i_k x_k\}$ is the optimal control strategy of MTOC.*
2. *If positive matrices $\{Q^i_\delta\}_{i=1}^S$ and $\{\tilde{P}_0, \tilde{P}_1, \ldots, \tilde{P}_{K-1}\}$ exist such that (12.14) and (12.15) hold with $\tilde{L}_k = \left[\hat{L}^{1T}_k, \hat{L}^{2T}_k, \ldots, \hat{L}^{ST}_k\right]^T$, then $\tilde{u}_k = -\tilde{L}_k = \begin{bmatrix}\hat{L}^{1T}_k & \hat{L}^{2T}_k & \cdots & \hat{L}^{ST}_k\end{bmatrix}^T x_k$ is the optimal control strategy of CTOC.*

Proof. The result is straightforward using Theorems 12.1 and 12.2.

According to Theorem 12.3, Problem 12.3 can be reformulated as the following convex optimization problems $\mathcal{P}_1$ and $\mathcal{P}_2$:

$$\begin{aligned}
&\mathcal{P}_1: \ \min \ \left| J^*_{\mathcal{G}_1}/\tilde{J}^*_{\mathcal{G}_0} - 1 \right| \\
&\text{s.t.} \ \left(\left\{ Q^i_\delta > 0 \right\}_{i=1}^S, \ \left\{ P^i_0 > 0, \ldots, P^i_{K-1} > 0 \right\}_{i=1}^S \right) \in \mathcal{C} \\
&\mathcal{C} := \left\{ \left(\left\{ Q^i_\delta \right\}_{i=1}^S, \ \left\{ P^i_0, \ldots, P^i_{K-1} \right\}_{i=1}^S \right) : \right. \\
&\left. L^i_k = \hat{L}^i_k, \ (12.8) \text{ and } (12.9) \text{ hold} \right\}
\end{aligned} \tag{12.17}$$

or

$$\begin{aligned}
&\mathcal{P}_2: \ \min \ \left| \tilde{J}^*_{\mathcal{G}_1}/\tilde{J}^*_{\mathcal{G}_0} - 1 \right| \\
&\text{s.t.} \ \left(\left\{ Q^i_\delta > 0 \right\}_{i=1}^S, \ \left\{ \tilde{P}_0 > 0, \ldots, \tilde{P}_{K-1} > 0 \right\} \right) \in \mathcal{C} \\
&\mathcal{C} := \left\{ \left(\left\{ Q^i_\delta \right\}_{i=1}^S, \ \left\{ \tilde{P}_0, \ldots, \tilde{P}_{K-1} \right\} \right) : \right. \\
&\left. \tilde{L}_k = \begin{bmatrix} \hat{L}^{1T}_k & \hat{L}^{2T}_k & \ldots & \hat{L}^{ST}_k \end{bmatrix}^T, \ (12.14) \text{ and } (12.15) \text{ hold} \right\}.
\end{aligned} \tag{12.18}$$

By solving $\mathcal{P}_1$ and $\mathcal{P}_2$, the players can achieve the desired strategies by just tuning the interface parameters Q^{i*}_δ. Furthermore, using the desired feedback gains from practical data to determine the optimality can be helpful to illustrate the goals of the players [455]. □

Remark 12.3. **1.** Note that $\mathcal{P}_1$ and $\mathcal{P}_2$ retain their convexity if weighting matrices R^i_δ and Q^i_δ are both viewed as decision variables. Thus, the weighting matrices R^i_δ and Q^i_δ can be determined simultaneously such that the players achieve desirable strategies.

2. The algorithm for the defense strategies can be seen as the pricing mechanism design [456] in game theory (i.e., by finding appropriate pricing parameters ($\{Q^{i*}_\delta\}_{i=1}^S$), the final outcome of the game can be driven to a desired target). On the other hand, the proposed pricing mechanism in this chapter can be regarded as an extension to the inverse LQR method [455], where the desired control strategies are known in advance, and the weighting matrix Q or R is to be determined.

12.4.2 Development of attack strategies

It should be noted that the traditional work on NCSs normally regards the packet dropout as a constraint [91] and seldom stands on the opposite side to exploit how to enhance the packet dropout rate to degrade the system performance. In this section, a strategic and resource-limited attacker is considered, and furthermore the corresponding attacking strategy is provided. This actually provides a worst case for the defender and is helpful in the defense mechanism design [448].

In what follows we provide Algorithm 1 to solve Problem 12.4, which is a standard dichotomy algorithm providing the optimal attacking strategy $\mathcal{G}^*$ such that the NCS with the MTOC (CTOC, respectively) structure is driven out of the safety zone with the smallest IoA $\alpha^i, i \in \mathbf{S}$.

Algorithm 1 The algorithm for optimal attacking strategy $\mathcal{G}^*$

Initialization: Set $a = 1; b = 0; S_o > 0; 0 < \varepsilon \ll 1$.

1: Denote $\alpha = \bar{\alpha}^1 = \bar{\alpha}^2 = \cdots = \bar{\alpha}^S$, and $\alpha = (a+b)/2$.

2: **while** $|b - a| > \epsilon$ **do**

3: Calculate the optimal defense strategy $(\{Q_\delta^{i*}\}_{i=1}^S)$ by solving (12.18) (resp. (12.19)).

4: Calculate $\left|J_{\mathcal{G}_1}^* / \tilde{J}_{\mathcal{G}_0}^* - 1\right|$ (resp. $\left|\tilde{J}_{\mathcal{G}_1}^* / \tilde{J}_{\mathcal{G}_0}^* - 1\right|$) using Q_δ^{i*}, and compare it with $\mathcal{S}_o$.

5: **if** $\left|J_{\mathcal{G}_1}^* / \tilde{J}_{\mathcal{G}_0}^* - 1\right| < \mathcal{S}_o$ So (resp. $\left|\tilde{J}_{\mathcal{G}_1}^* / \tilde{J}_{\mathcal{G}_0}^* - 1\right| < \mathcal{S}_o$) **then**

6: $a = \alpha, b = b$

7: **else if** $\left|J_{\mathcal{G}_1}^* / \tilde{J}_{\mathcal{G}_0}^* - 1\right| > \mathcal{S}_o$ (resp. $\left|\tilde{J}_{\mathcal{G}_1}^* / \tilde{J}_{\mathcal{G}_0}^* - 1\right| > \mathcal{S}_o$) **then**

8: $a = a, b = \alpha$

9: **end if**

10: Set $\alpha = (a+b)/2$.

11: **end while**

Remark 12.4. 1. The feasibility of Algorithm 1 is that the open-loop system $\delta x_k = A_\delta x_k$ should be unstable because the closed-loop system will be driven to instability again if all the control commands are lost (i.e., $\alpha^1 = \alpha^2 = \cdots = \alpha^S = 1$). Then we will have $J_{\mathcal{G}_1}^* / \tilde{J}_{\mathcal{G}_0}^* \to \infty$ ($\tilde{J}_{\mathcal{G}_1}^* / \tilde{J}_{\mathcal{G}_0}^* \to \infty$, respectively) and the constraint (12.7) will be feasible.

2. It can be deduced that the time complexity of the proposed dichotomy algorithm is $O(\log_2 n)$, where $n = 1/\varepsilon$, while the time complexity of the exhaustive method is $O(n)$. Thus, the proposed attacking algorithm saves more time for the attackers to discover the attacking strategy online. For example, if the safety zone $\mathcal{S}_o$ for the control system changes, the attacker can quickly adapt to this change and can determine the corresponding attacking strategy.

12.5 Validation results

In this section we aim to demonstrate the validity and applicability of the proposed method. For this purpose the proposed methodology is applied to the numerical simulations of the heating, ventilation, and air conditioning (HVAC) system, and the experimental verification on the ball and beam system is also provided.

12.5.1 Building model description

Consider the following dynamics of the HVAC system [456],

$$\rho v_i C_p \frac{d\mathcal{T}_i}{dt} = \sum_{j\neq i}^{S} h_{i,j} a_{i,j} (\mathcal{T}_j - \mathcal{T}_i) + T_0,$$

with $T_0 = h_{i,o} a_{i,o} (\mathcal{T}_\infty - \mathcal{T}_i) + \dot{m}_i C_p (\mathcal{T}_i^{sup} - \mathcal{T}_i)$, where $\mathcal{T}_i$ is the temperature in zone i, ρ is the density of air, C_p is the specific heat of air, and $\mathcal{T}_\infty$ is the outside air temperature. Parameter v_i is the volume of air in the ith zone, $a_{i,j}$ is the area of the wall between zone j and i, $a_{i,j}$ is the total area of the exterior walls and roof of zone i, $h_{i,j}$ and $h_{i,o}$ are respectively the heat transfer coefficients of the wall between zone j and i and the heat coefficient of the exterior walls, $\dot{m}_i$ is the mass flow rate of air into zone i, and $\mathcal{T}_i^{sup}$ is the supply air temperature for zone i. The specific values of the parameters can be found in [456]. Note that $\mathcal{T}_{di}$ is the set temperature of zone i. Then the temperature error of zone i is denoted as $\mathcal{T}_{ei} = \mathcal{T}_i - \mathcal{T}_{di}$. Define $x = \begin{bmatrix} x^{1T} & x^{2T} & \dots & x^{ST} \end{bmatrix}^T \triangleq \begin{bmatrix} \mathcal{T}_{e1} & \mathcal{T}_{e2} & \dots & \mathcal{T}_{eS} \end{bmatrix}^T$ to be the vector of zone temperatures and $\tilde{\mathbf{u}} = \begin{bmatrix} \tilde{u}^{1T} & \tilde{u}^{2T} & \dots & \tilde{u}^{ST} \end{bmatrix}^T \triangleq \begin{bmatrix} \mathcal{T}_1^{sup} & \mathcal{T}_2^{sup} & \dots & \mathcal{T}_S^{sup} \end{bmatrix}^T$ to be the control input (i.e., vector of supply air temperatures). Then the system dynamics are

$$\dot{x}_t = A_s x_t + \sum_{i=1}^{S} B_s^i \tilde{u}_t^i + d_s,$$

where

$$d_s := \begin{bmatrix} d_{s1}^T, & d_{s2}^T, & \dots, & d_{sS}^T \end{bmatrix}^T.$$

Matrices $A_s(i, j)$ and $B_s(i, j)$ are given as

$$A_s(i,j) = \begin{cases} -\left(\sum_{j\in\mathcal{N}_i} \frac{h_{i,j}a_{i,j}}{\rho v_i C_p} + \frac{\dot{m}_i}{\rho v_i} + \frac{h_{i,o}a_{i,o}}{\rho v_i C_p}\right), & i = j \\ \frac{h_{i,j}a_{i,j}}{\rho v_i C_p}, & j \in \mathcal{N}_i \text{ and } j \neq i \\ 0, & \text{otherwise} \end{cases}$$

$$B_s^i(j,1) = \begin{cases} \frac{\dot{m}_i}{\rho v_i}, & i = j \\ 0, & \text{otherwise} \end{cases}$$

where $\mathcal{N}_i$ denotes the neighborhood of player i. Let us denote $B_s = [B_s^1, B_s^2, \ldots, B_s^S] \in \mathbb{R}^{S\times S}$ and suppose that $\tilde{u}$ is composed of a feedback term u and a feedforward compensation term, which is $\tilde{u} = u - B_s^{-1}d_s$. Then we have $\dot{x} = A_s x + B_s(u - B_s^{-1}d_s) + d_s = A_s x + B_s u$, where $u = [u^{1T}, u^{2T}, \ldots, u^{ST}]^T$. Considering the packet dropout and assigning $T_S = 0.05$ s, we have the delta operator system

$$\delta x_k = A_\delta x_k + \sum_{i=1}^{S} \alpha_k^i B_\delta^i u_k^i.$$

12.5.2 Strategy design

We denote IoA $\alpha^i = 0.2$ and $R_\delta^i = I$ for all $i \in \mathbf{S}$. We set the weighting matrices $Q_\delta^i, i \in \mathbf{S}$ and Q_δ^K of the MTOC structure as identity matrix. The desired strategies are obtained using Theorem 12.2 with $\alpha^i = 0$, $Q_\delta^i = I, i \in \mathbf{S}$, and $Q_\delta^K = I$. Then the state trajectories of MTOC with and without compensation are shown in Figs. 12.1 and 12.2, respectively. It is evident that the control performance improves since all the players are induced to cooperate.

Next we stand on the side of the attacker and assign $\mathcal{S}_o = 0.38$ and $\varepsilon = 1 \times 10^{-6}$. By using Algorithm 1, we arrive at the results in Table 12.1, where we conclude that the optimal IoA for the attacker yields $\alpha^i = 0.4639$ if $\alpha^1 = \alpha^2 = \cdots = \alpha^S$.

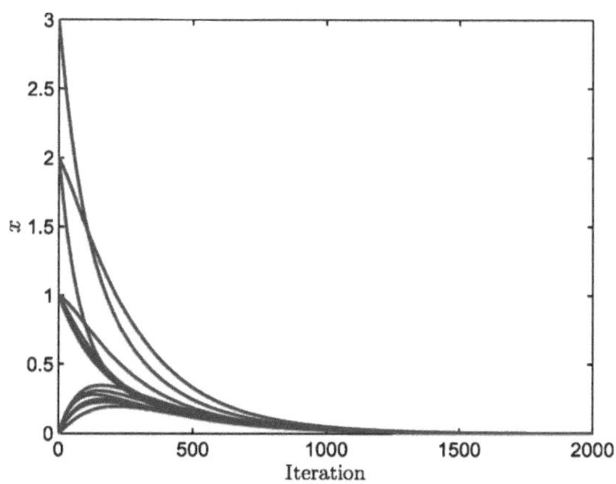

FIGURE 12.1 State trajectories of MTOC without compensation.

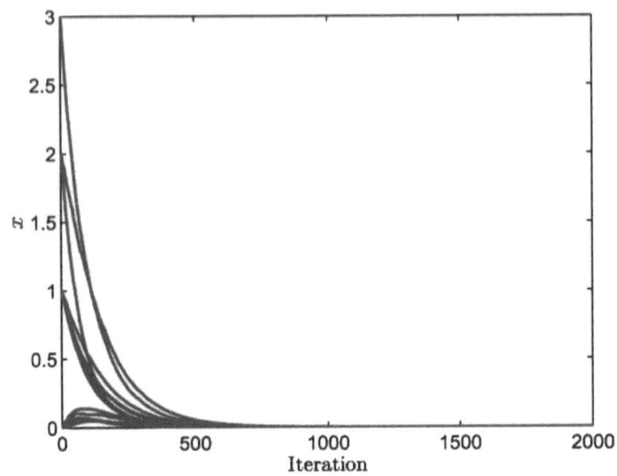

FIGURE 12.2 State trajectories of MTOC with compensation.

TABLE 12.1 Optimal values for network security.

Parameter	Value
Cost of CTOC $\tilde{J}^*$	74.9575
Cost with defense strategy of MTOC J^*	103.4415
Overall performance degradation $J^*_{\mathcal{G}_1}/\tilde{J}^*_{\mathcal{G}_0}$	1.3800
Optimal attack intense α^{i*}	0.4639

12.5.3 Robust study

In this section the robustness of the proposed defense algorithm is verified. Note that the proposed defense algorithm and control schemes all require the estimation of IoA which is not an easy task in practice. Therefore, it is necessary to test the robustness of the defense algorithm by checking whether it still works when the estimation of IoA is not accurate. As in Table 12.2, where J^* is the original cost of the MTOC system and J'^* is the cost of the MTOC system with defense strategies. We assume that the actual value of IoA is $\alpha^i = 0.2$, and the estimation error is described by a certain percentage of the actual value. From Table 12.2 we can see that the proposed defense mechanism still works well even when considering the estimation error.

TABLE 12.2 Comparisons for MTOC system with different estimation errors.

IoA	0	10%	20%	40%
J^*	117.1279	118.9997	120.9949	125.4236
J'^*	102.0808	102.0810	102.0832	102.0910

12.5.4 Comparative study

In this section the following four scenarios are considered for comparison:

1. Optimal: The optimal attack and defense strategies are used as proposed in this chapter.
2. Random: Both the attacker and defender set their strategies randomly, regardless of the existence of the other.
3. Unaware: The attacker does not know if there is a defender and chooses IoA randomly, but the defender chooses optimal strategy $Q^{i*}, i \in \mathbf{S}$ according to which IoA α^i the attacker chooses.
4. Misjudgement: The attacker believes that a defender exists, but the defense strategy is just selected randomly.

The comparison of the results is shown in Fig. 12.3, where we can see that the cases of "optimal" and "unaware" are better than "random" and "misjudgement". Thus, the proposed defense strategy can improve the system performance no matter what strategies the attacker adopts. On the other hand, it should be noted that the proposed attacking strategy can degrade the system performance more, especially when there are no defense mechanisms.

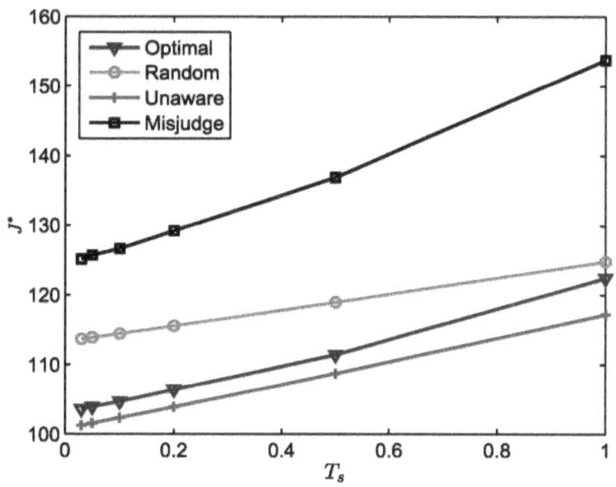

FIGURE 12.3 Comparison of the performance of the different scenarios.

12.6 Experiment verification

To further illustrate the validity and applicability, the proposed methodology is applied to the ball and beam system in Fig. 12.4. Let us denote $x = [\gamma\ \dot{\gamma}\ \theta\ \dot{\theta}]^T$, where γ and θ are the ball position and the beam angle, respectively. Then the state-space representation of the ball and beam system yields [312]

$$\dot{x}_t = A_s x_t + B_s u_t, \tag{12.19}$$

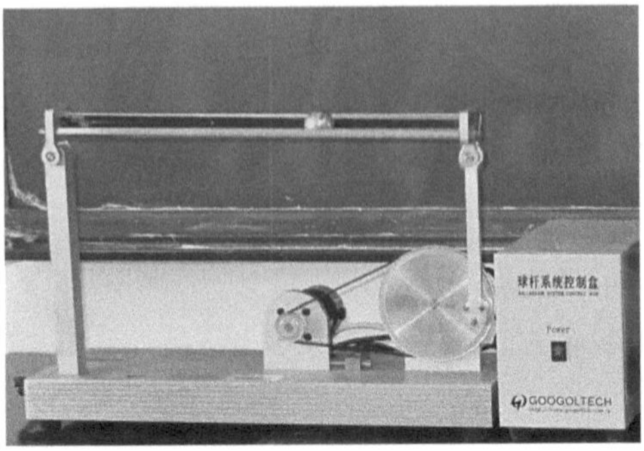

FIGURE 12.4 Ball and beam platform.

where

$$A_s = \begin{bmatrix} 0 & 1 & 0 & 0 \\ 0 & 0 & -7.007 & 0 \\ 0 & 0 & 0 & 1 \\ 0 & 0 & 0 & 0 \end{bmatrix}, \quad B_s = \begin{bmatrix} 0 \\ 0 \\ 0 \\ 1 \end{bmatrix}.$$

We set the initial value of the state as $x_0 = [-0.2\ 0\ 0\ 0]^T$ and the sampling period as $T_s = 0.02$ s. We aim to reselect the weight matrices of the cost function such that the controller obtained using the traditional LQR method [457] can adapt to the network environment. The original controller using the LQR method is $u_k^* = [-9.2713\ \ -8.47462\ \ 7.1400\ \ 7.2887]x_k$ without considering the packet dropout. The desired control strategy yields $u_k^* = [-9.3113\ \ -8.52352\ \ 7.3359\ \ 7.3519]x_k$, which can be obtained from Theorem 12.2 with $\alpha = 0.01$, $Q_1 = diag([1000\ \ 0\ \ 10\ \ 0])$, and $R_1 = [10]$. Then the following optimized weighting matrices can be obtained by solving $\mathcal{P}_2$ with $\bar{\alpha} = 1$:

$$\begin{aligned} Q_2 &= \begin{bmatrix} 181.0158 & 5.9030 & -48.0873 & -1.4436 \\ 5.9030 & 13.7252 & 2.6831 & -2.1865 \\ -48.0873 & 2.6831 & 29.8829 & -0.3409 \\ -1.4436 & -2.1865 & -0.3409 & 1.0521 \end{bmatrix} \\ R_2 &= \begin{bmatrix} 1.7921 \end{bmatrix}. \end{aligned}$$

The control results using LQR controller [457] with Q_1, R_1 and Q_2, R_2 are shown in Figs. 12.5 and 12.6, respectively. It is easy to see that the LQR controller with optimized weighting matrices achieves better control performance.

Next the control performance of the finite precision representation [458] of the system model (12.19) is exploited. Let us assign $T_s = 0.002$ s, $\alpha = 0.01$, $Q = diag([1000\ \ 0\ \ 10\ \ 0])$, and $R = [10]$. The finite precision representation

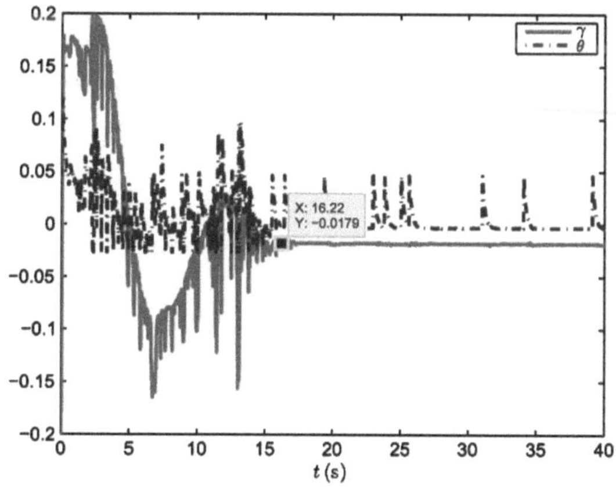

FIGURE 12.5 Results of the experiment using the LQR method with Q_1 and R_1.

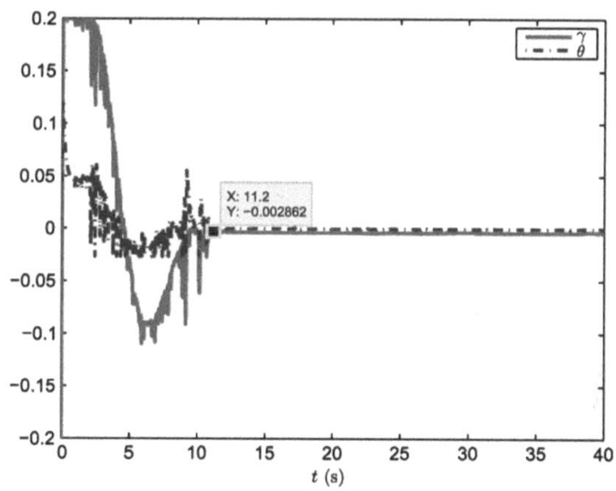

FIGURE 12.6 Results of the experiment using the LQR method with Q_2 and R_2.

of the system model (12.19) is derived with four significant digits in the traditional discrete domain and delta domain, respectively. Then we apply the control strategies obtained from ([459], Lemma 2) and Theorem 12.2 to these finite precision models, and the results are shown in Figs. 12.7 and 12.8, respectively. It is not difficult to see that the proposed delta domain control scheme performs better under the finite word-length constraint.

12.7 Notes

In this chapter the security issue and strategy design of NCSs under DoS attacks were analyzed. Two novel optimal control strategies were developed in the delta domain using the game theory approach. The attacker and defender possessing asymmetric information patterns were considered, and the respective algorithms

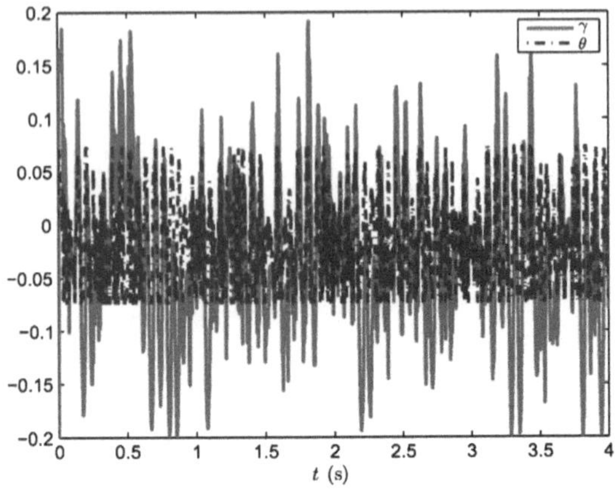

FIGURE 12.7 Results of the experiment using ([459], Lemma 2).

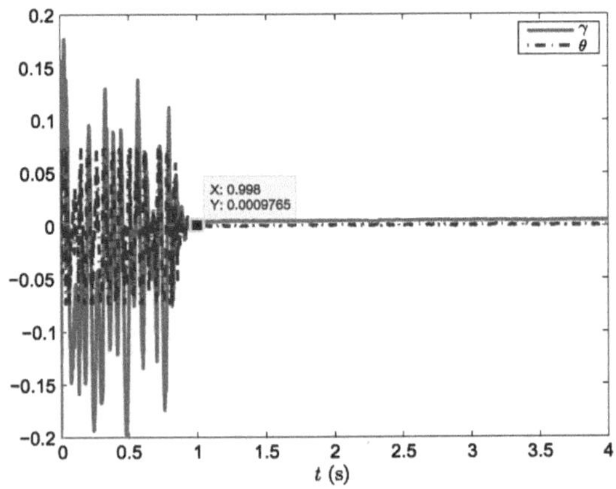

FIGURE 12.8 Results of the experiment using Theorem 12.2.

to develop the optimal defense and attack strategies were provided. The validity and advantage of the proposed methods were verified by numerical simulations and by practical experiments.

There are several ways that future work could be done based on the results of this chapter. There were some specific problems:

1. We can investigate the NCSs with the CTOC and MTOC structures over the infinite time horizon. For example, we only need to replace L_k^i by $\bar{L}^i$ and replace P_k^i and P_{k+1}^i by $\bar{P}^i$ in (12.8) and (12.9), and then the corresponding infinite-time horizon results for the MTOC structure can be obtained. However, this extension requires the stability analysis of the addressed system.
2. Another extension of this work is to exploit the pricing mechanisms in the cyber layer. The interactions of the DoS attacker and the cyber defender

(e.g., IDS [452]) can be captured by a cyber layer game model [113], whose output determines the packet dropout rate in the communication channel of the control system and further determines the control performance. Thus, the pricing mechanisms can be exploited, and by tuning the pricing parameters in the cyber layer game model the control system in the physical layer can achieve the desired performance.

3. The assumption made on the distribution of the stochastic variable α_k^i can be altered. The case of the cascading failure [105] can be considered where the link failure process $\{\alpha_k^i\}$ is correlated. A Markov chain can be employed to model this correlation.

Chapter 13

Secure estimation subject to cyber stochastic attacks

Contents

13.1 Estimation against stochastic cyber attacks

Cyber-physical systems (CPSs) are the integrations of control, communication, and computation so that the desired performance of physical processes is acquired. Recently, CPSs have become an interest of research due to their application in several fields, such as the generation and distribution of sustainable and blackout-free electricity [316], clean energy buildings, smart medical and healthcare systems, transportation networks, chemical process control, smart grids, water and gas distribution networks, and emergency management [38].

However, security threats have a high possibility of affecting CPSs and can be affected by several cyber attacks without providing any indication of failure. These attacks can cause disruption to the physical system. For example, the disarrangement of coordination packets in medium-access control layers could be the result malware injected by an adversary. Additionally, in order to destroy the normal operation, an attacker can obtain illegal access to supervision centers while obtaining the encryption key. That means that the system dynamics can be disturbed arbitrarily by the attacker, and when there is a lack of security protection either in hardware or software strategies the attacker has the capability of inducing disruptions [58].

Cloud Control Systems. https://doi.org/10.1016/B978-0-12-818701-2.00021-4

13.1.1 Introduction

In any control system, the communication among sensors, actuators, and controllers occurs through a common network medium. Lack of security protection during data transmission in these systems makes it vulnerable to attack by adversaries. As mentioned before, these attacks could lead the system to instability or drive the plant to undesired operations. This situation makes it very important to consider security when designing controllers and this question has attracted a great deal of attention in recent years. From a security control viewpoint, the two main types of cyber attacks affecting dynamical systems are 1) denial-of-service (DoS) attacks, which are strategies that are often used to occupy the communication resources in order to block the transmission of measurement or control signals and that cause the maximum possible deterioration of the system performance, and 2) deception attacks, also called false data injection (FDI) attacks, which are modifications of the data integrity for the transmitted packets among the cyber parts of a CPS.

Filtering of CPSs under cyber attack is one of the main issues in control engineering due to its importance, especially in power systems. This problem has been attracting a lot of research [460–463]. The nonlinear estimation problems were solved using a number of efficient algorithms. The extended Kalman estimation approach is the most famous algorithm among them thanks to its applicability to linearized systems with known Gaussian noise. Due to their excellent robustness against parameter uncertainties and disturbances, the H_∞ and robust estimators for nonlinear systems has also been attracting much attention [464,465]. In addition, the H_∞ estimator has been applied in several applications such as sector-bounded nonlinearities [466], randomly occurring nonlinear disturbances [467], nonlinear fractional transformations [468], and systems with affine nonlinearities [469].

In [470] the robust estimation problem was solved for discrete-time nonlinear systems including external stochastic disturbances, probabilistic missing measurements, and time delay. A secure estimator was proposed using stochastic analysis techniques for nonlinear stochastic discrete time-delay systems with random sensor saturation and deception attacks [134]. However, the proposed method considered constant probabilities of the saturation and the attacks.

This section presents two main contributions over the previous literature.

1. A secure estimator system is designed for discrete-time delayed systems that considers the two main types of attacks, DoS and deception attacks;
2. The occurrences of the DoS and deception attacks will be considered as Bernoulli distributed white sequences with variable conditional probabilities.

13.1.2 Problem formulation

The discrete-time nonlinear stochastic time delay system is described as

$$\begin{aligned} x(k+1) &= Ax(k) + A_d x(k-\tau_k^d) + Bf(x(k)) \\ &\quad + B_d f_d(x(k-\tau_k^d)) + w(k), \end{aligned} \tag{13.1}$$

where $x(k) \in \mathbb{R}^n$ is the plant's state vector and $w(k) \in \mathbb{R}$ is a zero-mean Gaussian white noise sequence with $\mathbb{E}w^2(k) \leq \delta_2$. A, A_d, B, and B_d are known real matrices with appropriate dimensions. The nonlinear functions $f : \mathbb{R}^{n_x} \to \mathbb{R}^{n_x}$ and $f_d : \mathbb{R}^{n_x} \to \mathbb{R}^{n_x}$ satisfying the bounded conditions

$$\begin{aligned} &[f(x) - K_1 x]^T [f(x) - K_2 x] \leq 0, \\ &[f_d(x) - T_1 x]^T [f_d(x) - T_2 x] \leq 0, \end{aligned} \tag{13.2}$$

where K_1, K_2, T_1, and T_2 are known real matrices with appropriate dimensions, and

$$K = K_1 - K_2, \quad T = T_1 - T_2, \tag{13.3}$$

where K and T are symmetric positive definite matrices.

In general we assume that the sensor measurement with random stochastic DoS and deception attacks is modeled as

$$\begin{aligned} y_f(k) &= \alpha(k)[Cx(k) + \beta(k)(-Cx(k) + \zeta(k))] \\ &\quad + (1-\alpha(k))Cx(k-\tau_k^m) + v(k), \end{aligned} \tag{13.4}$$

where τ_k^m stands for *measurement delay* caused by a DoS attack and its occurrence satisfying the Bernoulli distribution, $\zeta(k)$ is the signal that affects the system in the deception attack, and $\alpha(k)$ and $\beta(k)$ are Bernoulli distributed white sequences exhibiting the occurrence of forward DoS and deception attacks, respectively, with the following probabilities:

$$\begin{aligned} \rho_1(k) &= Prob\{\alpha(k) = 0\}, \ \hat{\rho}_1 = \mathbb{E}[\rho_1] \\ \rho_2(k) &= Prob\{\alpha(k) = 1, \beta(k) = 0\}, \ \hat{\rho}_2 = \mathbb{E}[\rho_2] \\ \rho_3(k) &= Prob\{\alpha(k) = 1, \beta(k) = 1\}, \ \hat{\rho}_3 = \mathbb{E}[\rho_3]. \end{aligned} \tag{13.5}$$

Remark 13.1. As noted from (13.4), there are three scenarios for the attacks: 1) DoS attack, when $\alpha(k) = 0$ and regardless the value of $\beta(k)$; 2) deception attack, when $\alpha(k) = 1$ and $\beta(k) = 1$; and 3) no attack, when $\alpha(k) = 1$ and $\beta(k) = 0$.

Remark 13.2. In the deception attack scenario, it is assumed that the injected false data sent by the attackers could be divided mathematically into two terms, as shown in (13.4), $-Cx(k)$, which cancels the original signal, and $\zeta(k)$, which is assumed to be an arbitrary bounded energy signal with the following characteristic:

$$\|\zeta(k)\|^2 \leq \delta_1^2. \tag{13.6}$$

The following estimator structure is applied based on the measurements mentioned before,

$$\hat{x}(k+1) = F\hat{x}(k) + Ny_f(k), \tag{13.7}$$

where $\hat{x}(k) \in \mathbb{R}^n$ is the estimate of the system states (13.1), and F and N are the parameters of the estimator.

Here τ_k^d and τ_k^m are assumed to be time-varying and with the following bounded condition:

$$\tau_d^- \leq \tau_k^d \leq \tau_d^+, \quad \tau_m^- \leq \tau_k^m \leq \tau_m^+. \tag{13.8}$$

Here τ_d^- and τ_d^+ are the minimum and maximum delays that occur in the system, respectively, and τ_m^- and τ_m^+ are the minimum and maximum measurement delays that occur due to the DoS attack.

By defining the estimation error as $e(k) = x(k) - \hat{x}(k)$, it is obtained that

$$\begin{aligned} e(k+1) \quad = \quad & Fe(k) + [A - F - \alpha(k)(1-\beta(k))NC]x(k) \\ & + A_d x(k-\tau_k^d) - (1-\alpha(k))NCx(k-\tau_k^m) \\ & + Bf(x(k)) + B_d f_d(x(k-\tau_k^d)) \\ & - \alpha(k)N\zeta(k) + w(k) - Nv(k). \end{aligned} \tag{13.9}$$

In addition, by using $\xi(k) = [x^T(k) \quad e^T(k)]^T$ and $\epsilon(k) = [w(k) \quad v(k)]^T$, systems (13.1) and (13.9) can be formulated as

$$\begin{aligned} \xi(k+1) \quad = \quad & \mathsf{A}_j\xi(k) + \mathsf{A}_\mathsf{d}\xi(k-\tau_k^d) + \mathsf{A}_{\mathsf{m}j}\xi(k-\tau_k^m) \\ & + \bar{\mathsf{B}}f(H\xi(k)) + \bar{\mathsf{B}}_\mathsf{d} f_d(H\xi(k-\tau_k^d)) \\ & + \mathsf{C}_j\zeta(k) + \mathsf{D}\epsilon(k), \end{aligned} \tag{13.10}$$

where $\{\mathsf{A}_j, \mathsf{A}_{\mathsf{m}j}, \mathsf{C}_j;\ j = 1, 2, 3\}$ and j is an index for one of the attack situations and their values as follows:

$$\begin{aligned} \mathsf{A}_1 \quad &= \quad \begin{bmatrix} A & 0 \\ A-F & F \end{bmatrix}, \ \mathsf{A}_2 = \begin{bmatrix} A & 0 \\ A-F-NC & F \end{bmatrix}, \\ \mathsf{A}_3 \quad &= \quad \begin{bmatrix} A & 0 \\ A-F & F \end{bmatrix}, \ \mathsf{A}_\mathsf{d} = \begin{bmatrix} A_d & 0 \\ A_d & 0 \end{bmatrix}, \\ \mathsf{A}_{\mathsf{m}1} \quad &= \quad \begin{bmatrix} 0 & 0 \\ -NC & 0 \end{bmatrix}, \ \mathsf{A}_{\mathsf{m}2} = \mathsf{A}_{\mathsf{m}3} = \mathbf{0}, \end{aligned}$$

$$\begin{aligned}
\bar{\mathrm{B}} &= \begin{bmatrix} B \\ B \end{bmatrix}, \ \bar{\mathrm{B}}_{\mathrm{d}} = \begin{bmatrix} B_d \\ B_d \end{bmatrix}, \\
H &= \begin{bmatrix} I & 0 \end{bmatrix}, \ \mathrm{C}_1 = \mathrm{C}_2 = \mathbf{0}, \\
\mathrm{C}_3 &= \begin{bmatrix} 0 \\ -N \end{bmatrix}, \ \mathrm{D} = \begin{bmatrix} I & 0 \\ I & -N \end{bmatrix}.
\end{aligned} \tag{13.11}$$

The aim of this section is to propose the estimator presented in (13.7) that guarantees the exponential stability in the mean square of system (13.10). The implemented method depends on the concepts of switched time-delay systems [376] and is inspired by [471]. For simplicity we introduce each probability ρ_j and its expected value $\mathbb{E}[\rho_j]$ for $j = 1, 2, 3$, as mentioned in (13.5).

Definition 13.1. Given the positive constant scalars $\delta_1, \delta_2, \delta_3$, the estimator (13.9) is said to be $\delta_1, \delta_2, \delta_3$ secure if, when $\mathbb{E}[\epsilon^2(k)] \leq \delta_3^2$, $\|\varsigma(k)\|^2 \leq \delta_1^2$, then $\mathbb{E}[\|e(k)\|^2] \leq \delta_2^2$ for all $k \geq \tau_d^+ + 1$.

13.1.3 Secure estimation design results

Theorem 13.1. *Given the positive scalars $\delta_1, \delta_2, \delta_3$ and the estimator gains F, N, the estimator (13.7) is $\delta_1, \delta_2, \delta_3$ secure if there exist positive definite matrices P, Q and positive scalars $\varepsilon_1, \varepsilon_2, \varepsilon_3, \varepsilon_4$ satisfying the inequalities*

$$\begin{cases} \bar{\Lambda} = \Lambda_{11} + \Lambda_{12}\Lambda_{22}\Lambda_{12}^T < 0 \\ max\left\{ \frac{z(r_0)}{\lambda_{min}(P)}, \frac{\theta^2 r_0^2}{\lambda_{min}(P)(r_0-1)} \right\} \leq \delta_3^2 \end{cases}, \tag{13.12}$$

where

$$\Lambda_{11} = \begin{bmatrix} \psi_1 & 0 & 0 & \psi_2 & 0 & 0 & 0 \\ * & \psi_3 & 0 & 0 & \psi_4 & 0 & 0 \\ * & * & -Q & 0 & 0 & 0 & 0 \\ * & * & * & -\varepsilon_2 I & 0 & 0 & 0 \\ * & * & * & * & -\varepsilon_4 I & 0 & 0 \\ * & * & * & * & * & -\varepsilon_1 I & 0 \\ * & * & * & * & * & * & -\varepsilon_3 I \end{bmatrix}$$

$$\begin{aligned}
\Lambda_{12} &= \begin{bmatrix} \bar{\mathrm{A}} & \mathrm{A_d} & \bar{\mathrm{A}}_{\mathrm{m}} & \bar{\mathrm{B}} & \bar{\mathrm{B}}_{\mathrm{d}} & \bar{\mathrm{C}} & \mathrm{D} \end{bmatrix}^T, \ \Lambda_{22} = P \\
\bar{\mathrm{A}} &= \hat{\rho}_1 \mathrm{A}_1 + \hat{\rho}_2 \mathrm{A}_2 + \hat{\rho}_3 \mathrm{A}_3 \\
\bar{\mathrm{A}}_{\mathrm{m}} &= \hat{\rho}_1 \mathrm{A}_{\mathrm{m}_1} + \hat{\rho}_2 \mathrm{A}_{\mathrm{m}_2} + \hat{\rho}_3 \mathrm{A}_{\mathrm{m}_3} \\
\bar{\mathrm{C}} &= \hat{\rho}_1 \mathrm{C}_1 + \hat{\rho}_2 \mathrm{C}_2 + \hat{\rho}_3 \mathrm{C}_3
\end{aligned}$$

$$\begin{aligned}
\psi_1 &= (\tau_d^+ - \tau_d^- + \tau_m^+ - \tau_m^- + 2)Q - P \\
&\quad -0.5\varepsilon_2(H^T K_1^T K_2 H + H^T K_2^T K_1 H) \\
\psi_2 &= 0.5\varepsilon_2 H^T (K_1^T + K_2^T) H \\
\psi_3 &= -Q - 0.5\varepsilon_4 (H^T T_1^T T_2 H + H^T T_2^T T_1 H) \\
\psi_4 &= 0.5\varepsilon_4 H^T (T_1^T + T_2^T) H \\
\theta^2 &= \varepsilon_1 \delta_1^2 + \varepsilon_2 \delta_2^2
\end{aligned}$$

and

$$\begin{aligned}
z(r_0) &= 2(\mu_1 + \mu_2)\delta_2^2 \\
\mu_1 &= \tau_d^+ \lambda_{max}(Q)(r_0^{\tau_d^+} - 1)(\tau_d^+ - \tau_d^- + 1) \\
&\quad + \tau_m^+ \lambda_{max}(Q)(r_0^{\tau_m^+} - 1)(\tau_m^+ - \tau_m^- + 1) \\
\mu_2 &= T_0 \, max\Big(\lambda_{max}(P), (\tau_d^+ - \tau_d^- + 1)\lambda_{max}(Q), \\
&\qquad (\tau_m^+ - \tau_m^- + 1)\lambda_{max}(Q) \Big) \\
T_0 &= max(\tau_d^+, \tau_m^+)
\end{aligned}$$

and $r_0 > 1$ *is the solution of*

$$\begin{aligned}
&-\lambda_{min}(-\bar{\Lambda}) r_0 + (r_0 - 1)\lambda_{max}(P) \\
&+ 2r_0(\tau_d^+ - \tau_d^- + 1)\lambda_{max}(Q)(r_0^{\tau_d^+} - 1) \\
&+ 2r_0(\tau_m^+ - \tau_m^- + 1)\lambda_{max}(Q)(r_0^{\tau_m^+} - 1) = 0.
\end{aligned} \tag{13.13}$$

Proof. To establish the main theorem a Lyapunov function is constructed,

$$V(k) = \sum_{i=1}^{5} V_i(k), \tag{13.14}$$

where

$$\begin{aligned}
V_1(k) &= \xi^T(k) P \xi(k), \; P > 0 \\
V_2(k) &= \sum_{i=k-\tau_k^d}^{k-1} \xi^T(i) Q \xi(i), \; Q = Q^T > 0 \\
V_3(k) &= \sum_{i=k-\tau_k^m}^{k-1} \xi^T(i) Q \xi(i)
\end{aligned}$$

$$V_4(k) = \sum_{\ell=-\tau_d^{+}+2}^{-\tau_d^{-}+1} \sum_{i=k+\ell-1}^{k-1} \xi^T(i) Q \xi(i)$$
$$V_5(k) = \sum_{\ell=-\tau_m^{+}+2}^{-\tau_m^{-}+1} \sum_{i=k+\ell-1}^{k-1} \xi^T(i) Q \xi(i). \tag{13.15}$$

Evaluating the difference of $V_1(k)$, we have

$$\begin{aligned}
&\mathbb{E}[\Delta V_1(k)] = \mathbb{E}\big[V_1(k+1) - V_1(k)\big] \\
&= \mathbb{E}\Big[\xi^T(k)(\bar{\mathrm{A}}^T P \bar{\mathrm{A}} - P)\xi(k) + 2\xi^T(k)\bar{\mathrm{A}}^T P \mathrm{A}_{\mathrm{d}}\xi(k-\tau_k^d) \\
&+2\xi^T(k)\bar{\mathrm{A}}^T P \bar{\mathrm{A}}_{\mathrm{m}}\xi(k-\tau_k^m) + 2\xi^T(k)\bar{\mathrm{A}}^T P \bar{\mathrm{B}} f(H\xi(k)) \\
&+2\xi^T(k)\bar{\mathrm{A}}^T P \bar{\mathrm{B}}_{\mathrm{d}} f_d(H\xi(k-\tau_k^d)) + 2\xi^T(k)\bar{\mathrm{A}}^T P \bar{\mathrm{C}}\zeta(k) \\
&+2\xi^T(k)\bar{\mathrm{A}}^T P \mathrm{D}\epsilon(k) + \xi^T(k-\tau_k^d)\mathrm{A_d}^T P \mathrm{A}_{\mathrm{d}}\xi(k-\tau_k^d) \\
&+2\xi^T(k-\tau_k^d)\mathrm{A_d}^T P \bar{\mathrm{A}}_{\mathrm{m}}\xi(k-\tau_k^m) \\
&+2\xi^T(k-\tau_k^d)\mathrm{A_d}^T P \bar{\mathrm{B}} f(H\xi(k)) \\
&+2\xi^T(k-\tau_k^d)\mathrm{A_d}^T P \bar{\mathrm{B}}_{\mathrm{d}} f_d(H\xi(k-\tau_k^d)) \\
&+2\xi^T(k-\tau_k^d)\mathrm{A_d}^T P \bar{\mathrm{C}}\zeta(k) \\
&+2\xi^T(k-\tau_k^d)\mathrm{A_d}^T P \mathrm{D}\epsilon(k) \\
&+\xi^T(k-\tau_k^m)\bar{\mathrm{A}}_{\mathrm{m}}^T P \bar{\mathrm{A}}_{\mathrm{m}}\xi(k-\tau_k^m) \\
&+2\xi^T(k-\tau_k^m)\bar{\mathrm{A}}_{\mathrm{m}}^T P \bar{\mathrm{B}} f(H\xi(k)) \\
&+2\xi^T(k-\tau_k^m)\bar{\mathrm{A}}_{\mathrm{m}}^T P \bar{\mathrm{B}}_{\mathrm{d}} f_d(H\xi(k-\tau_k^d)) \\
&+2\xi^T(k-\tau_k^m)\bar{\mathrm{A}}_{\mathrm{m}}^T P \bar{\mathrm{C}}\zeta(k) + 2\xi^T(k-\tau_k^m)\bar{\mathrm{A}}_{\mathrm{m}}^T P \mathrm{D}\epsilon(k) \\
&+f^T(H\xi(k))\bar{\mathrm{B}}^T P \bar{\mathrm{B}} f(H\xi(k)) \\
&+2f^T(H\xi(k))\bar{\mathrm{B}}^T P \bar{\mathrm{B}}_{\mathrm{d}} f_d(H\xi(k-\tau_k^d)) \\
&+2f^T(H\xi(k))\bar{\mathrm{B}}^T P \bar{\mathrm{C}}\zeta(k) + 2f^T(H\xi(k))\bar{\mathrm{B}}^T P \mathrm{D}\epsilon(k) \\
&+f_d^T(H\xi(k-\tau_k^d))\bar{\mathrm{B}}_{\mathrm{d}}^T P \bar{\mathrm{B}}_{\mathrm{d}} f_d(H\xi(k-\tau_k^d)) \\
&+2f_d^T(H\xi(k-\tau_k^d))\bar{\mathrm{B}}_{\mathrm{d}}^T P \bar{\mathrm{C}}\zeta(k) \\
&+2f_d^T(H\xi(k-\tau_k^d))\bar{\mathrm{B}}_{\mathrm{d}}^T P \mathrm{D}\epsilon(k) + 2\zeta^T(k)\bar{\mathrm{C}}^T P \bar{\mathrm{C}}\zeta(k) \\
&+2\zeta^T(k)\bar{\mathrm{C}}^T P \mathrm{D}\epsilon(k) + 2\epsilon^T(k)\mathrm{D}^T P \mathrm{D}\epsilon(k)\Big].
\end{aligned} \tag{13.16}$$

A straightforward computation gives

$$
\begin{aligned}
\mathbb{E}[\Delta V_2(k)] &= \mathbb{E}\Bigg[\sum_{i=k+1-\tau_{k+1}^d}^{k} \xi^T(i)Q\xi(i) - \sum_{i=k-\tau_k^d}^{k-1} \xi^T(i)Q\xi(i)\Bigg] \\
&= \mathbb{E}\Bigg[\xi^T(k)Q\xi(k) - \xi(k-\tau_k^d)Q\xi(k-\tau_k^d) \\
&\quad + \sum_{i=k+1-\tau_{k+1}^d}^{k-1} \xi^T(i)Q\xi(i) - \sum_{i=k+1-\tau_k^d}^{k-1} \xi(i)Q\xi(i)\Bigg].
\end{aligned}
\tag{13.17}
$$

In view of

$$
\begin{aligned}
&\sum_{i=k+1-\tau_{k+1}^d}^{k-1} \xi^T(i)Q\xi(i) \\
= &\sum_{i=k+1-\tau_{k+1}^d}^{k-\tau_k^d} \xi^T(i)Q\xi(i) + \sum_{i=k+1-\tau_k^d}^{k-1} \xi^T(i)Q\xi(i) \\
\le &\sum_{i=k+1-\tau_k^d}^{k-1} \xi^T(i)Q\xi(i) + \sum_{i=k+1-\tau_d^+}^{k-\tau_d^-} \xi^T(i)Q\xi(i),
\end{aligned}
\tag{13.18}
$$

we readily obtain

$$
\begin{aligned}
&\mathbb{E}[\Delta V_2(k)] \le \\
&\quad \mathbb{E}\Bigg[\xi^T(k)Q\xi(k) - \xi^T(k-\tau_k^d)Q\xi(k-\tau_k^d) \\
&\quad + \sum_{i=k+1-\tau_d^+}^{k-\tau_d^-} \xi^T(i)Q\xi(i)\Bigg].
\end{aligned}
\tag{13.19}
$$

Following a parallel procedure, we get

$$
\begin{aligned}
&\mathbb{E}[\Delta V_3(k)] \le \\
&\quad \mathbb{E}\Bigg[\xi^T(k)Q\xi(k) - \xi^T(k-\tau_k^m)Q\xi(k-\tau_k^m) \\
&\quad + \sum_{i=k+1-\tau_m^+}^{k-\tau_m^-} \xi^T(i)Q\xi(i)\Bigg].
\end{aligned}
\tag{13.20}
$$

Finally,

$$\mathbb{E}[\Delta V_4(k)] = \mathbb{E}\Bigg[\sum_{\ell=-\tau_d^+ +2}^{-\tau_d^- +1} [\xi^T(k)Q\xi(k)$$

$$-\xi^T(k+\ell-1)Q\xi(k+\ell-1)]\Bigg]$$

$$= \mathbb{E}\Bigg[(\tau_d^+ - \tau_d^-)\xi^T(k)Q\xi(k)$$

$$- \sum_{i=k+1-\tau_d^+}^{k-\tau_d^-} \xi^T(i)Q\xi(i)\Bigg] \tag{13.21}$$

$$\mathbb{E}[\Delta V_5(k)] = \mathbb{E}\Bigg[(\tau_m^+ - \tau_m^-)\xi^T(k)Q\xi(k)$$

$$- \sum_{i=k+1-\tau_m^+}^{k-\tau_m^-} \xi^T(i)Q\xi(i)\Bigg]. \tag{13.22}$$

On combining (13.16)–(13.22), while noting (13.2) and (13.6), we obtain

$$\mathbb{E}[\Delta V(k)] \le \mathbb{E}\Bigg[\xi^T(k)(\bar{\mathrm{A}}^T P\bar{\mathrm{A}} - P)\xi(k)$$
$$+2\xi^T(k)\bar{\mathrm{A}}^T P\mathrm{A_d}\xi(k-\tau_k^d)$$
$$+2\xi^T(k)\bar{\mathrm{A}}^T P\bar{\mathrm{A}}_\mathrm{m}\xi(k-\tau_k^m) + 2\xi^T(k)\bar{\mathrm{A}}^T P\bar{\mathrm{B}}f(H\xi(k))$$
$$+2\xi^T(k)\bar{\mathrm{A}}^T P\bar{\mathrm{B}}_\mathrm{d}f_d(H\xi(k-\tau_k^d)) + 2\xi^T(k)\bar{\mathrm{A}}^T P\bar{\mathrm{C}}\zeta(k)$$
$$+2\xi^T(k)\bar{\mathrm{A}}^T P\mathrm{D}\epsilon(k) + \xi^T(k-\tau_k^d)\mathrm{A_d}^T P\mathrm{A_d}\xi(k-\tau_k^d)$$
$$+2\xi^T(k-\tau_k^d)\mathrm{A_d}^T P\bar{\mathrm{A}}_\mathrm{m}\xi(k-\tau_k^m)$$
$$+2\xi^T(k-\tau_k^d)\mathrm{A_d}^T P\bar{\mathrm{B}}f(H\xi(k))$$
$$+2\xi^T(k-\tau_k^d)\mathrm{A_d}^T P\bar{\mathrm{B}}_\mathrm{d}f_d(H\xi(k-\tau_k^d))$$
$$+2\xi^T(k-\tau_k^d)\mathrm{A_d}^T P\bar{\mathrm{C}}\zeta(k)$$
$$+2\xi^T(k-\tau_k^d)\mathrm{A_d}^T P\mathrm{D}\epsilon(k)$$
$$+\xi^T(k-\tau_k^m)\bar{\mathrm{A}}_\mathrm{m}^T P\bar{\mathrm{A}}_\mathrm{m}\xi(k-\tau_k^m)$$
$$+2\xi^T(k-\tau_k^m)\bar{\mathrm{A}}_\mathrm{m}^T P\bar{\mathrm{B}}f(H\xi(k))$$
$$+2\xi^T(k-\tau_k^m)\bar{\mathrm{A}}_\mathrm{m}^T P\bar{\mathrm{B}}_\mathrm{d}f_d(H\xi(k-\tau_k^d))$$
$$+2\xi^T(k-\tau_k^m)\bar{\mathrm{A}}_\mathrm{m}^T P\bar{\mathrm{C}}\zeta(k) + 2\xi^T(k-\tau_k^m)\bar{\mathrm{A}}_\mathrm{m}^T P\mathrm{D}\epsilon(k)$$

$$
\begin{aligned}
&+f^T(H\xi(k))\bar{\mathrm{B}}^T P\bar{\mathrm{B}} f(H\xi(k)) \\
&+2f^T(H\xi(k))\bar{\mathrm{B}}^T P\bar{\mathrm{B}}_{\mathrm{d}} f_d(H\xi(k-\tau_k^d)) \\
&+2f^T(H\xi(k))\bar{\mathrm{B}}^T P\bar{\mathrm{C}}\zeta(k)+2f^T(H\xi(k))\bar{\mathrm{B}}^T P\mathrm{D}\epsilon(k) \\
&+f_d^T(H\xi(k-\tau_k^d))\bar{\mathrm{B}}_{\mathrm{d}}^T P\bar{\mathrm{B}}_{\mathrm{d}} f_d(H\xi(k-\tau_k^d)) \\
&+2f_d^T(H\xi(k-\tau_k^d))\bar{\mathrm{B}}_{\mathrm{d}}^T P\bar{\mathrm{C}}\zeta(k) \\
&+2f_d^T(H\xi(k-\tau_k^d))\bar{\mathrm{B}}_{\mathrm{d}}^T P\mathrm{D}\epsilon(k)+2\zeta^T(k)\bar{\mathrm{C}}^T P\bar{\mathrm{C}}\zeta(k) \\
&+2\zeta^T(k)\bar{\mathrm{C}}^T P\mathrm{D}\epsilon(k)+2\epsilon^T(k)\mathrm{D}^T P\mathrm{D}\epsilon(k) \\
&+\xi^T(k)((\tau_d^+-\tau_d^-+\tau_m^+-\tau_m^-+2)Q\xi(k) \\
&-\xi^T(k-\tau_k^d)Q\xi(k-\tau_k^d)-\xi^T(k-\tau_k^m)Q\xi(k-\tau_k^m) \\
&+\varepsilon_1(\delta_1^2-\zeta^T(k)\zeta(k))+\varepsilon_3(\delta_2^2-\epsilon^T(k)\epsilon(k)) \\
&-\varepsilon_2[f(H\xi(k))-K_1H\xi(k)]^T[f(H\xi(k))-K_2H\xi(k)] \\
&-\varepsilon_4[f_d(H\xi(k-d_k))-T_1H\xi(k-d_k)]^T \\
&\times[f(H\xi(k-d_k))-T_2H\xi(k-d_k)]\Bigg].
\end{aligned} \tag{13.23}
$$

Thus,

$$
\mathbb{E}[\Delta V(k)]\le\mathbb{E}\Big[\Omega^T(k)\bar{\Lambda}\Omega(k)+\theta^2\Big], \tag{13.24}
$$

where

$$
\begin{aligned}
\Omega(k)=\big[&\xi^T(k)\;\; \xi^T(k-\tau_k^d)\;\; \xi^T(k-\tau_k^m) \\
&f^T(H\xi(k))\;\; f_d^T(H\xi(k-d_k))\;\; \zeta^T(k)\;\; \epsilon^T(k)\big]^T.
\end{aligned} \tag{13.25}
$$

From (13.24), it is known that

$$
\mathbb{E}[\Delta V(k)]\le-\lambda_{min}(-\bar{\Lambda})\mathbb{E}\big[||\xi(k)||^2\big]+\theta^2. \tag{13.26}
$$

In addition, by referring to the definition of the energy-like functional $V(k)$, it is seen that

$$
\begin{aligned}
V(k)\le{}&\lambda_{max}(P)\mathbb{E}\big[||\xi(k)||^2\big] \\
&+\lambda_{max}(Q)(\tau_d^+-\tau_d^-+1)\sum_{i=k-\tau_d^+}^{k-1}\mathbb{E}\big[||\xi(i)||^2\big] \\
&+\lambda_{max}(Q)(\tau_m^+-\tau_m^-+1)\sum_{i=k-\tau_m^+}^{k-1}\mathbb{E}\big[||\xi(i)||^2\big].
\end{aligned} \tag{13.27}
$$

In addition, a scalar $r > 1$ is introduced, and from (13.26) and (13.27) it follows that

$$\begin{aligned}
&\mathbb{E}[r^{k+1}V(k+1)] - \mathbb{E}[r^k V(k)] \\
&= r^{k+1}\mathbb{E}[\Delta V(k)] + r^{k+1}\mathbb{E}[V(k)] - r^k\mathbb{E}[V(k)] \\
&\le r^{k+1}\Big[- \lambda_{min}(-\bar{\Lambda})\mathbb{E}\big[||\xi(k)||^2\big] + \theta^2\Big] \\
&\quad + r^k(r-1)\mathbb{E}[V(k)] \\
&\le a(r)r^k\mathbb{E}\big[||\xi(k)||^2\big] + b(r)\sum_{i=k-\tau_d^+}^{k-1} r^k\mathbb{E}\big[||\xi(i)||^2\big] \\
&\quad + c(r)\sum_{i=k-\tau_m^+}^{k-1} r^k\mathbb{E}\big[||\xi(i)||^2\big] + r^{k+1}\theta^2,
\end{aligned} \tag{13.28}$$

where

$$\begin{aligned}
a(r) &= -\lambda_{min}(-\bar{\Lambda})r + (r-1)\lambda_{max}(P), \\
b(r) &= (\tau_d^+ - \tau_d^- + 1)(r-1)\lambda_{max}(Q), \\
c(r) &= (\tau_m^+ - \tau_m^- + 1)(r-1)\lambda_{max}(Q).
\end{aligned}$$

For any integer $T \ge max(\tau_d^+, \tau_m^+) + 1$, summing up both sides of (13.28) from 0 to $T-1$ with respect to k yields

$$\begin{aligned}
&\mathbb{E}[r^T V(T)] - \mathbb{E}[V(0)] \\
&\quad \le a(r)\sum_{k=0}^{T-1} r^k\mathbb{E}\big[||\xi(k)||^2\big] + \frac{r(1-r^T)}{1-r}\theta^2 \\
&\qquad + b(r)\sum_{k=0}^{T_d-1}\sum_{i=k-\tau_d^+}^{k-1} r^k\mathbb{E}\big[||\xi(i)||^2\big] \\
&\qquad + c(r)\sum_{k=0}^{T_m-1}\sum_{i=k-\tau_m^+}^{k-1} r^k\mathbb{E}\big[||\xi(i)||^2\big].
\end{aligned} \tag{13.29}$$

Now the last two terms can be calculated as follows:

$$\begin{aligned}
&\sum_{k=0}^{T-1}\sum_{i=k-\tau_d^+}^{k-1} r^k\mathbb{E}\big[||\xi(i)||^2\big] \\
&\le \Bigg(\sum_{i=-\tau_d^+}^{-1}\sum_{k=0}^{i+\tau_d^+} + \sum_{i=0}^{T-\tau_d^+-1}\sum_{k=i+1}^{i+\tau_d^+} + \sum_{i=T-\tau_d^+}^{T-1}\sum_{k=i+1}^{T-1}\Bigg) r^k\mathbb{E}\big[||\xi(i)||^2\big]
\end{aligned}$$

$$
\begin{aligned}
&\leq \frac{r^{\tau_d^+}-1}{r-1}\sum_{i=-\tau_d^+}^{-1}\mathbb{E}[||\xi(i)||^2] \\
&+\frac{r(r^{\tau_d^+}-1)}{r-1}\sum_{i=0}^{T-1}r^i\mathbb{E}[||\xi(i)||^2] \\
&+\frac{r(r^{\tau_d^+-1}-1)}{r-1}\sum_{i=0}^{T-1}r^i\mathbb{E}[||\xi(i)||^2].
\end{aligned}
\tag{13.30}
$$

And similarly:

$$
\begin{aligned}
&\sum_{k=0}^{T-1}\sum_{i=k-\tau_m^+}^{k-1} r^k\mathbb{E}[||\xi(i)||^2] \\
&\quad\leq \frac{r^{\tau_m^+}-1}{r-1}\sum_{i=-\tau_m^+}^{-1}\mathbb{E}[||\xi(i)||^2] \\
&\quad+\frac{r(r^{\tau_m^+}-1)}{r-1}\sum_{i=0}^{T-1}r^i\mathbb{E}[||\xi(i)||^2] \\
&\quad+\frac{r(r^{\tau_m^+-1}-1)}{r-1}\sum_{i=0}^{T-1}r^i\mathbb{E}[||\xi(i)||^2].
\end{aligned}
\tag{13.31}
$$

Substituting (13.29), (13.30), and (13.31), we obtain

$$
\begin{aligned}
&\mathbb{E}[r^T V(T)] - \mathbb{E}[V(0)] \\
&\quad\leq \frac{r(1-r^T)}{1-r}\theta^2 \\
&\quad+\frac{b(r)(r^{\tau_d^+}-1)\tau_d^+}{r-1}\sup_{-\tau_d^+\leq i\leq 0}\mathbb{E}[||\xi(i)||^2] \\
&\quad+\frac{c(r)(r^{\tau_m^+}-1)\tau_m^+}{r-1}\sup_{-\tau_m^+\leq i\leq 0}\mathbb{E}[||\xi(i)||^2] \\
&\quad+g(r)\sum_{k=0}^{T-1}r^k\mathbb{E}[||\xi(k)||^2],
\end{aligned}
\tag{13.32}
$$

where

$$
g(r) = a(r) + \frac{2rb(r)(r^{\tau_d^+}-1)}{r-1} + \frac{2rc(r)(r^{\tau_m^+}-1)}{r-1}.
$$

Since $g(1) = -\lambda_{min}(-\bar{\Lambda}) < 0$ and $\lim_{r\to\infty} = +\infty$, there exists a scalar $r_0 > 1$ such that $g(r_0) = 0$. So, a scalar $r_0 > 1$ could be found such that

$$\begin{aligned}
&\mathbb{E}[r_0^T V(T)] - \mathbb{E}[V(0)] \\
&\leq \frac{r_0(1-r_0^T)}{1-r_0}\theta^2 \\
&+\frac{b(r_0)(r_0^{\tau_d^+}-1)\tau_d^+}{r_0-1}\sup_{-\tau_d^+\leq i\leq 0}\mathbb{E}\big[||\xi(i)||^2\big] \\
&+\frac{c(r_0)(r_0^{\tau_m^+}-1)\tau_m^+}{r_0-1}\sup_{-\tau_m^+\leq i\leq 0}\mathbb{E}\big[||\xi(i)||^2\big].
\end{aligned}\tag{13.33}$$

Noting that

$$\begin{aligned}
&\sup_{-\tau_d^+\leq i\leq 0}\mathbb{E}\big[||\xi(i)||^2\big] \\
&= \sup_{-\tau_d^+\leq i\leq 0}\mathbb{E}\big[||x(i)||^2+||e(i)||^2\big] \\
&\leq \sup_{-\tau_d^+\leq i\leq 0}\mathbb{E}\big[||x(i)||^2\big]+\sup_{-\tau_d^+\leq i\leq 0}\mathbb{E}\big[||e(i)||^2\big]\leq 2\delta_2^2,
\end{aligned}\tag{13.34}$$

and similarly that

$$\begin{aligned}
&\sup_{-\tau_m^+\leq i\leq 0}\mathbb{E}\big[||\xi(i)||^2\big] \\
&\leq \sup_{-\tau_m^+\leq i\leq 0}\mathbb{E}\big[||x(i)||^2\big]+\sup_{-\tau_m^+\leq i\leq 0}\mathbb{E}\big[||e(i)||^2\big]\leq 2\delta_2^2
\end{aligned}\tag{13.35}$$

$$\begin{aligned}
\mathbb{E}[r_0^T V(T)] &\geq \lambda_{min}(P)r_0^T\mathbb{E}\big[||\xi(T)||^2\big] \\
&\geq \lambda_{min}(P)r_0^T\mathbb{E}\big[||e(T)||^2\big]
\end{aligned}\tag{13.36}$$

and

$$\begin{aligned}
&\mathbb{E}[V(0)] \leq T_0\max\Big(\lambda_{max}(P), \\
&(\tau_d^+-\tau_d^-+1)\lambda_{max}(Q), (\tau_m^+-\tau_m^-+1)\lambda_{max}(Q)\Big) \\
&\times\sup_{-T_0\leq i\leq 0}\mathbb{E}\big[||\xi(i)||^2\big],
\end{aligned}\tag{13.37}$$

where $T_0 = \max(\tau_d^+, \tau_m^+)$, we have

$$\mathbb{E}\big[||e(T)||^2\big] \leq \frac{(r_0^T-1)\theta^2}{r_0^{T-1}(r_0-1)\lambda_{min}(P)}+\frac{z(r_0)}{r_0^T\lambda_{min}(P)}$$

$$
\begin{aligned}
&= r_0^{-T}\left[\frac{z(r_0)}{\lambda_{min}(P)} - \frac{\theta^2 r_0}{\lambda_{min}(P)(r_0-1)}\right] \\
&\quad + \frac{\theta^2 r_0}{\lambda_{min}(P)(r_0-1)} \\
&\le \max\left\{\frac{z(r_0)}{\lambda_{min}(P)}, \frac{\theta^2 r_0}{\lambda_{min}(P)(r_0-1)}\right\}. \qquad (13.38)
\end{aligned}
$$

Referring to (13.12), it can be shown that $\mathbb{E}\|e(T)\|^2 \le \delta_2^2$, which from Definition 13.1 implies that the estimation error system (13.9) is $\delta_1, \delta_2, \delta_3$ secure, and so the proof of Theorem 13.1 is complete. □

Corollary 13.1. *For the case of DoS attack only, given the positive scalars δ_2, δ_3, and given the estimator gains F, N, the estimator (13.7) is δ_2, δ_3 secure if there exist positive definite matrices P, Q and positive scalars $\varepsilon_2, \varepsilon_3, \varepsilon_4$ satisfying the inequalities*

$$
\begin{cases}
\bar{\Lambda}' = \Lambda'_{11} + \Lambda'_{12}\Lambda'_{22}\Lambda'^{T}_{12} < 0 \\
max\left\{\frac{z(r_0)}{\lambda_{min}(P)}, \frac{\theta^{2'} r_0^2}{\lambda_{min}(P)(r_0-1)}\right\} \le \delta_3^2
\end{cases}, \qquad (13.39)
$$

where

$$
\begin{aligned}
\Lambda'_{11} &= \begin{bmatrix}
\psi_1 & 0 & 0 & \psi_2 & 0 & 0 \\
* & \psi_3 & 0 & 0 & \psi_4 & 0 \\
* & * & -Q & 0 & 0 & 0 \\
* & * & * & -\varepsilon_2 I & 0 & 0 \\
* & * & * & * & -\varepsilon_4 I & 0 \\
* & * & * & * & * & 0 \\
* & * & * & * & * & -\varepsilon_3 I
\end{bmatrix} \\
\Lambda'_{12} &= \begin{bmatrix}\bar{\mathrm{A}} & \mathrm{A_d} & \bar{\mathrm{A}}_\mathrm{m} & \bar{\mathrm{B}} & \bar{\mathrm{B}}_\mathrm{d} & \mathrm{D}\end{bmatrix}^T, \ \Lambda'_{22} = P \\
\theta^{2'} &= \varepsilon_2 \delta_2^2.
\end{aligned}
$$

Proof. The proof of Corollary 13.1 follows by applying the same procedure as the proof of Theorem 3.3. □

Theorem 13.2. *Given the positive scalars $\delta_1, \delta_2, \delta_3$, if there exist positive definite matrices $P = diag\{P1, P2\}$, Q, matrices X, Y, and positive scalars $\varepsilon_1, \varepsilon_2, \varepsilon_3, \varepsilon_4$ satisfying the inequalities*

$$\begin{cases} \Lambda = \begin{bmatrix} \Lambda_{11} & \Lambda_3 \\ * & -\Lambda_{22} \end{bmatrix} < 0 \\ max\left\{ \frac{z(r_0)}{\lambda_{min}(P)}, \frac{\theta^2 r_0^2}{\lambda_{min}(P)(r_0-1)} \right\} \leq \delta_3^2 \end{cases}, \tag{13.40}$$

where

$$\begin{aligned} \Lambda_3 &= \begin{bmatrix} \tilde{\mathrm{A}} & P\mathrm{A_d} & \tilde{\mathrm{A}}_\mathrm{m} & P\bar{\mathrm{B}} & P\bar{\mathrm{B}}_\mathrm{d} & \tilde{\mathrm{C}} & \tilde{\mathrm{D}} \end{bmatrix}^T \\ \tilde{\mathrm{A}} &= \hat{\rho}_1\tilde{\mathrm{A}}_1 + \hat{\rho}_2\tilde{\mathrm{A}}_2 + \hat{\rho}_3\tilde{\mathrm{A}}_3 \\ \tilde{\mathrm{A}}_1 &= \begin{bmatrix} P_1 A & 0 \\ P_2 A - X & X \end{bmatrix}, \ \tilde{\mathrm{A}}_2 = \begin{bmatrix} P_1 A & 0 \\ P_2 A - X - YC & X \end{bmatrix} \\ \tilde{\mathrm{A}}_3 &= \begin{bmatrix} P_1 A & 0 \\ P_2 A - X & X \end{bmatrix}, \ \tilde{\mathrm{A}}_\mathrm{m} = \hat{\rho}_1 \begin{bmatrix} 0 & 0 \\ -YC & 0 \end{bmatrix} \\ \tilde{\mathrm{C}} &= \hat{\rho}_3 \begin{bmatrix} 0 \\ -Y \end{bmatrix}, \ \tilde{\mathrm{D}} = \begin{bmatrix} P_1 & 0 \\ P_2 & -Y \end{bmatrix}, \end{aligned} \tag{13.41}$$

and $r_0 > 1$ is the solution to the following equation:

$$\begin{aligned} &-\lambda_{min}(-\Lambda) r_0 + (r_0 - 1)\lambda_{max}(P) \\ &+ 2r_0(\tau_d^+ - \tau_d^- + 1)\lambda_{max}(Q)(r_0^{\tau_d^+} - 1) \\ &+ 2r_0(\tau_m^+ - \tau_m^- + 1)\lambda_{max}(Q)(r_0^{\tau_m^+} - 1) = 0, \end{aligned} \tag{13.42}$$

then the estimator system (13.7) is $\delta_1, \delta_2, \delta_3$ secure, and the estimator gain matrices F and N are calculated by the following equations:

$$F = P_2^{-1} X, \qquad N = P_2^{-1} Y. \tag{13.43}$$

Proof. Using the Schur complement, it is known that $\bar{\Lambda} < 0$ can be rewritten as

$$\bar{\Lambda} = \begin{bmatrix} \Lambda_{11} & \Lambda_{12} \\ * & -\Lambda_{22}^{-1} \end{bmatrix} < 0. \tag{13.44}$$

By multiplying the inequality (13.44) by $diag\{I, P\}$ on the right and left, we obtain

$$\Lambda = \begin{bmatrix} \Lambda_{11} & \Lambda_3 \\ * & -\Lambda_{22} \end{bmatrix} < 0, \tag{13.45}$$

where

$$\Lambda_3 = P\Lambda_{12}, \tag{13.46}$$

and by substituting $X = P_2F$ and $Y = P_2N$ it can be shown that $\bar{\Lambda} < 0$ is exactly the same of inequality (13.45), and that means the conditions in Theorem 3.3 are satisfied and the rest of the proof of Theorem 13.2 can be shown by following the proof of Theorem 13.1. □

Corollary 13.2. *For the case of DoS attack only, if there exist positive definite matrices $P = diag\{P1, P2\}$, Q, matrices X, Y, and positive scalars $\varepsilon_2, \varepsilon_3, \varepsilon_4$ with given positive scalars δ_2, δ_3 satisfying the inequalities*

$$\begin{cases} \Lambda' = \begin{bmatrix} \Lambda'_{11} & \Lambda'_3 \\ * & -\Lambda'_{22} \end{bmatrix} < 0 \\ max\left\{ \frac{z(r_0)}{\lambda_{min}(P)}, \frac{\theta^{2'} r_0^2}{\lambda_{min}(P)(r_0-1)} \right\} \leq \delta_3^2 \end{cases}, \tag{13.47}$$

where

$$\Lambda'_3 = \begin{bmatrix} \tilde{\mathrm{A}} & P\mathrm{A_d} & \tilde{\mathrm{A}}_\mathrm{m} & P\bar{\mathrm{B}} & P\bar{\mathrm{B}}_\mathrm{d} & \tilde{\mathrm{D}} \end{bmatrix}^T, \tag{13.48}$$

then the estimator system (13.7) is δ_2, δ_3 secure, and the estimator gain matrices F and N are calculated by the following equations:

$$F = P_2^{-1}X, \qquad N = P_2^{-1}Y. \tag{13.49}$$

Proof. The proof of Corollary 13.2 follows by applying the same procedure as the proof of Theorem 13.2. □

13.1.4 Illustrative example I

The effectiveness of the estimation methods presented in this chapter is shown by solving the following numerical examples. Following are the parameters of system (13.1):

$$A = \begin{bmatrix} -0.2 & 0.0 & 0.1 \\ 0.1 & -0.3 & 0.1 \\ 0.0 & 0.0 & -0.2 \end{bmatrix}, \ A_d = \begin{bmatrix} 0.1 & -0.1 & 0.0 \\ 0.1 & -0.2 & 0.0 \\ 0.0 & -0.2 & -0.1 \end{bmatrix},$$

$$B = \begin{bmatrix} 0.1 & 0.1 & 0.0 \\ 0.1 & 0.2 & 0.0 \\ 0.0 & 0.2 & 0.1 \end{bmatrix}, \ B_d = \begin{bmatrix} 0.1 & 0.0 & 0.1 \\ 0.1 & 0.2 & 0.0 \\ 0.1 & 0.0 & 0.1 \end{bmatrix}$$

$$C = \begin{bmatrix} 1 & 0 & 0 \\ 0 & 1 & 0 \end{bmatrix}. \tag{13.50}$$

The nonlinear functions (13.2) are selected as

$$\begin{aligned} f(x(k)) &= \text{sign}(x(k))\log(\text{sign}(x(k))x(k)+1); \\ f_d(x(k-d(k))) &= \text{sign}(x(k-d(k))) \\ &\quad \times\log(\text{sign}(x(k-d(k)))x(k-d(k))+1). \end{aligned} \tag{13.51}$$

In addition,

$$\begin{aligned} K_1 &= T_1 = \begin{bmatrix} 0.2 & 0.1 & 0 \\ 0.1 & 0.3 & 0 \\ -0.1 & 0.1 & 0.3 \end{bmatrix}, \\ K_2 &= T_2 = \begin{bmatrix} -0.2 & 0.1 & 0 \\ 0.1 & -0.3 & -0.1 \\ -0.1 & 0 & -0.3 \end{bmatrix}. \end{aligned} \tag{13.52}$$

Using YALMIP the gains of the estimator (13.7) were obtained:

$$\begin{aligned} F &= \begin{bmatrix} 0.0648 & -0.0783 & -0.0283 \\ 0.7386 & -0.4756 & -0.1054 \\ 1.0451 & -0.2533 & -0.4352 \end{bmatrix}, \\ N &= \begin{bmatrix} -0.4033 & -0.0801 \\ -0.3955 & -1.1062 \\ -1.2845 & -1.1494 \end{bmatrix}. \end{aligned} \tag{13.53}$$

Fig. 13.1 shows the time when the system is under attack, which is between 20 and 120 seconds. Three attack scenarios were considered and their estimation errors were obtained using MATLAB/Simulink as follows:

1. Filtering under DoS attack, Fig. 13.2;
2. Filtering under deception attack, Fig. 13.3;
3. Filtering without any attack, Fig. 13.4.

As shown in Figs. 13.2, 13.3, and 13.4, the designed estimator shows a small error in estimating the states under all possibilities of attack.

13.2 Resilience state estimation against integrity attacks

Cyber-physical security has received much research attention [472] in the last decade. Typically the sensors are vulnerable to integrity attacks since in most cases they are spatially distributed and cannot be fully protected. Compromised data detection via fault detection and isolation-based methods has been extensively studied [473,421,474,475]. However, in most of these works the system is assumed to be noiseless, which greatly favors the failure detector. In [127] the

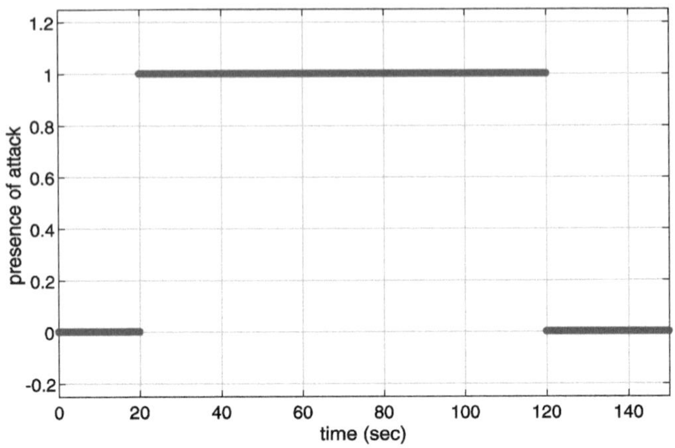

FIGURE 13.1 Time of attack.

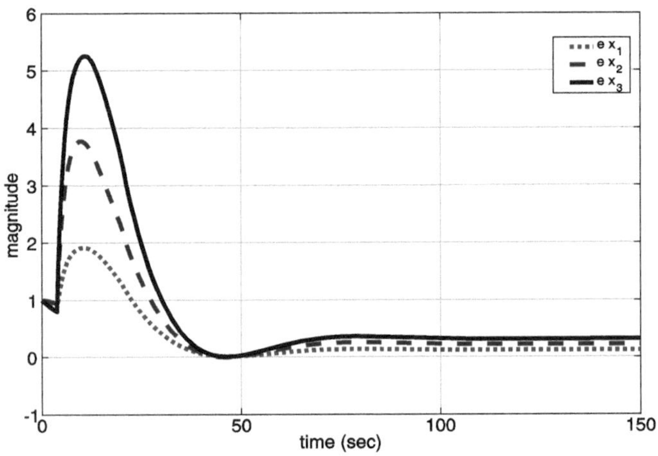

FIGURE 13.2 Error in estimation under DoS attack.

authors studied the worst bias an attack can cause through reachability analysis and ellipsoid approximation. In [476,475,477,478], they designed different robust estimators grounded on the so-called sparsity observability conditions.

If the fundamental sparsity observability conditions are violated, no resilient estimator exists; however, this is not enough. Instead of designing resilient estimators like [474,476,475,477,478], we conduct the research in the reverse way. We make an effort to answer the following questions: Given an estimator, what can we say about its resilience and performance? Are there any systematic analysis procedures we can use? We illustrate an analytical approach to solving the problem by taking a convex optimization-based estimator as an example. Our proposed approach can be applied to analyze the resilience of a class of com-

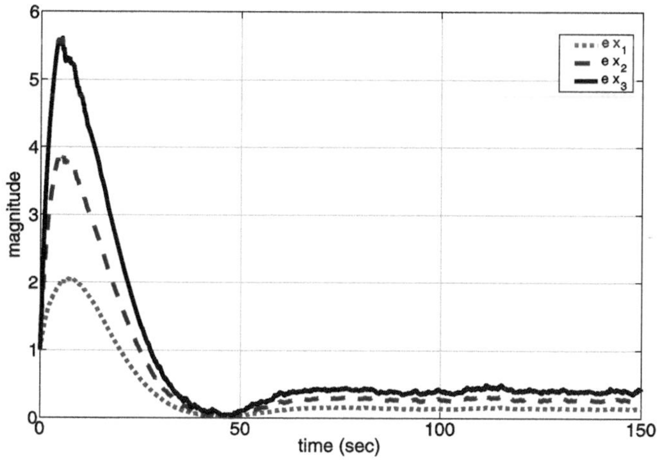

FIGURE 13.3 Error in estimation under deception attack.

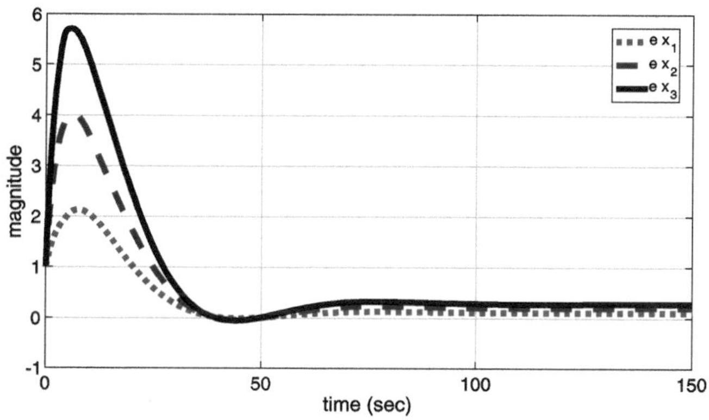

FIGURE 13.4 Error in estimation without attack.

monly used estimators and the estimation performance can be quantified, which is always ignored in the existing works.

13.2.1 Introduction

The significance of this section is threefold:

(i) We propose a framework for studying cyber-physical security problems by formally defining three key ingredients, (p,m)-sparse attacks, resilience, and translation invariance;

(ii) The employed approach for analyzing the resilience of an estimator is novel and generic for a class of estimators. Though we take an estimator with L1-penalty as an example, the convex optimization-based analysis applies to a large number of estimators. We analyze the necessary and sufficient conditions for resilience and the conclusion that benign sensors must be more than malicious sensors aligns with the results in different research scenarios [474,476,475,477,478];

(iii) The trade-off between minimum mean square error (MMSE) optimality and resilience has been thoroughly studied. The condition that the resilient estimator gives the MMSE estimate without attacks is given. A nontrivial upper bound on the gap between the resilient estimate and the MMSE estimate is also derived.

In the following the ith entry of the vector u is denoted as $u[i]$. The L_p norm of the vector u is denoted as $\|u\|_p$. If unspecified, $\|u\|$ means the L_2 norm of u by default. $\lfloor v \rfloor$ means the largest integer that is less than the scalar v. For a given set $\mathcal{X}$, $|\mathcal{X}|$ denotes its cardinality.

13.2.2 System model

Assume that m homogeneous sensors are measuring the following LTI system:

$$x(k+1) = Ax(k) + w(k). \tag{13.54}$$

The measurement equation for the ith sensor is given by

$$y_i(k) = Cx(k) + \varepsilon_i(k), i = 1, ..., m, \tag{13.55}$$

where $x(k) \in \mathbb{R}^n$ is the state, $y_i(k) \in \mathbb{R}^l$ is the measurement collected by the ith sensor, and $w(k) \in \mathbb{R}^n$ and $\varepsilon_i(k) \in \mathbb{R}^l$ are the process noise and measurement noise for the ith sensor, respectively. The noise values $w(k)$ and $\varepsilon_i(k)$ are Gaussian distributed (i.e., $w(k) \sim \mathcal{N}(0, Q)$, $\varepsilon_i(k) \sim \mathcal{N}(0, R)$). The noise values are assumed to be independent from each other across different time instants and sensors. We denote the tall measurement matrix $H \triangleq [C^T, C^T, ..., C^T]^T \in \mathbb{R}^{lm\times n}$ and $\sum \triangleq \text{diag}\,(R, ..., R)$, $R > 0$. The initial state $x(0)$ is Gaussian distributed with mean μ_0 and variance P_0, and is independent from all noise. Assume that (A, C) is observable and $(A, \sqrt{Q})$ is controllable. We denote the index set of the sensors as $\mathcal{S} \triangleq \{1, ..., m\}$.

The Kalman filter is well known to be the recursive MMSE estimator,

$$\begin{aligned}
\hat{x}_{KF}(k) &= (A - K(k)HA)\hat{x}_{KF}(k-1) + K(k)y(k), \\
P^-(k) &= AP(k-1)A^T + Q, P(k) = (I_n - K(k)H)P^-(k),
\end{aligned}$$

where the Kalman gain is given by

$$K(k) = P^-(k)H^T(HP^-(k)H^T + \sum)^{-1}. \tag{13.56}$$

The state error covariance $P^-(k)$ converges exponentially quickly to $\bar{P}$, which is obtained by solving the following discrete algebraic Riccati equation (DARE):

$$X = AXA^T - AXH^T(HXH^T + \sum)^{-1}HXA^T + Q. \tag{13.57}$$

Therefore, we assume the Kalman filter to be in the steady state, that is, $P(k) = (I_n - KH)\bar{P}$ and $K(k) = K$ from (13.56).

Due to the homogeneity of the sensors, we know that K can be written in the form of $[G, ..., G]$, $G \in \mathbb{R}^{n\times l}$. The Kalman filter can be equivalently rewritten as

$$\hat{x}_{KF}(k) = \frac{1}{m}\sum_{i\in\mathcal{S}}\hat{x}_i(k), \tag{13.58}$$

where

$$\hat{x}_i(k) = (A - KHA)\hat{x}_i(k-1) + mGy_i(k). \tag{13.59}$$

This means the estimation process at the estimator can be decomposed into m subprocesses, each of which involves measurements from only one sensor. This decomposition renders distributed estimation possible. To be specific, the sensor can locally compute $\hat{x}_i(k)$ based on its own measurements and then the information fusion of all local estimates occurs at the remote estimator. It is worth noting that this distributed estimation is more resilient to attacks than the centralized estimation (all sensors transmit raw measurements to a central estimator). Since each local estimate of one sensor encodes all its historical measurements, corruption of one local estimate at some time instant causes little damage to the estimation.

Even if the sensor lacks computational capability and can only transmit raw measurements, each local estimation process can be computed at the central estimator. Therefore, without loss of generality, we assume that each sensor computes a local estimate based on (13.59) and sends it to the estimator.

13.2.3 Attack model

The attacker launches an integrity attack to the sensory data in different ways. For example, it can change the physical environment to mislead the sensors, it can hack the onboard sensor chip, or it can manipulate the data packet during the sensor-to-estimator transmission. No matter how the attack is launched, we can rewrite the measurement equation as

$$z_i(k) = \hat{x}_i(k) + a_i(k), \tag{13.60}$$

where $z_i(k) \in \mathbb{R}^n$ is the *manipulated* local estimate and $a_i(k) \in \mathbb{R}^n$ is the attack vector. In other words, the attacker can change the local estimate of the ith sensor by $a_i(k)$. If the sensor is safe, then $a_i(k) = 0$. We define the local estimation error as $e_i(k) \triangleq \hat{x}_i(k) - x_i(k)$. Then we have

$$z_i(k) = x_i(k) + e_i(k) + a_i(k). \tag{13.61}$$

For concise notations, we denote

$$\hat{x}(k) \triangleq [\hat{x}_1(k)^T, \hat{x}_2(k)^T, ..., \hat{x}_m(k)^T]^T. \tag{13.62}$$

Similarly, we can define $z(k), e(k), a(k)$. For any index set $\mathcal{I} \subseteq \mathcal{S}$, we define the complement set as $\mathcal{I}^c \triangleq \mathcal{S} \backslash \mathcal{I}$. In our attack model, we assume that the attacker can only compromise at most p sensors but can arbitrarily choose $a_i(k)$. The index set of malicious sensors is assumed to be time invariant. Formally, a (p, m)-sparse attack can be defined as:

Definition 13.2 ((p,m)-sparse attack). A vector a is called a (p, m)-sparse attack if there exists an index set $\mathcal{I} \subset \mathcal{S}$ such that (i) $\|a_i(k)\| = 0, \forall \in \mathcal{I}^c$ and (ii) $|\mathcal{I}| \leq p$ both hold.

We define the collection of a possible index set of malicious sensors as $\mathbb{C} \triangleq \{\mathcal{I} : \mathcal{I} \subset \mathcal{S}, |\mathcal{I}| = p\}$. The set of all possible (p, m)-sparse attacks is denoted $\mathcal{A} = \mathcal{A}(k) \triangleq \bigcup_{\mathcal{I} \in \mathbb{C}} \{a(k) : \|a_i(k)\| = 0, i \in \mathcal{I}^c\}, \forall k$.

After introducing the (p, m)-sparse attack, we need to formally define what we mean by resilience.

Definition 13.3 (Resilience). An estimator that maps the measurements $z(k)$ to a state estimate $\hat{x}(k)$ is said to be resilient to the (p, m)-sparse attack if it satisfies the condition

$$\left\| g(\hat{x}(k)) - a(\hat{x}(k)) + a(k) \right\| \leq \mu(\hat{x}(k)), \forall a \in \mathcal{A}, \tag{13.63}$$

where $\mu : \mathbb{R}^{mn} \longmapsto \mathbb{R}$ is a real-valued mapping on $\hat{x}(k)$.

The resilience implies that the disturbance on the state estimate caused by an arbitrary attack is bounded. A trivial resilient estimator is $g(y) = 0$, which provides a very poor estimate. Therefore, another desirable property for an estimator is translation invariance, defined as follows, where $E \triangleq [I_n, ..., I_n]^T$:

Definition 13.4 (Translation invariance). An estimator g is translation invariant if $g(z + E_u) = u + g(z), \forall u \in \mathbb{R}^n$.

13.2.4 Generic resilient estimator

Apparently, the linear estimator (13.58) cannot give an estimate with bounded error even when only one estimate is arbitrarily manipulated. In other words, there is a conflict between the MMSE optimality and the resilience against attacks. One of the main tasks is to analyze an estimator that can achieve a desirable trade-off between MMSE optimality and resilience, and investigate the necessary and sufficient conditions for resilience to (p, m)-sparse attacks. To this end, a general estimator is proposed as

$$\hat{x}(k) \triangleq g(z(k)) = \arg\min_{\hat{x}(k)} \sum_{i \in \mathcal{S}} \varphi_i(z_i(k) - \hat{x}(k)), \tag{13.64}$$

where $\varphi_i : \mathbb{R}^n \longmapsto \mathbb{R}$. We note that to recover Kalman filter we can choose φ_i to be $L2$ norm. The candidate functions of φ_i may include the $L2$ norm or the least absolute shrinkage and selection operator (LASSO) [479], to name just a few. Though there are many important estimators as special cases of (13.64), the lack of an explicit form of (13.64) hinders the further discussion of the proposed approach. To facilitate conveying our idea, we mainly focus on the following concrete estimator in the rest of this chapter. The same methodology can be extended to other φ_i.

13.2.5 Resilient estimator with L1-penalty

We propose a resilient estimator in the presence of integrity attack [480],

$$\min_{\hat{x}, a, \hat{w}} \left\| [\|a_1\|_2 \ldots \|a_m\|_2]^T \right\|_0$$

subject to $z_i = H_i \hat{x} + \hat{w}_i + a_i$, $\hat{w} \in \omega$, where $\hat{w} = [\hat{w}_i^T, \ldots, \hat{w}_m^T]^T$ and where the authors assume that the noise is bounded and lies in a convex set ω. However, the minimization problem involves zero-norm, and is thus difficult to solve in general. A commonly adopted approach is to use L_1 relaxation to approximate zero-norm, which leads to the minimization problem,

$$\min_{\hat{x}(k), a, \hat{w}(k)} \sum_{i \in \mathcal{S}} \|\hat{w}_i(k)\|_2^2 + \lambda \sum_{i \in \mathcal{S}} \|a_i(k)\|_1, \tag{13.65}$$

subject to $z_i(k) = \hat{x}(k) + \hat{w}_i(k) + a_i(k), \forall i$.

If we define the function $F : \mathbb{R}^n \longmapsto \mathbb{R}$ as

$$F(u) \triangleq \min_{v \in \mathbb{R}^n} \|u - v\|_2^2 + \lambda \|v\|_1, \tag{13.66}$$

where v corresponds to $a_i(k)$ in the context, then we can easily prove that the optimization problem (13.65) can be rewritten as

$$\hat{x}(k) \triangleq g(z(k)) = \arg\min_{\hat{x}(k)} \sum_{i \in \mathcal{S}} F(z_i(k) - \hat{x}(k)). \tag{13.67}$$

It is easy to check that g is translation invariant. In fact, if $\hat{x}^* = g(z)$, then

$$g(z + Eu) = \arg\min_{\hat{x}(k)} \sum_{i \in \mathcal{S}} F[z_i - (\hat{x} - u)],$$

which implies that $g(z + Eu) - u = \hat{x}^*$. In the next section, we present the necessary and sufficient conditions for the resilience of the estimator (13.67) using a constructive method. For the concise notation, we omit the time index k if it is clear from the context.

13.2.6 Resilience analysis

We provide an answer to the following question in this section: Under what condition does the estimator in (13.67) satisfy the resilience requirement (13.63)? The first step of our analysis is to investigate the properties of F.

Before preceding to the main results, we give an explicit form of $F(u)$ given in (13.66). We can decompose $F(u)$ by $F(u) = \sum_{i=1}^{n} f(u[i])$, where $u[i]$ is the ith entry of u and $f(\tau) : \mathbb{R} \longmapsto \mathbb{R}$ is given by

$$f(\tau) \triangleq \min_{v \in \mathbb{R}} \quad (\tau - v)^2 + \lambda |v|. \tag{13.68}$$

We define the right-hand side of (13.68) as $\pi(v) \triangleq (\tau - v)^2 + \lambda|v|$. Applying the Karush–Kuhn–Tucker (KKT) conditions, we know that $0 \in \partial\tau(v^*) = -2\tau + 2v^* + \text{sgn}(v)\lambda$. Since $\pi(v)$ is not differentiable at $v = 0$, by calculating the subgradient we have that $v^* = 0$ if $|\tau| \leq \lambda/2$. For $(v^*) \neq 0$, by letting $0 = -2\tau + 2v^* + \text{sgn}(v^*)\lambda$, we obtain that

$$v^* = \begin{cases} \tau - \lambda/2, & \text{if } \tau > \lambda/2, \\ \tau + \lambda/2, & \text{if } \tau < -\lambda/2. \end{cases}$$

Therefore, we have f explicitly written as

$$v^* = \begin{cases} \tau^2, & \text{if } \tau \leq \lambda/2, \\ \lambda|\tau| - \lambda^2/2, & \text{if } \tau > \lambda/2. \end{cases} \tag{13.69}$$

In the next proposition we present some useful properties of f and F, which are easy to verify.

Proposition 13.1. *The following properties of f and F hold:*

1. *f and F are convex;*
2. *f and F are symmetric, $f(u) = f(-u)$;*
3. *f and F are non-negative and $f(0) = 0$.*

To obtain the necessary and sufficient conditions for resilience, we first need to show some findings on the derivative of F which are crucial for the derivation of the main results. To facilitate the analysis, we define two functions. For all $u, v \in \mathbb{R}^n$ and $t \in \mathbb{R}$, we define $h : \mathbb{R}^n \times \mathbb{R}^n \times \mathbb{R} \longmapsto \mathbb{R}$ as follows: $h(u, v, t) \triangleq F(v + tu)$. We define the mapping $\phi : \mathbb{R} \longmapsto \mathbb{R}$,

$$\phi(u) \triangleq \nabla F(u) = [\nabla f(u[1]), ..., \nabla f(u[n])]^T, \tag{13.70}$$

where

$$\nabla f(u[i]) = \begin{cases} 2u[i], & \text{if } |u[i]| \le \lambda/2, \\ \text{sgn}(u[i])\lambda, & \text{if } |u[i]| > \lambda/2. \end{cases} \tag{13.71}$$

Here sgn(·) is defined as follows: if $s = \text{sgn}(v)$, then $s[i] = \begin{cases} +1, & \text{if } v[i] \ge 0, \\ -1, & \text{if } v[i] < 0. \end{cases}$
Note that a useful equality in the sequel is $\partial h(u, v, t)/\partial t = \phi(v + tu)^T u$.

Lemma 13.1. *The following statements are true:*

1. *The limit below is well defined for all $u \in \{u \in \mathbb{R}^n : \|u\| < \infty\}$:*

$$C(u) \triangleq \lim_{t\to\infty} \partial h(u, 0, t)/\partial t = \lambda \|u\|_1 . \tag{13.72}$$

2. *The following point-wise limit holds:*

$$\lim_{t\to\infty} \partial h(u, v, t)/\partial t = C(u); \tag{13.73}$$

   moreover, the convergence is uniform on any compact set of (u, v).
3. *For any u, v, we have that*

$$\phi(v + u)^T u \le C(u). \tag{13.74}$$

Proof. **1.** It is easy to see that

$$\partial h(u, 0, t)/\partial t = \lambda \sum_{i=1}^{n} \text{sgn}(u[i])u[i] = \lambda \|u\|_1 .$$

2. We have that $\lim_{t\to\infty} \partial h(u, v, t)/\partial t = \lim_{t\to\infty} \phi(v + tu)^T u = \lambda \|u\|_1$. Due to the convexity of F, $\partial h(u, v, t)/\partial t$ is monotonically nondecreasing with respect to t. Furthermore, $C(u)$ is continuous since it is a norm. Therefore, by Dini's theorem [481], $\partial h(u, v, t)/\partial t$ converges uniformly to $C(u)$ on a compact set of (u, v).
3. From (13.70), we know that $\phi(v + u)^T u \le \sum_{i=1}^{n} \lambda |u[i]| \le \lambda \|u\|_1$. Therefore, we conclude that $\phi(v + u)^T u \le C(u)$ for any u, v. □

Remark 13.3. Intuitively speaking, we can interpret F as a potential field and the derivative of F as the force generated by each sensor. By (13.74), we know that the force from the potential field F along the u direction cannot exceed $C(u)$. On the other hand, Eq. (13.73) implies that this bound is achievable.

13.2.7 Necessary and sufficient conditions

Based on the convexity of F and the properties given in Lemma 13.1, we are now ready to give the sufficient conditions for the resilience of the estimator. The proof is constructive and general for (13.64) as long as φ_i is convex.

Many commonly used estimators based on convex optimization (i.e., L_p-norm minimization-based estimators and group LASSO estimators), can be decomposed with a convex φ_i.

Theorem 13.3. *(Sufficient condition). If $2p < m$, then the estimator g is resilient.*

Proof. To prove the result above we need to show that the optimal estimate $\tilde{x}$ is bounded by some constant dependent on $\tilde{x}$ if $2p < m$ holds. Equivalently, our goal is to prove that there exists a $\beta(\tilde{x})$, such that for any $t > \beta(\tilde{x}) \geq 0$, $\|u\| = 1, a \in \mathcal{A}$, the following inequality holds:

$$\sum_{i\in\mathcal{S}} \nabla F(z_i - tu)^T u = \sum_{i\in\mathcal{S}} \partial h(-u, z_i, t)/\partial t > 0. \tag{13.75}$$

As a result, any point satisfying $\|\hat{x}\| > \beta(\tilde{x})$ cannot be the solution of the optimization problem since there exists $\epsilon > 0$ such that $(\|\hat{x}\| - \epsilon)\hat{x}/\|\hat{x}\|$ is a better point. Therefore, we must have $\|g(z)\| \leq \beta(\tilde{x})$, and thus the estimator is resilient.

To prove (13.75) we first look at the benign sensors. We can always find a finite constant N_i depending on δ and $\tilde{x}_i$ such that for all $t \geq N_i(\delta, \tilde{x}_i)$, the following inequality holds

$$\partial h(-u, \tilde{x}_i, t)/\partial t \geq C(u) - \delta = \lambda - \delta \tag{13.76}$$

for any $\|u\| = 1$. We define $\beta(\tilde{x}) \triangleq \max_{1\leq i\leq m} N_i(\delta, \tilde{x}_i)$ and fix δ to be

$$\delta = (m - 2p)\lambda/m. \tag{13.77}$$

Hence, for $i = 1, ..., m$, if $t > \beta(\tilde{x})$ we know that

$$\sum_{i\in\mathcal{I}^c} \partial h(-u, \tilde{x}_i, t)/\partial t \geq (m - p)(\lambda - \delta), \forall \|u\| = 1. \tag{13.78}$$

We now consider malicious sensors. By Lemma 13.1 (iii), we know that for $i \in \mathcal{I}$, and any u

$$\phi(z_i - tu)^T tu = \phi(z_i - 2tu + tu)^T tu \leq C(tu) \quad \Rightarrow \quad \phi(z_i - tu)^T u \geq -\lambda.$$

Then we have

$$\sum_{i\in\mathcal{I}} \partial h(-u, z_i, t)/\partial t \geq -p\lambda, \forall \|u\| = 1. \tag{13.79}$$

Thus, from (13.77), (13.78) and (13.79), we know that

$$\sum_{i\in\mathcal{S}} \partial h(-u, z_i, t)/\partial t \geq (m - p)(\lambda - \delta) - p\lambda > 0,$$

which proves (13.75). □

Remark 13.4. The result is not surprising given the many other existing works [474,476,475,477,478]. This force analogy, however, provides very different insights from the existing works. It is shown that if the number of malicious sensors is less than the good sensors, then the estimator is resilient. The intuition is that the sum force injected by any p sensors from the potential field F along the u direction must be balanced by the sum force of the rest of the $m - p$ sensors, i.e., zero-sum. Otherwise, the optimal estimate must lie in the infinity due to unbalanced driving forces along u, and thus violates the resilience defined in (13.63). One of the advantages of this approach is that we can analytically quantify the estimation performance. The constructive proof of Theorem 13.1 sheds light on the derivation of a tight $\mu(\tilde{x})$, that is, we know that $\mu(\tilde{x}) = \beta(\tilde{x}) + 1$ is a good candidate, where $\beta(\tilde{x})$ is given before (13.77).

Remark 13.5. This sufficient condition is not generally true for inhomogeneous φ_i values in (13.64). The required ratio of good sensors will normally be higher than one-half to guarantee resilience. The proof of Theorem 13.1, however, applies for the case of inhomogeneous φ_i values as long as they are convex.

We next present a necessary condition for the resilience of the estimator.

Theorem 13.4 (Necessary condition). *If $2p > m$ holds, then the estimator is not resilient to the attack.*

Proof. The resilience of the estimator is equivalent to that of the optimal estimate c satisfying $\|\hat{x}\| \leq \mu(z)$ for all $a \in \mathcal{A}$. To this end, we prove that for any $r > 0$, there exists a y such that all $\hat{x}$ are satisfied cannot be the optimal solution of (13.67).

We first look at the compromised sensors. For every $\delta > 0$ we can always find a finite constant $N_i(\delta)$ such that for any $\tilde{x} \in \{\hat{x} : \|\hat{x}\| \leq r\}$ and for all $t > N_i$, the following inequality holds:

$$\partial h(u, z_i - \hat{x}, t)/\partial t \geq C(u) - \delta. \tag{13.80}$$

The inequality is due to the uniform convergence of $h(u, v, t)$ to $C(u)$ on $\{u\} \times \{v : v = z_i - \hat{x}, \|x\| \leq r\}$.

We choose $\delta = (2p - m)C(u)/m$ and $t = \max_{i \in \mathcal{I}} N_i(\delta, \tilde{x}_i)$ and $z_i = tu$ for all $i \in \mathcal{I}$, then we know for any $\|x\| \leq r$, $\sum_{i \in \mathcal{I}} \partial h(u, z_i - \hat{x}, t)/\partial t \geq pC(u) - p\delta$. Now we look at the benign sensors. From Lemma 13.1 (iii) we have

$$\partial h(u, \tilde{x}_i - \hat{x}, t)/\partial t \geq -C(u), \tag{13.81}$$

and from (13.80) and (13.81),

$$\sum_{i \in \mathcal{S}} \partial h(u, z_i - \hat{x}, t)/\partial t \geq (m - p)C(u) - pC(u) + p\delta > 0.$$

Thus, for such a z_i satisfying $y_i = \begin{cases} \tilde{x}_i, & \text{if } i \in \mathcal{I}^c \\ tu, & \text{if } i \in \mathcal{I} \end{cases}$, $\hat{x} + u$ is a better estimate than all $\hat{x}$ satisfying $\|x\| \le r$. Since r is an arbitrary positive real number, we can conclude that the estimator is not resilient. □

In the previous section we adopted a force analogy approach to studying the resilience of the estimator. Now we focus our attention on the performance of the proposed estimator. We deal with two questions in this section. The first is the sufficient condition that the estimator gives an MMSE estimate when there is no attack. The second is determining the maximum damage that an attacker can cause to the estimate (i.e., the upper bound of $\|g(\tilde{x}(k)) - g(\tilde{x}(k)) + a(k)\|$).

13.2.8 Performance evaluation without attacks

When no attacks are present, an MMSE estimate such as a Kalman filter provides is still preferable. Note that the proposed resilient estimator probabilistically provides an MMSE estimate. A sufficient condition for providing the MMSE estimate $\hat{x}_{KF}$ given in (13.58) is given as follows:

Proposition 13.2. *If $\tilde{x}$, defined in (13.62), falls into the set $\in \mathcal{G}$, where*

$$\mathcal{G} \triangleq \{\tilde{x} \in \mathbb{R}^{mn} : \max_{i \in \mathcal{S}} \left\|\tilde{x}_i - \hat{x}_{KF}\right\|_1 \le \lambda/2, \}, \tag{13.82}$$

then $\tilde{x} = \hat{x}_{KF}$ holds.

Proof. From (13.82) and (13.69), we know that $\hat{x}_{LS}$ is a solution of (13.67).

Now we characterize the probability density function (pdf) of $\tilde{x}$. We define the local estimation error covariance of the ith sensor and the local cross-estimation error covariance between the ith sensor and the jth sensor estimate as $P_{ii}(k) \triangleq \mathbb{E}[e_i(k)e_i(k)^T | y_i(1), ..., y_i(k)]$, $P_{ij}(k) \triangleq \mathbb{E}[e_i(k)e_j(k)^T | y_i(1), y_j(1), ..., y_j(k), y_j(k)]$. From (13.59) the error dynamics of the ith sensor estimate is thus given as follows:

$$e_i(k) = (A - KHA)e_i(k-1) + (mGC - I_n)w(k) + mG\epsilon_i(k). \tag{13.83}$$

Note that the local estimator for each sensor is a stable estimator since the spectral radius of $A - KHA$ is less than one [482]. It is easy to see that $P_{ii}(k)$ converges to $\bar{P}_{ii}$ at the steady state, where $\bar{P}_{ii}$ is the solution of the following Lyapunov equation of X:

$$\begin{aligned} X =& (A - KHA)X(A - KHA)^T \\ &+ (mGC - I_n)Q(mGC - I_n)^T + m^2GRG^T. \end{aligned} \tag{13.84}$$

Similarly, $P_{ij}(k)$ converges to $\bar{P}_{ij}$, where $\bar{P}_{ij}$ is the unique solution of the following Lyapunov equation of X:

$$X = (A - KHA)X(A - KHA)^T \\ + (mGC - I_n)Q(mGC - I_n)^T. \tag{13.85}$$

We denote $\Gamma = \{\bar{P}_{ij}\} \in \mathbb{R}^{nm \times nm}$. Now we know the pdf of $\tilde{x}$ (i.e., $\tilde{x} \sim \mathcal{N}(x, \Gamma)$), and thus the distribution of $\hat{x}_{KF}$. We can now compute the probability of generating the MMSE estimate:

$$Pr(\tilde{x} \in \mathcal{G}) = \int_{x \in \mathcal{G}} \mathcal{N}(x, \Gamma) d\tilde{x}. \tag{13.86}$$

The integration is not trivial and numerical methods can be used to approximate $Pr(\tilde{x} \in \mathcal{G})$. A closed-form solution to $Pr(\tilde{x} \in \mathcal{G})$ is left as an open question. Another interesting observation is that the larger λ is, the more likely the MMSE estimate is. □

13.2.9 Performance evaluation under attacks

We now consider the worst damage that an attacker can cause (i.e., the maximum deviation between the estimate under attack and that without attacks). If the necessary condition in Theorem 13.2 is violated, the estimator is not resilient and thus the deviation can be arbitrarily large. A more interesting question is how to obtain $\mu(\hat{x})$ in (13.63) for all possible attacks if the estimator is resilient.

Suppose the sufficient condition in Theorem 13.1 is satisfied. Let the resilient estimate without attacks be $\hat{x}_R = g(\hat{x})$.

Due to the translation invariance, we have

$$\|g(\tilde{x}) - g(\tilde{x} + a)\|_1 \\ = \|\hat{x}_R - g(\tilde{x} + a\|_1 = \|g(z - E\hat{x}_R)\|_1 \leq \mu(\tilde{x}).$$

We denote $\tilde{z}_i \triangleq z_i - \hat{x}_R$, $\tilde{z} = [\tilde{z}_1^T, ..., \tilde{z}_m^T]^T$ and $\breve{x}_i \triangleq \tilde{x}_i - \hat{x}_R$, $\breve{x} = [\breve{x}_1^T, ..., \breve{x}_m^T]^T$.

Similar to the proof of Theorem 13.1, there exists a value of β^* such that for any $\beta \in \{\beta : \|\beta\|_1 > \|\beta^*\|_1\}$ the following inequality holds:

$$\sum_{i \in \mathcal{S}} \phi(\tilde{z}_i - \beta)^T \text{sgn}(\beta) > 0. \tag{13.87}$$

In other words, we want to find a β^* for all j such that

$$\sum_{j=1}^{n} \sum_{i \in \mathcal{S}} \nabla f(\tilde{z}_i[j]) - \beta^*[j] \text{sgn}(\beta^*[j]) = 0, \tag{13.88}$$

where $\tilde{z}_i[j]$ and $\beta^*[j]$ are the jth entry of $\tilde{z}_i$ and β^*, respectively.

We define the two mappings $\underline{k}, \bar{k} : \mathbb{R}^m \times \mathbb{R} \times \mathbb{R} \longmapsto \mathbb{R}$ for any vector u and scalars p, m:

$$\begin{aligned}
&\underline{k}(u, p, m) \triangleq \\
&\quad \{u[i] : |\{u[j] : u[j] \leq u[i], j \neq i\}| = \lfloor (m-2p)/2 \rfloor + 1\}, \\
&\bar{k}(u, p, m) \triangleq \\
&\quad \{u[i] : |\{u[j] : u[j] \geq u[i], j \neq i\}| = \lfloor (m-2p)/2 \rfloor + 1\}.
\end{aligned}$$

We let $\zeta \triangleq [\breve{x}_1^T[j], ..., \breve{x}_m^T[j]]^T$. Then we denote $(\underline{\theta}_i, \bar{\theta}_i)$ to be

$$(\underline{\theta}_i, \bar{\theta}_i) = (\underline{k}(\zeta_j, p, m), \bar{k}(\zeta_j, p, m)), \; j = 1, ..., n. \tag{13.89}$$

Now we are ready to present the upper bound on the worst damage.

Theorem 13.5. *The upper bound of $\mu(\tilde{x})$ is shown as follows,*

$$\mu(\tilde{x}) = \left\| \beta^+ \right\|_1, \tag{13.90}$$

where $\beta^+ = max(|\underline{\theta}_j - \lambda/2|, |\bar{\theta}_j + \lambda/2|)$.

Proof. A sufficient condition for (13.88) is that for each j the following inequality holds:

$$\sum_{i \in \mathcal{S}} \nabla f(\tilde{z}_i[j]) - \beta[j] \mathrm{sgn}(\beta^*[j]) = 0. \tag{13.91}$$

We first show that $\beta^*[j]$ must lie in $[\underline{\theta}_j - \lambda/2, \bar{\theta}_j + \lambda/2]$. We prove this by contradiction. Suppose $\beta^*[j] < \underline{\theta}_j - \lambda/2$. For any possible $\mathcal{I}^c$, we then have

$$\sum_{i \in \mathcal{I}^c} \nabla f(\breve{x}_i[j]) - \beta[j] \mathrm{sgn}(\beta^*[j]) \geq (m-p)\lambda.$$

This is due to the fact that there are at least $m - p$ points of $\tilde{z}_i[j]$ that are $\lambda/2$ larger than $\beta^*[j]$ from (13.89) and that the maximum gradient is λ from (13.71).

On the other hand, from Lemma 13.1 (iii) we know that for any possible $\mathcal{I}$,

$$\left| \sum_{i \in \mathcal{I}} \nabla f(\tilde{z}_i[j]) - \beta[j] \mathrm{sgn}(\beta^*[j]) \right| \geq p\lambda.$$

Hence (13.91) cannot hold for $\beta^*[j] < \underline{\theta}_j - \lambda/2$. A similar argument applies for $\beta^*[j] > \bar{\theta}_j - \lambda/2$. Therefore, we know that $\beta^*[j] \in [\underline{\theta}_j - \lambda/2, \bar{\theta}_j + \lambda/2]$. If we take the maximum over $(|\underline{\theta}_j - \lambda/2|, |\bar{\theta}_j + \lambda/2|)$ as $\beta^*[j]$, then any β satisfying $\|\beta\| > \|\beta^+\|$ cannot be $g(z - E\hat{x}_R)$. □

13.2.10 Illustrative example II

We consider the system with the following parameters:

$$A = \begin{bmatrix} 0.95 & 0.1 \\ 0.1 & 1.01 \end{bmatrix}, Q = \begin{bmatrix} 1.5 & 1 \\ 1 & 2 \end{bmatrix}, C = \begin{bmatrix} 1 & 0 \\ 0 & 1 \end{bmatrix}, R = \begin{bmatrix} 2 & 1 \\ 1 & 1 \end{bmatrix}.$$

We assume there are $m = 5$ sensors. First we verify the necessary and sufficient conditions for resilience. Assume that the number of malicious sensors is $p = 2$. Each entry of the attack vector is generated with a Gaussian distribution (i.e., $\mathcal{N}(1000, 1)$). The traces of the expected estimation error covariance for $\lambda = 0.1, 1, 10$ are 1.1687, 1.1879, 4.4634, respectively. When $p = 3$, the traces of the expected estimation error covariance for $\lambda = 0.1, 1, 10$ are all around 485, which implies that the estimate is no longer reliable.

TABLE 13.1 Relationship between λ and the probability of recovering $\hat{x}_{KF}$ without attacks.

λ	1	2	3	5	10
$P_r(\hat{x} = \hat{x}_{KF})$	0.0002	0.0151	0.1088	0.5019	0.9843

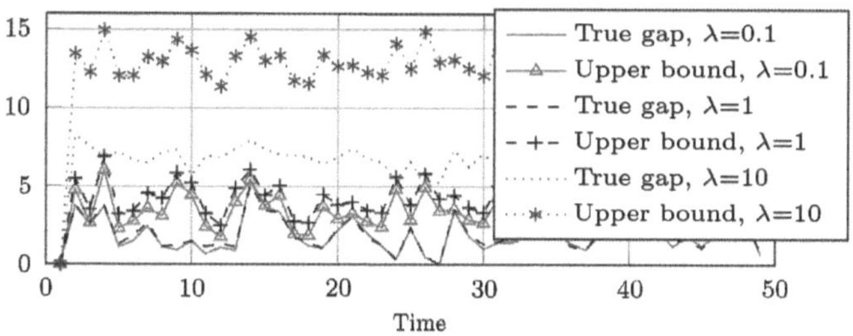

FIGURE 13.5 Upper bound $\mu(\hat{x})$ and the true gap vs. time.

In Table 13.1 we show the relationship between the penalty parameter λ and the probability of recovering the Kalman filter when there is no attack. It can been seen that the recovery probability increases with λ. On the other hand, we plot the upper bound $\mu(\tilde{x})$ given in Theorem 13.3 versus time in Fig. 13.5. Note that when $\lambda = 1$ or $\lambda = 0.1$ the upper bound on the deviation caused by attacks is smaller. The trade-off between resilience when the sensors are under attack and the MMSE optimality when the attacker is not present is clearly shown via different values of λ.

13.3 Notes

Cyber-physical systems (CPSs) are the integrations of computation, communication, and control that achieve the desired performance of physical processes.

Security threats have a high possibility of affecting CPSs, and they can be affected by several cyber attacks without providing any indication of failure. One important problem, especially in power systems, is the estimation problem of systems under cyber attacks. In this chapter a secure estimator for discrete-time delayed nonlinear systems considering both DoS and deception attacks was presented. The occurrences of the DoS and deception attacks is considered as Bernoulli distributed white sequences with variable conditional probabilities. First a sufficient condition was designed to obtain the required security level using the stochastic analysis techniques. Then a linear matrix inequality was solved to derive the gains of the estimator using YALMIP and MATLAB/Simulink. Finally the feasibility of the proposed estimation system was proved by solving a numerical example.

In the following section, we studied the resilient state estimation problem in the presence of integrity attacks. We proposed a resilient estimation framework and formulated a convex optimization problem with L_1 regulation to find the resilient estimate. The approach employed to analyze the resilience of an estimator is novel and generic for a class of estimators. We also showed the necessary and sufficient conditions for the estimator to be resilient against the (p, m)-sparse attack, which aligns with the results in the literature. This validates our analytical approach. The force analogy during the proof provides insights into the nature of resilience of an estimator. Furthermore, we analyzed the estimation performance without attacks and under attacks, which is always missing in the existing works.

Chapter 14

Cloud-based approach in data centers

Contents

14.1 Preliminaries

The past few years have witnessed a rapid development of communication technology and a significant improvement in computing capacity. These progresses have inspired computing systems to accurately control physical devices in remote areas in real time and in a reliable manner. This means that the cyber space and the physical space need to be tightly coupled with each other to accomplish the desired goals. To deal with these issues, cyber-physical systems (CPSs) have been proposed.

Recall that CPSs are a kind of emerging large-scale distributed systems [488], [489]. The new systems combine computing capacity and communication technology with control in a collaborative manner. CPSs are considered the next generation of engineered systems. However, there are several significant challenges to be addressed. CPSs require close interaction between the two distinct spaces. These interactions involve discrete dynamics in the cyber space and continuous dynamics in the physical space. In the past years the developments and design of the two spaces have moved in parallel. The intrinsic feature of a CPS necessitates efficient co-design methods between the two spaces at all

Cloud Control Systems. https://doi.org/10.1016/B978-0-12-818701-2.00022-6

levels. Using CPS approaches to design and implementing industrial goals has become more prevalent. The emerging applied fields include smart grids [490], intelligent transportation systems [491], and high confidence medical systems [492].

In this section, we survey recent CPS research projects related to cloud computing. As the definition in [493], cloud computing refers to the applications delivered as services over the Internet and to the hardware and systems software in the data centers that provide those services. In recent years this computing paradigm has been attracting extensive attention from both academia and industry, and has become more prevalent. We classify the related works into two categories: CPS approach to model data centers and cloud-based CPS applications. In the following we briefly introduce the related works according to the classification.

14.1.1 A modeling approach

The past decades have witnessed an unprecedented explosion of data space. The plethora of emerging applications require a large amount of data to accomplish the data-intensive computing. Due to the complicated requirements of massive storage and computation, data centers have been proposed that host a large number of servers for managing and processing the big data. The data center hardware and software is called "the cloud" [493]. Many providers of web services, such as Google, Facebook, and Amazon, offer various online services, such as web search, online gaming, social networks, to end-users based on the cloud. According to the reports [494], [495], in 2010 the power consumed by data centers all over the world was between 203 and 271 billion kilowatt hours of electricity. The energy consumed by data centers is mainly used for computation and cooling [496].

Definition 14.1. A data center (DC), also called a datacenter, is a repository that holds computing facilities such as servers, desks, switches, and routers, as well as supporting components, fire suppression facilities, power generators, backup equipment, and cooling systems.

Thus, a data center could be as complex as a dedicated building, for example in the United States [488], or as simple as a few machines that can be located within one room.

14.1.2 Architecture

The data center's architecture is a design blueprint that includes the physical and logical layout of the resources within a data center. Usually, it is a layered design that provides the architectural guidelines in data center development. The architecture includes the following:

- DC network architecture;

- o DC computing architecture (cloud computing);
- o DC security architecture.

14.1.3 Tier levels

Tier levels refer to the ability of data centers to maintain functionality during various failure scenarios such as power outage and storage desk corruption, among others. The higher the tier, the better the data center service, so higher tier levels indicate continuous data center operation with fault-tolerance systems that allow for uninterrupted use during emergencies. There are four commonly known tier levels, and a fifth tier standard has recently been added. Fig. 14.1 illustrates their features.

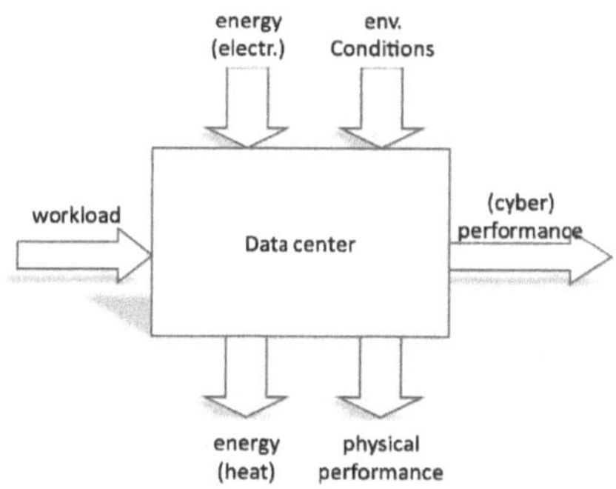

FIGURE 14.1 Data center model.

14.2 Modeling and control for energy efficiency

Data centers are facilities hosting a large number of servers dedicated to massive computation and storage. They can be used for several purposes, including interactive computation (e.g., web browsing), batch computation (e.g., renderings of images and sequences), or real-time transactions (e.g., banking). Data centers can be seen as a composition of information technology (IT) systems and support infrastructure. The IT systems provide services to the end-users, while the infrastructure supports the IT systems by supplying power and cooling. IT systems include servers, storage, and networking devices and middleware and software stacks, such as hypervisors, operating systems, and applications. The support infrastructure includes backup power generators, uninterruptible power supplies (UPSs), power distribution units (PDUs), batteries, and power supply units that generate and/or distribute power to the individual IT systems. The cooling technology (CT) systems, including server fans, computer room air conditioners (CRACs), chillers, and cooling towers, generate and deliver the cooling capacity to the IT systems [497], [498], [499], [500], [501], [502].

To provide the quality of service (QoS) required by the service level agreements, the IT control system can dynamically provision IT resources or actively manage the workloads through mechanisms such as admission control and workload balance. The IT systems consume power and generate heat whenever they are on. The power demand of the IT system can vary over time and is satisfied by the power delivery systems. The CT systems extract the heat to maintain the thermal requirements of the IT devices in terms of temperature and humidity. The IT, power, and cooling control systems have to work together to manage the IT resources, power, and cooling supply and demand.

The number of data centers is rapidly growing throughout the world, fueled by the increasing demand for remote storage and cloud computing services. Fig. 14.2 shows the increase in data center expenditures for power, cooling, and new servers from 1994 to 2010. The energy consumed for computation and cooling is dominating data center runtime costs [503], [504]. A report of the US Environmental Protection Agency (EPA) shows that data center power consumption doubled from 2000 to 2006, reaching a value of 60 TWh/y (terawatt hour/year) [505]. Historical trends suggest another doubling will occur by the end of 2019/2020.

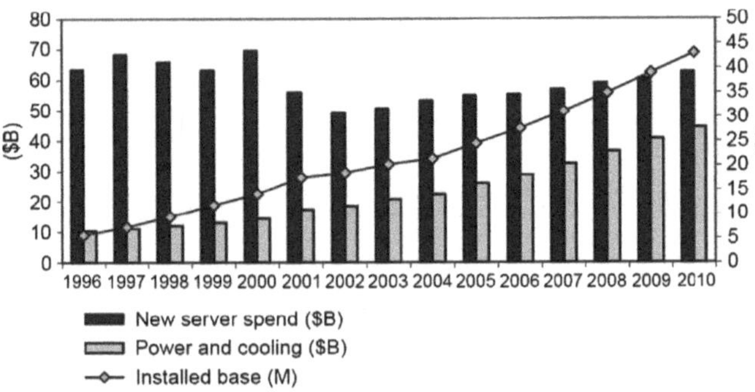

FIGURE 14.2 Data center spending trend [511].

As computational density has been increasing at multiple levels, from transistors on integrated circuits (ICs) to servers in racks and to racks in a room, the rate at which heat must be removed has increased, leading to nearly equal costs for operating the IT system and the CT system [506], [507]. Fig. 14.3 shows the measured and the expected growth of the power consumption density in data center equipment from 1994 until 2014 [508]. The available cooling capacity has in some cases become the limiting factor on the computational capacity [509]. Although liquid cooling is a promising alternative to air cooling, particularly for high-density data centers, this technology has not been widely adopted so far due to high costs and safety concerns [510].

The increase in data center operating costs is driving the innovation that will improve their energy efficiency. A measure of data center efficiency typically used in the industry is the power usage effectiveness (PUE), defined as the ratio of the total facility power consumption to the power consumption of the IT [512]. A PUE of 1.0 indicates that all of the data center power consumption is

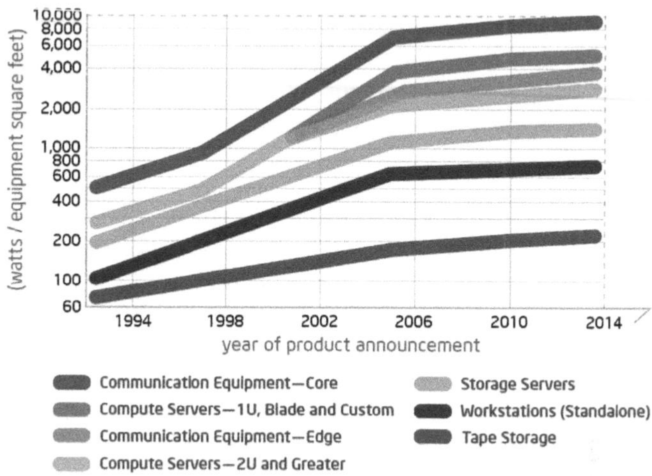

FIGURE 14.3 Power consumption per equipment footprint [508].

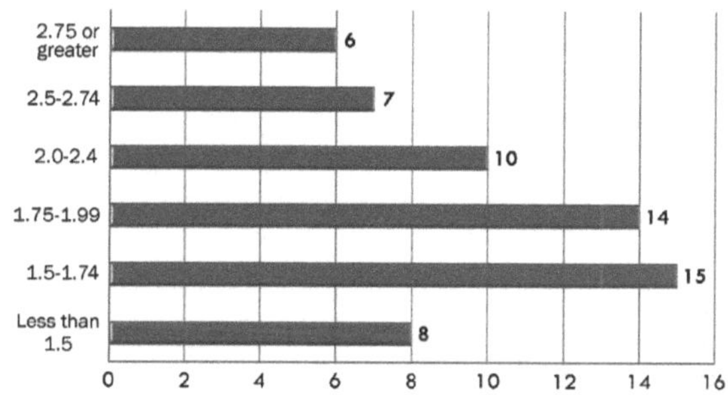

FIGURE 14.4 PUE measurements and number of respondents. The average value is 2.03 [512].

due to the IT. Fig. 14.4 shows the PUE values measured by 60 different data centers in 2007 [512]. Their average PUE value is 2.03; almost half of the total data center power consumption is consumed by the CT, which dominates the non-IT facility power consumption. Data centers based on state-of-the-art cooling and load balancing technology can reach PUE values of 1.1, which means that 90.9% of the total data center power consumption is consumed by IT.[1] A drawback of PUE is that it does not take into account IT equipment efficiency.

This section considers data centers as CPSs, with a focus on runtime management and operating costs. In what follows, we first review the current methods for controlling IT and CT in data centers, noting the degree to which they take into account both cyber and physical considerations. To evaluate the potential impact of coordinated CPS strategies at the data center level, a control-oriented model is introduced that represents the data center as two coupled

1. http://www.google.com/corporate/datacenter/efficiency-measurements.html.

networks: a computational network representing the cyber dynamics and a thermal network representing the physical dynamics. These networks are coupled through the influence of the IT on both networks: servers affect both the QoS delivered by the computational network and the generation of heat in the thermal network.

Then previous work on control in data centers is reviewed. Active IT management, power control, and cooling control systems can each have their own hierarchies [513], [514], and the interactions and coordination between the cyber and physical aspects of data centers also occur on multiple spatial and temporal scales. Depending on the metrics and the control variables that are used, we classify the work into three typical scales: server level, group level, and data center level.

At every level controllers can base their control actions considering only the state of the local controlled subsystem, or they can take into consideration the effects on other subsystems. In the first case the controller's actions are based on a local view of the system, whereas in the second case the controller's actions are based on a global view of the system. We classify the control approaches based on the scale of the controlled subsystem. For example, a controller at the server level manages the operations of a single server, even though the way the control actions are taken may derive from a global view of the data center.

14.2.1 Server level control

There are many control variables available at the server level for IT, power, and cooling management. The "server" in this case means all the IT devices, including the computing servers, the storage units, and the networking equipment. The computing resources, such as central processing unit (CPU) cycles, memory capacity, storage access bandwidth, and networking bandwidth, are all local resources that can be dynamically tuned, especially in a virtualized environment. Power control can be performed from either the demand side or the supply side, even at the server level. The power consumption of servers can be controlled by active management of the workload hosted by the server, for instance through admission control, load balance, and by workload migration or consolidation. On the other hand, power consumption can be tuned through physical control variables such as dynamic voltage and frequency scaling (DVFS) and through the on-off state control [515], [516], [517], [518], [519], [520], [521], [522], [523]. DVFS has already been implemented in many operating systems, for example the "CPU governors" in Linux systems. CPU utilization usually drives the DVFS controller, which adapts the power consumption to the varying workload.

Previous work has focused on how to deal with the trade-off between power consumption and IT performance. For instance, Varma et al. [524] discuss a control-theory approach to DVFS. Cho et al. [515] discuss a control algorithm that varies both the clock frequency of a microprocessor and the clock frequency of the memory. Leverich et al. [525] propose a control approach to reduce static power consumption of the chips in a server through dynamic per-core power gating control.

Cooling control at the server level is usually implemented through active server fan tuning to cool down the servers [526]. Similar to power management, the thermal status of the servers (e.g., the temperature of the processors) can also be affected by active control of the workload or power. As one example Mutapcic et al. [520] focus on the maximization of the processing capabilities of a multicore processor subject to a given set of thermal constraints. In another example Cohen et al. [519] propose strategies to control the power consumption of a processor via DVFS so as to enforce the given constraints on the chip temperature and on the workload execution.

14.2.2 Group level control

There are several reasons to control groups of servers rather than single servers. First, nowadays the IT workload most often runs on multiple nodes (e.g., a multi-tier application can span a set of servers). Second, when the metrics that drive the controllers are the performance metrics of the IT workloads, the control decision has to be made at the group level. One typical example is the workload migration-consolidation in the virtualized environment, which today is applied widely for sharing resources, improving resource utilization, and reducing power consumption [527], [528], [529], [530], [485], [531], [532]. Third, the IT and the infrastructure are usually organized into groups, for example the server enclosures that contain several servers cooled by a set of fans, the racks that have more than 40 servers each, the rows of racks, and the hot or cold aisles. The principal goal of group-level control is still to meet the cyber performance requirements while improving the physical resource utilization and energy efficiency. A few examples follow.

Padala et al. [533] propose a control algorithm to allocate IT resources among servers when they are divided among different tiers. Gandhi et al. [521], [522] focus on workload scheduling policies and transitions between the off and on states of servers as a means of minimizing the average response time of a data center given a certain power budget. A model-predictive control (MPC) approach [534], [535] is considered in the work of Aghajani et al. [523], where the goal of the control action is to dynamically adjust the number of active servers based on the current and predicted workload arrival rates. Wang et al. [530] propose a model-predictive controller to minimize the total power consumption of the servers in an enclosure subject to a given set of QoS constraints. Tolia et al. [529] discuss an MPC approach for coordinated workload performance, server power, and thermal management of a server enclosure, where a set of blade servers shares the cooling capacity of a set of fans. The fans are controlled through a multiple-input, multiple-output (MIMO) controller to minimize the aggregate fan power, while the location-dependent cooling efficiency of the fans is exploited so that the decisions of workload migration can result in the least total server and fan power consumption.

14.2.3 Data center level control

Depending on the boundaries defining the groups, group-level controllers have to be implemented on the data center scale in some cases. For instance, in a workload management system the workload may be migrated to the servers at any location in a data center. The other reason for the need of data center level control is that power and cooling capacity has to be shared throughout the data center. As one example the cooling capacity in a raised-floor air-cooled data center can be generated by a chiller plant, distributed to the IT systems through a set of CRAC units, the common plenum under the floor, and the open space above the floor. Sharing the capacity makes the cooling management in the first order a data center level control [536], [537], [538].

Quershi et al. [539] discuss the savings that can be achieved by migrating workload to the data center locations where the electricity cost is lowest. A similar problem is considered in the work of Rao et al. [540]. Tang et al. [541] discuss a control algorithm that allocates the workload among servers so as to minimize their peak inlet temperatures. Parolini et al. [542], [543] consider a data center as a node of the smart grid, where time-varying and power-consumption-dependent electricity price information is used to manage data center operations.

Bash et al. [536] discuss a control algorithm for the CRAC units. The proposed control approach aims at minimizing the amount of heat removed by each CRAC unit, while enforcing the thermal constraints of the IT. Anderson et al. [544] consider a MIMO robust control approach to the control of CT. Computational fluid dynamic (CFD) simulations are a widely used tool to simulate and predict the heat distribution in a data center. These simulations take a very long time to execute and cannot be used in real-time control applications. In [545], Toulouse et al. discuss an innovative approach to CFD that is able to perform fast simulations.

Chen et al. [546] propose a holistic workload, power, and cooling management framework in virtualized data centers through the exploration of the location-dependent cooling efficiency in the data center level. Workload migration and consolidation through virtual machine (VM) migration is taken as the main approach for active power management. On the thermal side, the rack inlet temperatures are under active control of a cooling controller. The controller dynamically tunes the cooling air flow rates and temperatures from the CRAC units, in response to the "hot spots" of the data center. On the cyber side, the authors consider multi-tier IT applications or workloads hosted in VMs that can span multiple servers, and are able to migrate VMs. The models are developed for application performance such as end-to-end response time. Based on the models and the predicted workload intensity, the computing resource demand of the workloads can be estimated. The demand of the VMs hosted by a server is then satisfied through dynamic resource allocation, if possible. Otherwise, the VMs may be migrated to other servers that have resources available. When possible, the VMs are consolidated onto fewer servers so that the idle ones can be turned off to save power. With the decision to migrate workload and turn servers on or off via the consolidation controller, the effect of the actions on the cooling power is taken into consideration through introduction of the local workload placement index (LWPI). As one index for the interaction between the cyber and physical systems, the LWPI indicates how much cooling capacity is available in a certain location of the data center, and how efficiently a server in that location can be cooled. Using the LWPI, the consolidation controller tends to migrate the workload to the servers that are more efficiently cooled than others, while turning off the idle servers that are located at "hot spots."

An idle server typically consumes about 60% of its peak power. Servers in data centers typically operate between 10% and 50% of their maximum utilization and often are completely idle [503], [547], [548], [549]. In order to maximize the server energy efficiency it would be desirable to operate them at 100% utilization by consolidating the workload onto a few servers and turning off the unused ones. The main problem related to turning off a server is the time required to turn it back on. This time, often called setup time, is on the order of a few minutes and it is typically not acceptable for interactive workloads [522], [523]. This problem can be solved by predicting the incoming workload and adjusting the computational power accordingly. Another potential drawback of using these techniques is the effect they may have on the cooling effort. Concentrating computation on a few servers while turning off others in general has the effect of generating hot spots for which the cooling system needs to do extra work. As air cooling cannot be targeted to a specific device, this method may in turn overcool the overall data center, leading to energy inefficiencies in the cooling system, thus potentially offsetting the savings achieved by reducing overall server power consumption.

14.3 A cloud control system model of data centers

As described in the previous section, the CT is managed at the data center level. At this level the thermal properties of the IT are managed as groups of components (racks of servers) aggregated into zones. From a CPS perspective the CT is purely physical, consuming power to remove heat from the zones, whereas zones have both cyber and physical characteristics. The IT processes the computational workload and also consumes power and generates heat. In this section we present a model of the data center level dynamics using common representations of these cyber and physical features, and the coupling between them. We use this model in the following sections to study the aspects of data center architectures and operating conditions that influence the potential impact of coordinated cloud (IT and CT) control strategies.

We model the data center as two networks. In the computational network nodes correspond to the cyber aspects of zones, which can process jobs at rates determined by the allocated resources, and connections represent pathways for moving jobs between zones. The thermal network includes nodes to represent the thermal aspects of zones along with nodes for the CT. Connections in the thermal network represent the exchange of heat between nodes as determined by the data center's physical configuration. The two networks are coupled by connections between the two nodes that represent each zone in the computational network and thermal network. These connections reflect the correlation between the computational performance of the resources allocated in the zone (a control variable) and the physical features of the zone (power consumption and generated heat).

14.3.1 Computational network

We model the computational network using a fluid approximation of the workload execution and arrival processes (i.e., the workload is represented by job flow rates rather than as discrete jobs). The proposed modeling approach represents a first-order approximation of a queuing system. The strength of the proposed approach resides in its simplicity, which allows an uncluttered discussion of the model and of the control approach. On the other hand, a certain amount of computational details of a data center are not included in the model. We consider the approximation provided by the proposed model adequate for the goal of the chapter. However, when additional details of the system are relevant for the system analysis, more refined approximations can be considered [550].

Let N be the number of nodes and let $l_i(\tau)$ denote the amount of workload in the ith node at time τ. From a queuing perspective, $l_i(\tau)$ represents the queue length of the ith node at time τ. The workload arrival rate at the data center is denoted by $\lambda^w(\tau)$. The relative amount of workload that is assigned to the ith node at time τ is denoted by $s_i(\tau)$. Variables $\{s_i(\tau)\}$ represent the workload scheduling (or allocation) action. The rate at which workload migrates from the ith computational node to the jth computational node at time τ is denoted by $\xi_{j,i}(\tau)$. The rate at which the workload departs from the ith node at time τ, after being executed on the node, is denoted by $\eta_i(\tau)$. Let $\mu_i(\tau)$ denote the desired workload execution rate at the ith node at time τ and let $\delta_{j,i}(\tau)$ denote the required migration rate of workload from the ith computational node to the jth computational node at time τ. Variables $\{\mu_i(\tau)\}$, $\{s_i(\tau)\}$, and $\{\delta_i(\tau)\}$ are controllable variables, whereas $\{l_i(\tau)\}$ are the state variables. Fig. 14.5 illustrates these variables that describe the ith computational node (a zone in the data center).

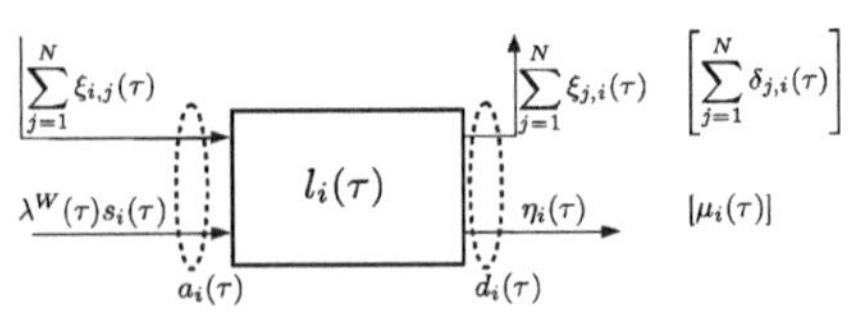

FIGURE 14.5 Input, state, and output variables of the ith computational node.

We define the following variables for $i = 1, \ldots, N$:

$$
\begin{aligned}
a_i(\tau) &= \lambda^W(\tau)s_i(\tau) + \sum_{j=1}^{N} \xi_{i,j}(\tau) \\
d_i(\tau) &= \eta_i(\tau) + \sum_{j=1}^{N} \xi_{j,i}(\tau) \\
\nu_i(\tau) &= \mu_i(\tau) + \sum_{j=1}^{N} \delta_{j,i}(\tau).
\end{aligned}
$$

Variable $\alpha_i(\tau)$ represents the total rate at which workload arrives at the ith node, variable $d_i(\tau)$ represents the total rate at which workload departs from the ith node, and variable $\nu_i(\tau)$ represents the desired total workload departure rate from the ith node. The evolution of the amount of workload at the ith computational node is given by

$$\dot{l}_i(\tau) = a_i(\tau) - d_i(\tau). \tag{14.1}$$

The relationship between the departure rate, the state, and the control variables at the ith node is given by

$$\eta_i(\tau) = \begin{cases} \mu_i(\tau), & \text{if } l_i(\tau) > 0 \text{ or } a_i(t) > \mu_i(t) \\ a_i(t), & \text{otherwise.} \end{cases} \tag{14.2}$$

The relationship between the workload migration rate, the state, and control variables at the ith node can be written as (14.3).

$$\xi_{j,i}(\tau) = \begin{cases} \delta_{j,i}(\tau), & \text{if } l_i(\tau) > 0 \text{ or } a_i(t) > \nu_i(t) \\ \frac{\delta_{j,i}(\tau)}{\sum_{j=1}^{N} \delta_{j,i}(\tau)} \left(a_i(\tau) - \eta_i(\tau)\right), & \text{otherwise.} \end{cases} \tag{14.3}$$

Eqs. (14.2) and (14.3) model the case where the ith node does not migrate workload to other computational nodes if the total rate of incoming workload is lower than or equal to the desired workload execution rate, that is, $\alpha_i(\tau) \leq \mu_i(\tau)$, and the queue length is 0, that is, $l_i(\tau) = 0$.

The model for the workload execution rates developed above is sufficient for our purposes, but we note that it can be extended to include different migration policies, workload classes, hardware requirements, and interactions among different types of workloads [542], [543], [551].

14.3.2 Thermal network

Let M be the number of nodes in the thermal network. The dynamics of the thermal network is characterized in terms of temperatures associated with each of the thermal nodes. For each node of the network we define two temperatures: the input temperature and the output temperature. The input temperature of the ith node represents the amount of heat received from the other thermal nodes and its value at time τ is denoted by $T_{in,i}(\tau)$. Variable $T_{in,i}(\tau)$ includes the recirculation and cooling effects due to all thermal nodes. The output temperature of the ith thermal node, denoted by $T_{out,i}(\tau)$, represents the amount of heat contained in the ith thermal node at time τ.

Following [484], we assume each input temperature is a linear combination of the output temperatures from all of the nodes, that is

$$T_{in,i}(\tau) = \sum_{j=1}^{M} \psi_{i,j} T_{out,j}(\tau), \qquad \text{for all } i = 1, \ldots, M, \tag{14.4}$$

where the coefficients $\{\psi_{i,j}\}$ are nonnegative and $\sum_{j=1}^{M} \psi_{i,j} = 1$ for all $i = 1, \ldots, M$. The values of the coefficients $\{\psi_{i,j}\}$ can be estimated following the procedure in [484]. We collect the input and output temperatures in the $M \times 1$ vectors $\mathbf{T}_{in}$ and $\mathbf{T}_{out}$, respectively. Consequently, their relationship can be written in vector form as

$$\mathbf{T}_{in}(\tau) = \Psi \mathbf{T}_{out}(\tau), \tag{14.5}$$

where $\{\psi_{i,j}\}$ are the components of the matrix Ψ.

Measurements taken on a server in our laboratory and discussed in [552] show that a linear time-invariant (LTI) system is a good approximation of the evolution of the outlet temperature of a server. Therefore, we model the evolution of the output temperatures of the thermal nodes for zones as

$$\dot{T}_{out,i}(\tau) = -\mathbf{k}_i T_{out,i}(\tau) + \mathbf{k}_i T_{in,i}(\tau) + \mathbf{c}_i \mathbf{p}_i(\tau), \tag{14.6}$$

where $1/\mathbf{k}_i$ is the time constant of the temperature of ith node, $\mathbf{c}_i$ is the coefficient that maps power consumption into output temperature variation, and $\mathbf{p}_i$ is the power consumption of the ith node at time τ.

The power consumption of the nodes representing a zone is proportional to the rate at which workload departs, after being executed, from the associated computational node

$$\mathbf{p}_i(\tau) = \alpha_i \eta_i(\tau), \tag{14.7}$$

where α_i is a non-negative coefficient. A linear model is chosen since we assume that lower-level controllers, for example, the one in [529], can be used to make the power consumption of a zone proportional to the amount of workload processed by the zone. This linear model can be extended to include more complicated functions that can account for the on-off state of every server.

In the CT we focus on the CRAC units, which are the primary power consumers. The output temperatures of the CRAC units are modeled as

$$\dot{T}_{out,i}(\tau) = -\mathbf{k}_i T_{out,i}(\tau) + \mathbf{k}_i \min\left\{T_{in,i}(\tau), T_{\mathrm{ref},i}(\tau)\right\}, \tag{14.8}$$

where $T_{ref,i}(\tau)$ represents the reference temperature of the CRAC node i and is assumed to be controllable. The "min" operator in (14.8) ensures that the node always provides output temperatures that are not greater than the input temperatures. As discussed by Moore et al. [553], the power consumption of a CRAC node is given by

$$\mathbf{p}_i(\tau) = \begin{cases} \mathbf{c}_i \frac{T_{in,i}(\tau) - T_{out,i}(\tau)}{\mathrm{COP}(T_{out,i}(\tau))}, & T_{in,i}(\tau) \geq T_{out,i}(\tau) \\ 0, & T_{in,i}(\tau) < T_{out,i}(\tau) \end{cases}, \tag{14.9}$$

where $\mathbf{c}_i$ is a coefficient that depends on the amount of air passing through the CRAC and the air heat capacity. The variable COP($T_{out,i}(\tau)$) is the Coefficient Of Performance of the CRAC unit modeled by the ith node, which is a function of the node's output temperature [553].

In order to provide a compact representation of the overall model we use vector notation. We use the $N \times 1$ vector $\mathbf{p}_{\mathcal{N}}(\tau)$ to denote the power consumption of the thermal nodes representing the zones, and use $\mathbf{p}_{\mathcal{C}}(\tau)$ and $T_{ref}(\tau)$ to denote respectively the power consumption and the reference temperatures of the thermal nodes representing CRAC units. The state of the thermal network is represented by the vector $T_{out}(\tau)$, the controllable input of the thermal network by the vector $T_{ref}(\tau)$, and the uncontrollable input of the thermal network by $\mathbf{p}_{\mathcal{N}}(\tau)$. Finally, the outputs of the thermal network are the vector of the thermal node input temperatures $T_{in}(\tau)$ and the vector $\mathbf{p}_{\mathcal{C}}(\tau)$. The vector $T_{in}(\tau)$ is a function of the network state and therefore, it is an output of the network. However, when we look at a single node, the input temperature becomes an uncontrollable input of the node. In this sense the input vector is an output of the thermal network and at the same time an uncontrollable input for each of the nodes.

14.3.3 Control strategies

Three control strategies are introduced in this section: *baseline, uncoordinated, and coordinated.* The control strategies are abstractions of three different control approaches that can be implemented at the data center level. The baseline strategy represents those control approaches where IT and CT are set so as to satisfy the QoS and the thermal constraints for the worst-case scenario, regardless of the actual computational and cooling demands of the data center. The uncoordinated strategy represents those control approaches where the efficiencies of IT and CT are considered in two separate optimization problems. The coordinated strategy represents those control approaches where the efficiencies IT and CT are controlled using a single optimization problem.

The goal of the control strategies is to minimize the total data center power consumption while satisfying both the QoS and the thermal constraints. The QoS constraint requires the workload execution rate of every zone to be greater than or equal to the workload arrival rate at the zone

$$\mu(\tau) \geq \text{diag}\left\{\mathbf{1}\lambda^{W}(\tau)\right\}\mathbf{s}(\tau), \qquad (14.10)$$

where diag$\{x\}$ is the diagonal matrix having the elements of the vector x on the main column and $\mathbf{1}$ is the vector of appropriate dimension whose elements are all 1. A more general formulation of the computational requirements can be obtained by considering the profit obtained by executing the workload with a certain QoS in the controller's cost function. In this case the goal of the control becomes the search of the best trade-off between minimizing the cost of powering the data center and maximizing the profit induced by executing the workload. The initial results in this direction can be found in [542].

The thermal constraints on IT devices are formulated in terms of upper bounds on the input temperature of the thermal nodes

$$\mathbf{T}_{in}(\tau) \leq \overline{\mathbf{T}_{in}}. \qquad (14.11)$$

Controllable variables are also subject to constraints. Constraints on the vector of workload execution rates are given by

$$\mathbf{0} \leq \mu(\tau) \leq \overline{\mu}, \qquad (14.12)$$

where the inequalities are applied component-wise. Controllers do not migrate workload among the zones, that is,

$$\delta(\tau) = \mathbf{0}, \qquad (14.13)$$

where $\mathbf{0}$ is the zero vector. The constraints on the vector of workload scheduling are given by

$$\mathbf{0} \leq \mathbf{s}(\tau) \leq \mathbf{1}, \qquad \mathbf{1}^{T}\mathbf{s}(\tau) \leq 1. \qquad (14.14)$$

The second constraint in (14.11) allows the data center controller to drop workload. The constraints on the vector of reference temperatures are given by

$$\underline{\mathbf{T}_{\text{ref}}} \leq \mathbf{T}_{\text{ref}}(\tau) \leq \overline{\mathbf{T}_{\text{ref}}}. \qquad (14.15)$$

14.3.4 Baseline controller

The baseline control approach is a reasonable control approach when the amount of power consumed by the CT is much less than the total power consumed by IT. A drawback of this control approach is that it cannot guarantee that the QoS and the thermal constraints are enforced.

For all $\tau \in \mathbb{R}$, the baseline controller sets the control variables to

$$\begin{aligned} \mu(\tau) &= \overline{\mu}, \qquad \delta(\tau) = 0, \\ \mathbf{s}(\tau) &= \mathbf{1}\frac{1}{N}, \qquad \mathbf{T}_{\text{ref}}(\tau) = \underline{\mathbf{T}_{\text{ref}}}. \end{aligned} \tag{14.16}$$

14.3.5 Uncoordinated controller

The uncoordinated controller represents a control strategy typically found in modern data centers where the management of the IT and the CT is assigned to two uncoordinated controllers. The uncoordinated controller is obtained by the composition of two controllers. The first controller deals with the optimization of the IT. The second controller manages the CT. Both controllers consider a predictive, discrete-time model of the data center. In this case the QoS and the thermal constraints are considered and enforced only at the beginning of every interval.

Let $\mathcal{T}$ be the time horizon considered by the two optimization problems. We define $\hat{\mu}(h|k)$ as the predicted value of the variable $\mu(\tau)$ at the beginning of the hth interval based on the information available up to the beginning of the kth interval, and define the set $\mathcal{M} = \{\hat{\mu}(k|k), \ldots, \hat{\mu}(k + \mathcal{T} - 1|k)\}$. Similarly, we define the variables $\hat{\mathbf{s}}(h|k)$, $\hat{\delta}(h|k)$, and $\hat{\mathbf{T}}_{ref}(h|k)$, and the sets $\mathcal{S} = \{\hat{\mathbf{s}}(k|k), \ldots, \hat{\mathbf{s}}(k+\mathcal{T}-1|k)\}$, $\mathcal{D} = \{\hat{\delta}(k|k), \ldots, \hat{\delta}(k+\mathcal{T}-1|k)\}$, and $\mathcal{T}_{ref} = \{\hat{\mathbf{T}}_{ref}(k|k), \ldots, \hat{\mathbf{T}}_{ref}(k+\mathcal{T}-1|k)\}$. The predicted value of the workload arrival rate at the data center during the hth interval, based on the information available up to the kth interval, is denoted $\hat{\lambda}(h|k)$. We use $\hat{p}_{\mathcal{N}}(h|k)$ to denote the expected average power consumption of the zones during the hth interval, based on the information available up to the kth interval. We use $\hat{p}_{\mathcal{C}}(h|k)$ to denote the expected average power consumption of the CRAC units during the hth interval, based on the information available up to the kth interval. At the beginning of every interval the first part of the uncoordinated controller solves the following optimization problem:

$$\min_{\mathcal{M},\mathcal{S},\mathcal{D}} \sum_{h=k}^{k+\mathcal{T}-1} \mathbf{1}^T \hat{p}_{\mathcal{N}}(h|k)$$

$$\textbf{s.t.} \quad \text{for all } h = k, \ldots, k + \mathcal{T} - 1$$

$$\text{computational dynamics}$$

$$\hat{\mu}(h|k) \geq \text{diag}\left\{\mathbf{1}\hat{\lambda}^W(h|k)\right\} \hat{\mathbf{s}}(h|k)$$

$$
\begin{aligned}
&\mathbf{0} \leq \hat{\mu}(h|k) \leq \overline{\mu}, \quad \hat{\delta}(h|k) = \mathbf{0} \\
&\mathbf{0} \leq \hat{\mathbf{s}}(h|k) \leq \mathbf{1}, \quad \mathbf{1}^T \hat{\mathbf{s}}(h|k) \leq 1 \\
&\hat{\mathbf{l}}(k|k) = \mathbf{l}(k).
\end{aligned} \tag{14.17}
$$

Based on the solution obtained for the optimization in (14.17), the second part of the uncoordinated controller generates and solves the following optimization problem:

$$
\begin{aligned}
\min_{\mathcal{T}_{ref}} \quad & \sum_{h=k}^{k+\mathcal{T}-1} \mathbf{1}^T \hat{p}_{\mathcal{C}}(h|k) \\
\textbf{s.t.} \quad & \text{for all } h = k, \ldots, k+\mathcal{T}-1 \\
& \text{thermal dynamics} \\
& \underline{\mathbf{T}_{\text{ref}}} \leq \hat{\mathbf{T}}_{\text{ref}}(h|k) \leq \overline{\mathbf{T}_{\text{ref}}} \\
& \hat{\mathbf{T}}_{\text{in}}(h+1|k) \leq \overline{\mathbf{T}_{\text{in}}} \\
& \hat{\mathbf{T}}_{\text{out}}(k|k) = \mathbf{T}_{\text{out}}(k).
\end{aligned} \tag{14.18}
$$

Since the uncoordinated controller considers the computational and thermal constraints in two different optimization problems, it cannot guarantee their enforcement. The uncoordinated controller manages variables related to both cyber and physical aspects of the data center, and therefore it is a cyber-physical controller. We call it uncoordinated because the management of the IT and CT are treated separately.

14.3.6 Coordinated controller

The coordinated control strategy is based on a discrete-time MPC approach and it manages the IT and CT resources as a single optimization problem. Under mild assumptions the coordinated controller is able to guarantee the enforcement of both the QoS and the thermal constraints [542]. The sets $\mathcal{M}$, $\mathcal{S}$, $\mathcal{D}$, and $\mathcal{T}_{ref}$ are defined as in the uncoordinated controller case. At the beginning of every interval the coordinated controller solves the following optimization problem:

$$
\begin{aligned}
\min_{\mathcal{M},\mathcal{S},\mathcal{D},\mathcal{T}_{\text{ref}}} \quad & \sum_{h=k}^{k+\mathcal{T}-1} \mathbf{1}^T \hat{p}_{\mathcal{N}}(h|k) + 1^T \hat{p}_{\mathcal{C}}(h|k) \\
\textbf{s.t.} \ & \text{for all } h = k, \ldots, k+\mathcal{T}-1 \\
& \text{computational dynamics} \\
& \text{thermal dynamics} \\
& \hat{\mu}(h|k) \geq \text{diag}\left\{\mathbf{1}\hat{\lambda}^W(h|k)\right\} \hat{\mathbf{s}}(h|k) \\
& \mathbf{0} \leq \hat{\mu}(h|k) \leq \overline{\mu}, \qquad \hat{\delta}(h|k) = \mathbf{0}
\end{aligned}
$$

$$
\begin{aligned}
&\mathbf{0} \le \hat{\mathbf{s}}(h|k) \le \mathbf{1}, \qquad \mathbf{1}^T \hat{\mathbf{s}}(h|k) \le 1 \\
&\underline{\mathbf{T}_{\text{ref}}} \le \hat{\mathbf{T}}_{\text{ref}}(h|k) \le \overline{\mathbf{T}_{\text{ref}}}, \qquad \hat{\mathbf{T}}_{\text{in}}(h+1|k) \le \overline{\mathbf{T}_{\text{ref}}} \\
&\hat{\mathbf{l}}(k|k) = \mathbf{l}(k), \qquad \hat{\mathbf{T}}_{\text{out}}(k|k) = \mathbf{T}_{\text{out}}(k).
\end{aligned} \tag{14.19}
$$

A drawback of the coordinated controller is the complexity of the optimization problem that has to be solved. Typically, the optimization problem is nonconvex and large. Local optimal solutions may yield control strategies that are worse than those obtained by an uncoordinated controller.

14.3.7 Simulation results I

We evaluate the long-term performance of the three control strategies for multiple constant workload arrival rates. The performances of the three controllers are evaluated in the ideal case, where the controllers have perfect knowledge of the data center and when the data center has reached its thermal and computational equilibrium. When the robustness of the control approaches is the focus of the simulation, then modeling errors and prediction errors should also be considered in the simulation. The simulations are developed using the TOMSym language with KNITRO as the numerical solver.[2]

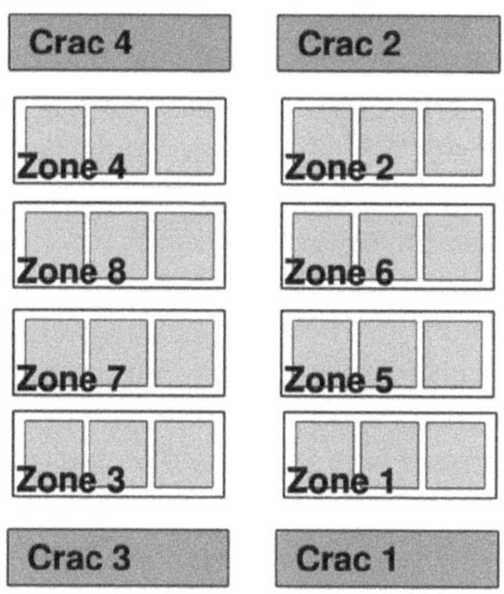

FIGURE 14.6 An example of a data center layout. Blue (light gray in print version) rectangles represent groups of three racks each and yellow (mid gray in print version) rectangles represent the CRAC units.

We consider the data center layout depicted in Fig. 14.6. The picture represents a small data center containing 32 racks and four CRAC units. The CT comprises four CRAC units that cool the servers through a raised-floor architecture. Racks are grouped into eight zones and every rack contains 42 servers. Under the same workload, servers in zones 5–8 consume 10% less power than

2. http://tomsym.com/and http://www.ziena.com/knitro.html.

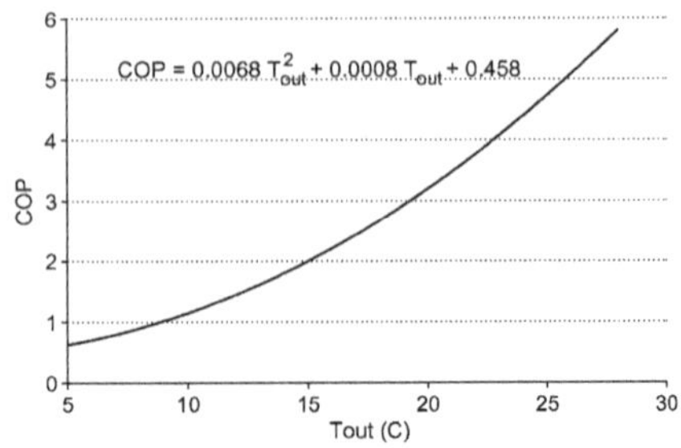

FIGURE 14.7 Coefficient of performance of the CRAC units for different output temperature values [553].

TABLE 14.1 Average and standard deviation values of $\sum_j \psi_{i,j}$. The coefficients refer to the graphs in Figs. 14.8 and 14.9.

	$j \rightarrow$	Zones 1–4		Zones 5–8		CRACs	
$\downarrow i$		*Avg.*	*Std.*	*Avg.*	*Std.*	*Avg.*	*Std.*
Zones	1–4	0.04	2.6e-6	0.03	2.2e-6	0.93	4.8e-6
Zones	5–8	0.05	9.9e-7	0.52	4.8e-5	0.43	4.8e-5
CRACs		0.63	2.0e-5	0.25	4.3e-5	0.12	2.3e-5

other servers, but they are not cooled as efficiently as the servers in zones 1–4, which are closer to the CRAC units. The maximum power consumption of the servers in zones 5–8 is 270 (W) and the maximum power consumption of the servers in zones 1–4 is 300 (W). It is assumed that every zone has a local controller that forces the zone to behave as postulated in the model, that is, the amount of power consumed by every zone is proportional to the workload execution rate. The CRAC units are identical and their efficiency (i.e., their COP) increases quadratically with respect to their output temperatures. The relationship between the output temperatures of the CRAC units and their COP is shown in Fig. 14.7. The coefficients relating input and output temperatures of zones and CRAC units are summarized in Table 14.1. We set the maximum allowed input temperature of every zone to 27°C (i.e., $\overline{\mathbf{T}_{in}} = 27$). This constraint reflects the environmental guidelines of the American Society of Heating, Refrigerating, and Air-Conditioning Engineers (ASHRAE) [554].

We define the average utilization of the data center as the mean values of the ratios $\eta_i(\tau)/\overline{\mu}_i$ for $i = 1, \ldots, 8$. When the average utilization is 0, then zones process no workload (i.e., $\eta(\tau) = \mathbf{0}$). When the average utilization is 1, then zones process the maximum amount of workload they can process (i.e., $\eta(\tau) = \overline{\mu}$). The behavior of the three controllers is not considered for small average utilization values since, at very low utilization values, nonlinearities of the IT and CT neglected in the proposed model become relevant. Fig. 14.8 shows the total data center power consumption obtained by the three controllers for different average utilization values.

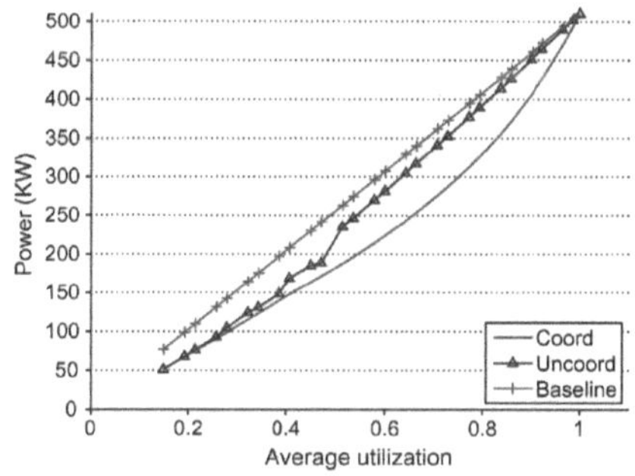

FIGURE 14.8 Average data center power consumption for different utilization values.

The total data center power consumption obtained by the baseline controller grows proportionally with the average utilization. The proportional growth is due to two factors. The first factor is the assumption that rack-level and server-level controllers make the power consumption of every zone grow proportionally with the amount of workload they process. The second factor is that the reference temperature of each CRAC unit is fixed and is always lower than or equal to its input temperature. In this case the efficiency of every CRAC unit is constant, and the power consumed by the CT grows proportionally with the amount of power that the zones consume.

The total data center power consumption obtained by the uncoordinated controller is always lower than that obtained by the baseline controller. This happens because the uncoordinated controller assigns as much workload as possible to the most energy-efficient servers (i.e., those located in zones 5–8) and it tries to maximize the efficiency of the CT by setting the reference value of every CRAC unit to the largest value that still enforces the thermal constraints.

The additional savings obtained by the coordinated controller are due to the coordinated management of the IT and CT resources. In particular, the coordinated controller, depending on the amount of workload that has to be processed by the zones, decides whether it is more efficient to allocate workload to the energy-efficient servers (in zones 5–8) or to the efficiently cooled servers (in zones 1–4).

The PUE values obtained by the three controller are shown in Fig. 14.9. In the baseline controller case the CRAC units operate at a constant minimum efficiency, and therefore the PUE values obtained by the baseline controller are constant. The uncoordinated and coordinated controllers are able to improve (i.e., to lower) the PUE obtained by the baseline controller. The PUE curves obtained by the uncoordinated and coordinated controllers are not smooth because for some values of the workload arrival rate, the controllers are unable to locate the true minimum of the optimization problems they solve. In these cases control actions are based on a local minimum that may be different from the control actions chosen for nearby values of the workload arrival rate.

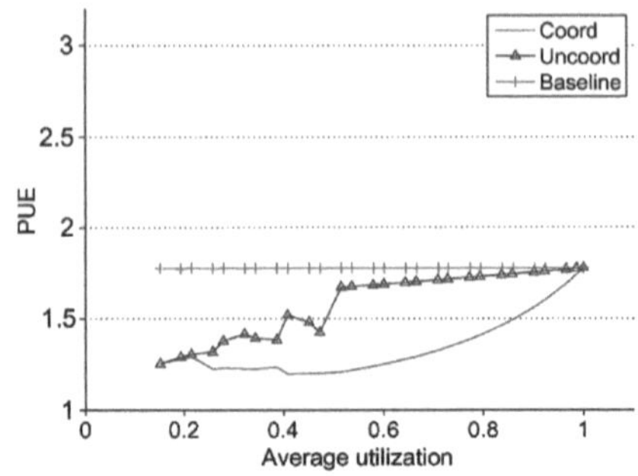

FIGURE 14.9 PUE values obtained by the baseline, uncoordinated, and coordinated controllers.

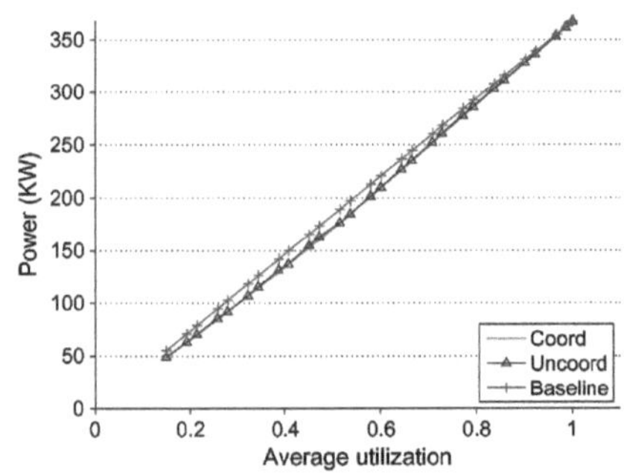

FIGURE 14.10 Average data center power consumption for different utilization values. All of the zones are efficiently cooled.

In the second simulation we focus on a case where the inlet temperatures of the servers in zones 1–4 equal the supplied air temperatures of the CRAC units, that is, $\sum_j \psi_{i,j} \simeq 1$, where i is the index of the zones 1–4 and j only represents the CRAC units. In addition, in the second simulation the servers in zones 1–4 are subject to less air-recirculation than in the first simulation case, and their inlet temperatures depends more on the output temperatures of the servers in zones 5–8. The total power consumption obtained by the uncoordinated equals the total power consumption obtained by the coordinated controller for every average utilization value, which means that there is no loss of performance in managing IT and CT separately. Fig. 14.10 shows the total data center power consumption obtained by the three controllers in the second simulation. Table 14.2 summarizes the values of the coefficients relating input and output temperatures of zones and CRAC units for this simulation. The other parameters did not change.

The third simulation describes a data center where almost half of the airflow cooling servers in zones 5–8 comes from other servers; in other words, the third simulation considers a data center case where some servers are poorly positioned with respect to the CRAC units. Table 14.3 summarizes the values of the coefficients relating input and output temperatures of the zones and CRAC

TABLE 14.2 Average and standard deviation values of $\sum_j \psi_{i,j}$. The coefficients refer to the graphs in Figs. 14.10 and 14.12.

	$j \rightarrow$	Zones 1–4		Zones 5–8		CRACs	
$\downarrow i$		*Avg.*	*Std.*	*Avg.*	*Std.*	*Avg.*	*Std.*
Zones	1–4	0	0	0	0	1	0
Zones	5–8	0.3	2.9e-5	0.4	8.0e-6	0.30	2.9e-5
CRACs		0.51	5.6e-5	0.34	3.4e-5	0.15	2.5e-5

TABLE 14.3 Average and standard deviation values of $\sum_j \psi_{i,j}$. The coefficients refer to the graphs in Figs. 14.11 and 14.13.

	$j \rightarrow$	Zones 1–4		Zones 5–8		CRACs	
$\downarrow i$		*Avg.*	*Std.*	*Avg.*	*Std.*	*Avg.*	*Std.*
Zones	1–4	0.08	0	0.08	0	0.84	0
Zones	5–8	0.08	7.0e-8	0.66	4.8e-5	0.26	4.8e-5
CRACs		0.57	5.4e-5	0.18	9.0e-5	0.25	4.1e-5

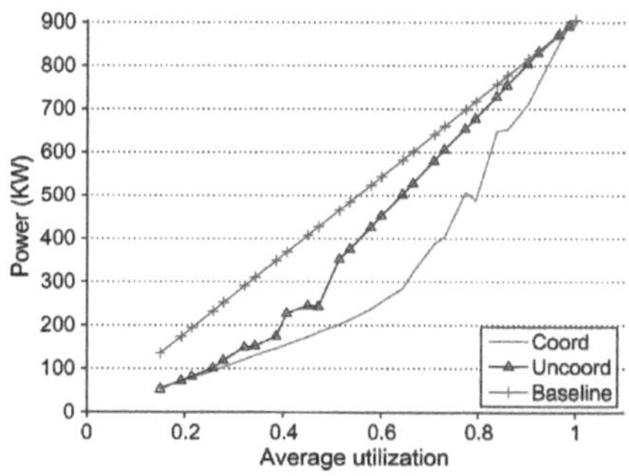

FIGURE 14.11 Average data center power consumption for different utilization values. There is wide variability in the cooling efficiency of the zones.

units for this simulation. The other parameters did not change. Fig. 14.11 shows the total data center power consumption obtained by the three controllers in the third simulation.

The PUE values obtained by the three controllers for these new cases are shown in Figs. 14.12 and 14.13. As shown in Fig. 14.13, when there is a large variability in the server cooling efficiency, the PUE strongly depends on the average utilization of the data center.

14.3.8 A cyber-physical index for data centers

For a given a data center, it would be useful to estimate a priori how much energy could be saved by using a coordinated controller rather than an uncoordinated one. To this end, we define relative efficiency as the area between

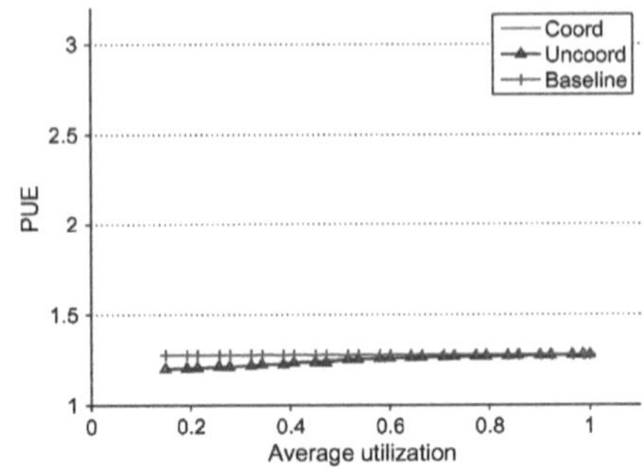

FIGURE 14.12 PUE values obtained by the baseline, uncoordinated, and coordinated controller. All of the zones are efficiently cooled.

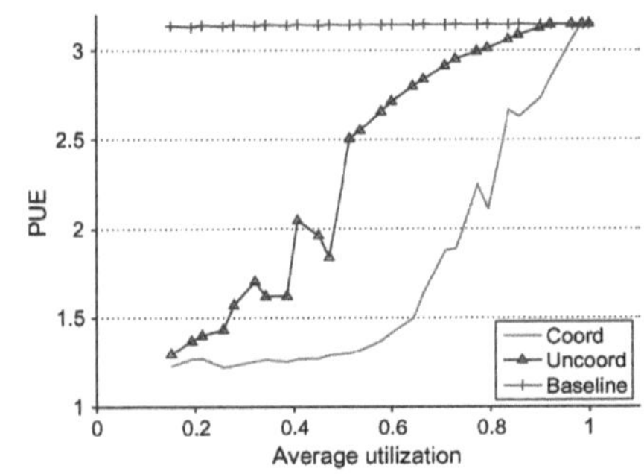

FIGURE 14.13 PUE values obtained by the baseline, uncoordinated, and coordinated controller. There is wide variability in the cooling efficiency of the zones.

the power consumption curve obtained by the uncoordinated controller and the power consumption curve obtained by the coordinated controller in Figs. 14.8, 14.10, and 14.11. With the appropriate weights the relative efficiency can be mapped against the average monthly or average yearly energy savings obtained by using a coordinated controller with respect to an uncoordinated controller.

Consider a data center at its thermal and computational equilibrium and assume that both the QoS and the thermal constraints are satisfied. Furthermore, we assume that every CRAC unit provides a certain amount of cooling: for every thermal node j modeling a CRAC unit $T_{ref,j} \leq T_{in,j}$, let i be the index of a cyber node representing a zone and let $T_{in,i}$ be its input temperature value at thermal equilibrium. Collecting the input temperatures of the thermal nodes representing zones into vector $\mathbf{T}_{in,\mathcal{N}}$, we define $\mathbf{\Psi}_{[\mathcal{N},\mathcal{C}]}$ as the matrix composed of the $\{\psi_{i,j}\}$ variables such that i is the index of a thermal node modeling a zone and j is the index of a thermal node modeling a CRAC unit. We also collect all of the output temperatures of the thermal nodes modeling zones into the vector $\mathbf{T}_{out,\mathcal{N}}$ and define $\mathbf{\Psi}_{[\mathcal{N},\mathcal{N}]}$ as the matrix composed of the $\{\psi_{i,j}\}$ variables such that i and j are the indexes of two thermal nodes modeling zones. From the above, we can write

$$\mathbf{T}_{in,\mathcal{N}} = \Psi_{[\mathcal{N},\mathcal{N}]}\mathbf{T}_{out,\mathcal{N}} + \Psi_{[\mathcal{N},\mathcal{C}]}\mathbf{T}_{\text{ref}}. \tag{14.20}$$

With a slight abuse of notation, we use the symbol diag $\{\alpha_i \mathbf{c}_i/\mathbf{k}_i\}$ to denote the diagonal matrix composed of the $\{\alpha_i \mathbf{c}_i/\mathbf{k}_i\}$ terms, where $\mathbf{k}_i$, $\mathbf{c}_i$, and α_i are the coefficients introduced in (14.6) and (14.7). Assuming every zone is processing a constant amount of workload and the matrix $(I - \boldsymbol{\Psi}_{[\mathcal{N},\mathcal{N}]})$ is invertible, (14.20) can be rewritten as

$$\begin{aligned} \mathbf{T}_{in,\mathcal{N}} &= \left(I - \Psi_{[\mathcal{N},\mathcal{N}]}\right)^{-1} \text{diag}\left\{\frac{\alpha_i \mathbf{c}_i}{\mathbf{k}_i}\right\}\eta + \Psi_{[\mathcal{N},\mathcal{C}]}\mathbf{T}_{\text{ref}} \\ &= \mathcal{L}\eta + \Psi_{[\mathcal{N},\mathcal{C}]}\mathbf{T}_{\text{ref}}, \end{aligned} \tag{14.21}$$

where $\mathcal{L} = (I - \boldsymbol{\Psi}_{[\mathcal{N},\mathcal{N}]})^{-1}\text{diag}\{\alpha_i \mathbf{c}_i/\mathbf{k}_i\}$ and $\eta = [\eta_1, \ldots, \eta_N]^T$ is the vector of workload departure rates from every zone at the equilibrium.

The variation in the input temperature of the thermal nodes with respect to a variation in the workload execution rate in the computational nodes or to a variation in the reference temperature vector can be written as

$$\frac{\partial \mathbf{T}_{in,\mathcal{N}}}{\partial \eta} = \mathcal{L}, \qquad \frac{\partial \mathbf{T}_{in,\mathcal{N}}}{\partial \mathbf{T}_{\text{ref}}} = \Psi_{[\mathcal{N},\mathcal{C}]}. \tag{14.22}$$

The physical meaning of the variables $\{\psi_{i,j}\}$ implies that the matrix $(I - \boldsymbol{\Psi}_{[\mathcal{N},\mathcal{N}]})$ is invertible when the input temperatures of all the thermal nodes representing a zone are affected by at least one thermal node representing a CRAC unit.

The inlet temperature of an efficiently cooled server largely depends on the reference temperature of the CRAC units and marginally on the execution rate of other servers. Let i be the index of a thermal node representing a zone. The ith node is efficiently cooled if[3]

$$\left\| \frac{\partial \mathbf{T}_{in,i}}{\partial \mathbf{T}_{\text{ref}}} \right\|_2 \gg \left\| \frac{\partial \mathbf{T}_{in,i}}{\partial \eta} \right\|_2.$$

We consider the vector $\mathbf{z} = [\mathbf{T}_{\text{ref}}\eta]^T$ and define the *relative sensitivity index* of the ith node as

$$\mathcal{S}_i = \left\| \frac{\partial \mathbf{T}_{in,i}}{\partial \mathbf{T}_{\text{ref}}} \right\|_2 \Big/ \left\| \frac{\partial \mathbf{T}_{in,i}}{\partial \mathbf{z}} \right\|_2.$$

When the relative sensitivity index of the ith zone equals 1, the input temperature of the ith zone uniquely depends on the reference temperature of the

3. We focus on L2-norm, but other norms can be considered.

CRAC nodes, whereas when the relative sensitivity index equals 0, then the input temperature of the ith zone only depends on the workload execution rate of the other zones.

A large variability among the relative sensitivity indexes can be exploited by a coordinated controller in order to improve the efficiency of a data center. We collect the relative sensitivity indexes in the vector **S** and define as cyber-physical index (CPI) of a data center the normalized standard variation of **S**

$$\mathrm{CPI} = k \cdot \mathrm{std}(\mathbf{S}), \tag{14.23}$$

where k is the normalizing coefficient and "std" is the standard deviation of the elements of the vector argument.

Fig. 14.14 shows the relative efficiency obtained by the coordinated controller for different values of the CPI. When the CPI values are higher than about 0.55, the uncoordinated controller is unable to find a cooling strategy that satisfies the thermal constraints for high values of the average data center utilization. When the CPI is almost 0, the relative efficiency of an uncoordinated controller is almost 0. As the CPI increases, the efficiency of a coordinated controller grows exponentially fast. The simulation cases discussed in the previous section correspond to a CPI of 0.33, 0.04, and 0.52, respectively.

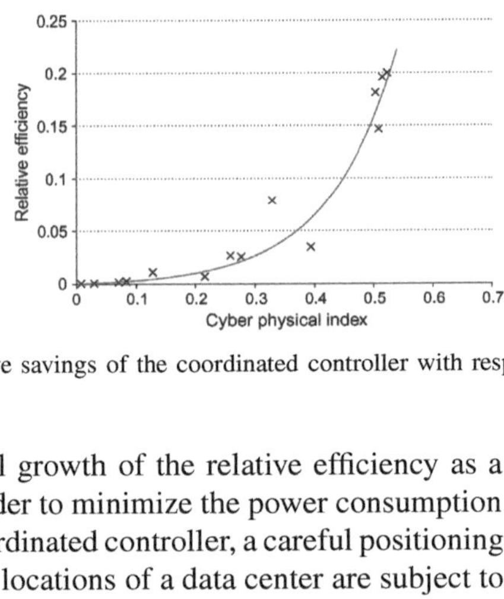

FIGURE 14.14 Relative savings of the coordinated controller with respect to the uncoordinated controller.

The exponential growth of the relative efficiency as a function of the CPI suggests that, in order to minimize the power consumption of a data center controlled by an uncoordinated controller, a careful positioning of the servers should be made. Different locations of a data center are subject to different cooling efficiency values. Given a data center, the relocation of some of its servers may move the CPI of the data center toward lower values so that an uncoordinated controller will be as efficient as a coordinated controller.

14.4 Dynamic server provisioning

The global census on data center trends [483] shows that data center power requirements grew globally to 38 GW (gigawatts) in 2012, about a 63% growth from 24 GW in 2011. A fast temperature evaluation model of data center thermal dynamics was developed in [484], which could help implement real-time control in data centers while considering the control of CRAC units. [485] discussed several server provisioning algorithms to dynamically turn servers on and off for the purpose of saving energy. [486] and [487] studied several dynamic provisioning and load patching algorithms.

In previous works [497], [498], a CPS model was developed to improve data center energy efficiency and the potential of energy saving was discussed. It should be noted that the details of server level dynamics were overlooked, and the models of servers were regarded as simple linear models at the data center level. Additionally, their work also lacks a discussion of the QoS, even though it is an important aspect in the control of data centers.

Dynamic server provisioning and load patching algorithms just took computational dynamics into consideration, regardless of the location of servers that were related to cooling efficiency. An unreasonable active server distribution can easily produce hot spots in data center, but implementing dynamic server provisioning while considering thermal dynamics could help prevent hot spots and achieve significant energy efficiency in both IT and CT systems.

This section considered a case where a data center does not perform cooling efficiently. In the precondition of satisfying QoS requirements, we focused on the energy saving offered by implementing dynamic server provisioning while considering data center thermal dynamics. As IT systems consume more than half of the total power consumption, dynamic server provisioning could largely decrease the power consumption of IT systems by closing idle servers on demand. At the same time different areas of data centers have different cooling efficiency, dynamically allocating tasks to server considering their thermal dynamics could achieve optimal efficiency. To maximize profits, several MPC approaches considering different situations are used in data centers to compare the energy efficiency performance. By optimizing active server numbers and reference temperatures of CRAC units, the MPC controllers achieved a significant improvement in energy efficiency.

14.4.1 Zone level model

A data center mainly consists of the IT system and the CT system. In the IT part, racks of servers are grouped into zones. In the CT part several CRAC units are used to cool down the data center. In the following, we consider connection service as the basic service of the data center; each server in a zone runs only one connection service application.

As presented in [485], in a front door architecture for connection intensive applications, every login request sent to the zone from end-users will reach a dispatch server first. The dispatch server picks a connection server and returns its IP address to the client. Then the client directly connects to the connection server. The connection server authenticates the user and if successful, a live transmission control protocol (TCP) connection will be maintained between the client and the connection server until the client logs off. The TCP connection is usually used to update user status (e.g., on-line, busy, off-line) and to redirect further activities such as chatting and multimedia conferencing to other back-end servers.

14.4.2 System dynamics

At an application level, each connection server has to enforce two major constraints: the maximum login rate and the maximum number of connections it can host. The login rate **L** is defined as the number of new connection requests sent to a connection server in one second. A limit on login rate $\mathbf{L}_{max}$ is used to protect the server. For the consideration of memory constraints and fault tolerance concerns, a limit $\mathbf{N}_{max}$ is considered on the total number of connections for each connection server.

This section assumes that a zone consists of $\mathbf{H}_{max}$ connection servers. Let $H_i(t)$ denote the number of available servers in Zone i, $N_\delta(t)$ the number of connections on Server δ, with $L_\delta(t)$ and $D_\delta(t)$ the login rate and departure rate, respectively. The dynamics of the δ-th server can be modeled as

$$\dot{N}_\delta(t) = L_\delta(t) - D_\delta(t), \tag{14.24}$$

which represents the relationship between the login rate $L_\delta(t)$ and the number of connections $N_\delta(t)$. The number of departures $D_\delta(t)$ is usually a part of $N_\delta(t)$, which varies widely with time. The login rate $L_\delta(t)$ dispatched to one of the available servers in Zone i is a part of total login rate $L_{Z,i}(t)$. A dispatch algorithm can be expressed as

$$L_\delta(t) = L_{Z,i}(t) p_\delta(t), \quad \delta = 1, ..., H_i(t), \tag{14.25}$$

where $p_\delta(t)$ is the fraction of the total login requests assigned to the Server δ,

$$0 \le p_\delta(t) \le 1, \ \sum_{\delta=1}^{H_i(t)} = 1,$$

and $H_i(t)$ is the number of available servers in Zone i.

14.4.3 Performance model

The performance model of a connection server is derived from the model described [485]. The key variables affecting CPU usage and power of connection of server δ are login rates L(t) and the number of active connections N(t). The linear model is

$$\hat{U}_\delta(t) = L_\delta(t) - D_\delta(t), \tag{14.26}$$

where $\hat{U}_\delta(t)$ denotes the CPU utilization percentage. The power consumption of a connection server increases almost linearly with CPU utilization, while the idle servers consume up to almost 60% of the peak power. The power consumption of Server δ is modeled as

$$P_{s,\delta}(t) = P_{s,0}(t) + \eta U_\delta(t), \tag{14.27}$$

where $P_{s,0}(t)$ is the power consumption of idle servers, and η is a positive coefficient.

14.4.4 Data center level model

At the data center level, as discussed in [497], the thermal properties of the IT system are managed as groups of components (racks of servers). Servers are aggregated into several zones, and each zone has both computational and thermal dynamics. The IT system processes the clients' requests and maintains active connections; it consumes power and generates heat at the same time. The CT system removes heat from zones by consuming power. In the following, we

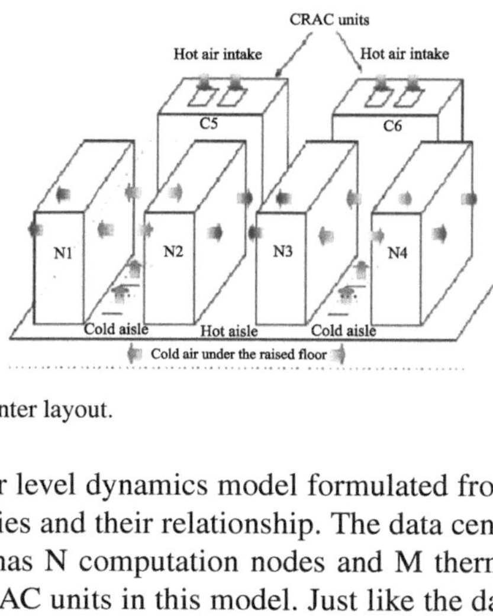

FIGURE 14.15 Data center layout.

show the data center level dynamics model formulated from the computational and thermal properties and their relationship. The data center level model studied in this chapter has N computation nodes and M thermal nodes, that is, N zones and M–N CRAC units in this model. Just like the data center considered in the following simulation (see Fig. 14.15), there are four computational nodes and six thermal nodes. In the following sections $i = 1, ..., N$ represents the number of each zone, and $i = N + 1, ..., M$ of each CRAC unit. The MPC method is utilized to save the potential energy by dynamically shutting down idle servers and changing the operation conditions of each component of the data center.

Computational and thermal models are expressed as

$$\dot{N}_{Z,i}(t) = L_{Z,i}(t) - D_{Z,i}(t), \tag{14.28}$$

$$L_{Z,i}(t) = L_{DC}(t)q_i(t), \quad i = 1, ..., N, \tag{14.29}$$

where Zones are modeled as a single computational node and Zone i has the following:

- A total login rate $L_{Z,i}(t) < L_{Z,max}(t)$. The limit $L_{Z,max}(t)$ is affected by the number of available servers $H_i(t)$ with $(L_{Z,max}(t) = H_i(t).L_{max})$. For a zone having H_{max} servers in all, $H_i(t) \leq H_{max}$;
- A total connection departure rate $D_{Z,i}(t)$ and a total number of active connections $N_{Z,i}(t)$;
- Given a data center with N computation nodes and M thermal nodes, the dynamics of Zone i can be expressed by (14.28);
- The total login rate $L_{Z,i}(t)$ is dispatched by the data center level controller. The dispatch algorithm is specified by (14.29) where $L_{DC}(t)$ is the total number of client requests sent to the data center, $q_i(t)$ the fraction of the total login requests assigned to Zone i such that

$$0 \leq q_i(t) \leq 1, \ \sum_{i=1}^{N} = 1.$$

The model of requests execution developed above can also be extended to include additional components, such as different workload dispatch policies, job classes, hardware requirements, and interactions among different workload classes.

From a thermal perspective the zones, CRAC units, and other support devices are modeled as thermal nodes. Since we focus here on the data center level, the slight effect of support devices on the thermal environment is neglected and we consider M thermal nodes in the data center. Following [484], each thermal node is associated with an output temperature and an input temperature. Thermal models are expressed by

$$Q_{in,i}(t) = \varrho.f_i.C_p.T_{in,i}(t), \tag{14.30}$$

$$Q_{out,i}(t) - Q_{in,i}(t) = \lambda_i.P_i(t), \tag{14.31}$$

where

- $T_{in,i}(t)$ is the input temperature;
- $Q_{in,i}(t)$ is the rate of the amount of heat brought into Node i, ϱ is the air density, f_i the flow rate of Node i, and C_p the specific heat of air;
- $Q_{out,i}(t)$ is the rate of the amount of heat taken away from node i;
- λ_i is the coefficient of the power consumption of Node i converted to the heat $\lambda_i \approx 1, \forall i = 1, ..., N$; $\lambda_i < 1, \forall i = N+1, ..., M$; and $P_i(t)$ is the power consumption.

The rate of the amount of input heat $Q_{in,i}(t)$ carried by inlet air flow is a mixture of supplied cold air flow from CRAC nodes and recirculated hot air from other server nodes, so there is

$$Q_{in,i}(t) = \sum_{j=1}^{M} \varphi_{j,i} \cdot Q_{out,i}(t), \quad i = 1, ..., M, \tag{14.32}$$

where the coefficient $\varphi_{j,i}$ is the percentage of the heat flow from node j to node i. The matrix $\Phi = [\varphi_{j,i}]_{M \times M}$ is defined as the cross-interference among all thermal nodes, with

$$\varphi_{j,i} \geq 0, \quad \sum_{j=1}^{M} \varphi_{j,i} = 1, i = 1, ..., M.$$

On considering (7) and (9), and substituting $\varrho f_i C_p$ with K_i, for node i we have

$$T_{in,i} = \frac{\sum_{j=1}^{M} \varphi_{j,i}.K_j.T_{out,j}}{K_i}, \quad i = 1, ..., M. \tag{14.33}$$

It is demonstrated in [558] that the revolution of the outlet temperature of a server node can approximate a liner-invariant system, the revolution of the output temperature of the thermal nodes for zones is then modeled by

$$\dot{T}_{out,i}(t) = -\alpha_i T_{out,i}(t) + \alpha_i T_{in,i}(t) + c_i.P_i(t), \tag{14.34}$$

where $i = 1, ..., N$, and $1/\alpha_i$ is the time constant of the temperature of node i, c_i is the coefficient that maps power consumption into output temperature variation, and $P_i(t)$ is the power consumption of node i.

Considering the zone level model (14.26), (14.27) and server On-Off state control. The power consumption of the nodes of zones can be modeled as

$$\begin{aligned} P_i(t) &= H_i(t).(P_{s,0} - 0.82) + 2.84 \times 10^{-4}.\eta.N_{Z,i} + 0.55.\eta L_{Z,i}, \\ i &= 1, ..., N, \end{aligned} \tag{14.35}$$

where $H_i(t)$ is the number of active servers in Zone i and is controlled by the data center level controller.

The CRAC units consume the primary power consumption of the CT system, and the output temperature of the CRAC units is modeled by

$$\dot{T}_{out,i}(t) = -\alpha_i T_{out,i}(t) + \alpha_i.\min\{T_{in,i}(t), T_{ref,i}(t)\}, \tag{14.36}$$

where $i = N + 1, ..., M$, and $T_{ref,i}(t)$ is the reference temperature of the CRAC node i. In the following it is assumed that $T_{ref,i}(t)$ is controllable. The "min" operator in (14.36) ensures that the supplied air temperatures from CRAC units are not greater than the input temperatures.

In [559] the power consumption of a CRAC node is modeled as

$$P_i(t) = \begin{cases} K_i \frac{T_{in,i}(t) - T_{out,i}}{COP(T_{out,i})}, & \text{if } T_{in,i}(t) \geq T_{out,i}(t) \\ 0, & \text{if } T_{in,i}(t) < T_{out,i}(t) \end{cases}, \tag{14.37}$$

where $i = N+1, ..., M$. As shown in [559], $COP(T_{out,i}(t))$ is the coefficient of performance of CRAC Node i and it is a function of the node's output temperature. In fact, $\lambda_i = -COP(T_{out,i}(t))$ for CRAC nodes. Notably, the relationship between the COP and the output temperatures of the CRAC units is expressed by

$$COP(T_{out,i}(t)) = 0.007.T_{out,i}^2(t) + 0.001.T_{out,i}(t) + 0.46. \tag{14.38}$$

For the purpose of computer simulation, the power consumption of the thermal nodes of the zones with the $M \times 1$ vector $\mathbf{P}_Z(t)$, with $\mathbf{T}_{ref}(t)$ and $\mathbf{P}_C(t)$ denoting the reference temperatures and power consumption of CRAC nodes, respectively. In the same way, we define vector $\mathbf{H}(t)$ as the number of available servers of zone nodes. The output temperature vector $\mathbf{T}_{out}(t)$ is the state of the thermal dynamics model, $\mathbf{T}_{ref}(t)$ is the controllable input, and $\mathbf{P}_Z(t)$ is the uncontrollable input. The input temperature $\mathbf{T}_{in}(t)$ and the power consumption of CRAC units $\mathbf{P}_C(t)$ are outputs of the thermal dynamics model. The evolution of the thermal dynamics can then be presented by

$$\dot{\mathbf{T}}_{out,i}(t) = A_{T,C}\ \mathbf{T}_{out,i}(t) + B_{T,C}[\mathbf{P}_Z^T(t)\ \ \mathbf{T}_{ref}(t)]^T, \tag{14.39}$$

where $A_{T,C}$ and $B_{T,C}$ can be derived from (14.34)–(14.36).

Now we direct our attention to a multilevel control strategy, the control architecture that is depicted in Fig. 14.16. The data center level controller collects the information from each data center component and provides the optimal set-point to the low-level controllers. The low-level controllers operate independently from each other in all parts of the data center; they work in zones called zone-level controller. The controller in each CRAC unit can keep the output air flow temperature up with the reference temperature.

14.4.5 Zone-level controller

The zone-level controller's main task is to dispatch login requests to the available servers in the zone and keep the number of connections on the servers as close as possible. If the upper-level controller wants to increase the number of active servers in Zone i, the corresponding zone-level controller will choose closed servers and turn them on. If the upper-level controller wants to decrease the number of active servers in Zone i, the corresponding zone-level controller will choose active servers and turn them off. The zone-level controller will migrate connections on chosen servers to other available servers before closing them.

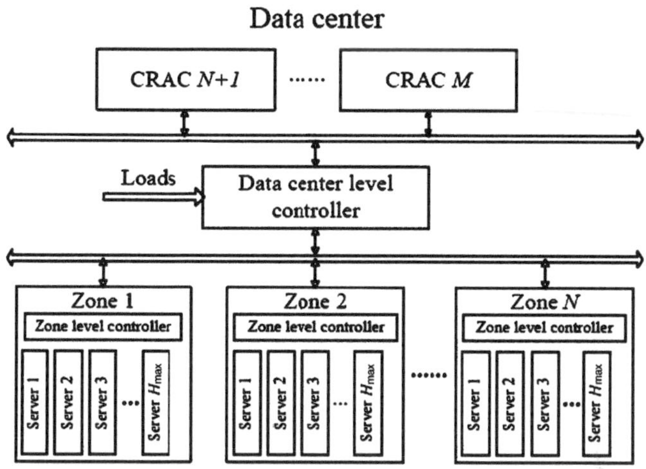

FIGURE 14.16 Multilevel control of a data center.

Some important issues should be noted:

Quality of service (QoS) metrics This section relies on **"Service not available" (SNA)** and **"Service-initiated disconnections" (SID)** as metrics of the QoS. With regards to SNA, users will receive errors if the number of connection servers is insufficient to serve new login requests, and those new login requests will be rejected. The SID will occur if a connection server with active users is turned off and users may experience a period of disconnection. When this happens, reconnections will create an artificial surge on the number of new connections, it will generate unnecessary SNAs. Neither error is allowed in order to protect the user's experience.

Load balancing The load balancing algorithm is considered in [485], where the controller dispatches the following portion of total loads of Zone i to Server δ,

$$p_\delta(t) \quad = \quad \frac{1}{H_i(t)} + \alpha\Big(\frac{1}{H_i(t)} - \frac{N_\delta(t)}{N_{Z,i}(t)}\Big), \tag{14.40}$$

where $\delta = 1, ..., H_i(t)$, $i = 1, ..., N$, and $\alpha > 0$. α is a parameter that can be tuned to change the dynamic behavior of the system. The algorithm tries to drive the number of connections quickly to uniform for high values of α. This algorithm assigns larger portions to servers with relatively small number of connections. By migrating connections internally the controller can take load off from Server δ, then $p_\delta(t)$ can be negative.

Server On-Off state control Because the main power consumption of the data center is caused by the IT system, the On-Off state control is used to save more energy. The number of active servers needs to handle the login rate and the total connection of zones. Since a server needs some time to turn on and turn off, we cannot change a server's status whenever we need to. The controller recalculates the number of servers that a zone needs for a period of time, and in the time horizon login rates and connections may change drastically. The constraints of active servers of Zone i are considered as

$$L_{max}(t)H_i(t) > \gamma_L L_{Z,i}(t), \quad \gamma_L > 1, \tag{14.41}$$
$$N_{max}(t)H_i(t) > \gamma_N N_{Z,i}(t), \quad \gamma_N > 1, \tag{14.42}$$

where γ_L and γ_N are prescribed parameters that are associated with the dynamics of $L_{Z;i}(t)$ and $N_{Z;i}(t)$, respectively. Note that the strategies used to dispatch loads in zones are also associated with those parameters. On considering the load balancing algorithm used here, when a server in zone j is newly turned on, the login rate assigned to it is $(1+\alpha)L_{Z;i}(t)/H_i(t)$, so $\gamma_L > 1+\alpha$.

When a server with active connections on it needs to be turned off, the connections will be migrated at a speed $V_m(t)$ and no new user requests will be dispatched to it. The time-continuous dynamics of the connection on Server δ which need to be turned off is represented as

$$\dot{N}_\delta = \begin{cases} -D_\delta(t) - V_m(t), & \text{if } N_\delta(t) > 0 \\ 0, & \text{if } N_\delta(t) = 0 \end{cases}. \tag{14.43}$$

This server will be turned off immediately when the number of connections on it is zero.

14.4.6 Data center level controller

A predictive, discrete-time model of the data center is considered by the data center level controller. A discrete-time MPC approach is used to optimize the energy efficiency of the total data center with the QoS and the thermal constraints enforced. We consider the optimization problem at a horizon $\Gamma \in \mathbb{N}$, and solve the optimal control problem once in every step. The loads that used to predict the states of the data center level model are obtained by a short-term load forecasting algorithm.

Load forecasting Let $L_{DC}(t)$ or $N_{DC}(t)$ be measured at regular time intervals and designated as the time series $y(t)$. A auto-regression (AR) model is used to predict the value of $y(t)$ over a period of T time units. The value of $\hat{y}(t)$ measurements is

$$\hat{y}(t) = \sum_{k=1}^{n} \sigma_k y(t-kT), \tag{14.44}$$

where n is the order of the AR model and $\{\sigma_k\}$ are the parameters of the AR model.

Optimization problem In the optimization problem we define the predicted value of the variable $\mathbf{N}_Z(t)$ as $\hat{\mathbf{N}}_Z(h|k)$ at the beginning of the hth interval, based on the information available up to the beginning of the kth interval, and similarly we define the variables $\hat{\mathbf{T}}_{in}(h|k)$, $\hat{\mathbf{T}}_{out}(h|k)$, and $\hat{\mathbf{T}}_{ref}(h|k)$. The expected value of $q_i(t)$ is denoted with $\hat{\mathbf{q}}(h|k)$ during the kth interval, based on the information available up to kth interval. Similarly, $\hat{\mathbf{P}}_N(h|k)$ denotes the expected average power consumption of the zones during the kth interval, $\hat{\mathbf{P}}_C(h|k)$ denotes the expected average power consumption of the CRAC units during the kth interval, and $\hat{\mathbf{H}}(h|k)$ denotes the expected active servers in zones during the kth interval. We define the sets

$$\begin{aligned}
\mathcal{H} &= \{\hat{\mathbf{H}}(k|k), \ldots., \hat{\mathbf{H}}(k+\Gamma-1|k)\} \\
\mathcal{Q} &= \{\hat{\mathbf{q}}(k|k), \ldots., \hat{\mathbf{q}}(k+\Gamma-1|k)\} \\
\mathcal{T}_{ref} &= \{\hat{\mathbf{T}}_{ref}(k|k), \ldots., \hat{\mathbf{T}}_{ref}(k+\Gamma-1|k)\} \\
\mathcal{N} &= \{\hat{\mathbf{N}}_Z(k|k), \ldots., \hat{\mathbf{N}}_Z(k+\Gamma-1|k)\} \\
\mathcal{L} &= \{\hat{\mathbf{L}}_Z(k|k), \ldots., \hat{\mathbf{L}}_Z(k+\Gamma-1|k)\}.
\end{aligned} \tag{14.45}$$

Among the different control strategies, we apply two methods:

- In the first method, called uncoordinated MPC, the controller has two independent solvers and considers a discrete-time model of the data center, and manages the IT and CT in two steps. In the first step, the controller solves the optimization problem at time k as:

$$\begin{aligned}
&\min_{\mathcal{H},\mathcal{Q},\mathcal{N},\mathcal{L}} \sum_{h=k}^{k+\Gamma-1} ||\hat{\mathbf{P}}_N(h|k)||_1 \\
&\forall\, h = k, \ldots, k+\Gamma-1 \\
&\textit{computational dynamics} \\
&\textit{subject to} \\
&0 \le \hat{\mathbf{H}}(h|k) \le \mathbf{H}_{max}, \;\; \mathbf{0} \le \hat{\mathbf{q}}(h|k) \le 1, \;\; \mathbf{1}^T\hat{\mathbf{q}}(h|k) = 1 \\
&\hat{\mathbf{L}}_Z(h|k) = diag\{\mathbf{1}\hat{L}_{DC}(h|k)\}\hat{\mathbf{q}}(h|k) \\
&\gamma_L\, \hat{\mathbf{L}}_Z(h|k) \le L_{max}\hat{\mathbf{H}}_Z(h|k), \quad \gamma_N\, \hat{\mathbf{N}}_Z(h|k) \le N_{max}\hat{\mathbf{H}}_Z(h|k) \\
&\gamma_N\, \hat{\mathbf{N}}_Z(h+1|k) \le N_{max}\hat{\mathbf{H}}_Z(h|k) \\
&\hat{\mathbf{N}}_Z(k|k) = \mathbf{N}_Z(k).
\end{aligned} \tag{14.46}$$

  The controller minimizes the power consumption of the IT system first, and then in the second step, optimizes the power consumption of the CT system as

$$\min_{\mathcal{T}_{ref}} \sum_{h=k}^{k+\Gamma-1} ||\hat{\mathbf{P}}_C(h|k)||_1$$

$$
\begin{aligned}
&\forall\, h = k, \ldots, k+\Gamma-1\\
&\textit{thermal dynamics}\\
&\textit{subject to}\\
&\mathbf{T}_{ref,min} \;\le\; \hat{\mathbf{T}}_{ref}(h|k) \le \mathbf{T}_{ref,max}\\
&\hat{\mathbf{T}}_{in}(h+1|k) \le \mathbf{T}_{in,max}\\
&\hat{\mathbf{T}}_{out}(k|k) \;=\; \mathbf{T}_{out}(k).
\end{aligned}
\tag{14.47}
$$

- The second method, called coordinated MPC, coordinates IT power consumption and CT power consumption minimization in one optimization problem. It is also based on a discrete-time MPC approach and minimizes the power consumption of the data center in each step. At time k, the controller has to solve the following optimization problem:

$$
\begin{aligned}
&\min_{\mathcal{H},\mathcal{T}_{ref},\mathcal{Q},\mathcal{N},\mathcal{L}} \sum_{h=k}^{k+\Gamma-1} ||\hat{\mathbf{P}}_N(h|k)||_1 + ||\hat{\mathbf{P}}_C(h|k)||_1\\
&\forall\, h = k, \ldots, k+\Gamma-1\\
&\textit{data center dynamics } (14.28)\text{–}(14.37)\\
&\textit{subject to}\\
&\mathbf{T}_{ref,min} \;\le\; \hat{\mathbf{T}}_{ref}(h|k) \le \mathbf{T}_{ref,max}\\
&\hat{\mathbf{T}}_{in}(h+1|k) \le \mathbf{T}_{in,max}\\
&0 \le \hat{\mathbf{H}}(h|k) \le \mathbf{H}_{max},\;\; \mathbf{0} \le \hat{\mathbf{q}}(h|k) \le 1,\;\; \mathbf{1}^T\hat{\mathbf{q}}(h|k) = 1\\
&\hat{\mathbf{L}}_Z(h|k) = diag\{\mathbf{1}\hat{L}_{DC}(h|k)\}\hat{\mathbf{q}}(h|k)\\
&\gamma_L\,\hat{\mathbf{L}}_Z(h|k) \le L_{max}\hat{\mathbf{H}}_Z(h|k),\quad \gamma_N\,\hat{\mathbf{N}}_Z(h|k) \le N_{max}\hat{\mathbf{H}}_Z(h|k)\\
&\gamma_N\,\hat{\mathbf{N}}_Z(h+1|k) \le N_{max}\hat{\mathbf{H}}_Z(h|k)\\
&\hat{\mathbf{T}}_{out}(k|k) \;=\; \mathbf{T}_{out}(k),\quad \hat{\mathbf{N}}_Z(k|k) = \mathbf{N}_Z(k).
\end{aligned}
\tag{14.48}
$$

Remark 14.1. Since the first strategy considers the optimization problem in two steps, the computational complexity of the uncoordinated MPC is expected to be lower than the second strategy. However, both control strategies are able to guarantee the enforcement of the QoS and the thermal constraints. The optimal solutions which were generated by the optimization solver, such as $\mathbf{T}_{ref}(k)$, $\mathbf{q}(k)$, and $\mathbf{H}(k)$, will be sent to the low-level controllers step by step.

14.4.7 Simulation results II

This section considers a data center layout like the one shown in Fig. 14.15. Server nodes $N1$–$N4$ represent a collection of two racks, each composed of 30 servers ($H_{max} = 60$). The servers in $N1$ and $N2$ are slightly more efficient than those in $N3$ and $N4$. The power consumption of idle servers in $N1$ and $N2$ is 150 kW, in $N3$ and $N4$ it is 180 kW. The coefficient matrix Φ of the thermal dynamics is

$$\Phi = \begin{bmatrix} 0.01 & 0.5 & 0 & 0.2 & 0.15 & 0.14 \\ 0 & 0.03 & 0 & 0.01 & 0.56 & 0.40 \\ 0 & 0.03 & 0.02 & 0.05 & 0.44 & 0.46 \\ 0 & 0.02 & 0.02 & 0.01 & 0.40 & 0.55 \\ 0.227 & 0.124 & 0.278 & 0.251 & 0.12 & 0.0 \\ 0.335 & 0.114 & 0.268 & 0.163 & 0 & 0.12 \end{bmatrix}. \tag{14.49}$$

Some places in the data center do not cool efficiently. Because about half of the output hot air of server node N1 is recycled to server node $N2$, $\varphi_{1,2} = 0.5$ is much larger than the amount around it. This means that when server node N1 has heavy loads, the input temperature of node $N2$ will be higher than the other server nodes. The simulations set the maximum allowed input temperature of every server node at 27°C, according to [554], the environmental guidelines for datacom equipment, expanding the recommended environmental envelope. All connection servers are equally in computational performance, the constraints on login rates and active connections are Lmax = 70/s and Nmax = 100, 000 (27). Before servers are shut down, the corresponding zone-level controller migrates the connections on them at a speed Vm = 100/s.

Three days of load pattern illustrated in Fig. 14.17 is taken into consideration. The login rates subject to a random noise uniformly distributed having zero mean and a variance proportional to the mean login rate. The total number of connections fluctuates throughout the day and night.

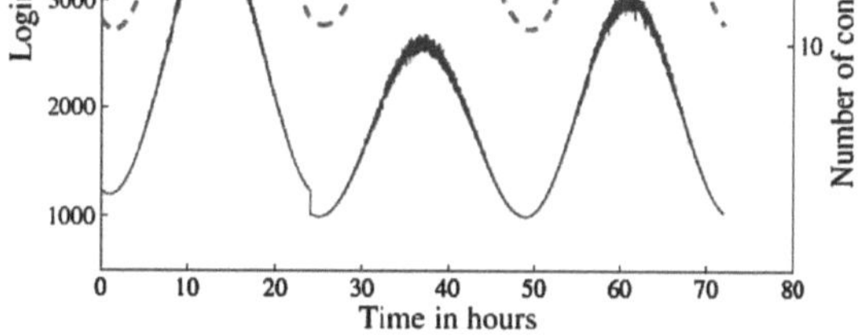

FIGURE 14.17 Load pattern over three days.

Simulations were developed under three different control scenarios. In the first scenario the servers in each zone are always active and the controller coordinated the energy efficiency of servers and the thermal dynamics in one optimization problem that we called constant MPC. Constant MPC is a simplified edition of the coordinated MPC scenario. It considers the same optimization problem as the coordinated MPC, but keeps all the servers active and ignores the On-Off state control. The controller dynamically dispatches jobs to each zone and adjusts the reference temperature for all CRAC units through the simulation. The second scenario uses the uncoordinated MPC scenario to dynamically adjust load dispatching and the reference temperature of CRAC units. The third scenario takes advantage of the coordinated MPC method by considering all the dynamics of the data center. Under these three scenarios the SID and SNA errors were prevented from influencing the user experience.

The MATLAB® optimization toolbox was used as the numerical solver. We assume that all zone nodes have the same initial state. The time step of the simulation is 30 s and the MPC controller solves the optimization problem every 20 min. The prediction horizon of the optimization problem is six steps (2 hours), and the control horizon is three steps (1 hour). The servers were forced to process all login requests, and no active connections were dropped due to the migration actions. The energy consumption values of the total data center under different control strategies are listed in Table 14.4.

TABLE 14.4 Comparing of energy consumption (kWh) under different control scenarios.

Control scenarios	IT energy	CT energy	Total energy	PUE
Constant MPC	3066.20	993.98	4060.18	1.32
Uncoordinated MPC	2293.34	771.43	3064.77	1.34
Coordinated MPC	2380.02	560.23	2940.25	1.24

Compared with the constant MPC, the coordinated MPC method reduced the energy consumption of the total data center to 27.58% and the uncoordinated MPC to 24.52% with enforced QoS and thermal constraints when considering dynamic server provisioning. No SNA and SID errors occurred in the simulation processes. The energy saving under coordinated MPC was mainly from the CT system, compared with the energy saving under uncoordinated MPC. The benefit from taking thermal dynamics into consideration in the optimization of active server numbers was that the PUE decreased to 1.24 under coordinated MPC. Uncoordinated MPC allocated login requests to zones regardless of thermal dynamics and had a relatively high PUE, almost to 1.34.

The total power consumption of the data center under different control scenarios is shown in Fig. 14.18.

Fig. 14.19 indicates that the variation in the number of active servers in nodes N1–N4 under MPC, reflects the number of login requests each zone processes. The uncoordinated MPC strategy dispatched loads to server node N1 and N2 first and N3 and N4 last, because the servers in nodes N1 and N2 have relatively high energy efficiency.

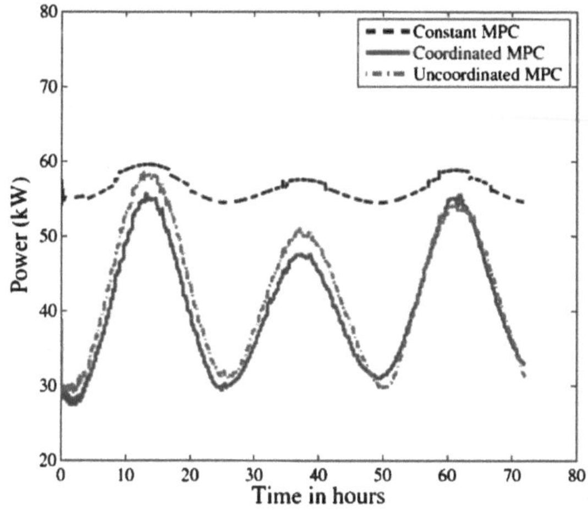

FIGURE 14.18 Total data center power consumption.

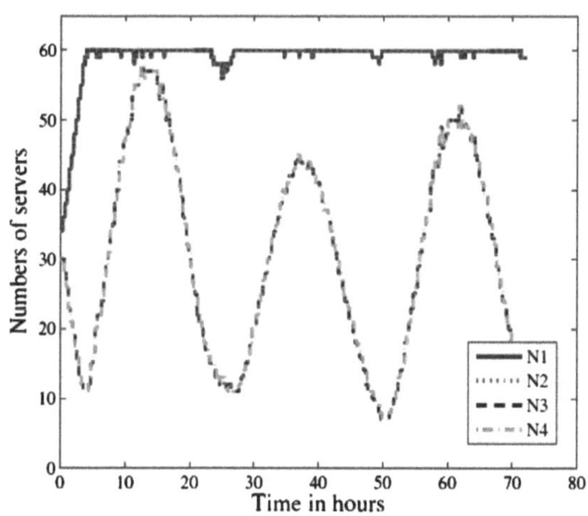

FIGURE 14.19 The number of active servers in zones N1–N4 under Uncoordinated MPC.

Fig. 14.20 indicates data center level controller using coordinated MPC dispatched loads to server Node N2 first and N1 last, where the servers in node N2 have relatively high energy efficiency and the output air flow recirculated less. Because about half of the output hot air from server node N1 is recirculated to server node N2, N1 has the last priority to process loads.

Fig. 14.21 demonstrates the average CRAC reference temperature under three control strategies. The average reference temperature varies a lot under coordinated MPC and is always higher than the uncoordinated MPC case. When compared with the uncoordinated MPC, the coordinated MPC scenario is able to maintain a higher level of cooling efficiency and is able to largely reduce the power consumption of the CRAC units. Although the uncoordinated MPC

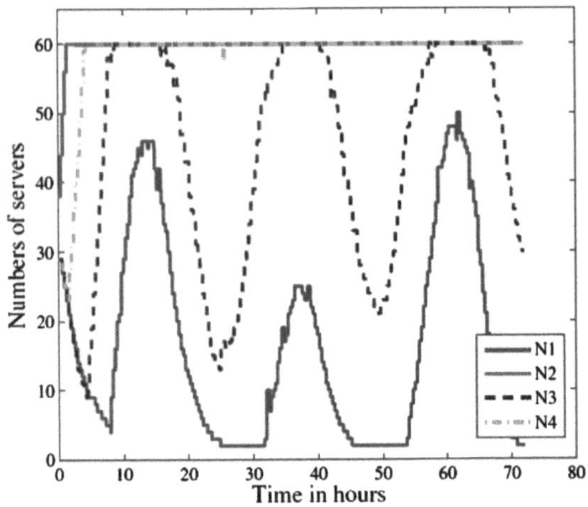

FIGURE 14.20 The number of active servers in zones N1–N4 under Coordinated MPC.

mostly reduces the energy consumption of the IT system, the average reference temperature under uncoordinated MPC is lower than under constant MPC because uncoordinated MPC uses N1 as one of the main zones for processing jobs. The recirculation of the output air flow of node N1 causes the CRAC units to turn down the reference temperatures to enforce the thermal constraints.

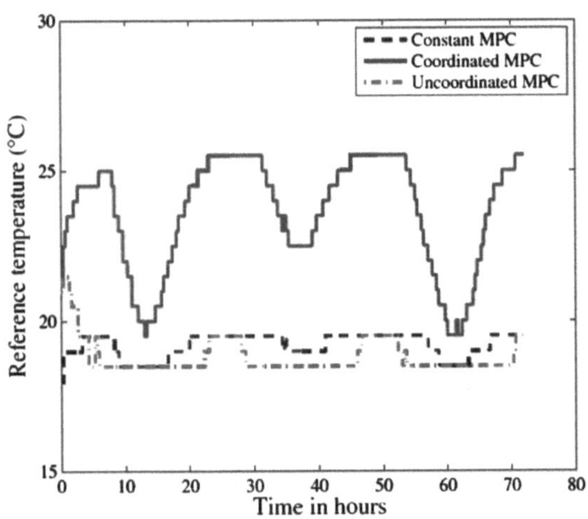

FIGURE 14.21 Average reference temperature.

Fig. 14.22 shows that coordinated MPC and uncoordinated MPC are able to save energy from both the IT system and the CT system. The power consumption of IT and CT have the same variation, and they changed more quickly compared with the constant MPC scenario. In simulations, three different control scenarios were able to enforce the inlet temperature constraints and prevented

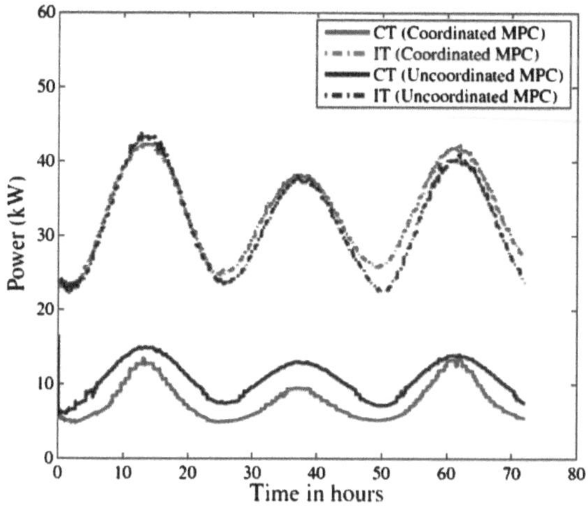

FIGURE 14.22 Power consumption of the IT system and the CT system.

SNA and SID errors. In the simulation processes, because of the computational complexity, the solver takes more time to solve the optimization problem (14.48) than the sum of optimization problems (14.46) and (14.47) in our simulation environment. The coordinated MPC and uncoordinated MPC both satisfy the requirement of real-time control, and after solving the optimization problem there is enough time to change the servers' states. We considered another case where the servers in N3 and N4 are slightly more efficient than those in N1 and N2. The power consumption of idle servers in N3 and N4 is 150 kW, in N1 and N2 is 180 kW. Then under either coordinated MPC or uncoordinated MPC the N1 always has the last priority to process login requests. The cooling of the data center will be much better than before. The energy consumption values of the total data center under different control strategies are listed in Table 14.5. Both Coordinated MPC and uncoordinated MPC obtain a significant energy saving compared with constant MPC. Considering the efficiency of the control algorithm, the uncoordinated MPC is preferred in this case.

TABLE 14.5 Comparing of energy consumption (kWh) of the total data center under different control scenarios.

Control scenarios	IT energy	CT energy	Total energy	PUE
Constant MPC	3066.20	1121.09	4187.29	1.36
Uncoordinated MPC	2208.65	599.12	2807.76	1.27
Coordinated MPC	2252.93	557.53	2810.46	1.24

The coordinated MPC scenario could achieve more energy saving than uncoordinated MPC when the data center's relatively high efficient server zones have large hot air recirculation. However, for a data center with relatively high server zone cooling efficiency, the uncoordinated MPC is preferred. By taking dynamic server provisioning into optimal control, both control strategies can obtain more than 24% energy reduction.

14.5 Notes

This section presented the control of data centers from a CPS perspective. A survey of the literature and current practice shows how energy efficiency has been improved at all levels by taking into account the coupling between the cyber and physical features of data center components. We developed a control-oriented model of the coupled cyber and physical dynamics in data centers to study the potential impact of coordinating the control of the IT with the control of the CT at the data center level. The simulation results show that the amount of savings that can be realized by coordinated control depends upon the amount of workload relative to the total data center capacity and the way the variations in the efficiency of servers are physically distributed relative to the physical distribution of cooling efficiency throughout the data center.

A new CPI is proposed to quantify this dependence of the potential impact of coordinated control on the distribution of cyber (computational) and physical (power and thermal) efficiency. The CPI can be used to assess the need for coordinated control for a given data center, or as a tool to evaluate alternative data center designs. Further research is needed to understand how servers with different efficiencies should be distributed to reduce the need for coordinated control. We are also investigating improvements in the CPI to better quantify additional features, such as the impact of different efficiencies on CRAC units.

More research is needed to develop strategies for coordinating data center control with the power grid. Initial results in this direction can be found in [540] and [542]. Data centers can play an important role in the smart grid because of their high power consumption density. A low-density data center can have a peak power consumption of 800 W/m2 (75 W/ft 2), whereas a high-density data center can reach 1.6 KW/m2 (150 W/ft 2) [508], [509], [555]. These values are much higher than residential loads, where the peak power consumption is about a few watts per square meter [556], [557].

Finally, we believe that the observations made in this section concerning data centers from a CPS perspective can offer insights into how to understand and control other large-scale CPSs. Many CPSs can be viewed as coupled cyber and physical networks, similar to the computational and thermal networks used in this chapter to model data centers. In CPS applications, it is important to understand the potential impact of coordinated control strategies versus uncoordinated strategies. Uncoordinated strategies offer the possibility of a "divide and conquer" approach to complexity, and in some cases the benefits of introducing more complex strategies to coordinate cyber and physical elements of a system may not be significant. The CPI defined for data centers offers one example of how to measure the potential impact of coordinated cyber and physical control. We expect that developing similar indices for other large-scale CPS applications could be of value.

References

[1] Y. Ali, Y. Xia, L. Ma, A. Hammad, Secure design for cloud control system against distributed denial of service attack, Control Theory and Technology 16 (1) (2018) 14–24.

[2] B. Kehoe, S. Patil, P. Abbeel, et al., A survey of research on cloud robotics and automation, IEEE Transactions on Automation Science and Engineering 12 (2) (2015) 398–409.

[3] K.D. Kim, P.R. Kumar, Cyber-physical systems: a perspective at the centennial, Proceedings of the IEEE 100 (2012) 1287–1308.

[4] I. Lee, O. Sokolsky, S. Chen, J. Hatcliff, E. Jee, B. Kim, A. King, M. Mullen-Fortino, S. Park, A. Roederer, K.K. Venkatasubramanian, Challenges and research directions in medical cyber–physical systems, Proceedings of the IEEE 100 (1) (2012) 75–90.

[5] I. Stojmenovic, Large scale cyber-physical systems: distributed actuation, in-network processing and Machine-to-Machine communications, in: Proc. 2nd Mediterranean Conference on Embedded Computing, MECO, 15–20 June 2013, pp. 21–24.

[6] K. Sampigethaya, R. Poovendran, Cyber-physical integration in future aviation information systems, in: Proc. IEEE/AIAA 31st Conference on Digital Avionics Systems, DASC, 14–18 October 2012 pp. 7C2-1–7C2-12.

[7] H. Li, L. Lai, H.V. Poor, Multicast routing for decentralized control of cyber physical systems with an application in smart grid, IEEE Journal on Selected Areas in Communications 30 (6) (2012) 1097–1107.

[8] X. Cao, P. Cheng, J. Chen, Y. Sun, An online optimization approach for control and communication codesign in networked cyberphysical systems, IEEE Transactions on Industrial Informatics 9 (1) (2013) 439–450.

[9] J. Lin, S. Sedigh, A. Miller, Towards integrated simulation of cyber-physical systems: a case study on intelligent water distribution, in: Proc. Eighth IEEE Int. Conference on Dependable, Autonomic and Secure Computing, DASC'09, 12–14 December 2009, pp. 690–695.

[10] N.N.P. Mahalik, K.K. Kim, A prototype for hardware-in-the-loop simulation of a distributed control architecture, IEEE Transactions on Systems, Man and Cybernetics. Part C, Applications and Reviews 38 (2) (2008) 189–200.

[11] Y. Liu, Y. Peng, B. Wang, S. Yao, Z. Liu, Review on cyber-physical systems, IEEE/CAA Journal of Automatica Sinica 4 (1) (January 2017) 27–40.

[12] C. Meng, T. Wang, W. Chou, S. Luan, Y. Zhang, Z. Tian, Remote surgery case: robot-assisted teleneurosurgery, in: IEEE Int. Conf. Robot. and Auto., ICRA'04 1 (Apr. 2004) 819–823.

[13] J.P. Hespanha, M.L. McLaughlin, G. Sukhatme, Haptic collaboration over the Internet, in: Proc. 5th Phantom Users Group Workshop, Oct. 2000.

[14] K. Hikichi, H. Morino, I. Arimoto, K. Sezaki, Y. Yasuda, The evaluation of delay jitter for haptics collaboration over the Internet, in: Proc. IEEE Global Telecomm. Conf. (GLOBECOM), vol. 2, Nov. 2002, pp. 1492–1496.

[15] S. Shirmohammadi, N.H. Woo, Evaluating decorators for haptic collaboration over the Internet, in: Proc. 3rd IEEE Int. Workshop Haptic, Audio and Visual Env. Applic., Oct. 2004, pp. 105–109.

[16] J. Deng, R. Han, S. Mishra, Secure code distribution in dynamically programmable wireless sensor networks, in: Proc. of ACM/IEEE IPSN, 2006, pp. 292–300.

[17] S. Munir, J. Stankovic, C. Liang, S. Lin, New cyber physical system challenges for human-in-the-loop control, in: 8th Int. Workshop on Feedback Computing, June 2013.
[18] G. Schirner, D. Erdogmus, K. Chowdhury, T. Padir, The future of human-in-the-loop cyber-physical systems, Computer 46 (1) (2013) 36–45.
[19] M.S. Mahmoud, Control and Estimation Methods Over Communication Networks, Springer-Verlag, UK, 2014.
[20] J.S. Baras, Security and trust for wireless autonomic networks: system and control methods, European Journal of Control 13 (2007) 105–133.
[21] T. Alpcan, T. Basar, Network Security: A Decision and Game Theoretic Approach, Cambridge Univ. Press, Cambridge, U.K., 2011.
[22] D. Wei, K. Ji, Resilient industrial control system: concepts, formulation, metrics, and insights, in: Proc. 3rd Int. Symp. Resilient Control Systems, 2010, pp. 15–22.
[23] Q. Zhu, T. Basar, Robust and resilient control design for cyberphysical systems with an application to power systems, in: Proc. 50th IEEE Conf. Decision Control European Control, 2011, pp. 4066–4071.
[24] Q. Zhu, T. Basar, A dynamic game-theoretic approach to resilient control system design for cascading failures, in: Proc. 1st Conf. High Confidence Networked Systems, CPS Week, Beijing, China, Apr. 16, 2012, pp. 41–46.
[25] Q. Zhu, L. Bushnell, T. Basar, Resilient distributed control of multi-agent cyber-physical systems, in: Proc. Workshop Control Cyber-Physical Systems, Baltimore, MD, Mar. 20–21, 2013, pp. 301–316.
[26] Y. Yuan, Q. Zhu, F. Sun, Q. Wang, T. Basar, Resilient control of cyber-physical systems against denial-of-service attacks, in: Proc. 6th Int. Symp. Resilient Control Systems, 2013, pp. 54–59.
[27] J. Stankovic, R. Rajkumar, Real-time operating systems, Real-Time Systems 28 (2–3) (2004) 237–253.
[28] K.J. Park, R. Zheng, X. Liu, Cyber-physical systems: milestones and research challenges, Computer Communications 36 (1) (2012) 1–7.
[29] R. Rajkumar, I. Lee, L. Sha, J. Stankovic, Cyber-physical systems: the next computing revolution, in: Int. Conf. Design Automation Conference, Anaheim, California, USA, 2010.
[30] A. Petrovski, P. Rattadilok, S. Petrovski, Designing a Context-Aware Cyber Physical System for Detecting Security Threats in Motor Vehicles, ACM, 2015.
[31] P. Maheshwari, Security issues of cyber physical system: a review, International Journal of Computer Applications (2016) 7–11.
[32] S.K. Khaitan, J.D. McCalley, Design techniques and applications of cyber physical systems: a survey, IEEE Systems Journal (2014).
[33] C. Konstantinou, M. Maniatakos, F. Saqib, S. Hu, J. Plusquellic, Y. Jin, Cyber-physical systems: a security perspective, in: Proc. 20th IEEE European Test Symposium, ETS, 2015.
[34] N. Adam, Workshop on Future Directions in Cyber-Physical Systems Security, Final Report, January 2010.
[35] R. Saltzman, A. Sharabani, Active Man in the Middle Attacks, A Security Advisory, A white-paper from IBM Rational Application Security Group, February 27, 2009.
[36] K.K. Venkatasubramanian, Security Solutions for Cyber-Physical Systems, Ph.D. Dissertation, Arizona State University, December 2009.
[37] A.E. Lee, Cyber physical systems: design challenges, in: Object Oriented Real-Time Distributed Computing (ISORC), Proc. IEEE Int. Symposium, 2008, pp. 363–369.
[38] K.D. Kim, P.R. Kumar, An overview and some challenges in cyber-physical systems, Journal of the Indian Institute of Science 93 (3) (2013) 341–352.
[39] J. Slay, M. Miller, Lessons learned from the maroochy water breach, in: E. Goetz, S. Shenoi (Eds.), Critical Infrastructure Protection, ser. IFIP Int. Federation for Information Processing, Springer Boston, Boston, MA, 2007, pp. 73–82, vol. 253, Ch. 6.
[40] R. Esposito, Hackers penetrate water system computers, ABC News [Online]. Available: http://abcnews.go.com/blogs/headlines/2006/10/hackers penetra/, 2006.

[41] A. Greenberg, Hackers cut cities power, Forbes [Online]. Available: http://www.forbes.com/2008/01/18/cyber-attack-utilities-tech-intel-cx ag 0118attack.html, 2008.
[42] J. Leyden, Polish teen derails tram after hacking train network, The Register [Online]. Available: http://www.theregister.co.uk/2008/01/11/tram hack/, 2008.
[43] S.E. Schechter, J. Jung, A.W. Berger, Fast detection of scanning worm infections, in: Proc. 7th Int. Symposium on Recent Advances in Intrusion Detection, 2004, pp. 59–81.
[44] N. Falliere, L.O. Murchu, E. Chien, W32.Stuxnet dossier, Symantec [Online]. Available: http://www.symantec.com/content/en/us/enterprise/media/security response/whitepapers/w32 stuxnet dossier.pdf.
[45] R. McMillan, Siemens: Stuxnet worm hit industrial systems, Computerworld [Online]. Available: http://www.computerworld.com/s/article/9185419/Siemens Stuxnet worm hit industrial systems, 2010.
[46] S. Cherry, How stuxnet is rewriting the cyberterrorism playbook, Computerworld [Online]. Available: http://spectrum.ieee.org/podcast/telecom/security/how-stuxnet-is-rewriting-the-cyberterrorism-playbook, 2010.
[47] A. Teixeira, D. Perez, H. Sandberg, K.H. Johansson, Attack models and scenarios for networked control systems, in: Proc. 1st Int. Conference on High Confidence Networked Systems, ACM, 2012, pp. 55–64.
[48] A. Teixeira, Toward Cyber-Secure and Resilient Networked Control Systems, Ph.D. diss., KTH Royal Institute of Technology, 2014.
[49] Bo Shen, An overview of cloud-related cyber-physical systems, International Journal of Data Science and Analytics 1 (2015) 8–13.
[50] H. Sandberg, A. Saurabh, K.H. Johansson, Cyberphysical security in networked control systems: an introduction to the issue, IEEE Control Systems 35 (1) (2015) 20–23.
[51] G. Wu, J. Sun, J. Chen, A survey on the security of cyber-physical systems, Control Theory and Technology 14 (1) (2016) 2–10.
[52] E. Molina, E. Jacob, Software-defined networking in cyber-physical systems: a survey, Computers & Electrical Engineering (2017).
[53] J. Lin, W. Yu, N. Zhang, X. Yang, H. Zhang, W. Zhao, A survey on Internet of things: architecture, enabling technologies, security and privacy, and applications, IEEE Internet of Things Journal (2017).
[54] B. Bordel, R. Alcarria, T. s Robles, D. Martin, Cyber-physical systems: extending pervasive sensing from control theory to the Internet of things, Pervasive and Mobile Computing 40 (2017) 156–184.
[55] J. Giraldo, E. Sarkar, A. Cardenas, M. Maniatakos, M. Kantarcioglu, Security and privacy in cyber-physical systems: a survey of surveys, IEEE Design & Test (2017).
[56] A. Humayed, J. Lin, F. Li, Bo Luo, Cyber-physical systems security–a survey, IEEE Internet of Things Journal (2017).
[57] Y. Ashibani, Q.H. Mahmoud, Cyber physical systems security: analysis, challenges and solutions, Computers & Security 68 (2017) 81–97.
[58] D. Ding, Q.L. Han, Y. Xiang, X. Ge, X.M. Zhang, A survey on security control and attack detection for industrial cyber-physical systems, Neurocomputing 275 (2018) 1674–1683.
[59] I. Shames, F. Farokhi, T.H. Summers, Security analysis of cyber-physical systems using H_2 norm, IET Control Theory & Applications 11 (11) (2017) 1749–1755.
[60] B. Kailkhura, Y.S. Han, S. Brahma, P.K. Varshney, Asymptotic analysis of distributed Bayesian detection with byzantine data, IEEE Signal Processing Letters 22 (5) (2015) 608–612.
[61] B. Kailkhura, Y.S. Han, S. Brahma, P.K. Varshney, Distributed Bayesian detection in the presence of byzantine data, IEEE Transactions on Signal Processing 63 (19) (2015) 5250–5263.
[62] A.S. Rawat, P. Anand, H. Chen, P.K. Varshney, Collaborative spectrum sensing in the presence of byzantine attacks in cognitive radio networks, IEEE Transactions on Signal Processing 59 (2) (2011) 774–786.

[63] O. Kosut, L. Jia, R.J. Thomas, L. Tong, Malicious data attacks on the smart grid, IEEE Transactions on Smart Grid 2 (4) (2011) 645–658.
[64] Bo Tang, J. Yan, S. Kay, H. He, Detection of false data injection attacks in smart grid under colored Gaussian noise, in: IEEE Conference on Communications and Network Security (CNS), 2016, pp. 172–179.
[65] W. Meng, W. Li, C. Su, J. Zhou, R. Lu, Enhancing trust management for wireless intrusion detection via traffic sampling in the era of big data, IEEE Access (2017).
[66] G. Hug, J.A. Giampapa, Vulnerability assessment of AC state estimation with respect to false data injection cyber-attacks, IEEE Transactions on Smart Grid 3 (3) (2012) 1362–1370.
[67] L. Liu, M. Esmalifalak, Q. Ding, V.A. Emesih, Z. Han, Detecting false data injection attacks on power grid by sparse optimization, IEEE Transactions on Smart Grid 5 (2) (2014) 612–621.
[68] C.H. Lo, N. Ansari, CONSUMER: a novel hybrid intrusion detection system for distribution networks in smart grid, IEEE Transactions on Emerging Topics in Computing 1 (1) (2013) 33–44.
[69] Y. Huang, J. Tang, Y. Cheng, H. Li, K.A. Campbell, Z. Han, Real-time detection of false data injection in smart grid networks: an adaptive CUSUM method and analysis, IEEE Systems Journal 10 (2) (2016) 532–543.
[70] R. Deng, G. Xiao, R. Lu, Defending against false data injection attacks on power system state estimation, IEEE Transactions on Industrial Informatics 13 (1) (2017) 198–207.
[71] Y. Mo, R. Chabukswar, B. Sinopoli, Detecting integrity attacks on SCADA systems, IEEE Transactions on Control Systems Technology 22 (4) (2014) 1396–1407.
[72] Y. Mo, B. Sinopoli, On the performance degradation of cyber-physical systems under stealthy integrity attacks, IEEE Transactions on Automatic Control 61 (9) (2016) 2618–2624.
[73] D.B. Rawat, C. Bajracharya, Detection of false data injection attacks in smart grid communication systems, IEEE Signal Processing Letters 22 (10) (2015) 1652–1656.
[74] Z. Guo, D. Shi, K.H. Johansson, L. Shi, Optimal linear cyber-attack on remote state estimation, IEEE Transactions on Control of Network Systems 4 (1) (2017) 4–13.
[75] J. Milosevic, T. Tanaka, H. Sandberg, K.H. Johansson, Analysis and mitigation of bias injection attacks against a Kalman filter, IFAC-Papers On-Line 50 (1) (2017) 8393–8398.
[76] I.S. Thaseen, C.A. Kumar, Intrusion detection model using fusion of chi-square feature selection and multi class SVM, Journal of King Saud University: Computer and Information Sciences 29 (4) (2017) 462–472.
[77] F. Pasqualetti, F. Dorfler, F. Bullo, Attack detection and identification in cyber-physical systems, IEEE Transactions on Automatic Control 58 (11) (2013) 2715–2729.
[78] H. Nishino, H. Ishii, Distributed detection of cyber attacks and faults for power systems, IFAC Proceedings 47 (3) (2014) 11932–11937.
[79] S. Amin, X. Litrico, S.S. Sastry, A.M. Bayen, Cyber security of water SCADA systems–part II: attack detection using enhanced hydrodynamic models, IEEE Transactions on Control Systems Technology 21 (5) (2013) 1679–1693.
[80] C. Alippi, S. Ntalampiras, M. Roveri, Model-free fault detection and isolation in large-scale cyber-physical systems, IEEE Transactions on Emerging Topics in Computational Intelligence 1 (1) (2017) 61–71.
[81] M. Housh, Z. Ohar, Model based approach for cyber-physical attacks detection in water distribution systems, in: World Environmental and Water Resources Congress, 2017, pp. 727–736.
[82] A.A. Yaseen, M. Bayart, Cyber-attack detection with fault accommodation based on intelligent generalized predictive control, IFAC-Papers On-Line 50 (1) (2017) 2601–2608.
[83] N. Hoque, H. Kashyap, D.K. Bhattacharyya, Real-time DDoS attack detection using FPGA, Computer Communications 110 (2017) 48–58.
[84] M. Semerci, A.T. Cemgil, B. Sankur, An intelligent cyber security system against DDoS attacks in SIP networks, Computer Networks (2018).
[85] P. Srikantha, D. Kundur, Denial of service attacks and mitigation for stability in cyber-enabled power grid, in: IEEE Innovative Smart Grid Technologies Conference (ISGT), IEEE Power & Energy Society, 2015, pp. 1–5.

[86] H. Beitollahi, G. Deconinck, A dependable architecture to mitigate distributed denial of service attacks on network-based control systems, International Journal of Critical Infrastructure Protection 4 (3–4) (2011) 107–123.
[87] J.H. Sarker, A.M. Nahhas, Mobile RFID system in the presence of denial-of-service attacking signals, IEEE Transactions on Automation Science and Engineering 14 (2) (2017) 955–967.
[88] M. Long, C.H. Wu, J.Y. Hung, Denial of service attacks on network-based control systems: impact and mitigation, IEEE Transactions on Industrial Informatics 1 (2) (May 2005) 85–96.
[89] C. De Persis, P. Tesi, Input-to-state stabilizing control under denial-of-service, IEEE Transactions on Automatic Control 60 (11) (2015) 2930–2944.
[90] S. Amin, A.A. Cardenas, S.S. Sastry, Safe and secure networked control systems under denial-of-service attacks, in: Int. Workshop on Hybrid Systems: Computation and Control, Springer, Berlin, Heidelberg, 2009, pp. 31–45.
[91] S. Amin, G.A. Schwartz, S.S. Sastry, Security of interdependent and identical networked control systems, Automatica 49 (1) (2013) 186–192.
[92] G.K. Befekadu, V. Gupta, P.J. Antsaklis, Risk-sensitive control under Markov modulated denial-of-service (dos) attack strategies, IEEE Transactions on Automatic Control 60 (12) (2015) 3299–3304.
[93] H. Zhang, Y. Qi, H. Zhou, J. Zhang, J. Sun, Testing and defending methods against DOS attack in state estimation, Asian Journal of Control 19 (4) (2017) 1295–1305.
[94] B. Sinopoli, L. Schenato, M. Franceschetti, K. Poolla, M.I. Jordan, S.S. Sastry, Kalman filtering with intermittent observations, IEEE Transactions on Automatic Control 49 (9) (2004) 1453–1464.
[95] K. Ding, S. Dey, D.E. Quevedo, L. Shi, Stochastic game in remote estimation under DoS attacks, IEEE Control Systems Letters 1 (1) (July 2017) 146–151.
[96] Y. Wu, Y. Li, L. Shi, A game-theoretic approach to remote state estimation in presence of a DoS attacker, IFAC-Papers On-Line 50 (1) (2017) 2595–2600.
[97] L. Zhao, G.H. Yang, Adaptive sliding mode fault tolerant control for nonlinearly chaotic systems against DoS attack and network faults, Journal of the Franklin Institute 354 (15) (2017) 6520–6535.
[98] D. Ding, Z. Wang, G. Wei, F.E. Alsaadi, Event-based security control for discrete-time stochastic systems, IET Control Theory & Applications 10 (15) (2016) 1808–1815.
[99] V.S. Dolk, P. Tesi, C. De Persis, W.P.M.H. Heemels, Output-based event-triggered control systems under denial-of-service attacks, in: IEEE 54th Annual Conf. Decision and Control (CDC), 2015, pp. 4824–4829.
[100] V.S. Dolk, P. Tesi, C. De Persis, W.P.M.H. Heemels, Event-triggered control systems under denial-of-service attacks, IEEE Transactions on Control of Network Systems 4 (1) (2017) 93–105.
[101] S. Feng, P. Tesi, C. De Persis, Towards stabilization of distributed systems under denial-of-service, in: IEEE 56th Annual Conf. Decision and Control (CDC), 2017, pp. 5360–5365.
[102] H.S. Foroush, S. Martinez, On event-triggered control of linear systems under periodic denial-of-service jamming attacks, in: IEEE 51st Annual Conf. Decision and Control (CDC), 2012, pp. 2551–2556.
[103] C. De Persis, P. Tesi, Resilient control under denial-of-service, in: Proceedings of the IFAC World Conference, Cape Town, South Africa, 2014, pp. 134–139.
[104] C. De Persis, P. Tesi, On resilient control of nonlinear systems under denial-of-service, in: IEEE 53rd Annual Conf. Decision and Control (CDC), 2014, pp. 5254–5259.
[105] Q. Zhu, T. Basar, Game-theoretic methods for robustness, security, and resilience of cyber-physical control systems: games-in-games principle for optimal cross-layer resilient control systems, IEEE Control Systems 35 (1) (2015) 46–65.
[106] Z. Han, D. Niyato, W. Saad, T. Basar, A. Hjorungnes, Game Theory in Wireless and Communication Networks: Theory, Models, and Applications, Cambridge University Press, 2012.
[107] W. Saad, Z. Han, H.V. Poor, T. Basar, Game-theoretic methods for the smart grid: an overview of microgrid systems, demand-side management, and smart grid communications, IEEE Signal Processing Magazine 29 (5) (2012) 86–105.

[108] M.H. Manshaei, Q. Zhu, T. Alpcan, T. Basar, J.P. Hubaux, Game theory meets network security and privacy, ACM Computing Surveys (CSUR) 45 (3) (2013) 25–45.
[109] H. Orojloo, M.A. Azgomi, A game-theoretic approach to model and quantify the security of cyber-physical systems, Computers in Industry 88 (2017) 44–57.
[110] H. Sun, C. Peng, T. Yang, H. Zhang, W. He, Resilient control of networked control systems with stochastic denial of service attacks, Neurocomputing 270 (2017) 170–177.
[111] K. Ding, Y. Li, D.E. Quevedo, S. Dey, L. Shi, A multi-channel transmission schedule for remote state estimation under DoS attacks, Automatica 78 (2017) 194–201.
[112] Y. Yuan, H. Yuan, L. Guo, H. Yang, S. Sun, Resilient control of networked control system under DoS attacks: a unified game approach, IEEE Transactions on Industrial Informatics 12 (5) (2016) 1786–1794.
[113] Y. Yuan, F. Sun, H. Liu, Resilient control of cyber-physical systems against intelligent attacker: a hierarchical Stackelberg game approach, International Journal of Systems Science 47 (9) (2016) 2067–2077.
[114] S. Liu, P.X. Liu, A.E. Saddik, A stochastic game approach to the security issue of networked control systems under jamming attacks, Journal of the Franklin Institute 351 (9) (2014) 4570–4583.
[115] Y. Li, L. Shi, P. Cheng, J. Chen, D.E. Quevedo, Jamming attacks on remote state estimation in cyber-physical systems: a game-theoretic approach, IEEE Transactions on Automatic Control 60 (10) (2015) 2831–2836.
[116] H. Zhang, P. Cheng, L. Shi, J. Chen, Optimal DoS attack scheduling in wireless networked control system, IEEE Transactions on Control Systems Technology 24 (3) (2016) 843–852.
[117] A. Benslimane, H.N. Minh, Jamming attack model and detection method for beacons under multichannel operation in vehicular networks, IEEE Transactions on Vehicular Technology 66 (7) (2017) 6475–6488.
[118] L. Peng, X. Cao, C. Sun, Y. Cheng, S. Jin, Energy efficient jamming attack schedule against remote state estimation in wireless cyber-physical systems, Neurocomputing 272 (2018) 571–583.
[119] S. Hu, D. Yue, X. Xie, X. Chen, X. Yin, Resilient event-triggered controller synthesis of networked control systems under periodic DoS jamming attacks, IEEE Transactions on Cybernetics 99 (2018) 1–11.
[120] D. Ding, Z. Wang, Q.L. Han, G. Wei, Security control for discrete-time stochastic nonlinear systems subject to deception attacks, IEEE Transactions on Systems, Man, and Cybernetics: Systems 48 (5) (2018) 779–789.
[121] D. Ding, Z. Wang, D.W. Ho, G. Wei, Observer-based event-triggering consensus control for multiagent systems with lossy sensors and cyber-attacks, IEEE Transactions on Cybernetics 47 (8) (2017) 1936–1947.
[122] S. Amin, X. Litrico, S. Sastry, A.M. Bayen, Cyber security of water SCADA systems–part I: analysis and experimentation of stealthy deception attacks, IEEE Transactions on Control Systems Technology 21 (5) (2013) 1963–1970.
[123] J. Hao, R.J. Piechocki, D. Kaleshi, W.H. Chin, Z. Fan, Sparse malicious false data injection attacks and defense mechanisms in smart grids, IEEE Transactions on Industrial Informatics 11 (5) (2015) 1–12.
[124] J. Wang, Y. Song, S. Liu, S. Zhang, Security in H_2-sense for polytopic uncertain systems with attacks based on model predictive control, Journal of the Franklin Institute 353 (15) (2016) 3769–3785.
[125] L. Ma, Z. Wang, Q.L. Han, H.K. Lam, Variance-constrained distributed filtering for time-varying systems with multiplicative noises and deception attacks over sensor networks, IEEE Sensors Journal 17 (7) (2017) 2279–2288.
[126] X. Liu, Z. Li, False data attack models, impact analyses and defense strategies in the electricity grid, The Electricity Journal 30 (4) (2017) 35–42.
[127] Y. Mo, E. Garone, A. Casavola, B. Sinopoli, False data injection attacks against state estimation in wireless sensor networks, in: IEEE 49th Conf. Decision and Control (CDC), 2010, pp. 5967–5972.

[128] Y. Liu, P. Ning, M.K. Reiter, False data injection attacks against state estimation in electric power grids, ACM Transactions on Information and System Security (TISSEC) 14 (1) (2011) 13–19.

[129] Z. Wang, D. Wang, B. Shen, F.E. Alsaadi, Centralized security-guaranteed filtering in multirate-sensor fusion under deception attacks, Journal of the Franklin Institute 355 (1) (2018) 406–420.

[130] W. Yang, L. Lei, C. Yang, Event-based distributed state estimation under deception attack, Neurocomputing 270 (2017) 145–151.

[131] N. Forti, G. Battistelli, L. Chisci, B. Sinopoli, Joint attack detection and secure state estimation of cyber-physical systems, preprint, arXiv:1612.08478, 2016.

[132] S. Mishra, Y. Shoukry, N. Karamchandani, S.N. Diggavi, P. Tabuada, Secure state estimation against sensor attacks in the presence of noise, IEEE Transactions on Control of Network Systems 4 (1) (2017) 49–59.

[133] L. Hu, Z. Wang, Q.L. Han, X. Liu, State estimation under false data injection attacks: security analysis and system protection, Automatica 87 (2018) 176–183.

[134] D. Wang, Z. Wang, B. Shen, F.E. Alsaadi, Security-guaranteed filtering for discrete-time stochastic delayed systems with randomly occurring sensor saturations and deception attacks, International Journal of Robust and Nonlinear Control 27 (7) (2017) 1194–1208.

[135] Z.H. Pang, G.P. Liu, Design and implementation of secure networked predictive control systems under deception attacks, IEEE Transactions on Control Systems Technology 20 (5) (2012) 1334–1342.

[136] L. Ma, Z. Wang, Y. Yuan, Consensus control for nonlinear multi-agent systems subject to deception attacks, in: IEEE 22nd Int. Conf. Automation and Computing (ICAC), 2016, pp. 21–26.

[137] T. Rhouma, K. Chabir, M.N. Abdelkrim, Resilient control for networked control systems subject to cyber/physical attacks, International Journal of Automation and Computing 15 (3) (2018) 345–354.

[138] X. Jin, W.M. Haddad, T. Yucelen, An adaptive control architecture for mitigating sensor and actuator attacks in cyber-physical systems, IEEE Transactions on Automatic Control 62 (11) (2017) 6058–6064.

[139] L. An, G.H. Yang, Improved adaptive resilient control against sensor and actuator attacks, Information Sciences 423 (2018) 145–156.

[140] X. Huang, D. Zhai, J. Dong, Adaptive integral sliding-mode control strategy of data-driven cyber-physical systems against a class of actuator attacks, IET Control Theory & Applications (2018).

[141] X. Huang, J. Dong, Reliable control policy of cyber-physical systems against a class of frequency-constrained sensor and actuator attacks, IEEE Transactions on Cybernetics (2018) 3432–3439.

[142] Y. Mo, B. Sinopoli, Secure control against replay attacks, in: IEEE 47th Annual Allerton Conf. Communication, Control, and Computing, 2009, pp. 911–918.

[143] Y. Mo, B. Sinopoli, Secure estimation in the presence of integrity attacks, IEEE Transactions on Automatic Control 60 (4) (2015) 1145–1151.

[144] H. Beikzadeh, H.J. Marquez, Multirate observers for nonlinear sampled-data systems using input-to-state stability and discrete-time approximation, IEEE Transactions on Automatic Control 59 (9) (2014) 2469–2474.

[145] H. Tan, B. Shen, Y. Liu, A. Alsaedi, B. Ahmad, Event-triggered multi-rate fusion estimation for uncertain system with stochastic nonlinearities and colored measurement noises, Information Fusion 36 (2017) 313–320.

[146] W. Chen, L. Qiu, Stabilization of networked control systems with multirate sampling, Automatica 49 (6) (2013) 1528–1537.

[147] W. Zhang, M.S. Branicky, S.M. Phillips, Stability of networked control systems, IEEE Control Systems 21 (1) (Feb 2001) 84–99.

[148] J. Hespanha, P. Naghshtabrizi, Y. Xu, A survey of recent results in networked control systems, Proceedings of the IEEE 95 (1) (Jan. 2007) 138–162.
[149] R.A. Gupta, M.Y. Chow, Networked control system: overview and research trends, IEEE Transactions on Industrial Electronics 57 (7) (July 2010) 2527–2535.
[150] L. Zhang, H. Gao, O. Kaynak, Network-induced constraints in networked control systems–a survey, IEEE Transactions on Industrial Informatics 9 (1) (Feb. 2013) 403–416.
[151] C. Zhang, Z. Cai, W. Chen, X. Luo, J. Yin, Flow level detection and filtering of low-rate DDoS, Computer Networks 56 (15) (2012) 3417–3431.
[152] Kaspersky, Kaspersky Internet security & anti-virus, http://www.kaspersky.com/, 2012, Russian Federation.
[153] C. Douligeris, A. Mitrokotsa, DDos attacks and defense mechanisms: classification and state-of-the-art, Computer Networks 44 (2004) 643–666.
[154] K.M. Prasad, A.R. Mohan, K.V. Rao, Dos and DDoS attacks: defense, detection and traceback mechanisms-a survey, Global Journal of Computer Science and Technology 14 (7) (2014).
[155] S. Specht, R.B. Lee, Distributed denial of service: Taxonomies of attacks, tools, and countermeasures, in: Proc. ISCA 17th Int. Conf. Parallel and Distributed Computing Systems (ISCA), San Francisco, USA, 15–17 September 2004, pp. 543–550.
[156] J.R. Collins, RAMEN - a Linux Worm, http://www.giac.org/paper/gsec/505/ramenlinux-worm/101193, SANS Institute, Maryland, USA, 2000.
[157] CERT, CERT Coordination Center, CERT Advisory CA-2001-19 'Code Red' Worm Exploiting Buffer Overflow in IIS Indexing Service DLL, http://www.cert.org/adviso-ries/CA-2001-19.html, Carnegie Mellon Software Engineering Institute, Pittsburgh, USA, 2001.
[158] J.J. Saman, D. Tipper, A survey of defense mechanisms against distributed denial of service (DDoS) flooding attacks, IEEE Communications Surveys and Tutorials 15 (4) (2013) 2046–2069.
[159] Y. Xie, S.-Z. Yu, A large-scale hidden semi-Markov model for anomaly detection on user browsing behaviors, IEEE/ACM Transactions on Networking 17 (1) (2009) 54–65.
[160] Y. Xie, S.-Z. Yu, Monitoring the application-layer DDoS attacks for popular websites, IEEE/ACM Transactions on Networking 17 (1) (2009) 15–25.
[161] N. Woolf, Retrieved from https://www.theguardian.com/technology/2016/oct/26/ddos-attack-dynmirai-botnet, October 26, 2016.
[162] T. Subramaniam, D. Bethany, Preventing distributed denial of service attacks in cloud environments, International Journal of Information Technology, Control and Automation 6 (2016) 23–32, https://doi.org/10.5121/ijitca.2016.6203.
[163] S. Sivamohan, R. Veeramani, K. Liza, S. Krishnaveni, B. Jothi, Data mining technique for DDoS attack in cloud computing, International Journal of Computer Technology and Applications 9 (2016) 149–156.
[164] M. Masdari, J. Marzie, A survey and taxonomy of DoS attacks in cloud computing, Security and Communication Networks 2 (2016) 3274–3751, https://doi.org/10.1002/sec.1539.
[165] A. Bonquet, B. Martine, A survey of denial-of-service and distributed denial of service attacks and defense in cloud computing, Future Internet 9 (2017) 1–9, https://doi.org/10.3390/fi9030043.
[166] A. Kaur, K. Anupama, A review on various attack detection techniques in cloud architecture, International Journal of Advanced Research in Computer Engineering & Technology 4 (2015) 3861–3867.
[167] S.G. Kene, P.T. Deepti, A review on intrusion detection techniques for cloud computing and security challenges, in: 2nd International Conference on Electronics and Communication Systems, Coimbatore, 26–27 February 2015, vol. 2, 2015, pp. 227–231.
[168] R.V. Deshmukh, K.D. Kailas, Understanding DDoS attack & its effect in cloud environment, Procedia Computer Science 49 (2015) 202–210, https://doi.org/10.1016/j.procs.2015.04.245.
[169] I. Sattar, et al., A review of techniques to detect and prevent distributed denial of service (DDoS) attack in cloud computing environment, International Journal of Computer Applications 115 (2015) 23–27.

[170] S. Navaz, Entropy based anomaly detection system to prevent DDoS attacks in cloud, International Journal of Computer Applications 15 (2013) 42–47.
[171] P. Ankita, K. Fenil, Survey on DDoS attack detection and prevention in cloud, International Journal of Engineering Technology, Management, and Applied Sciences 3 (2015) 43–47.
[172] C. Modi, P. Dhiren, B. Bhavesh, P. Avi, R. Muttukrishnan, A survey on security issues and solutions at different layers of cloud computing, Journal of Supercomputing 63 (2013) 561–592.
[173] Research Nielsen, Downtime Costs Auto Industry, 2006.
[174] A. Willig, K. Matheus, A. Wolisz, Wireless technology in industrial networks, Proceedings of the IEEE 93 (6) (2005) 1130–1151.
[175] W. Zhang, M. Branicky, Stability of networked control systems with time-varying transmission period, in: Allerton Conference on Communication, Control, and Computing, 2001.
[176] A. Saifullah, Y. Xu, C. Lu, Y. Chen, Real-time scheduling for WirelessHART networks, in: The 31st IEEE Real-Time Systems Symposium, 2010, pp. 150–159.
[177] M. Pajic, R. Mangharam, Embedded virtual machines for robust wireless control and actuation, in: RTAS'10: Proceedings of the 16th IEEE Real-Time and Embedded Technology and Applications Symposium, 2010, pp. 79–88.
[178] R. Alur, A. D'Innocenzo, K.H. Johansson, G.J. Pappas, G. Weiss, Compositional modeling and analysis of multi-hop control networks, IEEE Transactions on Automatic Control 56 (10) (2011) 2345–2357.
[179] G. Fiore, V. Ercoli, A. Isaksson, K. Landernas, M.D. Di Benedetto, Multi-hop multi-channel scheduling for wireless control in WirelessHART networks, in: IEEE Conference on Emerging Technology & Factory Automation, 2009, pp. 1–8.
[180] A. D'Innocenzo, G. Weiss, R. Alur, A. Isaksson, K. Johansson, G. Pappas, Scalable scheduling algorithms for wireless networked control systems, in: CASE'09: IEEE Int. Conf. Automation Science and Engineering, 2009, pp. 409–414.
[181] S. Graham, G. Baliga, P. Kumar, Abstractions, architecture, mechanisms, and a middleware for networked control, IEEE Transactions on Automatic Control 54 (7) (2009) 1490–1503.
[182] J. Hill, R. Szewczyk, A. Woo, S. Hollar, D. Culler, K. Pister, System architecture directions for networked sensors, SIGPLAN Notices 35 (11) (2000) 93–104.
[183] H. Kopetz, G. Bauer, The time-triggered architecture, Proceedings of the IEEE 91 (1) (2003) 112–126.
[184] R. Alur, A. D'Innocenzo, K.H. Johansson, G.J. Pappas, G. Weiss, Modeling and analysis of multi-hop control networks, in: RTAS '09: Proceedings of the 2009 15th IEEE Symposium on Real-Time and Embedded Technology and Applications, 2009, pp. 223–232.
[185] M. Welsh, G. Mainland, Programming sensor networks using abstract regions, in: NSDI'04: Proceedings of the 1st Conf. Symposium on Networked Systems Design and Implementation, 2004.
[186] C. Robinson, P. Kumar, Optimizing controller location in networked control systems with packet drops, IEEE Journal on Selected Areas in Communications 26 (4) (2008) 661–671.
[187] P. Jalote, Fault Tolerance in Distributed Systems, Prentice-Hall, Inc., 1994.
[188] P.A. Lee, T. Anderson, Fault tolerance - principles and practice, in: J.C. Laprie, A. Avizienis, H. Kopetz (Eds.), Springer Verlag, 1990.
[189] M. Pajic, A. Chernoguzov, R. Mangharam, Robust architectures for embedded wireless network control and actuation, ACM Transactions on Embedded Computing Systems 11 (4) (2012) 82.
[190] M. Pajic, S. Sundaram, G.J. Pappas, R. Mangharam, The wireless control network: a new approach for control over networks, IEEE Transactions on Automatic Control 56 (10) (2011) 2305–2318.
[191] C.N. Hadjicostis, R. Touri, Feedback control utilizing packet dropping network links, in: Proceedings of the 41st IEEE Conference on Decision and Control, 2002, pp. 1205–1210.
[192] M. Pajic, S. Sundaram, G.J. Pappas, R. Mangharam, Topological conditions for wireless control networks, in: Proceedings of the 50th IEEE Conference on Decision and Control, 2011, pp. 2353–2360.

[193] S. Sundaram, M. Pajic, C. Hadjicostis, R. Mangharam, G. Pappas, The wireless control network: monitoring for malicious behavior, in: Proceedings of the 49th IEEE Conference on Decision and Control, 2010, pp. 5979–5984.

[194] P. Levis, D. Culler, Mate: a tiny virtual machine for sensor networks, SIGARCH Computer Architecture News 30 (5) (2002) 85–95.

[195] P. Stanley-Marbell, L. Iftode, Scylla: a smart virtual machine for mobile embedded systems, in: WMCSA '00: Proceedings of the 3rd IEEE Workshop on Mobile Computing Systems and Applications, 2000, pp. 41–50.

[196] R. Muller, G. Alonso, D. Kossmann, A virtual machine for sensor networks, in: EuroSys '07: Proceedings of the 2nd ACM SIGOPS/EuroSys European Conference on Computer Systems, 2007, pp. 145–158.

[197] C.-C. Han, R. Kumar, R. Shea, E. Kohler, M. Srivastava, A dynamic operating system for sensor nodes, in: MobiSys '05: Proceedings of the 3rd Int. Conf. Mobile Systems, Applications, and Services, ACM, 2005, pp. 163–176.

[198] A. Dunkels, B. Gronvall, T. Voigt, Contiki - a lightweight and flexible operating system for tiny networked sensors, in: LCN '04: Proceedings of the 29th Annual IEEE Int. Conf. Local Computer Networks, 2004, pp. 455–462.

[199] S. Bhatti, J. Carlson, H. Dai, J. Deng, J. Rose, A. Sheth, B. Shucker, C. Gruenwald, A. Torgerson, R. Han, MANTIS OS: an embedded multi-threaded operating system for wireless micro sensor platforms, Mobile Networks and Applications 10 (4) (2005) 563–579.

[200] K. Lorincz, B.-r. Chen, J. Waterman, G. Werner-Allen, M. Welsh, Resource aware programming in the pixie OS, in: SenSys '08: Proceedings of the 6th ACM Conf. Embedded Network Sensor Systems, ACM, 2008, pp. 211–224.

[201] Q. Cao, T. Abdelzaher, J. Stankovic, T. He, The LiteOS operating system: towards Unix-like abstractions for wireless sensor networks, in: Proc. 7th ACM/IEEE Int. Conf. Information Processing in Sensor Networks, IPSN'08, 2008, pp. 233–244.

[202] M. Brown, S. Gilbert, N. Lynch, C. Newport, T. Nolte, M. Spindel, The virtual node layer: a programming abstraction for wireless sensor networks, SIGBED Review 4 (3) (2007) 7–12.

[203] R. Newton, G. Morrisett, M. Welsh, The regiment macroprogramming system, in: Proceedings of the 6th ACM/IEEE Int. Conf. Information Processing in Sensor Networks, IPSN'07, 2007, pp. 489–498.

[204] R. Gummadi, O. Gnawali, R. Govindan, Macro-programming wireless sensor networks using Kairos, in: Distributed Computing in Sensor Systems, Springer, Berlin, 2005, pp. 126–140.

[205] K. Gatsis, M. Pajic, A. Ribeiro, G.J. Pappas, Power-aware communication for wireless sensor-actuator systems, in: Proceedings of the 52th IEEE Conf. Decision and Control, 2013.

[206] V. Gupta, A.F. Dana, J. Hespanha, R.M. Murray, B. Hassibi, Data transmission over networks for estimation and control, IEEE Transactions on Automatic Control 54 (8) (2009) 1807–1819.

[207] M. Pajic, S. Sundaram, G.J. Pappas, Stabilizability over deterministic relay networks, in: Proceedings of the 52th IEEE Conf. Decision and Control, 2013.

[208] E.K. Conklin, E.D. Rather, FORTH Programmer's Handbook, FORTH Inc., 2007.

[209] M. Pajic, R. Mangharam, Embedded Virtual Machines, Tech. Rep., University of Pennsylvania, Sept. 2009.

[210] Simulink Documentation, MathWorks, 2012.

[211] NanoRK, Sensor RTOS – http://www.nanork.org, 2010.

[212] A. Rowe, R. Mangharam, R. Rajkumar, RT-link: a global time synchronized link protocol for sensor networks, Ad Hoc Networks 6 (8) (2008) 1201–1220.

[213] A. Schrijver, Theory of Linear and Integer Programming, John Wiley & Sons, 1998.

[214] HART Field Communication Protocol Specification, Rev 7, 2007.

[215] A. Cervin, J. Eker, B. Bernhardsson, K.E. Arzen, Feedback feedforward scheduling of control tasks, Real-Time Systems Journal 23 (1–2) (2002) 25–53.

[216] Z. Fu, Y. Mahajan, S. Malik, New features of SAT'04 version of zChaff, in: The Int. Conf. Theory and Applications of Satisfiability Testing, 2004.

[217] T. Bhme, F. Gring, J. Harant, Menger's theorem, Journal of Graph Theory 37 (1) (2001) 35–36.
[218] B. Yang, S. Zheng, E. Lu, Finding two disjoint paths in a network with α^+-min-sum objective function, in: Algorithms and Computation, in: Lecture Notes in Computer Science, 2005, pp. 954–963.
[219] J. Liu, Real-Time Systems, Prentice-Hall, Inc., 2000.
[220] L. Sha, R. Rajkumar, J. Lehoczky, K. Ramamritham, Mode change protocols for priority-driven preemptive scheduling, Real-Time Systems Journal 1 (3) (1989) 126–140.
[221] J. Real, A. Crespo, Mode change protocols for real-time systems: a survey and a new proposal, Real-Time Systems Journal 26 (2) (2004) 161–197.
[222] R. Mangharam, A. Rowe, R. Rajkumar, FireFly: a cross-layer platform for real-time embedded wireless networks, Real-Time Systems Journal 37 (3) (2007) 183–231.
[223] EVM website – http://mlab.seas.upenn.edu/evm, 2009.
[224] D.R. Lewin, Using Process Simulators in Chemical Engineering: A Multimedia Guide for the Core Curriculum, Wiley, 2009.
[225] D. Prett, M. Morari, The Shell Process Control Workshop, Butterworths, 1986.
[226] P. Seiler, R. Sengupta, Analysis of communication losses in vehicle control problems, in: Proceedings American Control Conference, 2001, pp. 1491–1496.
[227] N. Elia, Remote stabilization over fading channels, Systems & Control Letters 54 (3) (2005) 237–249.
[228] T. Schmid, P. Dutta, M.B. Srivastava, High-resolution, low-power time synchronization an oxymoron no more, in: Proceedings of the 9th ACM/IEEE Int. Conf. Information Processing in Sensor Networks, IPSN'10, 2010, pp. 151–161.
[229] Why WirelessHART?, White Paper, HART Communication Foundation, 2007.
[230] ISA100.11a: Wireless systems for industrial automation, process control and related applications, Standard, 2009.
[231] R.E. Skelton, T. Iwasaki, K. Grigoriadis, A Unified Algebraic Approach to Linear Control Design, CRC Press, 1998.
[232] J. Han, R. Skelton, An LMI optimization approach for structured linear controllers, in: Proc. 42nd IEEE Conf. Decision and Control, 2003, pp. 5143–5148.
[233] L. El Ghaoui, F. Oustry, M. Ait Rami, A cone complementarity linearization algorithm for static output-feedback and related problems, IEEE Transactions on Automatic Control 42 (8) (1997) 1171–1176.
[234] P. Antsaklis, A. Michel, Linear Systems, McGraw Hill, 1997.
[235] K.S. Pister, L. Doherty, Tsmp: time synchronized mesh protocol, in: Int. Symposium on Distributed Sensor Networks (DSN), 2008, pp. 391–398.
[236] S. Skogestad, I. Postlethwaite, Multivariable Feedback Control: Analysis and Design, Wiley, 1996.
[237] Real-time windows target - run simulink models on a PC in real time, http://www.mathworks.com/products/rtwt.MathWorks.
[238] CVX: Matlab Software for Disciplined Convex Programming, Version 2.0, http://cvxr.com/cvx, CVX Research, Inc., 2012.
[239] N. Sandell, P. Varaiya, M. Athans, M. Safonov, Survey of decentralized control methods for large scale systems, IEEE Transactions on Automatic Control 23 (2) (1978) 108–128.
[240] P. Cheng, A. Datta, L. Shi, B. Sinopoli (Eds.), Special Issue on Secure Control of Cyber Physical Systems, IEEE Transactions on Control of Network Systems (2017).
[241] H. Fawzi, P. Tabuada, S. Diggavi, Secure state-estimation for dynamical systems under active adversaries, in: Annual Allerton Conference on Communication, Control, and Computing, 2011.
[242] A. Teixeira, I. Shames, H. Sandberg, K.H. Johansson, A secure control framework for resource-limited adversaries, Automatica 51 (2015) 135–148.
[243] W. Xu, W. Trappe, Y. Zhang, T. Wood, The feasibility of launching and detecting jamming attacks in wireless networks, in: Proc. 6th ACM International Symposium on Mobile Ad Hoc Networking and Computing, ACM, 2015, pp. 46–57.

[244] X. Wang, M.D. Lemmon, Event-triggering in distributed networked control systems, IEEE Transactions on Automatic Control 56 (3) (2011) 586–601.
[245] M. Guinaldo, D.V. Dimarogonas, D. Lehmann, K.H. Johansson, Distributed event-based control for interconnected linear systems, in: Asynchronous Control for Networked Systems, Springer, 2015, pp. 149–179.
[246] C. De Persis, R. Sailer, F. Wirth, On a small-gain approach to distributed event-triggered control, IFAC Proceedings Volumes 44 (1) (2011) 2401–2406.
[247] C. De Persis, R. Sailer, F. Wirth, Event-triggering of large-scale systems without Zeno behavior, in: Proc. 20th International Symposium on Mathematical Theory of Networks and Systems MTNS2012, 2012.
[248] C. De Persis, R. Sailer, F. Wirth, On inter-sampling times for event-triggered large-scale linear systems, in: IEEE 52nd Annual Conference Decision and Control (CDC), 2013, pp. 5301–5306.
[249] C. De Persis, R. Sailer, F. Wirth, Parsimonious event-triggered distributed control: a Zeno free approach, Automatica 49 (7) (2013) 2116–2124.
[250] T. Liu, Z.-P. Jiang, D.J. Hill, Decentralized output-feedback control of large-scale nonlinear systems with sensor noise, Automatica 48 (10) (2012) 2560–2568.
[251] A. Cetinkaya, H. Ishii, T. Hayakawa, Networked control under random and malicious packet losses, IEEE Transactions on Automatic Control 61 (2016) 1–16.
[252] A. Gupta, C. Langbort, T. Basar, Optimal control in the presence of an intelligent jammer with limited actions, in: Proc. of the 49th IEEE Conference on Decision and Control, Atlanta, GA, 2010, pp. 1096–1101.
[253] V. Ugrinovskii, C. Langbort, Control over adversarial packet dropping communication networks revisited, arXiv:1403.5641, 2014.
[254] H. Shisheh Foroush, S. Martínez, On multi-input controllable linear systems under unknown periodic dos jamming attacks, in: SIAM Conference on Control and Its Applications, San Diego, CA, 2013.
[255] V. Dolk, P. Tesi, C. De Persis, W. Heemels, Event-triggered control systems under denial-of-service attacks, in: Proc. of the 54th IEEE Conference on Decision and Control, Osaka, Japan, 2015.
[256] S. Feng, P. Tesi, Resilient control under denial-of-service: robust design, Automatica 79 (2017) 42–51.
[257] M. Mazo, P. Tabuada, Decentralized event-triggered control over wireless sensor/actuator networks, IEEE Transactions on Automatic Control 56 (10) (2011) 2456–2461.
[258] S. Dashkovskiy, H. Ito, F. Wirth, On a small gain theorem for ISS networks in dissipative Lyapunov form, European Journal of Control 17 (4) (2011) 357–365.
[259] F. Forni, S. Galeani, D. Nesic, L. Zaccarian, Lazy sensors for the scheduling of measurement samples transmission in linear closed loops over networks, in: IEEE Conference on Decision and Control and European Control Conference, Atlanta, USA, 2010.
[260] X. Wang, M.D. Lemmon, Event-triggered broadcasting across distributed networked control systems, in: IEEE American Control Conference, 2008, pp. 3139–3144.
[261] S. Yin, X. Li, H. Gao, O. Kaynak, Data-based techniques focused on modern industry: an overview, IEEE Transactions on Industrial Electronics 62 (1) (Jan. 2015) 657–667.
[262] Z.H. Pang, G.P. Liu, D.H. Zhou, D.H. Sun, Data-based predictive control for networked nonlinear systems with network-induced delay and packet dropout, IEEE Transactions on Industrial Electronics 63 (2) (Feb. 2016) 1249–1257.
[263] J. Qiu, H. Gao, S.X. Ding, Recent advances on fuzzy-model-based nonlinear networked control systems: a survey, IEEE Transactions on Industrial Electronics 63 (2) (Feb. 2016) 1207–1217.
[264] D. Dzung, et al., Security for industrial communication systems, Proceedings of the IEEE 93 (6) (Jun. 2005) 1152–1177.
[265] W. Zeng, M.Y. Chow, A reputation-based secure distributed control methodology in D-NCS, IEEE Transactions on Industrial Electronics 61 (11) (Nov. 2014) 6294–6303.

[266] S. Huang, C.J. Zhou, S.H. Yang, Y.Q. Qin, Cyber-physical system security for networked industrial processes, International Journal of Automation and Computing 12 (6) (Dec. 2015) 567–578.

[267] A.A. Cardenas, S. Amin, S.S. Sastry, Secure control: towards survivable cyber-physical systems, in: Proc. 28th Int. Conf. Distrib. Comput. Syst. Workshop, 2008, pp. 495–500.

[268] Z.H. Pang, G.P. Liu, Z. Dong, Secure networked control systems under denial of service attacks, in: Proc. 18th IFAC World Congr., 2011, pp. 8908–8913.

[269] Y. Yuan, F. Sun, Q. Zhu, Resilient control in the presence of DoS attack: switched system approach, International Journal of Control, Automation, and Systems 13 (6) (Dec. 2015) 1423–1435.

[270] Z. Gao, C. Cecati, S.X. Ding, A survey of fault diagnosis and fault-tolerant techniques–part I: fault diagnosis with model-based and signal based approaches, IEEE Transactions on Industrial Electronics 62 (6) (Jun. 2015) 3757–3767.

[271] S. Yin, G. Wang, H. Gao, Data-driven process monitoring based on modified orthogonal projections to latent structures, IEEE Transactions on Control Systems Technology 24 (4) (October 2015) 1–8.

[272] G. Wang, S. Yin, Quality-related fault detection approach based on orthogonal signal correction and modified PLS, IEEE Transactions on Industrial Informatics 11 (2) (Apr. 2015) 398–405.

[273] Y. Mo, B. Sinopoli, False data injection attacks in control systems, in: Proc. 1st Workshop Secure Control Syst., 2010, pp. 1–6.

[274] K. Manandhar, X. Cao, F. Hu, Y. Liu, Detection of faults and attacks including false data injection attacks in smart grid using Kalman filter, IEEE Transactions on Control of Network Systems 1 (4) (Dec. 2014) 370–379.

[275] R. Niu, L. Huie, System state estimation in the presence of false information injection, in: Proc. IEEE Stat. Signal Process. Workshop, 2012, pp. 385–388.

[276] A. Teixeira, S. Amin, H. Sandberg, K.H. Johansson, S.S. Sastry, Cyber security analysis of state estimators in electric power systems, in: Proc. 49th IEEE Conf. Decision Control, 2010, pp. 5991–5998.

[277] C. Kwon, W. Liu, I. Hwang, Security analysis for cyber-physical systems against stealthy deception attacks, in: Proc. Amer. Control Conf., 2013, pp. 3344–3349.

[278] Z.H. Pang, G.P. Liu, D.H. Zhou, M.Y. Chen, Output tracking control for networked systems: a model-based prediction approach, IEEE Transactions on Industrial Electronics 61 (9) (Sep. 2014) 4867–4877.

[279] Y. Mo, T.H.-J. Kim, K. Brancik, D. Dickinson, H. Lee, A. Perrig, B. Sinopoli, Cyber-physical security of a smart grid infrastructure, Proceedings of the IEEE 100 (1) (Jan. 2012) 195–209.

[280] L. Schenato, B. Sinopoli, M. Franceschetti, K. Poolla, S. Sastry, Foundations of control and estimation over lossy networks, Proceedings of the IEEE 95 (1) (Jan. 2007) 163–187.

[281] Y. Liu, M. Reiter, P. Ning, False data injection attacks against state estimation in electric power grids, in: Proc. ACM Conf. Computer and Commun. Security, Chicago, IL, USA, 2009.

[282] G. Dan, H. Sandberg, Stealth attacks and protection schemes for state estimators in power systems, in: Proc. IEEE Int. Conf. Smart Grid Commun., Gaithersburg, MD, 2010, pp. 214–219.

[283] Y. Fujita, T. Namerikawa, K. Uchida, Cyber attack detection and faults diagnosis in power networks by using state fault diagnosis matrix, in: Proc. Eur. Control Conf., Zurich, Switzerland, 2013.

[284] W. Xu, K. Ma, W. Trappe, Y. Zhang, Jamming sensor networks: attack and defense strategies, IEEE Network 20 (3) (May-Jun. 2006) 41–47.

[285] D. Thuente, M. Acharya, Intelligent jamming in wireless networks with applications to 802.11b and other networks, in: Proc. 25th IEEE Commun. Soc. Military Commun. Conf., Washington, DC, USA, 2006.

[286] B. De Bruhl, P. Tague, Digital filter design for jamming mitigation in 802.15.4 communication, in: Proc. Int. Conf. Comput. Commun. Netw., Maui, HI, 2011.

[287] P. Tague, M. Li, R. Poovendran, Mitigation of control channel jamming under node capture attacks, IEEE Transactions on Mobile Computing 8 (9) (Sep. 2009) 1221–1234.
[288] G. Zhai, B. Hu, K. Yasuda, A. Michel, Stability analysis of switched systems with stable and unstable subsystems: an average dwell time approach, in: Proc. Amer. Control Conf., Chicago, IL, 2000, pp. 200–205.
[289] W. Zhang, L. Yu, Modelling and control of networked control systems with both network-induced delay and packet-dropout, Automatica 44 (2008) 3206–3210.
[290] X. Sun, G. Liu, D. Rees, W. Wang, Stability of systems with controller failure and time-varying delay, IEEE Transactions on Automatic Control 53 (10) (Oct. 2008) 2391–2396.
[291] W. Zhang, L. Yu, Stabilization of sampled-data control systems with control inputs missing, IEEE Transactions on Automatic Control 55 (2) (Feb. 2010) 447–452.
[292] P. Tabuada, Event-triggered real-time scheduling of stabilizing control tasks, IEEE Transactions on Automatic Control 52 (9) (Sep. 2007) 1680–1685.
[293] G. Befekadu, V. Gupta, P. Antsaklis, Risk-sensitive control under a class denial-of-service attack models, in: Proc. Amer. Control Conf., San Francisco, CA, USA, 2011.
[294] G. Befekadu, V. Gupta, P. Antsaklis, Risk-sensitive control under a Markov modulated denial-of-service attack model, in: Proc. IEEE Conf. Decision Control and Eur. Control Conf., Orlando, FL, 2011, pp. 5714–5719.
[295] E. Sontag, Input to state stability: basic concepts and results, in: Nonlinear and Optimal Control Theory, in: Lecture Notes in Math., vol. 1932, 2008, pp. 163–220.
[296] T. Strom, On logarithmic norms, SIAM Journal on Numerical Analysis 12 (1975) 741–753.
[297] J. Hespanha, A. Morse, Stability of switched systems with average dwell-time, in: Proc. 38th IEEE CDC, Orlando, FL, 1999, pp. 2655–2660.
[298] J. Lunze, D. Lehmann, A state-feedback approach to event-based control, Automatica 46 (2010) 211–215.
[299] W. Heemels, K. Johansson, P. Tabuada, An introduction to event-triggered and self-triggered control, in: Proc. 51th IEEE Conf. Decision and Control, Maui, HI, 2012, pp. 3270–3285.
[300] M. Abdelrahim, R. Postoyan, J. Daafouz, D. Nesic, Stabilization of nonlinear systems using event-triggered output feedback laws, in: Proc. 21th Int. Symp. Math. Theory of Netw. Syst., 2014.
[301] M. Mazo, A. Anta, P. Tabuada, An ISS self-triggered implementation of linear controllers, Automatica 46 (2010) 1310–1314.
[302] G. Seyboth, D. Dimarogonas, K. Johansson, Event-based broadcasting for multi-agent average consensus, Automatica 49 (2013) 245–252.
[303] C. Stocker, J. Lunze, Distributed event-based control of physically interconnected systems, in: Proc. IEEE Conf. Decision Control, Florence, Italy, 2013, pp. 7376–7383.
[304] C. De Persis, P. Frasca, Robust self-triggered coordination with ternary controllers, IEEE Transactions on Automatic Control 58 (12) (Dec. 2013) 3024–3038.
[305] G. Hardy, J. Littlewood, G. Polya, Inequalities, Cambridge Univ. Press, Cambridge, U.K., 1952.
[306] M.S. Mahmoud, M.M. Hamdan, U.A. Baroudi, Modeling and control of cyber-physical systems subject to cyber attacks: a survey of recent advances and challenges, Neurocomputing 338 (1) (January 2019) 101–115.
[307] M. Kogel, R. Findeisen, Distributed control of interconnected systems with lossy communication networks, IFAC Proceedings Volumes 46 (27) (Jan 2013) 363–368.
[308] W. Si, X. Dong, F. Yang, Decentralized adaptive neural control for interconnected stochastic nonlinear delay-time systems with asymmetric saturation actuators and output constraints, Journal of the Franklin Institute 355 (1) (Jan. 2018) 54–80.
[309] C.H. Xie, G.Y. Yang, Observer-based attack-resilient control for linear systems against FDI attacks on communication links from controller to actuators, International Journal of Robust and Nonlinear Control 28 (15) (Oct. 2018) 4382–4403.
[310] J. Liu, E. Tian, X. Xie, H. Lin, Distributed event-triggered control for networked control systems with stochastic cyber-attacks, Journal of the Franklin Institute (March 2018).

[311] L. Zha, E. Tian, X. Xie, Z. Gu, J. Cao, Decentralized event-triggered H_∞ control for neural networks subject to cyber-attacks, Information Sciences 457 (August 2018) 141–155.
[312] H. Yuan, Y. Xia, H. Yang, Y. Yuan, Resilient control for wireless networked control systems under DoS attack via a hierarchical game, International Journal of Robust and Nonlinear Control 28 (15) (2018) 4604–4623.
[313] Z. Song, Y. Liu, M. Tan, Robust pinning synchronization of complex cyber physical networks under mixed attack strategies, International Journal of Robust and Nonlinear Control 29 (2019) 1265–1278.
[314] A.Y. Lu, G.H. Yang, Input-to-state stabilizing control for cyber-physical systems with multiple transmission channels under denial of service, IEEE Transactions on Automatic Control 63 (6) (June 2018) 1813–1820.
[315] K.H. Johansson, The quadruple-tank process: a multivariable laboratory process with an adjustable zero, IEEE Transactions on Control Systems Technology 8 (3) (May 2000) 456–465.
[316] R.R. Rajkumar, I. Lee, L. Sha, J. Stankovic, Cyber-physical systems: the next computing revolution, in: Proc. 47th Design Automation Conference, ACM, 2010, pp. 731–736.
[317] H.H. Yuan, Y. Xia, Resilient strategy design for cyber-physical system under DoS attack over a multi-channel framework, Information Sciences 454 (2018) 312–327.
[318] Y.C. Sun, G.H. Yang, Periodic event-triggered resilient control for cyber-physical systems under denial-of-service attacks, Journal of the Franklin Institute 355 (13) (2018) 5613–5631.
[319] H. Ren, F. Deng, Mean square consensus of leader-following multi-agent systems with measurement noises and time delays, ISA Transactions 71 (2017) 76–83.
[320] Y. Xie, Z. Lin, Global leader-following consensus of a group of discrete-time neutrally stable linear systems by event-triggered bounded controls, Information Sciences 459 (2018) 302–316.
[321] Y. Cheng, V. Ugrinovskii, Event-triggered leader-following tracking control for multivariable multi-agent systems, Automatica 70 (2016) 204–210.
[322] Y. Pan, H. Werner, Z. Huang, M. Bartels, Distributed cooperative control of leader-follower multi-agent systems under packet dropouts for quadcopters, Automatica 70 (2016) 204–210.
[323] G. Wen, J. Huang, Z. Peng, Y. Yu, On pinning group consensus for heterogeneous multi-agent system with input saturation, Neurocomputing 207 (2016) 623–629.
[324] Q. Deng, J. Wu, T. Han, Q. Yang, X. Cai, Fixed-time bipartite consensus of multi-agent systems with disturbances, Neurocomputing 516 (2019) 37–49.
[325] Y. Han, W. Lu, T. Chen, Cluster consensus in discrete-time networks of multiagents with inter-cluster nonidentical inputs, IEEE Transactions on Neural Networks and Learning Systems 24 (4) (2013) 566–578.
[326] Y. Shang, Couple-group consensus of continuous-time multi-agent systems under Markovian switching topologies, Journal of the Franklin Institute 352 (11) (2015) 4826–4844.
[327] D. Xie, Q. Liu, L. Lv, S. Li, Necessary and sufficient condition for the group consensus of multi-agent systems, Applied Mathematics and Computation 243 (2014) 870–878.
[328] J. Yu, L. Wang, Group consensus in multi-agent systems with switching topologies and communication delays, Systems & Control Letters 59 (2010) 340–348.
[329] A. Lu, G. Yang, Distributed consensus control for multi-agent systems under denial of service, Information Services (2018) 95–107.
[330] C. Zhao, J. He, P. Cheng, J. Chen, Secure consensus against message manipulation attacks in synchronous networks, in: Proc. IFAC Cape Town, South Africa, 2014, pp. 1182–1187.
[331] Z. Feng, G. Hu, Distributed secure average consensus for linear multi-agent systems under DoS attacks, in: Proc. American Control Conference, 2017, pp. 2261–2268.
[332] X. Jin, W. Haddad, An adaptive control architecture for leader-follower multi-agent systems with stochastic disturbances and sensor and actuator attacks, International Journal of Control (2018) 1–9.
[333] Z. Feng, G. Hu, Distributed tracking control for multi-agent systems under two types of attacks, in: Proc. 19th World Congress IFAC, Cape Town, 2014, pp. 5790–5795.

[334] M.O. Oyedeji, M.S. Mahmoud, Couple-group consensus conditions for general first-order multiagent systems with communication delays, Systems & Control Letters (2018) 37–44.

[335] J. Qin, C. Yu, Cluster consensus control of generic linear multi-agent systems under directed topology with acyclic partition, Automatica 49 (April 2013) 2898–2905.

[336] Y. Feng, J. Lu, S. Xua, Y. Zou, Couple-group consensus for multi-agent networks of agents with discrete-time second-order dynamics, Journal of the Franklin Institute 350 (2013) 3277–3292.

[337] L. Ji, Q. Liu, X. Li, On reaching group consensus for linearly coupled multi-agent networks, Information Sciences 287 (July 2014) 1–12.

[338] Y. Shang, L^1 group consensus of multi-agent systems with switching topologies and stochastic inputs, Physics Letters A 337 (May 2013) 1582–1586.

[339] D. Xie, Q. Liu, L. Lv, S. Li, Necessary and sufficient conditions for group consensus of multi-agent systems, Applied Mathematics and Computation 337 (2014) 1582–1586.

[340] Y. Chena, J. Lua, F. Han, X. Yu, On the cluster consensus of discrete-time multi-agent systems, Systems & Control Letters 60 (2011) 517–523.

[341] G. Wang, Y. Shen, Second-order cluster consensus of multi-agent dynamical systems with impulsive effects, Communications in Nonlinear Science and Numerical Simulation 19 (2014) 3220–3228.

[342] Y. Han, W. Lu, T. Chen, Achieving cluster consensus in continuous-time networks of multi-agents with inter-cluster non-identical inputs, IEEE Transactions on Automatic Control 60 (3) (March 2015) 793–798.

[343] K. Chen, J. Wang, Y. Zhang, F.L. Lewis, Cluster consensus of heterogeneous linear multi-agent systems, IET Control Theory & Applications 60 (3) (April 2018) 793–798.

[344] H. Zhao, J.H. Park, Y. Zhang, Couple-group consensus for second-order multi-agent systems with fixed and stochastic switching topologies, Applied Mathematics and Computation 232 (2014) 595–605.

[345] G. Wen, Y. Yu, Z. Peng, H. Wang, Dynamical group consensus of heterogeneous multi-agent systems with input time delays, Neurocomputing 175 (2016) 278–286.

[346] Y. Gao, J. Yu, J. Shao, Y. Duan, Group consensus for multi-agent systems under the effect of coupling strengths among groups, IFAC Papers Online 48 (28) (2015) 449–454.

[347] H. Hu, W. Yu, Q. Xuan, C. Zhang, G. Xie, Group consensus for heterogeneous multi-agent systems with parametric uncertainties, Neurocomputing 142 (2014) 383–392.

[348] Y. Gao, J. Yu, J. Shao, M. Yu, Group consensus for second-order discrete-time multi-agent systems with time-varying delays under switching topologies, Neurocomputing 207 (2016) 805–812.

[349] J. Yua, J. Liua, L. Xiang, J. Zhoua, Group consensus in networked mechanical systems with communication delays, Procedia IUTAM 22 (2017) 107–114.

[350] H. Xia, T. Huang, J. Shao, J. Yu, Group consensus of multi-agent systems with communication delays, Neurocomputing 171 (2016) 1666–1673.

[351] Z. Li, Z. Ding, Distributed adaptive consensus and output tracking of unknown linear systems on directed graphs, Automatica 55 (March 2015) 12–18.

[352] Z. Li, Z.Q. Chen, Z. Ding, Distributed adaptive controllers for cooperative output regulation of heterogeneous agents over directed graphs, Automatica 68 (February 2016) 179–183.

[353] Z. Ding, Z. Li, Distributed adaptive consensus control of nonlinear output-feedback systems on directed graphs, Automatica 72 (July 2016) 46–52.

[354] Y. Lv, Z. Li, Z. Duana, J. Chen, Distributed adaptive output feedback consensus protocols for linear systems on directed graphs with a leader of bounded input, Automatica 74 (October 2016) 308–314.

[355] Y. Zhang, S. Li, Adaptive near-optimal consensus of high-order nonlinear multi-agent systems with heterogeneity, Automatica 85 (September 2017) 426–432.

[356] J. Li, D.W.C. Ho, J. Li, Adaptive consensus of multi-agent systems under quantized measurements via the edge Laplacian, Automatica 92 (April 2018) 217–224.

[357] X. Niu, Y. Liu, Y. Man, Distributed adaptive consensus of nonlinear multi-agent systems with unknown coefficients, IFAC-PapersOnline 48 (28) (2015) 915–920.

[358] K. Chen, J. Wang, Y. Zhang, Z. Liu, Adaptive consensus of nonlinear multi-agent systems with unknown backlash-like hysteresis, Neurocomputing 175 (2015) 698–703.

[359] G. Wang, C. Wang, L. Li, Q. Du, Distributed adaptive consensus tracking control of higher-order nonlinear strict-feedback multi-agent systems using neural networks, Neurocomputing 214 (2016) 269–279.

[360] H. Rezaee, F. Abdollahi, Adaptive consensus control of nonlinear multiagent systems with unknown control directions under stochastic topologies, IEEE Transactions on Neural Networks and Learning Systems 29 (8) (2017).

[361] J. Sun, Z. Geng, Y. Lv, Adaptive output feedback consensus tracking for heterogeneous multi-agent systems with unknown dynamics under directed graphs, Systems & Control Letters 87 (2016).

[362] W. Ren, R.W. Beard, Distributed Consensus in Multi-Vehicle Cooperative Control, Springer-Verlag, London, 2008.

[363] A Systems View of the Modern Grid, National Energy Technology Laboratory (NETL), U.S. Department of Energy (DOE), 2007.

[364] NISTIR 7628: Guidelines for Smart Grid Cyber Security, National Institute for Standards and Technology, Aug. 2010.

[365] GAO-04-354: Critical Infrastructure Protection Challenges and Efforts to Secure Control Systems, U.S. Government Accountability Office (GAO), Mar. 2004.

[366] NERC Critical Infrastructure Protection (CIP) Reliability Standards, North American Electric Reliability Corporation, 2009.

[367] N. Falliere, L. Murchu, E. Chien, BW32. Stuxnet Dossier, Version 1.3, Symantec, Nov. 2010.

[368] S. Baker, S. Waterman, G. Ivanov, Crossfire: critical infrastructure in the age of cyber war, in: McAfee, 2009.

[369] GAO-11-117: Electricity Grid Modernization: Progress Being Made on Cybersecurity Guidelines, but Key Challenges Remain to be Addressed, U.S. Government Accountability Office (GAO), Jan. 2011.

[370] K. Stouffer, J. Falco, K. Scarfone, BNIST SP 800-82: Guide to Industrial Control Systems (ICS) Security, Tech. Rep., National Institute of Standards and Technology, Sep. 2008.

[371] D. Kundur, X. Feng, S. Liu, T. Zourntos, K. Butler-Purry, Towards a framework for cyber attack impact analysis of the electric smart grid, in: Proc. 1st IEEE Int. Conf. Smart Grid Commun., Oct. 2010, pp. 244–249.

[372] S. Li, Y.Y. lmaz, X.D. Wang, Quickest detection of false data injection attack in wide-area smart grids, IEEE Transactions on Smart Grid 6 (6) (Nov. 2015) 2725–2735.

[373] M. Esmalifalak, G. Shi, Z. Han, L.Y. Song, Bad data injection attack and defense in electricity market using game theory study, IEEE Transactions on Smart Grid 4 (1) (Mar. 2013) 160–169.

[374] K.C. Sou, H. Sandberg, K.H. Johansson, On the exact solution to a smart grid cyber-security analysis problem, IEEE Transactions on Smart Grid 4 (2) (Jun. 2013) 856–865.

[375] M. Esmalifalak, H. Nguyen, R. Zheng, Z. Han, Stealth false data injection using independent component analysis in smart grid, in: Proc. Int. Conf. Smart Grid Communications, Brussels, Belgium, 2011, pp. 244–248.

[376] M.S. Mahmoud, Switched Time-Delay Systems, Springer-Verlag, New York, 2010.

[377] E. Bayraktaroglu, C. King, X. Liu, G. Noubir, R. Rajaraman, B. Thapa, On the performance of IEEE 802.11 under jamming, in: Proc. IEEE Infocom 08, 2008, pp. 1265–1274.

[378] S. Bhattacharya, T. Basar, Graph-theoretic approach for connectivity maintenance in mobile networks in the presence of a jammer, in: IEEE Int. Conference on Decision and Control, Atlanta, USA, December 2010.

[379] A. Cardenas, S. Amin, B. Sinopoli, A. Giani, A. Perrig, S.S. Sastry, Challenges for securing cyber physical systems, in: Workshop on Future Directions in Cyber-Physical Systems Security, DHS, July 2009.

[380] A.G. Fragkiadakis, V.A. Siris, N. Petroulakis, Anomaly based intrusion detection algorithms for wireless networks, in: WWIC, 2010, pp. 192–203.
[381] C.V. Loan, The sensitivity of the matrix exponential, SIAM Journal on Numerical Analysis 14 (6) (1977) 971–981.
[382] F. Pasqualetti, R. Carli, F. Bullo, Distributed estimation and false data detection with application to power networks, personal communications, 2019.
[383] M. Zhu, S. Martnez, Attack-resilient distributed formation control via online adaptation, in: IEEE International Conference on Decision and Control, Orlando, USA, December 2011.
[384] S.A. Salinas, P. Li, Privacy-preserving energy theft detection in microgrids: a state estimation approach, IEEE Transactions on Power Systems 31 (2) (Mar. 2016) 883–894.
[385] A. Alimardani, F. Therrien, D. Atanackovic, J. Jatskevich, E. Vaahedi, Distribution system state estimation based on nonsynchronized smart meters, IEEE Transactions on Smart Grid 6 (6) (Nov. 2015) 2919–2928.
[386] S. Meliopoulos, R.K. Huang, E. Polymeneas, G. Cokkinides, Distributed dynamic state estimation: fundamental building block for the smart grid, in: Proc. Power & Energy Society General Meeting, Denver, CO, USA, 2015, pp. 1–6.
[387] E. Marris, Upgrading the grid, Nature 454 (2008) 570–573.
[388] S.M. Amin, For the good of the grid, IEEE Power & Energy Magazine 6 (6) (Nov./Dec. 2008) 48–59.
[389] Guidelines for Smart Grid Cyber Security, Draft NISTIR 7628, Jul. 2010.
[390] Understanding the Benefits of the Smart Grid, NETL, Jun. 2010.
[391] High-Impact, Low-Frequency Event Risk to the North American Bulk Power System, US-DOE, NERCH, Jun. 2010.
[392] J. Vijayan, Stuxnet renews power grid security concerns, Computerworld 26 (Jul. 2010).
[393] F. Cleveland, Cyber security issues for advanced metering infrastructure (AMI), in: Proc. Power Energy Soc. Gen. Meeting–Conv. Delivery Electr. Energy, 21st Century, Apr. 2008.
[394] S.M. Amin, Securing the electricity grid, The Bridge 40 (2010) 13–20.
[395] P. McDaniel, S. McLaughlin, Security and privacy challenges in the smart grid, IEEE Security & Privacy 7 (3) (May/Jun. 2009) 75–77.
[396] H. Khurana, M. Hadley, N. Lu, D.A. Frincke, Smart-grid security issues, IEEE Security & Privacy 8 (1) (Jan./Feb. 2010) 81–85.
[397] L. Xie, Y. Mo, B. Sinopoli, False data injection attacks in electricity markets, in: Proc. IEEE Int. Conf. Smart Grid Commun., Oct. 2010, pp. 226–231.
[398] NIST Framework and Roadmap for Smart Grid Interoperability Standards, Release 1.0, NIST Special Publication 1108 NIST, Jan. 2010.
[399] R. Goebel, R. Sanfelice, A. Teel, Hybrid dynamical systems, IEEE Control Systems 29 (2) (2009) 28–93.
[400] Advanced Metering Infrastructure (AMI), EPRI, Feb. 2007.
[401] A. Kerckhoffs, La cryptographie militairie, Journal des Sciences Militaires IX (1883) 5–38.
[402] S.S.S.R. Depuru, L. Wang, V. Devabhaktuni, N. Gudi, Smart meters for power grid–Challenges, issues, advantages and status, in: Proc. IEEE/PES Power Syst. Conf. Expo, 2011.
[403] P. Huitsing, R. Chandia, M. Papa, S. Shenoi, Attack taxonomies for the Modbus protocols, International Journal of Critical Infrastructure Protection 1 (Dec. 2008) 37–44.
[404] E. Barker, D. Branstad, S. Chokhani, M. Smid, A Framework for Designing Cryptographic Key Management Systems, NIST DRAFT Special Publication 800-130, Jun. 2010.
[405] H. Lee, J. Kim, W. Lee, Resiliency of network topologies under path-based attacks, IEICE Transactions on Communications E89-B (Oct. 2006) 2878–2884.
[406] D. Seo, H. Lee, A. Perrig, Secure and efficient capability-based power management in the smart grid, in: Proc. Int. Workshop Smart Grid Security Commun., May 2011, pp. 119–126.
[407] R.L. Pickholtz, D.L. Schilling, L.B. Milstein, Theory of spread spectrum communications–a tutorial, IEEE Transactions on Communications 30 (5) (May 1982) 855–884.
[408] S. McLaughlin, D. Podkuiko, A. Delozier, S. Mizdzvezhanka, P. McDaniel, Embedded firmware diversity for smart electric meters, in: Proc. USENIX Workshop Hot Topics in Security, 2010.

[409] M. LeMay, G. Gross, C. Gunter, S. Garg, Unified architecture for large-scale attested metering, in: Proc. Annu. Hawaii Int. Conf. Syst. Sci., Jan. 2007.

[410] M. LeMay, C.A. Gunter, Cumulative attestation kernels for embedded systems, in: Proc. Eur. Symp. Res. Comput. Security, Sep. 2009, pp. 655–670.

[411] A. Seshadri, A. Perrig, L. van Doorn, P. Khosla, SWATT: software-based attestation for embedded devices, in: Proc. IEEE Symp. Security Privacy, May 2004, pp. 272–282.

[412] A. Shah, A. Perrig, B. Sinopoli, Mechanisms to provide integrity in SCADA and PCS devices, in: Proc. Int. Workshop Cyber-Physical Syst. Challenges Appl., Jun. 2008.

[413] M. Shahidehpour, F. Tinney, Y. Fu, Impact of security on power systems operation, Proceedings of the IEEE 93 (11) (Nov. 2005) 2013–2025.

[414] A. Abur, A.G. Exposito, Power System State Estimation: Theory and Implementation, CRC Press, Boca Raton, FL, 2004.

[415] H. Sandberg, A. Teixeira, K.H. Johansson, On security indices for state estimators in power networks, in: Proc. 1st Workshop Secure Control Syst., 2010.

[416] O. Kosut, L. Jia, R.J. Thomas, L. Tong, Limiting false data attacks on power system state estimation, in: Proc. 44th Annu. Conf. Inf. Sci. Syst., 2010.

[417] N. Falliere, L.O. Murchu, E. Chien, W32.stuxnet Dossier, Tech. Rep., Symantec Corporation, 2011.

[418] Y. Chen, S. Kar, J.M.F. Moura, Dynamic attack detection in cyber-physical systems with side initial state information, arXiv e-prints, Mar. 2015.

[419] A.A. Cardenas, S. Amin, Z.-S. Lin, Y.-L. Huang, C.-Y. Huang, S. Sastry, Attacks against process control systems: risk assessment, detection, and response, in: Proc. 6th ACM Symposium on Information, Computer and Communications Security, ASIACCS 11, New York, NY, USA, 2011, pp. 355–366.

[420] T.T. Tran, O.S. Shin, J.H. Lee, Detection of replay attacks in smart grid systems, in: Int. Conference Computing, Management and Telecommunications (ComManTel), Jan 2013, pp. 298–302.

[421] F. Pasqualetti, F. Dörfler, F. Bullo, Cyber-physical attacks in power networks: models, fundamental limitations and monitor design, in: 50th IEEE Conference Decision and Control and European Control Conference, Dec. 2011, pp. 2195–2201.

[422] H. Fawzi, P. Tabuada, S. Diggavi, Secure estimation and control for cyber-physical systems under adversarial attacks, IEEE Transactions on Automatic Control 59 (6) (June 2014) 1454–1467.

[423] Y. Liu, P. Ning, M.K. Reiter, False data injection attacks against state estimation in electric power grids, in: Proc. 16th ACM Conference on Computer and Communications Security, CCS 09, New York, NY, USA, 2009, pp. 21–32.

[424] W.-H. Chen, Disturbance observer based control for nonlinear systems, IEEE/ASME Transactions on Mechatronics 9 (4) (Dec 2004) 706–710.

[425] Y. Mo, B. Sinopoli, False data injection attacks in control systems, in: Preprints of the 1st Workshop on Secure Control Systems, 2008.

[426] C.Z. Bai, F. Pasqualetti, V. Gupta, Security in stochastic control systems: fundamental limitations and performance bounds, in: American Control Conference (ACC), July 2015, pp. 195–200.

[427] W.H. Chen, J. Yang, L. Guo, S. Li, Disturbance-observer-based control and related methods – an overview, IEEE Transactions on Industrial Electronics 63 (2) (Feb 2016) 1083–1095.

[428] J.C. Willems, Dissipative dynamical systems part ii: linear systems with quadratic supply rates, Archive for Rational Mechanics and Analysis 45 (5) (1972) 352–393.

[429] Y. Yan, P. Antsaklis, Stabilizing nonlinear model predictive control scheme based on passivity and dissipativity, in: 2016 American Control Conference (ACC), July 2016, pp. 4476–4481.

[430] M. Xia, P.J. Antsaklis, V. Gupta, Passivity indices and passivation of systems with application to systems with input/output delay, in: 53rd IEEE Conference on Decision and Control, Dec. 2014, pp. 783–788.

[431] X. Chen, S. Komada, T. Fukuda, Design of a nonlinear disturbance observer, IEEE Transactions on Industrial Electronics 47 (2) (Apr 2000) 429–437.
[432] P. Colaneri, J.C. Geromel, A. Astolfi, Stabilization of continuous-time switched nonlinear systems, Systems & Control Letters 57 (1) (2008) 95–103.
[433] T. Basar, P. Bernhard, A General Introduction to Minimax (Hinfinity Optimal) Designs, Birkhäuser Boston, Boston, MA, 2008, pp. 1–32.
[434] J. Zhao, D.J. Hill, Dissipativity theory for switched systems, IEEE Transactions on Automatic Control 53 (4) (May 2008) 941–953.
[435] L. Guo, Q. Liao, S. Wei, Y. Huang, A kind of bicycle robot dynamic modeling and nonlinear control, in: The 2010 IEEE Int. Conference on Information and Automation, June 2010, pp. 1613–1617.
[436] A. Cardenas, S. Amin, S. Sastry, Research challenges for the security of control systems, in: 3rd USENIX Workshop on Hot Topics in Security, Jul. 2008, Article 6.
[437] A. Cardenas, S. Amin, S. Sastry, Secure control: towards survivable cyber-physical systems, in: First International Workshop on Cyber-Physical Systems, Jun. 2008.
[438] V. Gligor, A note on denial-of-service in operating systems, IEEE Transactions on Software Engineering (1984) 320–324.
[439] W. Liu, C. Kwon, I. Hwang, Cyber security analysis for state estimators in air traffic control systems, in: AIAA Conference on Guidance, Navigation, and Control, Aug. 2012.
[440] J.J. Gertler, Survey of model-based failure detection and isolation in complex plants, IFAC Proceedings Series 7 (1987).
[441] D. Titterton, J. Weston, Strapdown Inertial Navigation Technology, AIAA, 2004.
[442] H. Li, M.Y. Chow, Z. Sun, Optimal stabilizing gain selection for networked control systems with time delays and packet losses, IEEE Transactions on Control Systems Technology 17 (5) (Sep. 2009) 1154–1162.
[443] K.J. Park, J. Kim, H. Lim, Y. Eun, Robust path diversity for network quality of service in cyber-physical systems, IEEE Transactions on Industrial Informatics 10 (4) (Nov. 2014) 2204–2215.
[444] F. Pasqualetti, F. Dorfler, Control-theoretic methods for cyberphysical security: geometric principles for optimal cross-layer resilient control systems, IEEE Control Systems Magazine 35 (1) (Feb. 2015) 110–127.
[445] R.S. Smith, Covert misappropriation of networked control systems: presenting a feedback structure, IEEE Control Systems Magazine 35 (1) (Feb. 2015) 82–92.
[446] Y.W. Law, T. Alpcan, M. Palaniswami, Security games for risk minimization in automatic generation control, IEEE Transactions on Power Systems 30 (1) (Jan. 2015) 223–232.
[447] Y. Wu, X. He, S. Liu, L. Xie, Consensus of discrete-time multiagent with adversaries and time delays, International Journal of General Systems 43 (3) (2014) 402–411.
[448] H. Zhang, P.Cheng.L. Shi, J.M. Chen, Optimal denial-of-service attack scheduling with energy constraint, IEEE Transactions on Automatic Control 60 (11) (Nov. 2015) 3023–3028.
[449] M. Zhu, S. Martinez, On the performance analysis of resilient networked control systems under replay attacks, IEEE Transactions on Automatic Control 59 (3) (Mar. 2014) 804–808.
[450] L. Li, B. Hu, M.D. Lemmon, Resilient event triggered systems with limited communication, in: Proc. IEEE Conf. Decis. Control, 2012, pp. 6577–6582.
[451] S. Amin, A.A. Cardenas, S.S. Sastry, Hybrid Systems: Computation and Control, Springer, Berlin, Germany, 2009.
[452] Y. Yuan, F. Sun, Data fusion-based resilient control system under DoS attacks: a game theoretic approach, International Journal of Control, Automation, and Systems 13 (3) (2015) 513–520.
[453] Y. Li, L. Shi, P. Cheng, J. Chen, D.E. Quevedo, Jamming attack on cyber-physical systems: a game-theoretic approach, in: Proc. Annu. Int. Conf. Cyber Technol. Autom. Control Intell. Syst, 2013, pp. 252–257.
[454] Y. Xia, M. Fu, H. Yang, G.P. Liu, Robust sliding-mode control for uncertain time-delay systems based on delta operator, IEEE Transactions on Industrial Electronics 56 (9) (Sep. 2009) 3646–3655.

[455] M.C. Priess, R. Conway, J. Choi, J.M. Popovich, C. Radcliffe, Solutions to the inverse LQR problem with application to biological systems analysis, IEEE Transactions on Control Systems Technology 23 (2) (Mar. 2015) 770–777.
[456] S. Coogan, L. Ratliff, D. Calderone, C. Tomlin, S. Sastry, Energy management via pricing in LQ dynamic games, in: Proc. Amer. Control Conf., 2013, pp. 443–448.
[457] D.P. Bertsekas, Dynamic Programming and Optimal Control, Athena Scientific, Belmont, MA, USA, 1995.
[458] P. Suchomski, A j-lossless coprime factorization approach to H_∞ control in delta domain, Automatica 38 (10) (2002) 1807–1814.
[459] E. Garone, B. Sinopoli, A. Goldsmith, A. Casavola, LQG control for MIMO systems over multiple erasure channels with perfect acknowledgment, IEEE Transactions on Automatic Control 57 (2) (Feb. 2012) 450–456.
[460] T. Sui, K. You, Minyue Fu, Stability of Kalman estimation with multiple sensors involving lossy communications, IFAC Proceedings 47 (3) (2014) 116–121.
[461] X.M. Zhang, Q.L. Han, Event-based H_∞ estimation for sampled-data systems, Automatica 51 (2015) 55–69.
[462] J. Zhang, Z. Wang, D. Ding, X. Liu, H_∞ state estimation for discrete-time delayed neural networks with randomly occurring quantizations and missing measurements, Neurocomputing 148 (2015) 388–396.
[463] M. Lyu, B. Yuming Bo, Variance-constrained resilient $H\infty$ estimation for time-varying nonlinear networked systems subject to quantization effects, Neurocomputing 267 (2017) 283–294.
[464] K. Ito, K. Xiong, Gaussian estimators for nonlinear estimation problems, IEEE Transactions on Automatic Control 45 (5) (2000) 910–927.
[465] S. Julier, J. Uhlmann, H. Durrant-Whyte, A new method for the nonlinear transformation of means and covariances in estimators and estimators, IEEE Transactions on Automatic Control 45 (3) (2000) 477–482.
[466] Z. Wang, Y. Liu, X. Liu, H_∞ estimation for uncertain stochastic time-delay systems with sector-bounded nonlinearities, Automatica 44 (5) (2008) 1268–1277.
[467] H. Dong, Z. Wang, D.W.C. Ho, H. Gao, Robust H_∞ estimation for Markovian jump systems with randomly occurring nonlinearities and sensor saturation: the finite-horizon case, IEEE Transactions on Signal Processing 59 (7) (2011) 3048–3057.
[468] S. Zhou, J. Lam, H_∞ estimation for systems with delays and time-varying nonlinear parameters, Circuits, Systems, and Signal Processing 29 (4) (2010) 601–627.
[469] M. Basin, S. Elvira-Ceja, E. Sanchez, Mean-square H_∞ estimation for stochastic systems: application to a 2DOF helicopter, Signal Processing 92 (3) (2012) 801–806.
[470] G. Wei, Z. Wang, H. Shu, Robust estimation with stochastic nonlinearities and multiple missing measurements, Automatica 45 (3) (2009) 836–841.
[471] M.S. Mahmoud, S.Z. Selim, P. Shi, M.H. Baig, New results on networked control systems with non-stationary packet dropouts, IET Control Theory & Applications 6 (15) (2012) 2442–2452.
[472] A.A. Cárdenas, S. Amin, S. Sastry, Research challenges for the security of control systems, in: HotSec, 2008.
[473] F. Pasqualetti, A. Bicchi, F. Bullo, Consensus computation in unreliable networks: a system theoretic approach, IEEE Transactions on Automatic Control 57 (1) (2012) 90–104.
[474] H. Fawzi, P. Tabuada, S. Diggavi, Security for control systems under sensor and actuator attacks, in: IEEE 51st Annual Decision and Control Conference (CDC), 2012, pp. 3412–3417.
[475] M.S. Chong, M. Wakaiki, J.P. Hespanha, Observability of linear systems under adversarial attacks, in: American Control Conference (ACC), 2015, pp. 2439–2444.
[476] S. Mishra, Y. Shoukry, N. Karamchandani, S. Diggavi, P. Tabuada, Secure state estimation: optimal guarantees against sensor attacks in the presence of noise, in: IEEE Int. Symposium on Information Theory (ISIT), 2015, pp. 2929–2933.

[477] Y. Shoukry, P. Tabuada, Event-triggered state observers for sparse sensor noise/attacks, IEEE Transactions on Automatic Control 61 (8) (2016) 2079–2091.
[478] Y. Mo, R.M. Murray, Multi-dimensional state estimation in adversarial environment, in: IEEE 34th Chinese Control Conference (CCC), 2015, pp. 4761–4766.
[479] R. Tibshirani, Regression shrinkage and selection via the lasso, Journal of the Royal Statistical Society, Series B, Methodological (1996) 267–288.
[480] M. Pajic, J. Weimer, N. Bezzo, P. Tabuada, O. Sokolsky, I. Lee, G.J. Pappas, Robustness of attack-resilient state estimators, in: ICCPS'14: ACM/IEEE 5th Int. Conference on Cyber-Physical Systems (with CPS Week 2014), 2014, pp. 163–174.
[481] W. Rudin, et al., Principles of Mathematical Analysis, McGraw-Hill, New York, 1976.
[482] B.D. Anderson, J.B. Moore, Optimal Filtering, Prentice-Hall, Englewood Cliffs, 1979.
[483] http://www.computerweekly.com/news/2240164589/Datacentre-power-demand-grew-63-in-2012-Global-datacentre-census.
[484] Q. Tang, T. Mukherjee, S.K.S. Gupta, P. Cayton, Sensor-based fast thermal evaluation model for energy efficient high-performance data centers, in: The Fourth Int. Conf. Intelligent Sensing and Information Processing (ICISIP), 2006, pp. 203–208.
[485] G. Chen, W. He, J. Liu, S. Nath, L. Rigas, L. Xiao, F. Zhao, Energy-aware server provisioning and load dispatching for connection-intensive Internet services, in: USENIX Symposium on Networked Systems Design and Implementation, vol. 8, 2008, pp. 337–350.
[486] R.P. Doyle, J.S. Chase, O.M. Asad, W. Jin, A. Vahdat, Model-based resource provisioning in a web service utility, in: USENIX Symposium on Internet Technologies and Systems, vol. 4, 2003, pp. 5–15.
[487] J.S. Chase, D.C. Anderson, P.N. Thakar, A.M. Vahdat, R.P. Doyle, Managing energy and server resources in hosting centers, ACM SIGOPS Operating Systems Review 35 (5) (2001) 103–116.
[488] K. Gai, M. Qiu, H. Zhao, X. Sun, Resource management in sustainable cyber-physical systems using heterogeneous cloud computing, IEEE Transactions on Sustainable Computing 3 (2) (2018) 60–72.
[489] L. Parolini Luca, et al., A cyber–physical systems approach to data center modeling and control for energy efficiency, Proceedings of the IEEE 100 (1) (2012) 254–268.
[490] M. Al-Ayyoub, et al., Resilient service provisioning in cloud-based data centers, Future Generations Computer Systems 86 (2018) 765–774.
[491] A. Banerjee, et al., Cooling-aware and thermal-aware workload placement for green HPC data centers, in: IEEE Int. Green Computing Conference, 2010.
[492] T. Mukherjee, et al., Spatio-temporal thermal-aware job scheduling to minimize energy consumption in virtualized heterogeneous data centers, Computer Networks 53 (17) (2009) 2888–2904.
[493] M. Wahlroos, et al., Utilizing data center waste heat in district heating–impacts on energy efficiency and prospects for low-temperature district heating networks, Energy 140 (2017) 1228–1238.
[494] X. Zhang, et al., Cooling energy consumption investigation of data center IT room with vertical placed server, Energy Procedia 105 (2017), Elsevier.
[495] J.C. Ni, X. Bai, A review of air conditioning energy performance in data centers, Renewable & Sustainable Energy Reviews 67 (2017) 625–640.
[496] V.D. Tobias, C. De Persis, P. Tesi, Optimized thermal-aware job scheduling and control of data centers, IEEE Transactions on Control Systems Technology 99 (2018) 1–12.
[497] R.C. Chu, R.E. Simons, M.J. Ellsworth, R.R. Schmidt, V. Cozzolino, Review of cooling technologies for computer products, IEEE Transactions on Device and Materials Reliability 4 (4) (2004) 568–585.
[498] T.J. Breen, E.J. Walsh, J. Punch, A.J. Shah, C.E. Bash, From chip to cooling tower data center modeling: part I: influence of server inlet temperature and temperature rise across cabinet, in: Proc. 12th IEEE Intersoc. Conf. Thermal Thermomech. Phenomena Electron. Syst., Jun. 2010, pp. 1–10.

[499] E.J. Walsh, T.J. Breen, J. Punch, A.J. Shah, C.E. Bash, From chip to cooling tower data center modeling: part II: influence of chip temperature control philosophy, in: Proc. 12th IEEE Intersoc. Conf. Thermal Thermomech. Phenomena Electron. Syst., Jun. 2010, pp. 1–7.

[500] C.D. Patel, R.K. Sharma, C. Bash, M. Beitelmal, Thermal considerations in cooling large scale high compute density data centers, in: Proc. Intersoc. Conf. Thermal Thermomech. Phenomena Electron. Syst., May 2002, pp. 767–776.

[501] M.H. Beitelmal, C.D. Patel, Model-Based Approach for Optimizing a Data Center Centralized Cooling System, Tech. Rep. HPL-2006-67, Hewlett Packard Labs., Apr. 2006.

[502] M.K. Patterson, D. Fenwick, The State of Data Center Cooling, White Paper, Intel Corporation, Mar. 2008.

[503] X. Fan, W.-D. Weber, L.A. Barroso, Power provisioning for a warehouse-sized computer, in: Proc. Int. Symp. Comput. Architect., Jun. 2007, pp. 13–23.

[504] J. Hamilton, Cost of power in large-scale data centers [Online]. Available: http://perspectives.mvdirona.com, Nov. 2008.

[505] Report to Congress on Server and Data Center Energy Efficiency, ENERGY STAR Program, Tech. Rep., U.S. Environmental Protection Agency, Aug. 2007.

[506] C.D. Patel, A.J. Shah, Cost Model for Planning, Development and Operation of a Data Center, Tech. Rep., Internet Syst. Storage Lab., HP Labs., Palo Alto, CA, Jun. 2005.

[507] S. Rahman, Power for the Internet, IEEE Computer Applications in Power 14 (4) (2001) 8–10.

[508] M. Patterson, D. Costello, P.F. Grimm, M. Loeffler, Data Center TCO; a Comparison of High-Density and Low-Density Spaces, White Paper, Intel Corporation, 2007.

[509] R.K. Sharma, C.E. Bash, C.D. Patel, R.J. Friedrich, J.S. Chase, Balance of power: dynamic thermal management for Internet data centers, IEEE Internet Computing 9 (2005) 42–49.

[510] Hewlett-Packard, HP Modular Cooling System: Water Cooling Technology for High-Density Server Installations, Tech. Rep., Hewlett-Packard, 2007.

[511] J. Scaramella, Worldwide Server Power and Cooling Expense 2006–2010 Forecast, Int. Data Corporation (IDC), Sep. 2006.

[512] The Green Grid, The Green Grid Data Center Power Efficiency Metrics: PUE and DCiE, White Paper, Tech. Committee, 2007.

[513] X. Zhu, D. Young, B. Watson, Z. Wang, J. Rolia, S. Singhal, B. McKee, C. Hyser, D. Gmach, R. Gardner, T. Christian, L. Cherkasova, 1000 islands: integrated capacity and workload management for the next generation data center, in: Proc. Int. Conf. Autonom. Comput., Jun. 2008, pp. 172–181.

[514] R. Raghavendra, P. Ranganathan, V. Talwar, Z. Wang, X. Zhu, No "power" struggles: coordinated multi-level power management for the data center, in: Proc. Architect. Support Programming Lang. Oper. Syst., Mar. 2008, pp. 48–59.

[515] Y. Cho, N. Chang, Energy-aware clock-frequency assignment in microprocessors and memory devices for dynamic voltage scaling, IEEE Transactions on Computer-Aided Design of Integrated Circuits and Systems 26 (6) (2007) 1030–1040.

[516] H. Aydin, D. Zhu, Reliability-aware energy management for periodic real-time tasks, IEEE Transactions on Computers 58 (10) (Oct. 2009) 1382–1397.

[517] P. Choudhary, D. Marculescu, Power management of voltage/frequency island-based systems using hardware-based methods, IEEE Transactions on Very Large Scale Integration (VLSI) Systems 17 (3) (2009) 427–438.

[518] J. Kim, S. Yoo, C.-M. Kyung, Program phase-aware dynamic voltage scaling under variable computational workload and memory stall environment, IEEE Transactions on Computer-Aided Design of Integrated Circuits and Systems 30 (1) (2011) 110–123.

[519] Z. Jian-Hui, Y. Chun-Xin, Design and simulation of the CPU fan and heat sinks, IEEE Transactions on Components and Packaging Technologies 31 (4) (2008) 890–903.

[520] A. Mutapcic, S. Boyd, S. Murali, D. Atienza, G. De Micheli, R. Gupta, Processor speed control with thermal constraints, IEEE Transactions on Circuits and Systems. I, Regular Papers 56 (9) (2009) 1994–2008.

[521] A. Gandhi, M. Harchol-Balter, R. Das, C. Lefurgy, Optimal power allocation in server farms, in: Proc. ACM SIGMETRICS, 2009, pp. 157–168.

[522] A. Gandhi, M. Harchol-Balter, I. Adan, Server farms with setup costs, Performance Evaluation 67 (2010) 1123–1138.

[523] M. Aghajani, L. Parolini, B. Sinopoli, Dynamic power allocation in server farms: a real time optimization approach, in: Proc. 49th IEEE Conf. Decision Control, Dec. 2010, pp. 3790–3795.

[524] A. Varma, B. Ganesh, M. Sen, S.R. Choudhury, L. Srinivasan, B. Jacob, A control-theoretic approach to dynamic voltage scheduling, in: Proc. Int. Conf. Compilers Architect. Synthesis Embedded Syst., Oct. 2003, pp. 255–266.

[525] J. Leverich, M. Monchiero, V. Talwar, P. Ranganathan, C. Kozyrakis, Power management of datacenter workloads using per-core power gating, Computer Architecture Letters 8 (2) (2009) 48–51.

[526] R. Mahajan, C. pin Chiu, G. Chrysler, Cooling a microprocessor chip, Proceedings of the IEEE 94 (8) (Aug. 2006) 1476–1486.

[527] G.K. Thiruvathukal, K. Hinsen, K. Laufer, J. Kaylor, Virtualization for computational scientists, Computing in Science & Engineering 12 (4) (2010) 52–61.

[528] P. Padala, X. Zhu, Z. Wang, S. Singhal, K.G. Shin, Performance Evaluation of Virtualization Technologies for Server Consolidation, White Paper, Hewlett-Packard, 2007.

[529] N. Tolia, Z. Wang, P. Ranganathan, C. Bash, M. Marwah, X. Zhu, Unified power and cooling management in server enclosures, in: Proc. InterPACK, pp. 721–730, Jul. 2009.

[530] X. Wang, Y. Wang, Coordinating power control and performance management for virtualized server clusters, IEEE Transactions on Parallel and Distributed Systems 22 (2) (2011) 245–259.

[531] H. Jin, L. Deng, S. Wu, X. Shi, X. Pan, Live virtual machine migration with adaptive, memory compression, in: Proc. IEEE Int. Conf. Cluster Comput. Workshops, Aug. 2009, pp. 1–10.

[532] F. Ma, F. Liu, Z. Liu, Live virtual machine migration based on improved pre-copy approach, in: Proc. IEEE Int. Softw. Eng. Service Sci. Conf., 2010, pp. 230–233.

[533] P. Padala, K.G. Shin, X. Zhu, M. Uysal, Z. Wang, S. Singhal, A. Merchant, K. Salem, Adaptive control of virtualized resources in utility computing environments, in: SIGOPS Oper. Syst. Rev., vol. 41, Mar. 2007, pp. 289–302.

[534] J.B. Rawlings, Tutorial overview of model predictive control, IEEE Control Systems Magazine 20 (3) (Jun. 2000) 38–52.

[535] R. Scattolini, Architectures for distributed and hierarchical model predictive control – a review, Journal of Process Control 19 (5) (2009) 723–731.

[536] C.E. Bash, C.D. Patel, R.K. Sharma, Dynamic thermal management of air cooled data centers, in: Proc. 10th Intersoc. Conf. Thermal Thermomech. Phenomena Electron. Syst., no. 29, May 2006, pp. 445–452.

[537] C. Bash, G. Forman, Cool job allocation: measuring the power savings of placing jobs at cooling-efficient locations in the data center, in: USENIX Annu. Tech. Conf., no. 29, Jun. 2007, pp. 363–368.

[538] E. Ferrer, C. Bonilla, C. Bash, M. Batista, Data Center Thermal Zone Mapping, White Paper, Hewlett-Packard, 2007.

[539] A. Qureshi, R. Weber, H. Balakrishnan, J. Guttag, B. Maggs, Cutting the electric bill for Internet-scale systems, in: Proc. ACM SIGCOMM Conf. Data Commun., Aug. 2009, pp. 123–134.

[540] L. Rao, X. Liu, L. Xie, W. Liu, Minimizing electricity cost: optimization of distributed Internet data centers in a multi-electricity-market environment, in: Proc. 29th IEEE Int. Conf. Comput. Commun., Mar. 2010, pp. 1–9.

[541] Q. Tang, S.K.S. Gupta, G. Varsamopoulos, Energy-efficient thermal-aware task scheduling for homogeneous high-performance computing data centers: a cyber-physical approach, IEEE Transactions on Parallel and Distributed Systems 19 (11) (2008) 1458–1472.

[542] L. Parolini, B. Sinopoli, B.H. Krogh, Model predictive control of data centers in the smart grid scenario, in: Proc. 18th Int. Fed. Autom. Control (IFAC) World Congr., Aug. 2011.
[543] L. Parolini, N. Tolia, B. Sinopoli, B.H. Krogh, A cyber-physical systems approach to energy management in data centers, in: Proc. 1st Int. Conf. Cyber-Phys. Syst., Apr. 2010, pp. 168–177.
[544] M. Anderson, M. Buehner, P. Young, D. Hittle, C. Anderson, J. Tu, D. Hodgson, MIMO robust control for HVAC systems, IEEE Transactions on Control Systems Technology 16 (3) (2008) 475–483.
[545] M.M. Toulouse, G. Doljac, V.P. Carey, C. Bash, Exploration of a potential-flow-based compact model of air-flow transport in data centers, in: Proc. Amer. Soc. Mech. Eng. Conf., Nov. 2009, pp. 41–50.
[546] Y. Chen, D. Gmach, C. Hyser, Z. Wang, C. Bash, C. Hoover, S. Singhal, Integrated management of application performance, power and cooling in data centers, in: Proc. 12th IEEE/IFIP Netw. Oper. Manage. Symp., Apr. 2010, pp. 615–622.
[547] L.A. Barroso, U. Holzle, The case for energy-proportional computing, Computer 40 (12) (2007) 33–37.
[548] A. Hawkins, Unused Servers Survey Results Analysis, White Paper, The Green Grid, 2010.
[549] C. Lefurgy, X. Wang, M. Ware, Server-level power control, in: Proc. 4th Int. Conf. Autonom. Comput., Jun. 2007, p. 4.
[550] H. Kobayashi, B.L. Mark, System Modeling and Analysis: Foundations of System Performance Evaluation, Prentice-Hall, Englewood Cliffs, NJ, 2008.
[551] L. Parolini, E. Garone, B. Sinopoli, B.H. Krogh, A hierarchical approach to energy management in data centers, in: Proc. 49th IEEE Conf. Decision Control, Dec. 2010, pp. 1065–1070.
[552] L. Parolini, B. Sinopoli, B.H. Krogh, Reducing data center energy consumption via coordinated cooling and load management, in: Proc. Workshop Power Aware Comput. Syst., Dec. 2008, pp. 14–18.
[553] J. Moore, J. Chase, P. Ranganathan, R. Sharma, Making scheduling BCoo: temperature-aware workload placement in data centers, in: Proc. USENIX Annu. Tech. Conf., Apr. 2005, pp. 61–75.
[554] American Society of Heating, Refrigerating and Air-Conditioning Engineers (ASHRAE), Environmental Guidelines for Datacom Equipment. Expanding the Recommended Environmental Envelope, Tech. Rep., American Society of Heating, Refrigerating and Air-Conditioning Engineers (ASHRAE), 2008.
[555] R.A. Greco, High Density Data Centers Fraught With Peril, Slides, EYP Mission Critical Facilities Inc., 2003.
[556] D. Meisegeier, M. Howes, D. King, J. Hall, Potential Peak Load Reductions From Residential Energy Efficient Upgrades, White Paper, ICF International, 2002.
[557] National Association of Home Builders (NAHB) Research Center, Inc., Review of Residential Electrical Energy Use Data, White Paper, NAHB Research Center Inc., 2001.
[558] L. Parolini, B. Sinopoli, B. Krogh, Reducing data center energy consumption via coordinated cooling and load management, in: Proc. 2008 Conference on Power Aware Computing and Systems, HotPower, vol. 8, 2008, pp. 4–14.
[559] J. Moore, J. Chase, P. Ranganathan, R. Sharma, Making scheduling "cool": temperature-aware workload placement in data centers, in: USENIX Annual Technical Conference, 2005, pp. 61–75.

Index

J

K

L

M

MIX
Papier | Fördert gute Waldnutzung
FSC® C083411

Kolibri 360 GmbH
Ferdinand-Jühlke-Straße 7
99095 Erfurt, Deutschland
produktsicherheit@kolibri360.de